Combinatorics and Graph Theory '95

SSICC'95

Proceedings of the Summer School and International Conference on Combinatorics

Combinatorics and Graph Theory '95

Hefei, China, 25 May – 5 June 1995

Vol. 1

Editor

Ku Tung-Hsin

Academia Sinica, Hefei

World Scientific

Singapore • New Jersey • London • Hong Kong

Published by

World Scientific Publishing Co. Pte. Ltd.

P O Box 128, Farrer Road, Singapore 9128

USA office: Suite 1B, 1060 Main Street, River Edge, NJ 07661

UK office: 57 Shelton Street, Covent Garden, London WC2H 9HE

COMBINATORICS AND GRAPH THEORY '95

ISBN 981-02-2317-X

This book is printed on acid-free paper.

Printed in Singapore by Uto-Print

Summer School and International Conference on Combinatorics Hefei, China 25 May -- 5 June 1995

Presidium of the Conference

Honorary Chairman
P. Erdos (Hungary)
L.C. Hsu (Dalian: Dalian University of Technology)
Z.X. Wan (Beijing: Institute of System Science Academia Sinica)

Chairman
T.H.Ku (Hefei: Hefei Branch of Academia Sinica)

Vice Chairman
Y.Alavi (USA)
N.Biggs (England)
M.Deza (France)
A.Dress (Germany)
D.Z.Du (Beijing: Institute of Applied Mathematics Academia Sinica)
G.P.Egorychev (Russia)
C.K.Gupta (Canada)
F.Harary (USA)
F.K.Hwang (USA)
M.Iri (Japan)
S.Kageyama (Japan)
Ko–Wei Lih (Taipei: Academia Sinica)
H.Y.Lin (Wuhan: Huazhong University of Technology)
G.Z.Liu (Jinan: Shandong University)
Y. P. Liu (Beijing: Institute of Applied Mathematics Academia Sinica)
E.K.Lloyd (England)
K.C.Lu (Beijing: Qinghua University)
N.Polat (France)
A.Recski (Hungary)
R.W.Robinson (USA)
J.Schoenheim (Israel)
V.T.Sos (Hungary)
Hugo S.H.Sun (USA)
G.Tallini (Italy)
J.H.van Lint (The Netherlands)
J. F. Wang (Beijing: Institute of Applied Mathematics Academia Sinica)
R.J.Wilson (England)
J.Y.Yan (Beijing: University of Science and Technology of China)
H.P.Yap (Singapore)
J.Zhong (Guangzhou: Huanan Normal University)
L.Zhu (Suzhou: Suzhou University)

Academic Consaltant
L.T.Xie (Jinan: Shandong University}

List of Referees

Y.Alavi	**N.Biggs**	**Hui–Zhong Cao**
Yan–Xun Chang	**M.Deza**	**Bei–Liang Du**
Ding–Zhu Du	**Hong–Bing Fan**	**Jing–Zhen Gao**
F.Harary	**S.Kageyama**	**Ku Tung–Hsin**
Jiong–Sheng Li	**Ko–Wei Lih**	**Hua–Yi Lin**
Yi–Xun Lin	**Gui–Zhen Liu**	**Yan–Pei Liu**
E.Keith Lloyd	**Kai–Cheng Lu**	**Jing–Zhong Mao**
J.Schöenheim	**Hao Shen**	**V.T.Sós**
Rong–Guo Sun	**Jian–Fang Wang**	**Zhi–Jian Wang**
Zhe–Min Wang	**Hong–zeng Wei**	**Ming–Yuan Xia**
Bao–Gang Xu	**Min–Yi Yue**	**Bing–Xin Zhao**
Han–Ding Zheng	**Ji Zhong**	**Lie Zhu**

PREFACE

The Summer School and International Conference on Combinatorics (SSICC '95) was held in Hefei from May 25 to June 5, 1995, under the auspices of the Hefei Branch of Academia Sinica, some universities and scientific research institutions, and the international groups of combinatorics. More than 100 participants, representing a variety of nationalities and interests, attended the conference.

This conference focused on current research on combinatorics and was divided into three parts: Combinatorial Enumeration and Designs, Graphs and Hypergraphs, and Combinatorial Optimization. For each part, there were three series of instructional lectures which were followed by contributed talks.

This proceedings contains a selection of papers presented at the conference. It is hoped to provide a valuable work describing current research interests in the combinatorial community.

We are grateful to the contributors for their cooperation, many of them kindly submitted their papers in advance which enabled them to be published in the first volume of this proceedings.

We would like to thank the Third World Academy of Sciences, the National Natural Science Foundation of China, Academia Sinica, the Hefei Branch Research Center of Combinatorial Mathematics and Computer Science, Sino-US joint venture Anhui KCDNT High Technology Company for their financial supports. We are also grateful to World Scientific for their efficiency and skill in publishing this proceedings within a short time. Finally, thanks are also due to all referees, especially Prof. G. Z. Liu and her colleagues for their efficiency in refereeing the papers for the conference.

Hononary Chairmen of SSICC '95
P. Erdös
Leetsch C. Hsu
Z. X. Wan

CONTENTS

Equitable Labelings of Forests

Christian Barrientos, Universidad Católica de Valparaíso and Western Michigan University
Italo J. Dejter, Universidad de Puerto Rico, Río Piedras
Héctor Hevia[1], Universidad Católica de Valparaíso

In memory of Sergio Ruiz

ABSTRACT

A *difference labeling* of a graph G is an assignment of distinct integers to the vertices of G. If f is a difference labeling of G, the *weight* of an edge $uv \in E(G)$ is the integer $|f(u) - f(v)|$. A labeling f of a graph G is said to be *k-equitable* if for each weight h induced by f there are exactly k edges of G that have weight h. Equitable labelings of graphs are generalizations of the well-known graceful labelings of graphs and, they were introduced recently by Bloom and Ruiz. A graph G of size q is said to be *equitable* if G is k-equitable for each divisor k of q, $k \neq q$. It is known that cycles are equitable. We show that linear forests are another family of equitable graphs. We prove that forests of even size are 2-equitable. Let w be a positive integer. For $k = 3$ and $k = 4$, we show that if F is a forest of size kw then, F is k-equitable if and only if $\Delta(F) \leq 2w$. We discuss the k-equitability of forests for $k \geq 5$ and characterize all those caterpillars of diameter 2 that are k-equitable, for all possible value of k.

[1]Research supported in part by FONDECYT under Project 1941219 (94)

1. Introduction and basic results

The terminology and notation are taken from [5]. In general, graphs considered here have no isolated vertices. A *difference labeling* of a graph G is an assignment of distinct integers to the vertices of G. In this paper the term *labeling* is used to abbreviate "difference labeling". Assume that f is a labeling of G. The *weight* of an edge $uv \in E(G)$ is the integer $|f(u) - f(v)|$. A labeling of a graph G is called *proper* if all the labels of the vertices of G are non-negative. Observe that every labeling f of a graph G can be transformed into a proper labeling of G that preserves the weights induced by f. Thus, properties that hold for labelings of graphs can be also stated for proper labelings of graphs.

A labeling f of a graph G is *k-equitable* if for each weight h induced by f there are exactly k edges of G that have weight h. In this case, the graph G is said to be *k-equitable.* For example, in Figure 1 we show a 2-equitable labeling of $K_{2,3}$. Equitable labelings of graphs were introduced by Bloom and Ruiz in [3].

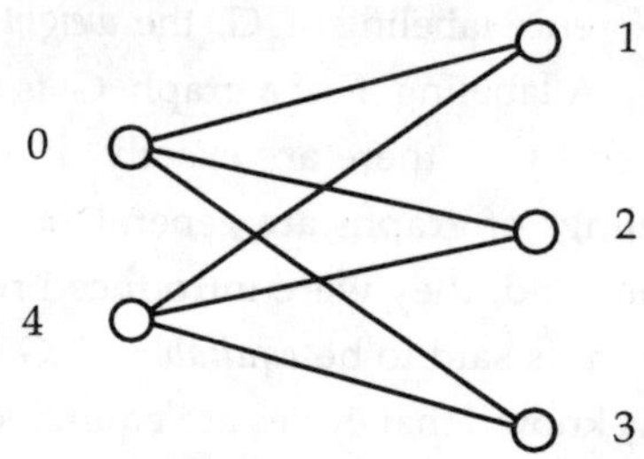

Figure 1

Suppose that f is a k-equitable labeling of a graph G and let h denote the weight of an edge of G. Let S_h be the subgraph of G induced by the edges of G that have weight h and assume that u and v are two adjacent vertices of S_h. Then,

$$|f(u) - f(v)| = h \iff f(u) = f(v) + h \text{ or } f(u) = f(v) - h.$$

So, $\deg_{S_h} u \le 2$ and hence, $\Delta(S_h) \le 2$. Observe that each component C of S_h can be embedded in the real line L by corresponding every vertex $u \in V(C)$ with that point of L whose coordinate is f(u). Therefore, C is a path and S_h is a linear forest. Let W denote the set of the weights induced by f. Then, it follows from the previous observation that $\{S_h\}_{h \in W}$ is a decomposition of G into linear forests of size k. With the help of this observation, we establish the next two results.

Lemma 1 Let G be a graph of size $q > 0$. Then, G is q-equitable if and only if G is a linear forest.

Proof $\Rightarrow$) Since G is q-equitable, there exists an equitable labeling of G in which all the edges of G have the same weight h. Therefore $G \cong S_h$, which is a linear forest.
$\Leftarrow$) Use consecutive integers to label the vertices of each component of the linear forest G. ▯

Assume that the graph G has size q and let f be a k-equitable labeling of G that induces exactly w distinct weights on the edges of G. Then $q = kw$ and, as we state next, the maximum degree of G is bounded above by 2w.

Lemma 2 (of the degree constraint) Let G be a k-equitable graph of size q. Then

$$\Delta(G) \le 2w,$$

where $w = \frac{q}{k}$.

Proof Let f be a k-equitable labeling of G and assume that W is the set of the weights induced by f on the edges of G. Then, $|W| = w = \frac{q}{k}$. For $h \in W$, let S_h denote the subgraph of G induced by the edges of weight h. Let $v \in V(G)$. Since the collection $\{S_h\}_{h \in W}$ of subgraphs of G is a decomposition of G into linear forests, $\deg_{S_h} v \le 2$ and hence, $\deg_G v \le 2w$, for each vertex $v \in V(G)$. ▯

Let G be a graph of size q. Then G is said to be *equitable* if G is k-equitable for each divisor k of q, $k \neq q$. In [11], J. Wojciechowski proved that cycles are equitable. (Bloom and Ruiz had mentioned this property in [3], and C. Barrientos gave a proof of this fact in [1].) In the following section we present another family of equitable graphs.

2. Linear forests are equitable

Suppose that G is a graph of size q and let f be a labeling of G. If the set of the labels of G is a subset of {0, 1, 2, ... , q} and the set of the weights is {1, 2, ... , q} then, the labeling f is called *graceful* and G is said to be a *graceful graph*. Graceful labelings of graphs have been studied extensively since they were introduced by Rosa [9] and Golomb [7]. To this respect, it is still unknown whether all trees are graceful. (See the related surveys [2], [4], and [6].)

Let G be a graph of size q and suppose that f is a k-equitable labeling of G that induces exactly w distinct weights. If the set of the w different weights induced by f is {1, 2, ... ,w} then, f is said to be *complete*. Thus, each graceful labeling of a graph G is a complete 1-equitable labeling of G. In Figure 2 we give a complete 1-equitable labeling of the *nongraceful* graph $K_2 \cup K_3$. This shows that not every complete 1-equitable labeling of a graph G is a graceful labeling.

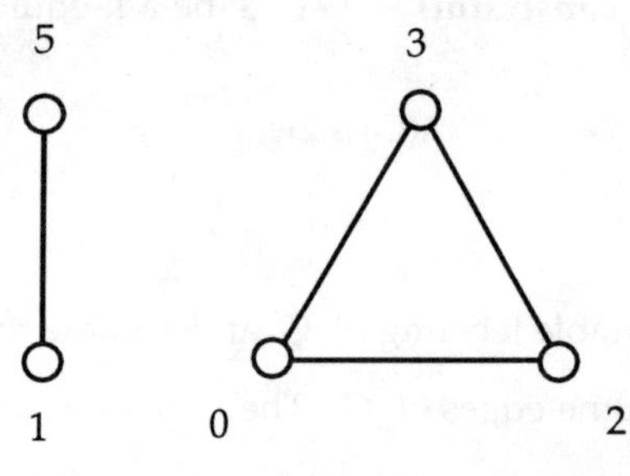

Figure 2

A labeling of a graph G is said to be *optimal* if the labels of the vertices of G are consecutive integers. Thus, if T is a tree then, T is graceful if and only if there exists an optimal complete 1-equitable labeling of T.

It is well-known that paths are graceful graphs. (See [9].) For example, if $P : v_0, v_1, \dots, v_n$ is a path of size n then, the labelings γ_n^0 and γ_n^1 defined below are two graceful labelings of P.

$$\gamma_n^0(v_{2i+1}) = i \qquad 0 \le i \le \left\lfloor \frac{n-1}{2} \right\rfloor$$

$$\gamma_n^0(v_{2i}) = n - i \qquad 0 \le i \le \left\lceil \frac{n-1}{2} \right\rceil$$

$$\gamma_n^1(v_j) = n - \gamma_n^0(v_j) \qquad\qquad 0 \le j \le n.$$

In Figure 3, we show the labelings γ_n^0 and γ_n^1 for $n = 4$ and $n = 5$.

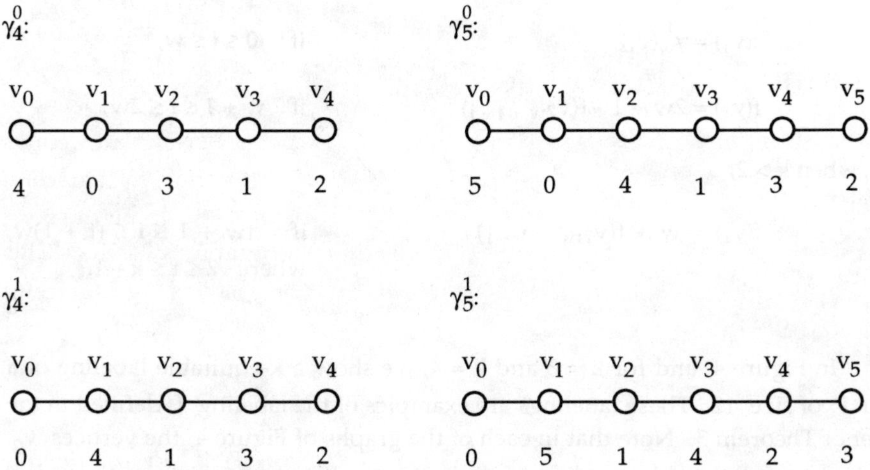

Figure 3

The existence of the graceful labeling γ_n^0 implies the existence of an optimal complete 1-equitable labeling of the path P_{n+1}. A generalization of this result is now presented.

Theorem 3 For each positive integer $n \ge 1$ and for each divisor k of n there exists an optimal complete k-equitable labeling of the path P_{n+1}.

Proof Since the result is clear for $k = 1$ and $k = n$, we may assume $1 < k < n$. Let $w = \frac{n}{k}$. We use a recursive definition to describe a complete k-equitable labeling f of the path $P: v_0, v_1, \cdots, v_n$.

Case 1: $w \equiv 0 \pmod 2$.

$$f(v_i) = \gamma_w^0(v_i) \qquad\qquad \text{if} \quad 0 \le i \le w,$$

$$f(v_i) = 2w - f(v_{2w+1-i}) \qquad\qquad \text{if} \quad w + 1 \le i \le 2w,$$

and, when $k > 2$,

$$f(v_i) = w + f(v_{2tw+1-i}) \qquad \text{if} \quad tw + 1 \le i \le (t+1)w, \quad \text{where } 2 \le t \le k-1.$$

Case 2: $w \equiv 1 \pmod 2$.

$$f(v_i) = \gamma_w^1(v_i) \qquad \text{if} \quad 0 \le i \le w,$$

$$f(v_i) = 2w + 1 - f(v_{2w+1-i}) \qquad \text{if} \quad w + 1 \le i \le 2w,$$

and, when $k > 2$,

$$f(v_i) = w + f(v_{2tw+1-i}) \qquad \text{if} \quad tw + 1 \le i \le (t+1)w, \quad \text{where } 2 \le t \le k-1.$$

□

In Figure 4, and for $k = 3$ and $k = 4$, we show a k-equitable labeling of a path P of size 12. These labelings are examples of the labeling f defined in the proof of Theorem 3. Note that in each of the graphs of Figure 4, the vertices v_0, $v_1, \dots, v_w$ of P have been placed forming a row. The remaining vertices of P are drawn below forming $k - 1$ rows, each containing the vertices $v_{tw+2}, v_{tw+3}, \dots, v_{(t+1)w+1}$, $1 \le t \le k-1$, respectively. In these drawings of P we may also observe that the k-labeling f repeats, level by level, the pattern of labels that the graceful labelings γ_n^0 and γ_n^1 present.

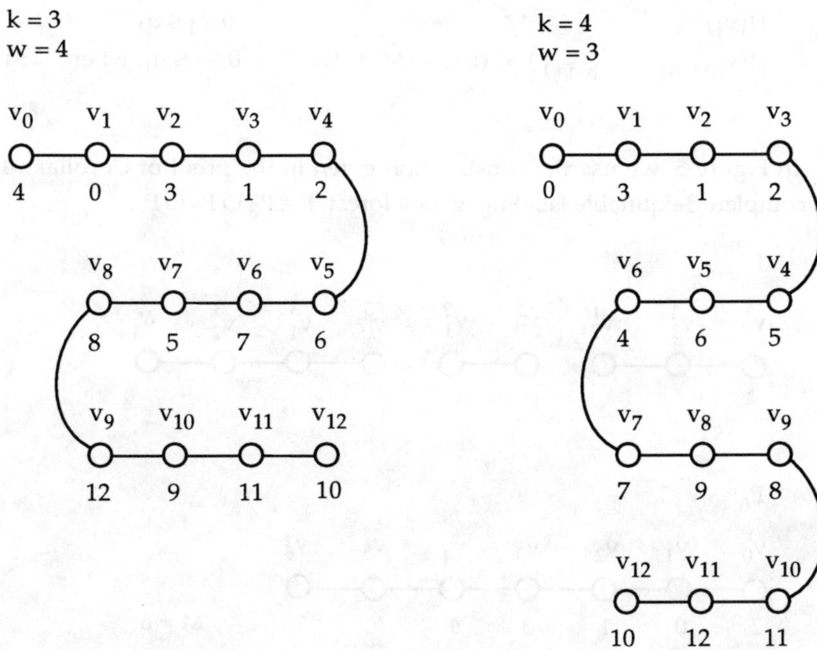

Figure 4

From Theorem 3, it follows that paths are optimal equitable graphs. This result can be partially extended to linear forests, as we show next.

Corollary 4 Let F be a linear forest of size q and suppose that k divides q. Then, there exists a complete k-equitable labeling of F.

Proof Suppose that $F = \bigcup_{i=1}^{m} P_{q_i+1}$ is a linear forest of size q which is isomorphic to the (disjoint) union of m paths $P_{q_i+1} : v_0^{(i)}, v_1^{(i)}, \ldots, v_{q_i}^{(i)}$, $1 \le i \le m$. (So, $q = \sum_{i=1}^{m} q_i$.) We may assume that $q_i > 0$, for each $i = 1, 2, \ldots, t$. Let k be a divisor of q. First, we find a proper complete k-equitable labeling f of the path $P : v_0, v_1, \ldots, v_q$. Let $M = \max \{f(v_i) \mid i = 0, 1, \ldots, q\}$. We use f to define a complete k-equitable labeling g of F.

$$g(v_j^{(i)}) = \begin{cases} f(v_j) & 0 \le j \le q_1 \\ f(v_{q_1+q_2+\dots+q_{i-1}+j}) + (i-1)(M+1) & 0 \le j \le q_i, \text{ where } 2 \le i \le m. \end{cases}$$

□

In Figure 5, we use the construction given in the proof of Corollary 4 to find a complete 3-equitable labeling g of a forest $F \cong P_3 \cup P_2 \cup P_4$.

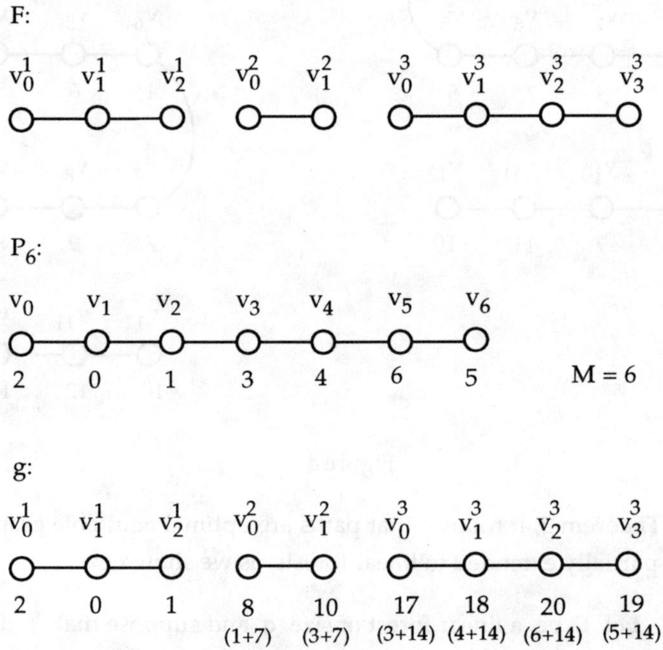

Figure 5

The existence of an optimal complete k-equitable labeling of a graph G is also related to the so-called Skolem-graceful graphs introduced by Lee and Shee in [8].

A graph G of order p and size q is said to be *Skolem-graceful* if there exists a labeling f of G such that $f(V(G)) = \{1, 2, \dots, p\}$ and the set of weights induced by f on the edges of G is $\{1, 2, \dots, q\}$.

Note that a graph G is a Skolem-graceful graph if and only if there exists an optimal complete 1-equitable labeling of G. Using a result from Skolem [10], it follows that the graph nP_2 is Skolem graceful if and only if $n \equiv 0$ or 1 (mod

4). We now state the existence of an optimal k-equitable labeling of the graph nP_2, for each divisor k of n.

Theorem 5 Let $G \cong nP_2$ and suppose that k is a divisor of the size n of G. Then, there exists an optimal k-equitable labeling of G.

In Figure 6 we show an optimal 3-equitable labeling of the graph $12\,P_2$. This example may also suggests a procedure to obtain such a labeling, in the general case stated in Theorem 5.

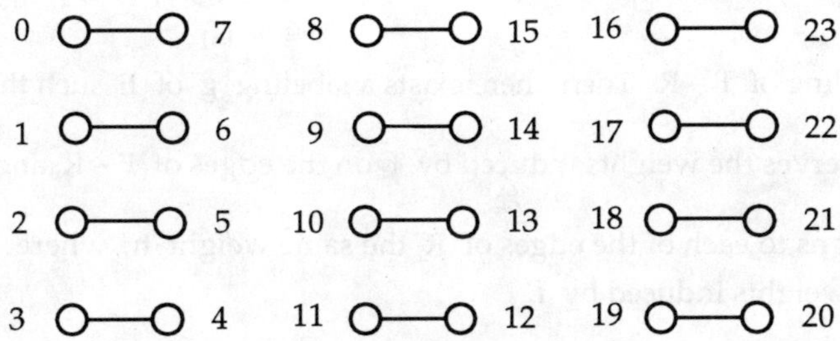

Figure 6

3. The k-equitability of forests, for $k = 2$, 3, and 4.

Let T be a tree of order at least 3 and, suppose that the deletion of all the end-vertices of T transforms T into a path P. Then, the tree T is called a *caterpillar* and the path P is called the *spine* of T. Let P: $v_1, v_2, \dots, v_n$ be the spine of the caterpillar T and, for $i = 1, 2, \dots, n$, let d_i denote the number of end-vertices of T adjacent to v_i. Since the numbers $d_1, d_2, \dots, d_n$ characterize T, we also denote T by $S_{d_1, d_2, \dots, d_n}$. In particular, if $n = 1$ the caterpillar S_{d_1} is isomorphic to the star K_{1,d_1} and, if $n = 2$ the caterpillar S_{d_1, d_2} is called a *double star.*

In [3], Bloom and Ruiz proved that all graphs are 1-equitable. They also proved that the complete graph K_p, $p \geq 2$, is not k-equitable for each $k > 1$. We may obtain this last result, observing that if f is a k-equitable labeling of a complete graph G then, the weight of the edge joining the vertex with the maximum label and the vertex with the minimum label is unique, which implies that $k = 1$.

The smallest graph of even size that is not 2-equitable is K_4. In this section we prove that each forest of even size is 2-equitable. Moreover, for $k = 3$ and for $k = 4$, we give necessary and sufficient conditions for the k-equitability of a forest whose size is divisible by k.

Let F be a forest. A subset $R \neq \emptyset$ of edges of F is called a *linear connector* of F if for each component T of F the edges of $R \cap E(T)$ lie in a single path.

The next lemma plays a main role in the proofs of the results of this section.

Lemma 6 Let R be a linear connector of a forest $F = \bigcup_{i=1}^{m} T_i$. Suppose that f is a proper labeling of $F - R$. Then, there exists a labeling g of F such that:

(i) g preserves the weights induced by f on the edges of $F - R$ and,

(ii) g assigns to each of the edges of R the same weight h, where h is none of the weights induced by f.

Proof Without loss of generality, we may assume that for some r, $1 \le r \le m$, $t_i = |E(T_i) \cap R| > 0$ if $1 \le i \le r$ and, when $r < m$, $E(T_i) \cap R = \emptyset$ if $r < i \le m$. For $i = 1, 2, \ldots, r$, let $e_j^{(i)} = v_{2j-1}^{(i)} v_{2j}^{(i)}$, $1 \le j \le t_i$, be the edges in $E(T_i) \cap R$. Since the edges $e_1^{(i)}, e_2^{(i)}, \ldots, e_{t_i}^{(i)}$ lie in a single path of T_i, it follows that the vertices $v_{2j}^{(i)}$ and $v_{2j+1}^{(i)}$ belong to the same component of $F - R$, for all $j = 1, 2, \ldots, t_i - 1$. (We remark that it may be possible that $v_{2j}^{(i)} = v_{2j+1}^{(i)}$, for some i, $1 \le i \le r$, and for some j, $1 \le j \le t_i - 1$.)
Let $C_0^{(i)}$ be the component of $F - R$ that contains the vertex $v_1^{(i)}$, $1 \le i \le r$, and denote the component of $F - R$ that contains the vertex $v_{2j}^{(i)}$ by $C_j^{(i)}$, $j = 1, 2, \ldots, t_i$. In Figure 7, we depict the component T_i of F and the components $C_j^{(i)}$ of $F - R$ obtained from F by the deletion of the edges of $E(T_i) \cap R$.

T_i:

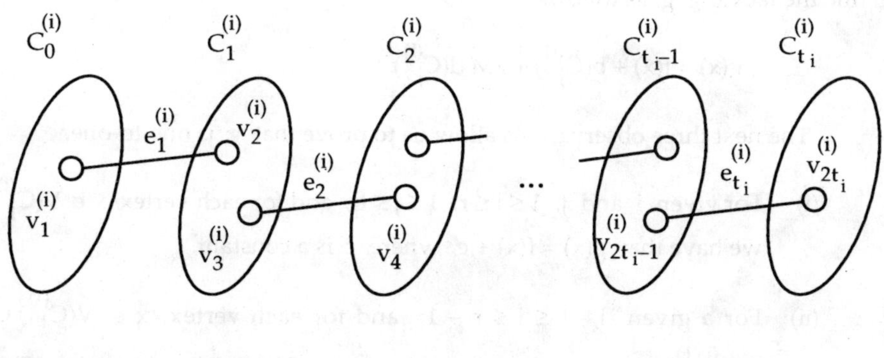

Figure 7

Let $M = \max\{f(x) \mid x \in V(F)\}$. As a matter of convenience, we define $b(e_j^{(i)}) = f(v_{2j-1}^{(i)}) - f(v_{2j}^{(i)})$, for $1 \le j \le t_i$ and $1 \le i \le r$. Assume that x is a vertex of the forest F. Now, we proceed to define the labeling g.

Case 1: $x \in V(T_i)$ for some $i > r$. In this case, define $g(x) = f(x)$.

Case 2: $x \in V(T_i)$ for some i, $1 \le i \le r$. In order to define g, let

$$b(C_j^{(i)}) = \sum_{z=1}^{i-1}\sum_{y=1}^{t_z} b(e_y^{(z)}) + \sum_{y=1}^{j} b(e_y^{(i)}), \text{ and}$$

$$d(C_j^{(i)}) = \sum_{z=1}^{i-1} t_i + j,$$

for each i, $1 \le i \le r$ and for each j, $0 \le j \le t_i$.

(In the definitions given above, take sums equal to 0 if the upper limit of the sum is less than the corresponding lower limit.)

Note that

$$b(C_j^{(i)}) - b(C_{j-1}^{(i)}) = b(e_j^{(i)}) \text{ and } d(C_j^{(i)}) - d(C_{j-1}^{(i)}) = 1, \tag{1}$$

for each $1 \le i \le r$, $1 \le j \le t_i$; and

$$b(C_{t_i}^{(i)}) = b(C_0^{(i+1)}) \text{ and } d(C_{t_i}^{(i)}) = d(C_0^{(i+1)}), \tag{2}$$

for each $1 \le i \le r-1$.

Suppose that $x \in V(C_j^{(i)})$ for some j, $0 \le j \le t_i$. In this second case, we define the labeling g as follows.

$$g(x) = f(x) + b(C_j^{(i)}) + 2M\, d(C_j^{(i)}).$$

The next three observations allow us to prove that g is one-to-one.

(i) For given i and j, $1 \le i \le r$, $1 \le j \le t_i$, and for each vertex $x \in V(C_j^{(i)})$ we have that $g(x) = f(x) + c$, where c is a constant.

(ii) For a given i, $1 \le i \le r-1$, and for each vertex $x \in V(C_{t_i}^{(i)}) \cup V(C_0^{(i+1)})$, we have that $g(x) = f(x) + c'$, where c' is a constant. (This follows directly from (2).)

(iii) For given i and j, $1 \le i \le r$, $1 \le j \le t_i$, let $x_{j-1} \in V(C_{j-1}^{(i)})$ and $x_j \in V(C_j^{(i)})$. Then,

$$g(x_j) > g(x_{j-1}).$$

To see this, we use (1) to obtain

$$\begin{aligned} g(x_j) - g(x_{j-1}) &= (f(x_j) - f(x_{j-1})) + b(e_j^{(i)}) + 2M \\ &= (f(x_j)) - f(x_{j-1})) + (f(v_{2j-1}^{(i)}) - f(v_{2j}^{(i)})) + 2M. \qquad (3) \end{aligned}$$

Now we show that $g(x_j) > g(x_{j-1})$. Since $x_{j-1} \in V(C_{j-1}^{(i)})$ and $v_{2j}^{(i)} \in V(C_j^{(i)})$, it follows that $x_{j-1} \ne v_{2j}^{(i)}$. In a similar manner we obtain that $x_j \ne v_{2j-1}^{(i)}$. From equation (3) and since f is a proper labeling of G, it follows that if $x_{j-1} = v_{2j-1}^{(i)}$ or $x_j = v_{2j}^{(i)}$ then $g(x_j) - g(x_{j-1}) \ge M$. Thus, we may assume that $x_{j-1} \ne v_{2j-1}^{(i)}$ and $x_j \ne v_{2j}^{(i)}$. Since $|f(u) - f(v)| = M$ for at most two vertices u, v of F, we conclude that $g(x_j) - g(x_{j-1}) \ge 1$.

Observation (i) also shows that g preserves the weights induced by f on the edges of $F - R$.

We now prove that g assigns the same weight $2M$ to each edge of R. Let $e \in R$. Then for some i, $1 \le i \le r$, and for some s, $1 \le s \le t_i$, we have that e

$= e_s^{(i)} = v_{2s-1}^{(i)} v_{2s}^{(i)}$. Note that $v_{2s}^{(i)} \in V(C_s^{(i)})$ and $v_{2s-1}^{(i)} \in V(C_{s-1}^{(i)})$. Thus, using (1) we obtain that the weight induced by g on the edge e is

$$|g(v_{2s}^{(i)}) - g(v_{2s-1}^{(i)})| =$$

$$|(f(v_{2s}^{(i)}) + b(C_s^{(i)}) + 2M\, d(C_s^{(i)})) - (f(v_{2s-1}^{(i)}) + b(C_{s-1}^{(i)}) + 2M\, d(C_{s-1}^{(i)}))| =$$

$$|(b(C_s^{(i)}) - b(C_{s-1}^{(i)}) - b(e_s^{(i)})) + 2M\,(d(C_s^{(i)}) - d(C_{s-1}^{(i)}))| = 2M. \qquad \square$$

A *k-linear connector* of a forest F is a linear connector of F that has cardinality k. Note that if F has a component whose diameter is d, then F has a k-linear connector for each $k \le d$.

We next prove that every forest of even size is 2-equitable.

Theorem 7 Every forest of even size is 2-equitable.

Proof Let F be a forest of size $2n$, $n \ge 1$. The proof is by induction on n. If $n = 1$ then, F is a linear forest of size 2 and, by Lemma 1, F is 2-equitable. Let $t \ge 1$ and assume that each forest of size $2t$ is 2-equitable. Observe that each forest of size at least 2 has a 2-linear connector. Let R be a 2-linear connector of a forest G of size $2(t+1)$. Then, by inductive hypothesis, $G - R$ is 2-equitable and, by an application of Lemma 6, it follows that G is 2-equitable. $\square$

We now show that, for $k = 3$ and $k = 4$, the degree constraint given in Lemma 2 is a sufficient condition for the k-equitability of a forest whose size is divisible by k.

Theorem 8 For $k = 3$ and for $k = 4$, let F be a forest of size $q = kw$. Then, the forest F is k-equitable if and only if $\Delta(F) \le 2w$.

Proof $\Rightarrow$) The necessity follows from Lemma 2.
$\Leftarrow$) Let F be a forest of size kw such that $\Delta(F) \le 2w$. We use induction on w to prove that F is k-equitable.

Suppose that $w = 1$. Then F is a linear forest of size k. By Lemma 1, F is k-equitable. Assume that every forest F of size kw, $w \ge 1$, for which $\Delta(F) \le 2w$ is k-equitable. Let G be a forest of size $q = k(w+1)$ such that $\Delta(G) \le 2(w+1)$. To show that G is k-equitable it is convenient to divide the rest of the proof into two cases.

Case 1: $k = 3$. Since G has size at least 6, it follows that in G there are at most two vertices of degree at least $\frac{2q}{3} - 1$ and, at most one of these two vertices may have degree equal to $\frac{2q}{3}$. Now, we proceed to choose a convenient 3-linear connector R of G. We consider three subcases.

Subcase 1.1: There exist two distinct vertices u, v in G that have degree at least $\frac{2q}{3} - 1$. Assume that $\deg u \geq \deg v$. If u and v belong to the same component, let $u, v_1, v_2, \ldots, v_n = v$ be a u-v path in G, and let $e_1 = uv_1$. We choose $R = \{e_1, e_2, e_3\}$, where e_2 and e_3 are two edges different from e_1 that are incident with u and v, respectively. If u and v belong to distinct components of G then, let $R = \{e_1, e_2, e_3\}$ where e_1 and e_2 are two distinct edges incident with u and, e_3 is an edge incident with v.

Subcase 1.2: There exists exactly one vertex u in G with degree at least $\frac{2q}{3} - 1$. If there is a path P of length 3 in G that contains u as one of its central vertices then, take $R = E(P)$. Otherwise, the component C of u is a star of size at most $\frac{2q}{3}$ and G must have another component C' with positive size. Let $e \in E(C')$ and let P be a path of length 2 in C. Then, we choose $R = E(P) \cup \{e\}$.

Subcase 1.3: $\Delta(G) \leq \frac{2q}{3} - 2$. It is straightforward to see that G has a 3-linear connector. In this subcase we choose any 3-linear connector R of G.

Observe that in each of these three subcases the forest $F = G - R$ has size kw and $\Delta(F) \leq \frac{2q}{3} - 2 = 2w$. Therefore, by inductive hypothesis, F is 3-equitable. By an application of Lemma 6 we obtain that G is 3-equitable.

Case 2: $k = 4$. First, we observe that if G has more than two vertices of degree at least $\frac{q}{2} - 1$, then $q = 8$ and G is isomorphic to one of the seven forests of Figure 8. It is straightforward to see that these forests are 4-equitable. Therefore, we may assume that G has at most two vertices of degree at least $\frac{q}{2} - 1$. We now proceed to choose a 4-linear connector R of G. Again, we consider three subcases.

Subcase 2.1: There exist two distinct vertices u, v in G that have degree at least $\frac{q}{2}-1$.

Assume that u and v belong to the same component C of G. If there is a path P of length 4 that contains u and v as internal vertices then, choose $R = E(P)$. Otherwise, the component C is isomorphic to a double star of size at most $q-1$ and G must have another component C' with positive size. Let $e \in E(C')$ and let P be a path of length 3 in C. Then, we choose $R = E(P) \cup \{e\}$.

If u and v belong to distinct components of G then, let P_u and P_v be two paths of length 2 that have u and v as their corresponding central vertices. We choose $R = E(P_u) \cup E(P_v)$.

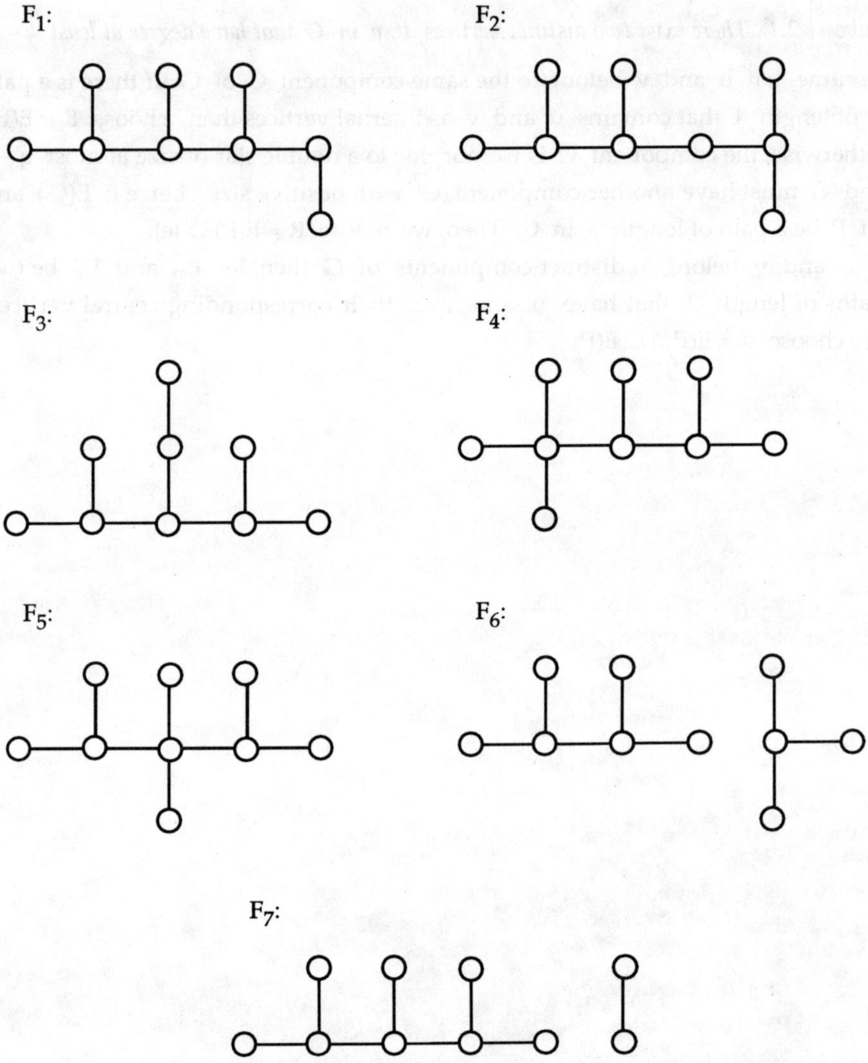

Figure 8

Subcase 2.2: There exists exactly one vertex u *in* G *with degree at least* $\frac{q}{2} - 1$.

Assume that there is a path P of length 4 in G that contains u as one of its internal vertices. Then, we choose $R = E(P)$. Otherwise, the component C of u is a tree of size at most $q-3$ and we can easily find in G a 4-linear connector R that contains two edges incident with u.

Subcase 2.3: $\Delta(G) \le \frac{q}{2} - 2$. It is straightforward to see that G has a 4-linear connector. In this subcase we choose any 4-linear connector R of G.

Similarly to Case 1, we observe that in each of these three subcases the forest $F = G - R$ has size kw and $\Delta(F) \le \frac{q}{2} - 2 = 2w$. Therefore, by inductive hypothesis, F is 4-equitable. By an application of Lemma 6 we obtain that G is 4-equitable. □

Corollary 9 For $k = 3$ and $k = 4$, let F be a forest of size $q = kw$ such that $\Delta(F) \le 2w$. Then F can be decomposed into w linear forests of size k.

Two observations related to Lemma 6 now follow.

Lemma 6 may be used to prove the k-equitability of certain classes of graphs. For example, we can construct a tree T from $m + 1$ paths P_{n_i}, $1 \le i \le m$, $m \ge 1$, by choosing vertices $u^{(n_i)} \in V(P_{n_i})$ and $v^{(n_{i+1})} \in V(P_{n_{i+1}})$, for each $i = 1, 2, \ldots, m$, which are then correspondingly joined by the edges $u^{(n_i)}v^{(n_{i+1})}$. We call the tree T a *fence* of width m. (In Figure 9, we show a fence whose width is 5.) From Lemma 6 and Corollary 4, we obtain the following result.

Theorem 10 Let T be a fence of width k whose size is divisible by k. Then, T is k-equitable.

Note that the fence shown in Figure 9 is 5-equitable. Theorem 10 also shows the existence of k-equitable trees that are not isomorphic to paths, for each positive integer k.

Lemma 6 can be generalized to an arbitrary graph G, extending the definition of a linear connector R to a set of bridges of G such that, for each component C of G, the edges of $R \cap E(C)$ lie in a single path.

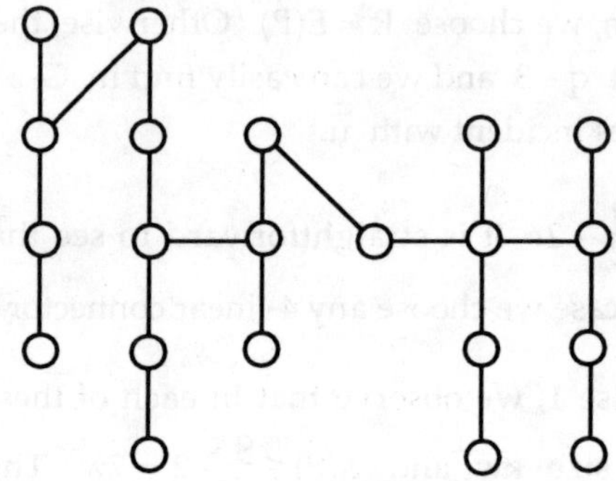

Figure 9

4. On the k-equitability of caterpillars, $k \geq 5$.

For $k \geq 5$, the degree constraint given in Lemma 2 is not a sufficient condition to assure the k-equitability of a forest whose size is divisible by k. For example, the caterpillar T of size 10 shown in Figure 10 is not 5-equitable (see Lemma 11) nevertheless, T satisfies the degree constraint $\Delta(T) \leq 2w = 4$.

T:

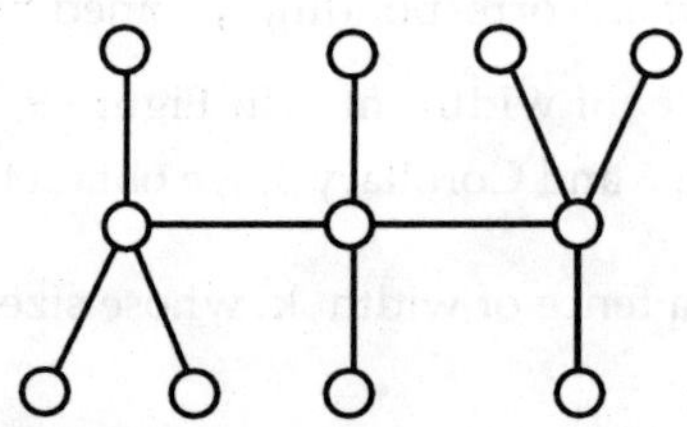

Figure 10

Let f be a k-equitable labeling of a graph G. We already know that the subgraph S_h induced by the edges of G with weight h is a linear forest. For each edge $e \in E(G)$, define $q_L(e)$ as the cardinality of a largest set of edges of G that induces a linear forest that contains e. Then,

$$k \leq \min_{e \in E(G)} \{q_L(e)\}.$$

In particular, if T is a caterpillar whose spine has order $n \geq 2$ then $q_L(e) \leq 2n - 1$, for each edge e that belongs to the spine of T. Thus, in this case, we obtain

$$k \leq 2n - 1.$$

With respect to this last inequality, Joseph McCanna observed (personal communication to one of us) that if $n \geq 3$ then $k < 2n - 1$.

We state these results on caterpillars in the following lemma.

Lemma 11 Let T be a k-equitable caterpillar whose spine has order $n \geq 2$. Then,

$$k \leq \begin{cases} 3 & \text{if } n = 2 \\ 2n-2 & \text{if } n \geq 3. \end{cases}$$

Proof We only have to prove that if $n \geq 3$ then, $k \neq 2n - 1$. Let f be a k-equitable labeling of T and let $n \geq 3$. Suppose, to the contrary, that $k = 2n - 1$. Since $n \geq 3$, there are two adjacent edges e_1 and e_2 in the spine of T whose induced weights are different. For $i = 1, 2$, let $e_i = v_i\, v_{i+1}$ and denote the weight of the edge e_i by a_i. Since $k = 2n - 1$, it follows that for $i = 1, 2$ the linear forest induced by the edges that have weight a_i is isomorphic to $(n-2)\, P_3 \cup P_4$, where e_i is the central edge of the component isomorphic to P_4. The subgraph H of T induced by the edges of weight a_1 or a_2 incident with some of the vertices v_1, v_2, or, v_3 is shown in Figure 10 (a). It is now straightforward to see that the distribution of weights shown in Figure 10 (a) forces the labels of two distinct vertices of H to be equal, which is a contradiction. (See the two basic cases in Figures 11 (b) and (c).) ▯

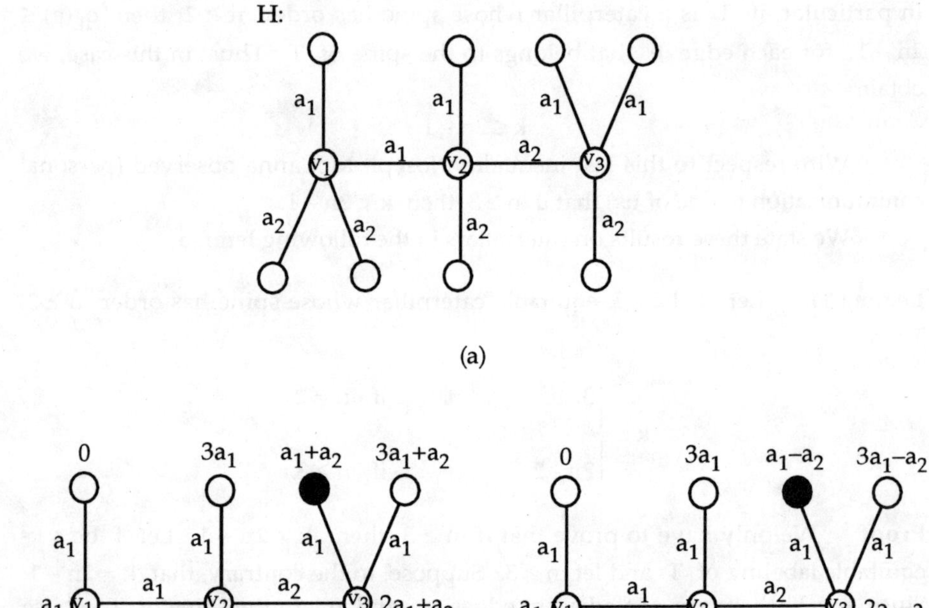

Figure 11

Let W be the set of the weights induced by a k-equitable graph G and let S_h be the subgraph of G induced by the edges of G that have weight h. Then, the collection of subgraphs $\{S_h \mid h \in W\}$ is a decomposition of G into linear forests of size k, called a *common weight decomposition* of G (see [3]). Thus, each k-equitable labeling of G determines a decomposition of G into linear forests of equal size. The caterpillar T in Figure 9 shows that the existence of such a decomposition of G does not guarantee the existence of a corresponding k-equitable labeling of G.

With respect to decompositions of caterpillars into linear forests of equal size, we believe that the following conjecture is true.

Conjecture Let T be a caterpillar of size kw. Then, T has a decomposition into w linear forests of size k if and only if $\Delta(G) \leq 2w$.

5. Equitable labelings of caterpillars

In this section we present results on optimal and complete k-equitable labelings of caterpillars of diameter $d \leq 4$.

Theorem 12 Let T be a caterpillar of even size q. Then, there exists an optimal complete 2-equitable labeling f of T.

Proof The construction of f is a generalization of the well-known construction given by Rosa in [9] to prove that caterpillars are graceful. First, we find two connected edge-disjoint subgraphs T_1 and T_2 of T, each having size $\frac{q}{2}$. Let v be the unique common vertex of T_1 and T_2. Note that for $i = 1, 2$, the vertex v is an end-vertex of the spine of T_i or v is adjacent to such a vertex. In any case, using Rosa's construction, we may find a graceful labeling f_i of T_i, $i = 1, 2$, for which $f_1(v) = \frac{q}{2}$ and $f_2(v) = 0$. We define the labeling f of T as follows.

$$f(x)=\begin{cases} f_1(x) & \text{if } x \in V(T_1) \\ f_2(x) + \frac{q}{2} & \text{if } x \in V(T_2). \end{cases} \qquad \square$$

Following the procedure given in the proof of Theorem 12, in Figure 12 we show some steps in the construction of an optimal complete 2-equitable labeling f of a caterpillar T of even size.

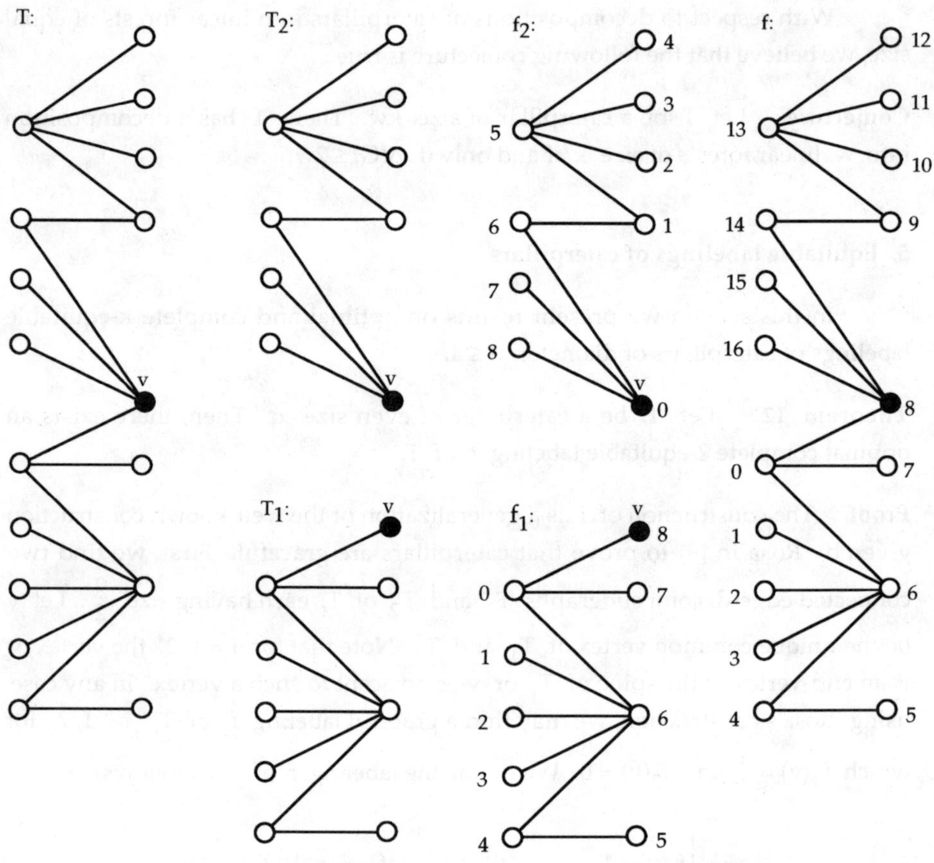

Figure 12

Let T be a k-equitable caterpillar. If the spine of T has order 1 (that is, if $T \cong K_{1,n}$, $n \geq 2$) then, by Lemma 2, $k \leq 2$ and, for these two possible values of k, there exist k-equitable labelings of T that are optimal and complete. (See Theorem 12.)

If the spine of a k-equitable caterpillar T has order 2 (that is, if T is a double star) then, by Lemma 10, $k \leq 3$. For $k = 2$, again Theorem 12 assures the existence of an optimal complete 2-equitable labeling of T. If $k = 3$, it is not true in general that T has a complete 3-equitable labeling. In Figure 13 we show the smallest 3-equitable double star that does not posses a complete 3-equitable labeling.

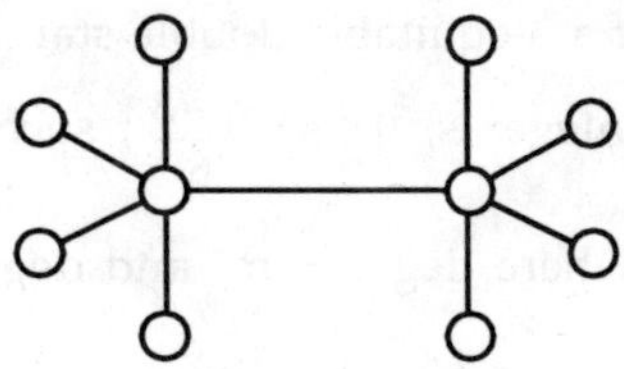

Figure 13

The following lemma will be used to describe a 3-equitable labeling of each 3-equitable double star.

Lemma 13 Let q be the size of the double star $S_{m,n}$, $1 \le m \le n$, and assume that 3 divides q. Then, $S_{m,n}$ is 3-equitable if and only if there exists an integer s, $0 \le s \le \frac{q-3}{6}$, such that $m = \frac{q}{3} + s$ and $n = \frac{2q}{3} - s - 1$.

Proof Let $C = S_{m,n}$. Denote the vertices of the spine of C by u and v, where $\deg u = m + 1$ and $\deg v = n + 1$.
$\Rightarrow$) Suppose that C is 3-equitable. Let $\mathcal{D}$ be a common weight decomposition of C induced by a 3-equitable labeling of C. Observe that the graph D of $\mathcal{D}$ that contains the unique edge of the spine of C must be isomorphic to P_4. On the other hand, each of the graphs in $\mathcal{D} - \{D\}$ is isomorpic to $P_2 \cup P_3$. Note that for each $H \in \mathcal{D} - \{D\}$, the vertex of degree 2 of H is either u or v. Assume that there are exactly s graphs in $\mathcal{D} - \{D\}$ for which u is the vertex of degree 2, and suppose that there are exactly t graphs in $\mathcal{D} - \{D\}$ for which v is the vertex of degree 2. Then,

$$m = 2s + t + 1 \quad \text{and} \quad n = s + 2t + 1. \tag{4}$$

Since the number of graphs of $\mathcal{D}$ is $\frac{q}{3}$, we have

$$\frac{q}{3} = s + t + 1. \tag{5}$$

Combining (4) and (5) we obtain

$$m = \frac{q}{3} + s \quad \text{and} \quad n = \frac{2q}{3} - s - 1,$$

and from $m \le n$, we deduce that $0 \le s \le \frac{q-3}{6}$.

$\Leftarrow$) Clearly, C satisfies the degree constraint. Therefore, by Theorem 7, C is 3-equitable. $\square$

Let q be the size of a 3-equitable double star $C = S_{m,n}$, $1 \le m \le n$. By Lemma 13, there exists an integer s, $0 \le s \le \frac{q-3}{6}$, such that $m = \frac{q}{3} + s$ and $n = \frac{2q}{3} - s - 1$. In Figure 14, where $\deg u = m$ and $\deg v = n$, we describe a 3-equitable labeling of C whose weights constitute the set $\{1, 3, \ldots, \frac{2q}{3} - 1\}$. (When $s = 0$, the bold vertices and their incident edges should be omitted from this figure.)

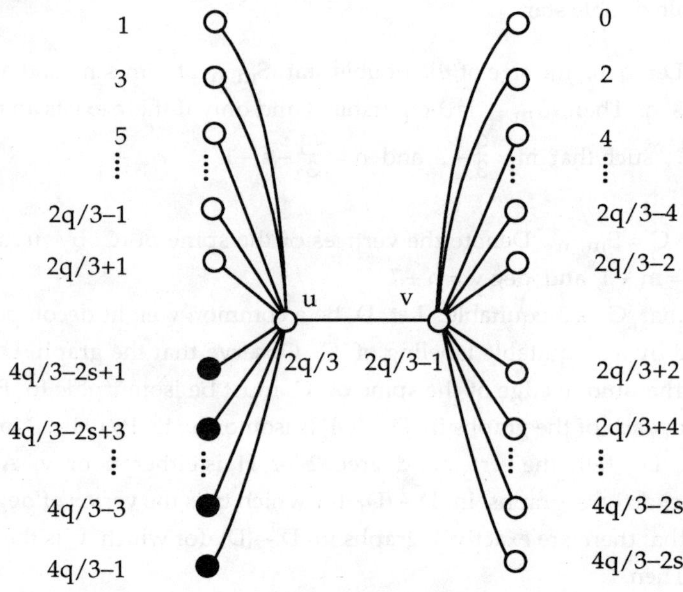

Figure 14

Using Lemma 13, we easily obtain the following characterization.

Theorem 14 Assume that 3 divides the size q of the double star $S_{m,n}$, $1 \le m \le n$. Then, $S_{m,n}$ is 3-equitable if and only if $\frac{q}{3} \le m \le \left\lfloor \frac{q-1}{2} \right\rfloor$.

In particular, we know of an infinite family of double stars that posses optimal complete 3-equitable labelings.

Theorem 15 Let $m \geq 1$. Then, there exists an optimal complete 3-equitable labeling of the double star $S_{m,\,2m-1}$.

Proof Let $C = S_{m,\,2m-1}$. The result is clear if $m = 1$. Let $m \geq 2$. It is straightforward to see that the labeling of C given in Figure 15 satisfies the required conditions. (In Figure 15, deg $u = m + 1$ and deg $v = 2m$.) □

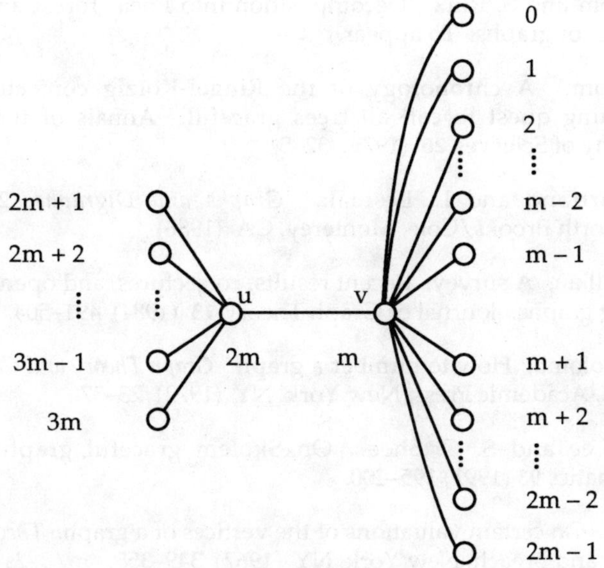

Figure 15

BIBLIOGRAPHY

[1] C. Barrientos, Rotulaciones de grafos en $\mathbb{Z}$. Licenciate Thesis, Universidad Católica de Valparaíso (1990).

[2] J. C. Bermond, Graceful graphs, radio antennae and French windmills. *Graph Theory and Combinatorics.* Pitman, London (1979) 13–37.

[3] G. Bloom and S. Ruiz. Decomposition into linear forest and difference labelings of graphs. To appear.

[4] G. Bloom. A chronology of the Ringel-Kotzig conjecture and the continuing quest to call all trees graceful. Annals of the New York Academy of Sciences **26** (1979) 32–51.

[5] G. Chartrand and L. Lesniak. *Graphs and Digraphs* 2nd. edition. Wadsworth Brooks/Cole, Monterey, CA (1986).

[6] J. A. Gallian. A survey: Recent results, conjectures, and open problems in labeling graphs. Journal of Graph Theory **13** (1984) 491–504.

[7] S. W. Golomb, How to number a graph. *Graph Theory and Computing*. R. C. Read, Academic Press, New York, NY (1972) 23–37.

[8] S. M. Lee and S. C. Shee. On Skolem graceful graphs. Discrete Mathematics **93** (1991) 195–200.

[9] A. Rosa. On certain valuations of the vertices of a graph. *Theory of Graphs*. Gordon and Breach, New York, NY (1967) 349–355.

[10] T. Skolem. On certain distributions of integers in pairs with given differences. Math. Scand. **5** (1957) 57–68.

[11] J. Wojciechowski. Equitable labelings of cycles. Journal of Graph Theory **17** (1993) 531–547.

An Application of the Marriage Theorem to Invariant Theory

Kenneth A. Berman

Department of Electrical & Computer Engineering & Computer Science
University of Cincinnati, Cincinnati OH

Abstract

In this paper we employ the Konig-Hall Matching Theorem (Marriage Theorem) for bipartite graphs to obtain an alternate proof of a combinatorial lemma from invariant theory, known as Kempe's Lemma. Kempe's Lemma is utilized (see [2]) in Hilbert's constructive proof of the existence of a finite generating set for the set of covariants of binary forms of degree n over a field of characteristic zero.

We begin by recalling some pertinent facts from invariant theory to be found in Kung and Rota ([2]). Consider a binary form $f(x,y)$ of degree n in the variables x and y over a field K of characteristic zero, i.e.,

$$f(x,y) = \sum_{k=0}^{n} \binom{n}{k} a_k x^k y^{n-k},$$

where the coefficient of a_k belongs to K, $k = 1,2,\ldots,n$. A linear change of variables (c_{ij}) is a transformation of the variables x and y given by

$$\begin{bmatrix} x \\ y \end{bmatrix} = \begin{bmatrix} c_{11} & c_{12} \\ c_{21} & c_{22} \end{bmatrix} \begin{bmatrix} \bar{x} \\ \bar{y} \end{bmatrix}$$

such that the determinant $c_{11}c_{22} - c_{12}c_{21}$ is nonzero. Under a linear change of variables the binary form $f(x,y)$ is transformed into another binary form $\bar{f}(\bar{x},\bar{y})$ in the new variables $\bar{x}$ and $\bar{y}$ defined by

$$\bar{f}(\bar{x},\bar{y}) = \sum_{k=0}^{n} \binom{n}{k} a_k (c_{11}\bar{x} + c_{12}\bar{y})^k (c_{21}\bar{x} + c_{22}\bar{y})^{n-k}.$$

After expanding and regrouping the terms we obtain

$$\bar{f}(\bar{x},\bar{y}) = \sum_{k=0}^{n} \binom{n}{k} \bar{a}_k \bar{x}^k \bar{y}^{n-k}.$$

A nonconstant polynomial $I(A_0,A_1,\ldots,A_n,X,Y)$ in the variables $A_0,A_1,\ldots,A_n,X,Y$

is a *covariant of binary forms of degree n* if for all binary forms $f(x,y)$ of degree n and all linear changes of variables (c_{ij})

$$I(\overline{a}_0,\overline{a}_1,\ldots,\overline{a}_n,\overline{x},\overline{y}) = (c_{11}c_{22} - c_{12}c_{21})^g \; I(a_0,a_1,\ldots,a_n,x,y)$$

for some nonnegative integer g called the *index* of I. An *invariant* is a covariant in which the variables X and Y do not occur.

The binary form $f(x,y)$ may be expressed as a product of linear forms, i.e.,

$$f(x,y) = (r_1x - s_1y)(r_2x - s_2y) \ldots (r_nx - s_ny).$$

Let A be the alphabet $\{1,2,\ldots,n,u\}$. The *difference* of homogeneous roots (r_i,s_i) and (r_j,s_j), represented by the *bracket* $[i\ j]$, is defined by

$$[i\ j] = \det \begin{bmatrix} r_i & r_j \\ s_i & s_j \end{bmatrix} = r_is_j - r_js_i.$$

The brackets $[i\ u]$ and $[u\ i]$ are defined by $[u\ i] = -\,[i\ u]$ and

$$[i\ u] = \det \begin{bmatrix} r_i & y \\ s_i & x \end{bmatrix} = r_ix - s_iy.$$

A *bracket monomial* is a product of brackets. A bracket monomial M is *regular of degree d* if every letter different from u occurs in exactly d brackets of M. The letter u may occurs in any number of brackets.

The *symmetrization operator S* is defined to act on a bracket monomial $M = [i\ j][k\ l]\ldots[p\ u]$ as follows:

$$<S|M> = \sum_{\pi} [\pi(i)\ \pi(j)][\pi(k)\ \pi(l)]\ldots[\pi(p)\ u],$$

where the summation is over all permutations π of $A - \{u\}$. We call $<S|M>$ a *symmetrized bracket monomial.* If M is regular then $<S|M>$ can be expressed in a unique way as a polynomial in the variables $a_0,a_1,\ldots,a_n,x,y$. This follows form the fundamental theorem of symmetric functions. See Kung and Rota ([2]). In fact $<S|M>$ is a covariant of index g, where g is the number of brackets of M not containing the letter u. Thus, any linear combination of symmetrized regular bracket monomials of the same index is a covariant. Remarkably the converse is true, that is, every covariant can be expressed as a linear combination of symmetrized regular bracket monomials. For a proof of this see Kung and Rota ([2]).

A set $\{M_1, M_2, ...\}$ of regular bracket monomials is a generating set for the set of regular bracket monomials formed with the alphabet A if every regular bracket monomial formed with A can be expressed as a linear combination of products of monomials from $\{M_1, M_2, ...\}$. A set $\{I_1, I_2, ...\}$ of covariants is a generating set for the set of covariants of binary forms of degree n if every covariant of binary forms of degree n can be expressed as a linear combination of covariants from $\{I_1, I_2, ...\}$. The following result is due to Hilbert.

Theorem (Hilbert). Let $\{E_1, E_2, ..., E_m\}$ be a finite generating set for the set of regular bracket monomials formed with the alphabet $A = \{1,2,...,n,u\}$. Then the set

$$\langle S | E_1^{e_1} E_2^{e_2} ... E_m^{e_m} \rangle \qquad 0 \le e_i \le n!\text{-}1,\ e_i \ne 0 \text{ from some } i,$$

$$\langle S | E_1^{n!} \rangle \qquad 1 \le i \le m$$

is a finite generating set for the set of covariants of binary forms of degree n.

A finite generating set for the set of regular bracket monomials formed with A is obtained from the following lemma of Kempe. A *elemental* bracket monomial is a regular bracket monomial of degree one or a regular bracket monomial of degree two which is not the product of two regular bracket monomials of degree one.

Kempe's Lemma. Every regular bracket monomial can be expressed as a linear combination with integer coefficients of elemental bracket monomials.

In [2] Rota and Kung establish this result using circular straightening, a concept dating back to Kempe. In this paper we give a short proof of the following slightly stronger result, using results about matchings and 2-factors in graphs. A *matching* in a graph G is a set of independent edges. A *perfect* matching is a matching that contains every vertex. A *2-factor* is a subgraph having degree 2 at every vertex.

Theorem 1. Let M be a regular bracket monomial of degree d formed with the alphabet $A = \{1,2,...,n,u\}$. If n is even or u occurs in at least d brackets then M can be expressed as a linear combination with integer coefficients of products of regular bracket monomials of degree one. On the other hand, if n is odd and u occurs in fewer than d brackets then M can be expressed as a linear combination with integer coefficients of products of regular bracket monomials of degree one containing u and regular bracket monomials of degree two not containing u.

We proof Theorem 1 with the aid of the following lemma. For G a multigraph and u a vertex of G, let $G - u$ denote the multigraph obtained from G by deleting vertex u and all incident edges.

Lemma 2. Let G be a multigraph with vertex set $A = \{1,2,...,n,u\}$ such that every vertex different from u has degree d.

(i) If n is even and $G - u$ is bipartite then G is decomposable into subgraphs having degree one at every vertex different from u.

(ii) If n is odd, u has degree less than d and A can be bipartitioned into classes X and Y of equal cardinality with $u \in X$ such that no two vertices of X are joined with an edge of G, then G is decomposable into perfect matchings and subgraphs not containing u having degree two at every vertex.

Proof. We first prove statement (i). By the Konig-Hall Matching theorem (see [1]), a bipartite multigraph with maximum degree Δ is decomposable (i.e., can be edge partitioned) into Δ matchings. By hypothesis $G - u$ is bipartite having maximum degree at most d. Therefore, $G - u$ is decomposable into d matchings $\mu_1, \mu_2, ..., \mu_d$, some of which may be *empty*, i.e., contain no edges. If a vertex $v \in V - \{u\}$ has degree d in $G - u$, i.e., v is not adjacent to u in G, then $\mu_1, \mu_2, ..., \mu_d$ all contain v. But, if $v \in V - \{u\}$ has degree $d - k$ in $G - u$, i.e., uv is an edge of G with multiplicity k, then v is contained in all but k of the matchings $\mu_1, \mu_2, ..., \mu_d$. To each of the matchings not containing v add the edge uv. Adding edges in this way for every vertex v that is adjacent to u yields a decomposition of G into subgraphs having degree one at every vertex different from u. This proves statement (i) of Lemma 2.

We now prove statement (ii) of Lemma 2. Let Q denote the set of edges of G having both edges in Y. By hypothesis $G - Q$ (the multigraph obtained from G

by deleting all the edges from Q) is bipartite having maximum degree d. Again, by the Konig-Hall Matching theorem $G - Q$ is decomposable into d matchings. Now every vertex from X has degree d in $G - Q$ except possibly u, and X contains exactly half the vertices. Therefore, the matchings containing u, say $\mu_1, \mu_2, \ldots, \mu_k$ are all perfect matchings. Removing these perfect matchings and deleting vertex u, we obtain an r-regular (where $r = d - k$) multigraph on n vertices. Since (by hypothesis) n is odd, it follows that r is even. But by a result of Petersen (see [1]) any r-regular multigraph, where r is even, is decomposable into 2-factors. This completes the proof of statement 2. ∎

Proof of Theorem 1. Let M be a regular bracket monomial of degree d formed with the alphabet $A = \{1,2,\ldots,n,u\}$. With M we associate the multigraph G on vertex set A, such that two vertices are joined with an edge of G whenever they occur together in a bracket of M.

Case 1: n is even. Arbitrarily divide $A - \{u\}$ into two classes X and Y of equal cardinality. Let P and Q denote the set of edges having both ends in X and both ends in Y, respectively. Let P_u and Q_u denote the set of edges joining u to a vertex from X and to a vertex from Y, respectively. Since every vertex different from u has degree d, it immediately follows that

$$2|P| + |P_u| = 2|Q| + |Q_u|.$$

Assume without loss of generality that $|P| \le |Q|$. Let P' be the set consisting of all the edges from P together with $|Q| - |P|$ edges chosen arbitrarily from P_u. Then $|P'| = |Q|$. If P' and Q are null sets then $G - u$ is bipartite. Otherwise, chose an edge x_1x_2 from P' and an edge y_1y_2 from Q. By the definition of the multigraph G the brackets $[x_1\ x_2]$ and $[y_1\ y_2]$ belong to M. Set $M' = M/([x_1\ x_2][y_1\ y_2])$. We now employ the following syzygy

$$[x_1\ x_2][y_1\ y_2] = [x_1\ y_2][y_1\ x_2] - [x_1\ y_1][y_2\ x_2].$$

This syzygy for 2×2 determinants is easily verified (see [2]). By the syzygy we have

$$M = M_1 - M_2,$$

where $M_1 = [x_1\ y_2][y_1\ x_2]M'$ and $M_2 = [x_1\ y_1][y_2\ x_2]M'$. Let G_1 and G_2 denote the multigraphs corresponding to M_1 and M_2, respectively. Observe that G_1 and

G_2 have degree d at every vertex different from u and that the number of edges in P' (and Q) has been reduced by one. By repeatedly applying the syzygy we can express M as a linear combination with integer coefficients of bracket monomials $M_1, M_2, \ldots, M_m$ such that the multigraph G_i corresponding to M_i has degree d at every vertex different from u and $G - u$ is bipartite. But, by Lemma 2, G_i is decomposable into subgraphs having degree one at every vertex different from u. Hence, M_i can be expressed as a product of regular bracket monomials of degree one.

Case 2: n is odd. First assume that the letter u occurs in at least d brackets of M. Arbitrarily choose a set of d brackets that contain u can replace u in each of these brackets with the letter n+1 to obtain a regular bracket monomial M' formed with the alphabet $\{1,2,\ldots,n+1,u\}$. By Case 1, M' can be expressed as a linear combination with integer coefficients of products of regular bracket monomials of degree one. Replacing the letter n+1 with the letter u in each bracket where the letter n+1 occurs yields an expression for M as a linear combination with integer coefficients of products of regular bracket monomials of degree one.

Now assume that the letter u occurs in fewer than d brackets of M. Then, the multigraph G corresponding to M has degree d at every vertex different from u and degree less than d at u. Arbitrarily divide the vertices of G into two classes X and Y of equal cardinality with u contained in X. Let P and Q denote the set of edges with both ends in X and both ends in Y, respectively. Since u has degree less than d and every other vertex has degree d it follows that $|P| < |Q|$. Thus, we can pair every edge of P with an edge of Q so that no two edges of P paired with the same edge of Q. As in Case 1 we can repeatedly apply the syzygy, each time taking a pair of edges, to express M as a linear combination with integer coefficients of bracket monomials $M_1, M_2, \ldots, M_m$ such that the multigraph G_i corresponding to M_i has the following properties: every vertex of G_i has degree d except u which has degree less than d and no two vertices from X are joined with an edge of G_i. By Lemma 2, G_i is decomposable into perfect matchings and subgraphs not containing u having degree two at every vertex different from u. Hence, M_i can be expressed as the product of regular bracket monomials of degree one containing the letter u and regular bracket monomials of degree two not containing u.

This completes the proof of the Theorem.

References

[1] J.A. Bondy and U.S.R. Murty, *Graph Theory and Applications*, Elsevier North Holland, New York, N.Y., 1980.

[2] J.P.S. Kung and G.-C. Rota, Invariant Theory of Binary Forms, *Bulletin (New Series) of the American Math. Soc.* **10** (1984), 27-85.

On the Optimal Strongly Connected Orientations of City Annular Street Graphs $H_{n_1 \times n_2}$, I: $n_1=2$, $n_2 \equiv 0 \pmod 2$. *

Cai Junliang and Xu Yonghua

(*Taiyuan Heavy Machinery Institute, Taiyuan, Shanxi*, 030024)

Abstract. The problem of finding strongly connected orientations (one — way street assignment) for graphs which arise from annular cities is studied. Specifically, the annular street graphs consisting of n_1 annular streets and n_2 avenues being spokes heading outward from the center are studied. We seek optimal strongly connected orientations according to several important criteria of optimality. Here, we solve the problem for $n_1=2$ and $n_2 \equiv 0 \pmod 2$.

Key words graph theory, strongly connected, orientation, traffic, transportation, annulus, optimization

1. Background

Motivated by the desire to make traffic move more efficiently, we sometimes seek to make every street in a traffic network one — way. In graph — theoretic terms, assuming all streets are two—way to begin with, we seek to find an orientation of an undirected graph, and in particular we seek a strongly connected orientation. Strongly connected orientations of graphs have been studied by Robbins[6], Roberts[7, 8], Chvatal and Thomassen [5], Boesch and Tindell [2], Atallah [1], Chung et al. [4], and Vishkin [13]. While it is easy to find some strongly connected orientations, the problem of finding a most efficient one according to various definitions of efficiency is difficult. Roberts and Xu[9] introduced the idea of studying this problem for specific graphs of interest in traffic, in particular, the city street or grid graphs consisting of n_1+1 east—west avenues and n_2+1 north—south streets. Under a reasonable definition of efficiency, they solved the problem of finding an optimal strongly connected orientations for n_1, n_2 sufficiently large, i.e., $n_1, n_2 \geqslant 4$. Roberts and Xu[10,11,12] began the study of this problem for n_1 or n_2 small, specifically $n_1=1$, $n_2=2$ and $n_1=3$, respectively. Here, we solve the same problem for other types of city street networks, specifically, the annular street graphs.

---

* The project supported by National Natural Science Foundation of China

2. City annular street graphs and criteria for optimality.

For $n_1 \geqslant 1$, $n_2 \geqslant 1$, let $H = H_{n_1 \times n_2} = (V, E)$ be $n_1 \times n_2$ city annular street graph defined as follows. H has vertices v_{ij}, $i = 0, 1, \cdots, n_1 - 1$, $j = 0, 1, 2, \cdots, n_2 - 1$, and an edge $\{V_{kp}, V_{rs}\}$ if and only if $\{|k - r|, |q - s|\} = \{0, 1(\bmod(n_2 - 1))\}$. For each edge of $H_{n_1 \times n_2}$ we evaluate a weight as follows: $|\{V_{ij}, V_{(i+1)j}\}| = r$ and $|\{V_{ij}, V_{i(j+1)}\}| = iar$, where $i = 0, 1, 2, \cdots, n_1 - 1, j = 0, 1, 2, \cdots, n_2 - 1$, $a = 2\pi / n_2$, $r > 0$ $(in_2 = i0)$. The graph $H_{n_1 \times n_2}$ can of course be drawn in the plane. Let C_i, $i = 0, 1, 2, \cdots, n_1 - 1$, be n_1 concentric circles, of which annular streets are composed. The radius of C_i is ir. Let $A_0, A_1, \cdots A_{n_2 - 1}$ be n_2 equipartition points on C_0 and B_0, $B_1, \cdots$, $B_{n_2 - 1}$ be n_2 correspondent points on $C_{n_1 - 1}$ that the straight line A_jB_j, $j = 0, 1, 2, \cdots, n_2 - 1$, passes through the centre of circle C_0. Thus, n_2 spoke avenues are composed of line segments A_jB_j, $j = 0, 1, 2, \cdots, n_2 - 1$, where V_{ij} is the point of intersection of C_i and A_iA_j. Here, we handle the case $n_1 = 2$ and $n_2 \equiv 0 (\bmod 2)$. For convenience, we usually will write j' for vertex V_{1j} and j for vertex V_{2j}, and n for n_2 in $H_{2 \times n}$. Assume that n vertices $0, 1, \cdots, n - 1$ on C_2 and n vertices $0', 1', \cdots, (n - 1)'$ on C_1 both are arranged in a clockwise direction (see Fig. 1).

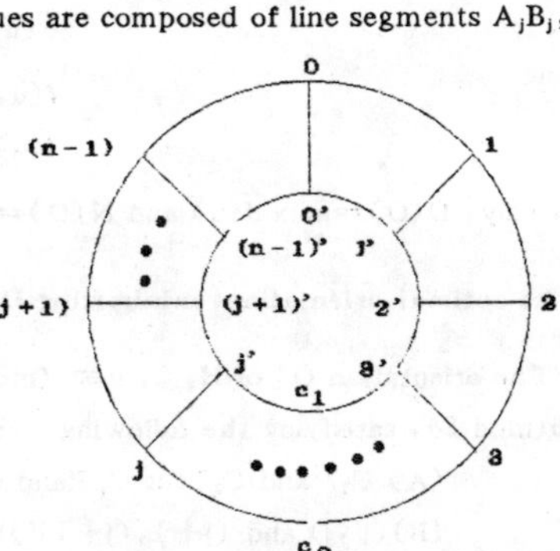

FIG. 1 $H_{2 \times n}$

Suppose O is a strongly connected orientation of $H_{2 \times n}$. In the directed graph H(O), let O denote the set of directed edges of H(O), too. In what follows, let $\{u, v\}$ denote an edge in H, and (u, v) denote a directed edge in O, and $p(u, v)$ denote some path from u to v. A simple closed path is called a cycle. In H(O) any (directed) path (path in this paper will always mean directed path) will be regarded as the subset of O. Let C_1^+, C_1^-, C_2^+, respectively, C_2^- denote the cycle $p_1(o'o') = (o'1'2' \cdots (n - 2')(n - 1)'o')$, $p_2(o'o') = (o'(n - 1)'(n - 2)' \cdots 2'1'0')$. $p_1(0, 0) = (012 \cdots (n - 2)(n - 1)0)$, respectively, $p_2(0, 0) = (0(n - 1)(n - 2) \cdots 210)$. A vertex w on $p(u, v)$, $w \neq u$ and v, is called a turning point if its two adjacent points on $P(u, v)$ are located on C_1 and C_2, respectively. We use the convention that using $(u \cdots w \cdots z \cdots v)$ to denote a path $p(u, v)$, we only indicate the turning points $w, z, \cdots$, if it contains. Let $d(u, v)$ denote the weighted length of the shortest chain in H from u to v, and $\bar{d}(u, v)$ denote the weighted length of the shortest path from u to v in O. j is called an internal knot on C_2 when $(j - 1, j), (j + 1, j) \in 0$; j is called an external knot on C_2 when $(j, j - 1), (j, j + 1) \in 0$. The internal and external knots on C_1 are defined analogously. The number of knots on C_i is denoted by h_i, h_i is even, $i = 1, 2$, and internal knots and external knots on $C_i, i = 1, 2$, are alternate, obviously.

Among the functions which one seeks to minimize are the following functions, which are discussed in Roberts[7,8], Chvatal and Thomassen[5], and Roberts and Xu[9–12] except N

(O).

$$D(O)=\max\{\bar{d}(u,v):u,v\in V\}$$
$$N(O)=\max\{\bar{d}(u,v)-d(u,v):u,v\in V\}$$
$$M(O)=\sum_{u\in V}(\max\{\bar{d}(u,x):x\in V\}-\max\{d(u,x):x\in V\})$$
$$A(O)=\sum_{u,v\in V}\bar{d}(u,v)$$

For $H_{2\times n}$, $n\equiv 0 \pmod 2$, we shall find strongly connected orientations that both mininize D and N. We do not consider the problem of minimizing M and A. We handle the other case in subsequent paper [3].

It is useful to introduce the notations

$$\bar{d}(u)=\max_{v\in V}\bar{d}(u,v)$$
$$f(u,v)=\bar{d}(u,v)-d(u,v)$$
$$\bar{f}(u)=\max_{v\in V}f(u,v)$$

Obviously, $D(O)=\max_{u\in V}\bar{d}(u)$ and $N(O)=\max_{u\in V}\bar{f}(u)$.

3. The optimal orientations minimizing D(O).

The orientation O^* of $H_{2\times n}$, $n\equiv o \pmod 2$. O^* is a strongly connected orientation of $H_{2\times n}$, $n\equiv 0 \pmod 2$, satisfying the following conditions

(A) C_1^+ and C_2^- (or C_1^- and C_2^+)$\subset O^*$;

(B) (j',j) and $(j+1,(j+1)')\in O^*$, $j=0,2,\cdots,(n-2)$.

In the following discuss, we use the notation $[x]$ to denote the part of integer of x.

Lemma 1. Let $k=2[\lambda]$, $\delta=\delta(\lambda)=\lambda-[\lambda]$, where $\lambda=(n-0,5-a^{-1})/3$. Suppose $n\geqslant 10$. Then

$$D(O^*)=\begin{cases}2(n+1-k)ar, & 0<\delta<1/3,\\ (k+5)ar+2r, & 1/3<\delta<1.\end{cases} \qquad (1)$$

Proof. Without loss of generality, suppose $C_1^+\subset O^*$. Clearly, by symmetry of $H(O^*)$, $\bar{d}(O)-\bar{d}(n-1)=2ar>o$ and $\bar{d}((n-1)')-\bar{d}(0')=ar>0$.

It follows that

$$D(O^*)=\max_{u\in V}\bar{d}(u)=\max\{\bar{d}(0),\bar{d}(n-1),\bar{d}(0'),\bar{d}((n-1)')\}$$

$$=\max\{\bar{d}(0),d((n-1)')\}.$$

To prove this lemma, it suffices to show that

(I)

$$\bar{d}(O)=\begin{cases}2(n+1-k)ar, & 0<\delta<1/3,\\ (k+5)ar+2r, & 1/3<\delta<1.\end{cases}$$

(II) $\bar{d}((n-1)')<\bar{d}(0)$

both hold.

Proof(I). Case 1. $0<\delta<1/3$. We have

$$d_1=\overline{d}(0\cdots(k-1))=2(n+1-k)ar$$

and

$$d_2=\overline{d}(0(n-1)(n-1)'\cdots k'k(k-1))=(k+5)ar+2r$$
$$>(2(\lambda-1/3)+5)ar+2r>2(n+1-k)ar=d_1.$$

Hence,
$$\overline{d}(0,k-1)=\min_{w\in V}\overline{d}(1\cdots w\cdots(k-1))=d_1. \tag{2}$$

Note that

$$d_3=\overline{d}(0(n-1)(n-1)'\cdots(k-2)'(k-2))$$
$$=(k+1)ar+2r=(2[\lambda]+1)ar+2r$$
$$=(2\lambda+1+2a^{-1}-2\delta)ar=2(n-k-3\delta)ar<d_1.$$

Thus,
$$\max_{0\leqslant j\leqslant n-1}\overline{d}(0,j)\leqslant d_1. \tag{3}$$

Let $k_1=k+2[0.5a^{-1}]$. A simple counting argument shows that

$$d_4=\overline{d}(0(n-1)(n-1)'\cdots k_1')=(k_1+3)ar+r$$
$$<(2\lambda+3+2a^{-1})ar=2(n+1-2\lambda)ar<d_1$$

and
$$d_5=\overline{d}(0\cdots(k_1+1)(k_1+1)'(k_1+2)')=2(n-1-k_1)ar+r+ar$$
$$<d_1+(a-1)r<d_1.$$

Then
$$\max_{0\leqslant j\leqslant n-1}\overline{d}(0,j')<d_1. \tag{4}$$

By (2), (3) and (4),

$$\overline{d}(0)=d_1=2(n+1-k)ar,\ 0<\delta<1/3. \tag{5}$$

Case 2. $1/3<\delta<1$. The argument similar to that of Case 1 shows that

$$b_1=\overline{d}(0(n-1)(n-1)'\cdots k'k(k-1))=(k+5)ar+2r.$$

and
$$b_2=\overline{d}(0\cdots(k-1))=2(n+1-k)ar>2(n+1-2(\lambda-1/3))ar$$
$$>(k+5)ar+2r=b_1.$$

Hence,
$$\overline{d}(0,k-1)=\min_{w\in V}\overline{d}(0\cdots w\cdots(k-1))=b_1. \tag{6}$$

Note that
$$b_3=\overline{d}(0\cdots(k+1))=2(n-1-k)ar<2(n-1-2(\lambda-1))ar$$
$$=(k+3)ar+2r<b_1.$$

Then
$$\max_{0\leqslant j\leqslant n-1}\overline{d}(0,j)\leqslant b_1. \tag{7}$$

Let $k_2=k+2+2[0.5a^{-1}]$. Thus,

$$b_4=\overline{d}(0(n-1)(n-1)'\cdots k_2')=(k_2+3)ar+r<(k+5+a^{-1})ar+r=b_1$$

and
$$b_5=\overline{d}(0\cdots(k_2+1)(k_2+1)'(k_2+2))=2(n-k_2-1)ar+r+ar$$
$$<(2n-1-2k-2a^{-1})ar+r<b_1+(a-1)r<b_1.$$

Thus,
$$\max_{0\leqslant j\leqslant n-1}\overline{d}(0,j')<b_1. \tag{8}$$

By (6), (7) and (8),

$$\overline{d}(0)=b_1=(k+5)ar+2r,\qquad 1/3<\delta<1. \tag{9}$$

It follows from (5) and (9) that (I) holds.

Proof(II). Case 1. $0<\delta<1/3$. We have

$$x_1=\bar{d}((n-1)'\cdots k_1'k_1(k_1-1))=(k_1+3)ar+r<d_1$$

and $$x_2=\bar{d}((n-1)'0'0\cdots(k_1+1))=2(n-1-k_1)ar+ar+r<d_1.$$

Hence, $$\max_{0\leqslant j\leqslant n-1} d((n-1)',j)<d_1=\bar{d}(0). \tag{10}$$

Let $k_3=k+2+2[a^{-1}]$. Note that

$$x_3=\bar{d}((n-1)'\cdots k_3')=(k_3+1)ar=(k+3+2[a^{-1}])ar$$
$$<2(n+1-k)ar=d_1$$

and $$x_4=\bar{d}((n-1)'0'0\cdots(k_3+1)(k_3+1)'(k_3+2)')$$
$$=2(n-1-k_3)ar+2ar+2r<2(n-k-a^{-1})ar<d_1.$$

Thus, $$\max_{0\leqslant j\leqslant n-1} \bar{d}((n-1)',j')<d_1=\bar{d}(0). \tag{11}$$

By (10) and (11), it follows that (II) holds when $0<\delta<1/3$.

Case 2. $1/3<\delta<1$. We have

$$y_1=\bar{d}((n-1)'\cdots k_2'k_2(k_2-1))=(k_2+3)ar+r<b_1$$

and $$y_2=\bar{d}((n-1)'0'0\cdots(k_2+1))=2(n-1-k_2)ar+ar+r<b_1.$$

Hence, $$\max_{0\leqslant j\leqslant n-1} \bar{d}((n-1)',j)<b_1=\bar{d}(0). \tag{12}$$

Let $k_4=k+4+2[a^{-1}]$. Note that

$$y_3=\bar{d}((n-1)'\cdots k_4')=(k_4+1)ar=(k+5+2[a^{-1}])ar<b_1$$

and $$y_4=\bar{d}((n-1)'0'0\cdots(k_4+1)(k_4+1)'(k_4+2)')$$
$$=2(n-1-k_4)ar+2ar+2r$$
$$=2(n-k-4-2[a^{-1}])ar+2r<2(n-2\lambda-2a^{-1})ar+2r$$
$$<(k+5)ar+2r=b_1.$$

Thus, $$\max_{0\leqslant j\leqslant n-1} \bar{d}((n-1)',j',)<b_1=\bar{d}(0). \tag{13}$$

By (12) and (13), it follows that (II) holds when $1/3<\delta<1$. Therefore, (II) holds.

Theorem 1. Suppose $n\geqslant 12$. If O is a strongly connected orientation of H_{2xn}, $n\equiv 0$ (mode 2), then $D(O)\geqslant D(O^*)$ More over, $D(O)=D(O^*)$ iff $H(O)$ and $H(O^*)$ are isomorphic.

Proof. Assume that $H(O)$ and $H(O^*)$ are nonisomorphic. It suffices to show that $D(O)>D(O^*)$.

First, suppose that there exists a vertex $j\in\{0,1,\cdots,n-1\}$ in V so that $(j,(j+1))$ and $(j',(j+1)')\in 0$. Then

$$D(O)\geqslant\bar{d}(j+1,j)\geqslant(n-1)ar+2r>D(O^*).$$

Thus, in the following argument, we assume that $((j+1)',j')\in O$ when $(j,j+1)\in O$ or vice versa.

Second, suppose that $h_1=h_2=0$. Then , by the assumption mentioned above, C_1 and C_2 have opposite orientations. We might as well assume that $C_1^+\subset O$. By nonisomorphism of $H(O)$ and $H(O^*)$, there exist two adjacent edges $\{j,j'\}$and $\{j+1,(j+1)'\}$ that either $(j,j'),(j+1,(j+1)')\in O$, or $(j',j),((j+1)',j+1)\in O$. Without loss of generality, we assume that

$$((n-1)',n-1),(0',0)\in O. \tag{14}$$

Let $k=2[\lambda](\geqslant 6)$, where $\lambda=(n-0.5-a^{-1})/3$. Now we may assume

$$(n-2,\ (n-2)'),\ ((k-2)',k-2)\in O \tag{15}$$

For suppose (15) fails, then

$$\begin{aligned} D(O) &\geqslant \bar{d}(0,k-2) \geqslant \min\{2(n+2-k)ar,\ (k+7)ar+2r\} \\ &= \min\{2(n+1-k)ar,\ (k+5)ar+2r\}+2ar \\ &= D(O^*)+2ar > D(O^*), \end{aligned}$$

a contradiction. By (14) and (15), we can assume

$$((k-3)',k-3)\in 0. \tag{16}$$

If (16) fails, then

$$D(0)\geqslant \bar{d}(0,k-3)=\min\{2(n+3-k)ar,\ (k+6)ar+2r\}>D(0^*).$$

By (14) and (15), it follows that $D(0) > D(0^*)$ if there are three adjacent edges $\{j-1,(j-1)'\}$, $\{j,j'\}$ and $\{j+1,(j+1)'\}$ with the same orientations from C_1 to C_2 or C_2 to C_1. Thus, by (14), (15) and (16), we may assume

$$(k-1,(k-1)'),\ (k-4,(k-4)')\in 0 \tag{17}$$

In the present case, when $0<\delta<1/3$, we have

$$\begin{aligned} D(0) &\geqslant \bar{d}(0,k-4)=\min\{2(n+4-k)ar,(k+5)ar+2r\} \\ &= (k+5)ar+2r>D(0^*). \end{aligned}$$

When $1/3<\delta<1$, we have

$$\begin{aligned} D(0) &\geqslant \bar{d}(0,k-1)\geqslant \min\{2(n+1-k)ar,(k+8)ar+2r\} \\ &= 2(n+1-k)ar>D(0^*). \end{aligned}$$

Finally, suppose that $h_2=h>0$. Let $j_1, j_2, \cdots, j_h$ be all knots on C_2 and $0=j_1<j_2<\cdots<j_h\leqslant n-1$. Without loss of generality, we assume that $j_2, j_4, \cdots, j_h$ and $j_1, j_3, \cdots, j_{h-1}$ are the internal knots and external knots on C_2, respectively. By the assumption of the first section of the proof, then $j_1', j_3', \cdots, j'_{h-1}$ and $j_2', j_4' \cdots j_h'$ are all internal knots and all external knots on C_1, respectively. We conclude that $h_1=h_2=h>0$. Note that strong connectedness implies that

$$\begin{aligned} \{&(j_1',j_1),(j_3',j_3),\cdots,(j'_{h-1},j_{h-1}), \\ &(j_2,j_2'),\ (j_4,j_4'),\cdots(j_h,j_h')\}\subset 0 \end{aligned} \tag{18}$$

Without loss of generality, we now assume $\bar{d}(0',(n/2)')=\bar{d}(0'01\cdots j_2j_2'\cdots(n/2)')$. Let $j_s<n/2\leqslant j_{s+1}$ and $1\leqslant s\leqslant h$, where $j_{h+1}\equiv 0$. When $s=1$, we have

$$D(0)\geqslant \bar{d}(0',(n/2)')\geqslant nar+2r\geqslant D(0^*).$$

When $3\leqslant s\leqslant h$,

$$D(0)\geqslant \bar{d}(0',(n/2)')\geqslant (n/2+2)ar+4r>D(0^*).$$

Thus, assume $s=2$. If $h\geqslant 5$, then we have

$$D(0)\geqslant \bar{d}(j_5',(j_5-n/2)')>D(0^*).$$

Here, the argument is analogous as above. Now we only need to consider the case: $s=2$ and $h\in\{2,4\}$.

Case 1. Suppose that $s=2$ and $h=4$. Then $0=j_1<j_2<n/2\leqslant j_3<j_4\leqslant n-1$. Let $p=j_3-n/2$. We may assume $0\leqslant p<j_2$. Otherwise, a similar argument as the case $s\neq 2$ leads us to the conclusion that

$$D(0)\geqslant \bar{d}(j_3',p')>D(0^*).$$

By reduction to absurdity, we now assume that $D(0) \leqslant D(0^*) = d$.

Subcase 1. Suppose $0 < p < j_2 < n/2 < j_3 < j_4 \leqslant n-1$. Then we have

$$\bar{d}(0', (n/2)') = (n/2 + j_2)ar + 2r \leqslant D(0) \leqslant d$$

and $$\bar{d}(j_3', p') = (n/2 + j_3 - j_2)ar + 2r \leqslant D(0) \leqslant d.$$

Hence. $$n + 1 + 2a^{-1} - da^{-1}r^{-1} \leqslant j_2 \leqslant da^{-1}r^{-1} - 2a^{-1} - n/2. \tag{19}$$

If $j_4 - j_3 \leqslant n/2 - j_2$, then, by (19),

$$\begin{aligned} D(0) \geqslant \bar{d}(0', n/2) &\geqslant \min\{n/2 + j_2, n/2 + (n - j_4) + (j_3 - n/2)\}ar + 3r \\ &= \min\{n/2 + j_2,\ n - j_4 + j_3\}ar + 3r = (n/2 + j_2)ar + 3r \\ &\geqslant (3n/2 + 1)ar + 5r - d > d, \end{aligned}$$

a contradiction. If $j_4 - j_3 \geqslant n/2 - j_2 + 1$, then, by (19),

$$\begin{aligned} D(0) \geqslant \bar{d}(j_3', p) &\geqslant \min\{n/2 + (j_4 - j_3) + p, n/2 + (j_3 - j_2)\}ar + 3r \\ &\geqslant \min\{n/2 + (n/2 - j_2 + 1) + 1,\ n/2 + (n/2 + 1 - j_2)\}ar + 3r \\ &= (n + 1 - j_2)ar + 3r \geqslant (3n/2 + 1)ar + 5r - d > d, \end{aligned}$$

again a contradiction.

Subcase 2. Suppose $p = 0$, i.e., $j_3 = n/2$. Then we have

$$\bar{d}(0', n/2) = (n/2 + j_2)ar + 3r \leqslant D(0) \geqslant d.$$

That is $j_2 \leqslant da^{-1}r^{-1} - 3a^{-1} - n/2$. Thus,

$$\begin{aligned} D(0) \geqslant \bar{d}((n/2)', 1) &\geqslant \min\{n - 1 - j_2, j_4 + 2\}ar + 3r \\ &\geqslant \min\{(3n/2 - 1)ar + 6r - d, (n/2 + 3)ar + 3r\} > d, \end{aligned}$$

a contradiction.

Case 2. Suppose $s = 2$ and $h = 2$. Then $0 < j_2 < n/2$. Suppose $n \geqslant 14$. We have

$$d \geqslant \bar{d}(0', n/2) \geqslant (n/2 + j_2)ar + 3r.$$

Then $j_2 = 1$. For $j \in (n/2, n/2 - 1, n/2 - 2)$, it follows that $(j', j) \in 0$. Otherwise,

$$d \geqslant \bar{d}(0', j) \geqslant (j + 4)ar + 3r \geqslant (n/2 + 2)ar + 3r > d$$

a contradiction. Thus, we have

$$\begin{aligned} d \geqslant \bar{d}(n/2, 0) &\geqslant \min\{n/2 + 9, n - j_2\}ar + 2r \\ &= \min\{n/2 + 9, n - 1\}ar + 2r > d, \end{aligned}$$

still a contradiction.

Suppose $n = 12$. Then $d = 14ar$. It follows that $j_2 \leqslant 3$ since $d \geqslant \bar{d}(0', 5) \geqslant (5 + j_2)ar + 3r$. If $j_2 = 2, 3$, then $(j', j) \in 0$ for $j \in \{4, 5\}$. Otherwise,

$$d \geqslant \bar{d}(0', j) \geqslant (j + 3 + j_2)ar + 3r > d,$$

a contradiction. In the present case, we have

$$d \geqslant \bar{d}(5, 10) \geqslant 11ar + 2r > d,$$

again a contradiction. If $j_2 = 1$, then $(5', 5) \in 0$ since $\bar{d}(0', 5) \leqslant D(0) \leqslant d$. Thus, $(11', 11) \in 0$ because $\bar{d}(5, 11) \leqslant D(0) \leqslant d$. Then, $d \geqslant D(0) \geqslant \bar{d}(11, 3) > d$, a contradiction.

Let $\tilde{O}$ be a strongly connected orientation of $H_{2\times n}$ and satisfy $h_1 = 0$, $h_2 = n$.

Lemma 2. Suppose $n = 4, 6, 8, 10$. Then $D(\tilde{O}) = (n - 1)ar + 2r$.

Proof. It is easily verified for $\tilde{O}$ directly.

Theorem 2. Suppose $n\geqslant 4$, $n\equiv 0 \pmod 2$. Let $\hat{0}\in\{0^*, \tilde{O}\}$. If 0 is a strongly connected orientation of $H_{2\times n}$, then $D(0)\geqslant D(\hat{0})$. Moreover, we have

(i) Suppose $n\geqslant 12$, $D(0)=D(0^*)$ iff $H(0)\cong(0^*)$;

(ii) Suppose $n=10$. $D(0)=D(\hat{0})$ iff $H(0)\cong H(\hat{0})$;

(iii) Suppose $n=4,6,8$. $D(0)=D(\tilde{O})$ iff $H(0)\cong H(\tilde{O})$.

Proof. (i) follows by Lemma 1 and Theorem 1. The proofs of (ii) and (iii) are omitted.

4. The optimal orientation minizing N(0).

Let $q=2[\mu]$, $\varepsilon=\mu-[\mu]$, where $\mu=(n+2.5+a^{-1})/3$.

Lemma 3. Suppose $n\geqslant 10$, $n\equiv 0 \pmod 2$. Then

$$N(0^*)=\begin{cases}(n+5-q)ar+2r, & 0<\varepsilon<1/3.\\(2q+2-n)ar, & 1/3<\varepsilon<1.\end{cases} \tag{20}$$

Proof. Suppose $C_1^+\subset 0^*$. By symmetry of $H(0^*)$, $N(0^*)=\max_{u\in V}\bar{f}(u)=\max\{\bar{f}(0),\bar{f}(n-1),\bar{f}(0'),\bar{f}((n-1)')\}$. For any $u\in V$, by the definition of $H(0^*)$, we have $\bar{d}(0,u)=\bar{d}(n-1,u)+2ar$ and $d(0,u)=d(n-1,u)\pm iar$, $i=1,2$. Then

$$\begin{aligned}\bar{f}(0)-\bar{f}(n-1)&=\max_{u\in V}f(0,u)-\max_{v\in V}f(n-1,v)\\&=\max_{u\in V}\{\bar{d}(0,u)-d(0,u)\}-\max_{v\in V}\{\bar{d}(n-1,v)-d(n-1,v)\}\\&=2ar\pm iar\geqslant 0,\ i=1,2.\end{aligned}$$

Then $\bar{f}(0)\geqslant\bar{f}(n-1)$ follows, By the same reasoning, $\bar{f}((n-1)')\geqslant\bar{f}(0')$.

We conclude that $N(O^*)=\max\{\bar{f}(0),\ \bar{f}((n-1)')\}$.

It is easy to verify that

$$\bar{d}((n+1)',j')=\begin{cases}(j+1)ar, & 0\leqslant j<q-1,\\ min\{(j+1)ar,\ (2n+1-2j)ar+2r\}, & j=q-1,\\ min\{(j+1)ar,\ 2(n+2-j)ar+2r\}, & j=q,\\ 2(n+2-j)ar+2r,\ j\equiv 0(mod\ 2), & q<j<n-1,\\ (2n+3-2j)ar+2r,\ j\equiv 1(mod\ 2), & q<j<n-1,\end{cases}$$

where $q=2[(n+2.5+a^{-1})/3]>n/2+1$, and

$$\bar{d}((n-1)',j')=\begin{cases}(j+1)ar, & 0\leqslant j<n/2,\\(n-1-j)ar, & n/2\leqslant j<n-1.\end{cases}$$

Then we have

$$f((n-1)',j')=\begin{cases}0, & 0\leqslant j<n/2,\\ (2j+2-n)ar, & n/2\leqslant j<q-1,\\ min\{(2j+2-n)ar,\ (n+2-j)ar+2r\}, & j=q-1,\\ min\ \{(2j+2-n)ar,\ (n+5-j)ar+2r\}, & j=q,\\ (n+5-j)ar+2r,\ j\equiv 0(mod\ 2), & q<j<n-1,\\ (n+3-j)ar+2r,\ j\equiv 1(mod\ 2), & q<j<n-1.\end{cases}$$

and $\max\limits_{0\leqslant j\leqslant n-1} f((n-1)',j')=f((n-1)',q')$

$$=\min\{(2q+2-n)ar,\ (n+5-q)ar+2r\}. \tag{21}$$

It is easily verified that

$$\bar{d}((n-1)',j)=\begin{cases}(j+1)ar+r,\ j\equiv 0(mod\ 2), & 0\leqslant j<t-1,\\ (j+3)ar+r,\ j\equiv 1(mod\ 2), & 0\leqslant j<t-1,\\ min\ \{\ (j+4)ar+r,(2n+1-2j)ar+r\ \}, & j=t-1,\\ (2n+1-2j)ar+r\ , & t\leqslant j\leqslant n-1,\end{cases}$$

where $t=2[(n+1)/3]>n/2+1$, and

$$d((n-1)',\ j)=\begin{cases}(j+1)ar+r, & 0\leqslant j<n/2,\\ (n-1-j)ar+r, & n/2\leqslant j<n-1.\end{cases}$$

Hence, we have

$$f((n-1)',\ j)=\begin{cases}0, & j\equiv 0(mod\ 2),\ 0\leqslant j<n/2\\ 2ar, & j\equiv 1(mod\ 2),\ 0\leqslant j<n/2\\ (2j+2-n)ar & j\equiv 0(mod\ 2),\ n/2\leqslant j<t-1\\ (2j+4-n)ar, & j\equiv 1(mod\ 2),\ n/2\leqslant j<t-1\\ min\{(2j+5-n)ar,(n+2-j)ar\}, & j=t-1\\ (n+2-j)ar, & t\leqslant j\leqslant n-1\end{cases}$$

and $\max\limits_{0\leqslant j\leqslant n-1} f((n-1)',j)=f((n-1)',t-1)$

$$=\min\{(2t+3-n)ar,\ (n+3-t)ar\}. \tag{22}$$

By (21) and (22), it is easy to verify that

$$\max_{0\leqslant j\leqslant n-1} f((n-1)'.\ j')\geqslant \max_{0\leqslant j\leqslant n-1} f((n-1)',j).$$

We conclude that

$$\bar{f}((n-1)')=\max_{u\in V} f((n-1)',u)=\max_{0\leqslant j\leqslant n-1} f((n-1)',j'),$$

that is ,

$$\overline{f}((n-1)')=\begin{cases}(n+5-q)ar+2r, & 0<\varepsilon<1/3,\\ (2q+2-n)ar, & 1/3<\varepsilon<1.\end{cases} \tag{23}$$

Similarily, it is easily verified that

$$\overline{d}(0,j)=\begin{cases}(j+3)ar+2r & j\equiv 0(mod\ 2),\ 0<j<k-1,\\ (j+6)ar+2r, & j\equiv 1\ (mod\ 2),\ 0<j<k-1,\\ min\{2(n-j)ar,\ (j+6)ar+2r\}, & j=k-1,\\ min\{2(n-j)ar,\ (j+3)ar+2r\}, & j=k,\\ 2(n-j)ar\quad , & k<j\leqslant n-1,\end{cases}$$

where $k=2[(n-0.5-a^{-1})/3]$, and

$$d(0,j)=\begin{cases}2jar & 0<j<2a^{-1},\\ jar+2r, & 2a^{-1}<j\leqslant n/2,\\ (n-j)ar+2r, & n/2\leqslant j<n-2a^{-1},\\ 2(n-j)ar, & n-2a^{-1}<j\leqslant n-1.\end{cases}$$

Thus, the values of $f(0,j)=\overline{d}(0,j)-d(0,j)$ are shown in the following table.

$f(0,j)$	$n/2<k-1$	$n/2=k-1$	$n/2=k$	$n/2>K$
$(3(\eta+1)-j)ar+2r$	$0<j<2a^{-1}$			
$3(\eta+1)ar$	$2a^{-1}<j<n/2$	$2a^{-1}<j<k-1$		
$(2j+3(\eta+1)-n)ar$	$n/2\leqslant j<k-1$			
$\min\{(n-j)ar-2r,(2j+6-n)\}ar$	$j=k-1$			
$\min\{(2n-3j)ar-2r,6ar\}$			$j=k-1$	
$\min\{(n-j)ar-2r,(2j+3-n)ar\}$	$j=k$			
$\min\{(2n-3j)ar-2r,3ar\}$			$j=k$	
$(n-j)ar-2r$	$k<j<n-2a^{-1}$			
$(2n-3j)ar-2r$				$k<j\leqslant n/2$
$(n-j)ar-2r$			$n/2\leqslant j<n-2a^{-1}$	
0	$n-2a^{-1}<j\leqslant n-1$			

where $j\equiv\eta(mod\ 2)$, $\eta=0,1$. By the table, we have

$$\max_{0\leqslant j\leqslant n-1} f(0,j) = f(0,k-1)$$
$$= \begin{cases} \min\{(n+1-k)ar - 2r,\ (2k+4-n)ar\}, & n/2 \leqslant k-1, \\ \min\{(2n+3-3k)ar - 2r,\ 6ar\}, & n/2 \geqslant k, \end{cases} \quad (24)$$

and

$$\max_{0\leqslant j\leqslant n-1} f(0,j') = \max_{0\leqslant j\leqslant n-1} \bar{f}((n-1)',j) \leqslant \bar{f}((n-1)').$$

By (23) and (24), it follows that

$$\bar{f}((n-1)') \geqslant f(0,k-1) = \max_{0\leqslant j\leqslant n-1} f(0,j).$$

Hence,

$$\bar{f}(0) = \max_{u\in V} f(0,u)$$
$$= \max\{\max_{0\leqslant j\leqslant n-1} f(0,j'),\ \max_{0\leqslant j\leqslant n-1} f(0,j)\} \leqslant \bar{f}((n-1)').$$

We conclude that $N(O^*) = \bar{f}((n-1)')$, and (20) holds by (23).

Theorem 3. Suppose $n\geqslant 10$, $n\equiv 0 \pmod 2$. If 0 is a strongly connected orientation of $H_{2\times n}$, then $N(0)\geqslant N(0^*)$.

Proof. In $H(0)$, if there exists a vertex $j\in\{0,1,\cdots,n-1\}$ so that $(j,j+1)$, $(j',(j+1)')\in 0$, then $N(0)\geqslant f(j+1,j)\geqslant (n-3)ar+2r\geqslant N(O^*)$. Thus, if $(j,j+1)\in 0$. then we assume that $((j+1)',j')\in 0$ or vice versa.

Case 1. Suppose $n\geqslant 10$, $h_1=h_2=0$ and $C_1^+, C_2^-\subset 0$. If $H(O)\cong H(O^*)$, then $N(O)=N(O^*)$. Now assume that $H(O)$ and $H(O^*)$ are nonisomorphic. Hence, there exist two adjacent edges $\{j,j'\}$ and $\{j+1,(j+1)'\}$ that either (j,j'), $(j+1,(j+1)')\in 0$, or (j',j), $((j+1)',j+1)\in 0$. We now assume that $(n-2,(n-2)')$, $(n-1,(n-1)')\in 0$ without loss of generality. If there is a vertex $j\in\{q,q+1,q+2\}$ so that $(j',j)\in 0$, then

$$\bar{d}((n-2)',j') \geqslant \min\{(2n+5-2j)ar+2r,\ (j+2)ar\}$$

and $$d((n-2)',j') = (n-2-j)ar.$$

Hence,

$$N(0) \geqslant f((n-2)',j') \geqslant \min\{(n+7-j)ar+2r, (2j+4-n)ar\}$$
$$\geqslant \min\{(n+5-q)ar+2r, (2q+2-n)ar\} = N(O^*).$$

So we now assume that $(j,j')\in 0$ for any $j\in\{q,q+1,q+2\}$. Let $s=2q+1-n$, where $0<s<q$. Then

$$\bar{d}(q',s') \geqslant \min\{(2q-2s+9)ar+2r,\ (n-q+s)ar\}$$
$$= \min\{(2n+7-2q)ar+2r, (q+1)ar\}$$

and $$d(q',s') = (q-s)ar = (n-1-q)ar.$$

It follows that

$$N(O) \geqslant f(q',s') \geqslant \min\{(n+8-q)ar+2r,\ (2q+2-n)ar\} \geqslant N(O^*).$$

Case 2. Suppose $n\geqslant 16$ and $h_1=h_2=h>0$. Let $j_1', j_2' \cdots, j_h'$ and $j_1, j_2, \cdots, j_h$ are knots on C_1 and on C_2, respectively, of which the definitions are the same as that of (18). Let $j_s<n/2\leqslant j_{s+1}$, $1\leqslant s\leqslant h$ $(j_{h+1}\equiv 0)$. Then $\bar{d}(0',(n/2)') = \min\{\bar{d}(0'0\cdots j_s\cdots(n/2)(n/2)'), \bar{d}(0'0\cdots j_h\cdots j_{s+1}$

$\cdots(n/2)')\}$. Without loss of generality, we assume that $\bar{d}(0',(n/2)')=\bar{d}(0'0\cdots j_s\cdots(n/2)(n/2)')$. Otherwise, we reverse the numbers of vertices of $H_{2\times n}$ on C_i, $i=1,2$. Let e denote $N(O^*)$ and b denote $(ea^{-1}r^{-1}-2a^{-1})$.

Let

$$\hat{b}=\begin{cases} b-1 & 0<\varepsilon<1/3, \\ [b], & 1/3<\varepsilon<1. \end{cases}$$

Then we have

$$\bar{d}(0',(n/2)')\geqslant\begin{cases}(n/2+3)ar+6r, & s\geqslant 5, \\ (n/2+2)ar+4r, & s=3,4, \\ nar+2r, & s=1,\end{cases}$$

and $\quad d(0',(n/2)')=nar/2$.

Thus,

$$N(0)\geqslant f(0',(n/2)')\geqslant\begin{cases}3ar+6r\geqslant e, & s\geqslant 5, \\ 2ar+4r\geqslant e, & s=3,4, \\ nar/2+2r\geqslant e, & S=1.\end{cases} \tag{25}$$

Now we suppose $s=2$. Then we may assume that $h=2,4$. Otherwise, it is easy to verify that $N(O)\geqslant f(j_5',(j_5-n/2)')\geqslant e$. By reduction to absurdity, we assume that $N(O)<N(O^*)=e$.

(i) Suppose $h=2$. We have $0<j_2<n/2$. Note that $e>N(O)\geqslant f(0',j_2')=j_2ar+2r$. Thus, $1\leqslant j_2\leqslant\hat{b}$. Let

$$q_1=(2n-j_2-i)/3, \tag{26}$$

where $2n-j_2\equiv i(\mathrm{mod}\ 3)$, then $q_1>n/2$. Suppose that $((n-j)',n-j)\in 0, 0\leqslant j\leqslant t$, and $(n-t-1,(n-t-1)')\in 0$, where $(n',n)=(0',0)$. When $t\leqslant q_1-j_2$, i.e., $q_1-t\geqslant j_2$, by (26), we have

$$\begin{aligned} e&>N(0)\geqslant f((n-t-1)',(q_1-t)') \\ &\geqslant\min\{n+2-q_1,\ 2q_1+j_2+2-n\}ar+2r \\ &=(n+6+j_2-2i)ar/3+2r\geqslant e, \end{aligned}$$

a contrdiction. Now suppose $t\geqslant q_1-j_2+1$, i.e., $j_2\geqslant q_1+1-t$. Then

$$\begin{aligned} e&>f(0',(n-1)')\geqslant\min\{3t+1,n-2+j_2\}ar+2r \\ &\geqslant\min\{3t+1,n-1\}ar+2r=(3t+1)ar+2r. \end{aligned}$$

Hence, $q_1-j_2+1\leqslant t\leqslant(\hat{b}-1)/3$. By (26), we have

$$n/2+1-\hat{b}/4-i/4\leqslant j_2\leqslant\hat{b}. \tag{27}$$

Note that $n-q_1\geqslant 2q_1+j_2-n$. Thus, by (27), we have

$$e>f(0',q_1')\geqslant\min\{n-q_1,2q_1+j_2-n\}ar+2r$$
$$=(2q_1+j_2-n)ar+2r=(n+j_2-2i)ar/3+2r\geqslant e,$$

a contradiction.

(ii) Suppose $h=4$. Let $p=j_3-n/2$. Then we have $0\leqslant p\leqslant j_2$. Otherwise, an argument similar to that of (25) shows that $e>N(O)\geqslant f(j_3',p')\geqslant e$, again a contradiction.

Claim 1. $1\leqslant j_2\leqslant\hat{b}$, $j_3-j_2\leqslant\hat{b}$, $j_4-j_3\leqslant\hat{b}$ and $n-j_4\leqslant\hat{b}$, (28)

where $0=j_1\leqslant p<j_2<n/2\leqslant j_3<j_4\leqslant n-1$, $p=j_3-n/2$, $\hat{b}\leqslant n/2-2$.

Verification. It is easily verified that $\hat{b}\leqslant n/2-2$ when $n\geqslant 16$. Note that $e>f(0',j_2')=j_2ar+2r$. Then $j_2\leqslant\hat{b}$. Using the analogous argument, we can verify the others.

Claim 2. Assume that $\overline{d}(0',j_3')=\overline{d}(0'0\cdots j_2j_2'\cdots j_3')$, then $j_3-j_2=1$. Verification. Suppose $j_3-j_2>1$. If there exists a vertex $j\in\{j_2+1,j_2+2,\cdots,j_3-1\}$ so that $(j,j')\in 0$, then, by (25) and (28), we have $e>f(0',j)\geqslant(j_2+3)ar+2r$ and $e>f((j_4+1)',(j_3-1)')\geqslant(n-4+j_2-2(j_4-j_3))ar+2r$.

Hence,

$$j_2\leqslant\hat{b}-3, \tag{29}$$

and

$$j_4-j_3\geqslant(n-4-\hat{b}+j_2)/2. \tag{30}$$

By (25), (28), (29) and (30), we have

$$e>f((j_2+1)',(n-1)')\geqslant(n-2j_2+j_4-j_3-4)ar+2r$$
$$\geqslant(n-4-3j_2/2+(n-4-\hat{b})/2)ar+2r$$
$$\geqslant(3n-4\hat{b}-3)ar/2+2r\geqslant e,$$

a contradiction. If $(j',j)\in 0$ for any $j\in\{j_2+1,j_2+2,\cdots,j_3-1\}$, then, by (25) and (28), we have

$$e>f((j_4-1)',1')\geqslant(n-4-2(n-j_4)+j_3-j_2)ar+2r$$

and

$$e>f(j_3-1,j_3)$$
$$\geqslant\min\{(3(j_3-j_2)-4)ar+2r,(2n-4+j_3-j_2-j_4)ar+4r\}$$
$$\geqslant\min\{(3(j_3-j_2)-4)ar+2r,(n-1)ar+4r\}.$$

Thus,

$$n-j_4\geqslant(n-4-\hat{b}+(j_3-j_2))/2 \tag{31}$$

and

$$j_3-j_2\leqslant(\hat{b}+4)/3. \tag{32}$$

By (25), (28), (31) and (32), we have

$$e>f((j_2-1)',(j_3+1)')\geqslant(n-4-2(j_3-j_2)+n-j_4)ar+2r$$
$$\geqslant(n-4-3(j_3-j_2)/2+(n-4-\hat{b})/2)ar+2r$$
$$\geqslant(3n-2\hat{b}-16)ar/2+2r\geqslant e,$$

a contradiction.

Claim 3. Assume that $\overline{d}(0',j_3')=\overline{d}(0'0\cdots j_4j_4'\cdots j_3')$. Then $j_4-j_3=1$. Verification. Suppose

$j_4-j_3>1$. We shall reach a contradiction. If there exists a vertex $j\in\{j_3+1,j_3+2,\cdots,j_4-1\}$ so that $(j,j')\in 0$, then, by (25) and (28), we have $e>f(o',j)\geqslant(n-j_4+3)ar+2r$ and

$$e>f((j_2-1)',(j_3+1)')\geqslant(n-4-2(j_3-j_2)+n-j_4)ar+2r.$$

Thus,

$$n-j_4\leqslant\hat{b}-3 \tag{33}$$

and

$$j_3-j_2\geqslant(2n-4-\hat{b}-j_4)/2. \tag{34}$$

By (25), (28), (33) and (34), we have

$$\begin{aligned}e&>f((j_4-1)',1')\geqslant(n-4-2(n-j_4)+j_3-j_2)ar+2r\\&\geqslant(n-4-3(n-j_4)/2+(n-4-\hat{b})/2)ar+2r\\&\geqslant(3n-4\hat{b}-3)ar/2+2r\geqslant e,\end{aligned}$$

a contradiction. If $(j',j)\in 0$ for any $j\in\{j_3+1,j_3+2,\cdots,j_4-1\}$, then, by (25) and (28), we have

$$e>f(j_3+1,j_3)\geqslant(3(j_4-j_3)-4)ar+2r$$

and

$$e>f((j_2+1)',(n-1)')\geqslant(n-4-2j_2+j_4-j_3)ar+2r.$$

Thus,

$$j_4-j_3\leqslant(\hat{b}+4)/3 \tag{35}$$

and

$$j_2\geqslant(n-4-\hat{b}+j_4-j_3)/2. \tag{36}$$

By (25), (28), (35) and (36), we have

$$\begin{aligned}e&>f((j_4+1)',(j_3-1)')\geqslant(n-4-2(j_4-j_3)+j_2)ar+2r\\&\geqslant(n-4-3(j_4-j_3)/2+(n-4-\hat{b})/2)ar+2r\\&\geqslant(3n-2\hat{b}-16)ar/2+2r\geqslant e,\end{aligned}$$

a contradiction. By Claim 2, $j_3-j_2=1$ when $\overline{d}(0',j_3')=\overline{d}(0'0\cdots j_2j_2'\cdots j'_3)$. That is $j_3=n/2=j_2+1$, hence we have $\overline{d}(j_3',0')=\overline{d}(j_3'j_3\cdots j_2j_2'\cdots 0')$. Then $j_2-j_1=j_2=1$. By Claim 1, $n=4$, a contradiction. By Claim 3, $j_4-j_3=1$ when $\overline{d}(0',j_3')=\overline{d}(0'0\cdots j_4j_4'\cdots j_3')$. Then we have $\overline{d}(j_3',0')=\overline{d}(j_3'j_3\cdots j_4j_4'\cdots 0')$ and $j_4=n-1=j_3+1$. By (28), we have $n-2-\hat{b}=j_3-\hat{b}\leqslant j_2\leqslant\hat{b}$, i.e., $\hat{b}\geqslant n/2-1$, contradiction $\hat{b}\leqslant n/2-2$, when $n\geqslant 16$.

Case 3. Suppose $10\leqslant n\leqslant 14$. and $h_1=h_2=h>0$. By counting directly, we have

$$e=N(0^*)=\begin{cases}8ar, & n=10,\\10ar, & n=12,\\7ar+2r, & n=14.\end{cases}$$

Without loss of generality, we assume that $j_2=\max\{j_{t+1}-j_t:1\leqslant t\leqslant h\ (j_{h+1}\equiv 0)\}$. If $j_2=1$, then $h=n$. Hence, $e>N(O)\geqslant f(4',(n-1)')\geqslant 3ar+6r>e$, a contradiction. If $j_2=2$, then $h\geqslant n/$

2. Thus,

$$e > N(0) \geqslant \begin{cases} f((j_h - 1)', 1') \geqslant 4ar + 4r > e, & (1', 1) \in 0, \\ f(1', (j_3 + 1)') \geqslant 4ar + 4r > e, & (1, 1')0 \end{cases}$$

a contradiction. If $j_2 \geqslant 3$, then

$$e > N(0) \geqslant \begin{cases} f(0', 1') \geqslant 7ar + 2r \geqslant e, & (1', 1),\ (2', 2) \in 0, \\ f(1', 2') \geqslant 7ar + 2r \geqslant e, & (1, 1'),\ (2', 2) \in 0, \\ f(2', 3') \geqslant 7ar + 2r \geqslant e, & (1, 1'),\ (2, 2') \in 0 \end{cases}$$

a contradiction. In the present case, we assume $(1', 1)$, $(2, 2') \in 0$. If $j_2 \geqslant 4$, then

$$e > N(0) \geqslant \begin{cases} f(2', 3') \geqslant 7ar + 2r \geqslant e, & (3', 3) \in 0, \\ f(3', 4') \geqslant 7ar + 2r \geqslant e, & (3, 3') \in 0, \end{cases}$$

a contradiction. So we assume that

$$j_2 = 3 \text{ and } (1', 1),\ (2, 2') \in 0 \tag{37}$$

If $j_h \leqslant n-3$, then using the argument similar to that of (37) we have $((n-1)', n-1) \in 0$. Thus, $e > N(0) \geqslant f(1, (n-1)') \geqslant 7ar + 2r \geqslant e$, a contradiction. Therefore,

$$j_h \geqslant n-2. \tag{38}$$

It follows immediatelly that

$$j_3 \leqslant 5 \tag{39}$$

by substituting j_3 and j_2 for j_1 and j_h, respectively. By (37), (38) and (39), for

$$10 \leqslant n \leqslant 14, \text{ we have}$$

$$e > N(0) \geqslant f((j_h - 1)', 1') \geqslant 4ar + 4r \geqslant e,$$

a contradiction.

References

[1] M. J. Atallah, Parallel strong orientation of an undirected graph, Inform. Process. Lett. 18 (1984) 37—39.

[2] F. Boesch and R. Tindell, Robbins', theorem for mixed multigraphs, Amer. Math. Monthly 87 (1980) 716—719.

[3] J. Cai and Y. Xu, On the optimal strongly connected orientations of city annular street graphs $H_{n_1 \times n_2}$ II. In preparation.

[4] F. R. K. Chung, M. R. Garey and R. E. Tarjan, Strongly connected orientations of mixed multigraphs, Networks 15 (1985) 477—484

[5] V. Chvatal and C. Thomassen, Distances in orientations of graphs, J. Combin. Theory B 24 (1978) 61—75.

[6] H. E. Robbins, A theorem on graphs, with an application to a problem of traffic control, Amer. Math. Monthly 46 (1939) 281—283.

[7] F. S. Roberts, Discrete Mathematical Models, with Applications to Social, Biological, and Environmental Problems (Prentice— Hall, Englewood Cliffs, N J, 1976).

[8] F. S. Roberts, Graph Theory and its Applications to problems of Society, NSF—CBMS Monograph 29 (SIAM, Philadelphia, PA, 1978).

[9] F. S. Roberts and Y. Xu, On the optimal strongly connected orientations of city street graphs I: Large grids, SIAM J. Discrete Math. 1 (1988) 199—222

[10] F. S. Roberts and Y. Xu, On the optimal strongly connected orientatins of city street graphs Ⅱ: Two east—west avenues or north—south streets, Networks 19 (1989) 221—233

[11] F. S. Roberts and Y. Xu, On the optimal strongly connected orientations of city street graphs Ⅲ: Three east—week avenues or north—south streets, Networks 22 (1992) 109—143

[12] F. S. Roberts and Y. Xu, On the optimal strongly connected orientations of city street graphs Ⅳ: Four east— west avenues or north—south streets, Discrete Applied Math. 49 (1994) 331—356

[13] U. Vishkin, On efficient parallel strong orientation. Inform. Process. Lett. 20(1985) 235—240.

On Some Modular Properties for Stirling Numbers of the Second Kind

Chen Wen Li

(Southwest Normal Universty, Chongqing, 630715)

Abstract For fixed n or m, Stirling numbers of the second kind, $S(n,m)$, have some modular properties. In this paper we discuss some its interesting modular properties for n and m.

Key words Stirling numbers of the second kind, Modular properties.

1. Introduction

The polynomials x^n can be represented as sums of facyorial polynomials $[x]_m$ with cofficients $S(n,m)$,

$$x^n = \sum_{1\leqslant m\leqslant n} S(n,m)[x]_m, \tag{1}$$

where the number $S(n, m)$ is the Stirling number of second kind. It satisfies the recurrence relation

$$S(n,r)=S(n-1,r-1)+rS(n-1,r). \tag{2}$$

The following lemma come from [1] by, using binomial inversion for (1),

Lemma 1 For all $n\geqslant m\geqslant 1$,

$$m!\ S(n,m)= \sum_{0\leqslant k\leqslant m-1} (-1)^k\binom{n}{m}(m-k)^n, \tag{3}$$

In particular, $S(n,1)=S(n,n)=1$, and

$$S(n,2)=2^{n-1}-1 \tag{4}$$

In this paper we discuss some interstng modular properties of $S(n,m)$.

2. Modular properties of $S(n,m)$ for n.

For fixed n we discuss modular properties of $S(n,m)$.

Theorem 1 A positive integer n is prime number if and only if, for all $2\leqslant m\leqslant n-1$

$$S(n,m)\equiv 0 \qquad (\text{mod } n) \tag{5}$$

Proof Suppose that n is a prime number. Using Fermat's little theorem to (2) and combinatorial identities [2]

$$\sum_{0\leqslant k\leqslant n} (-1)^k\binom{n}{k}k^j=0 \qquad (0\leqslant j\leqslant n-1)$$

we have

$$m!\ S(n,m)=\sum_{0\leqslant k\leqslant m-1}(-1)^k\binom{m}{k}(m-k)^n\equiv\sum_{0\leqslant k\leqslant m}(-1)^k\binom{m}{k}(m-k)\equiv 0.\ (\text{mod } n)$$

Since $2\leqslant m\leqslant n-1$ and n is a prime number, we know that $(n,m)=1$. Therefore, for all $2\leqslant m\leqslant n-1$

$$S(n,m)\equiv 0 \qquad (\text{mod } n).$$

Suppose that (5) is true, we show that for all $2\leqslant m\leqslant n-1$

$$m^{n-1}\equiv 1 \qquad (\text{mod } n)\ . \qquad (6)$$

We use induction on m. When $m=2$ the congruence (6) is true by (4).

New assume, that the congruence (6) is true for natural numberk, $2\leqslant k\leqslant m-1\leqslant n-1$, then when $k=m$ we have

$$\begin{aligned}0&\equiv\sum_{0\leqslant k\leqslant m-1}(-1)^k\binom{m}{k}(m-k)n && (\text{mod } n)\\ &\equiv\{\sum_{0\leqslant k\leqslant m}(-1)^k\binom{m}{k}(m-k)\}+1+m^n-1 && (\text{mod } n)\\ &\equiv\{\sum_{0\leqslant k\leqslant m}(-1)^k\binom{m}{k}(m-k)\}+m^n-1\equiv m^n-1 && (\text{mod } n).\end{aligned}$$

So the congruence (6)is true, for all $1\leqslant m\leqslant n-1$.

By (6) we know $(n,m)=1$ is true, for all $2\leqslant m\leqslant n-1$. Therefore n is a prime number □

Using recurrence relation (2) and Theorem 1, we obtained following

Corollary 1 A positive integer n is prime number if and only if

$$S(n+1,3)\equiv S(n+1,4)\equiv\cdots\equiv S(n+1,n)\equiv 0 \qquad (\text{mod } p) \qquad (7)$$

$$S(n+1,1)\equiv S(n+1,2)\equiv S(n+1,n+1)\equiv 1 \qquad (\text{mod } p)$$

Proof Suppose n is a prime number, using (7) and (2) for all $3\leqslant m\leqslant n-1$ we have

$$S(n+1,m)=S(n,m-1)+m\cdot S(n,m)\equiv 0. \qquad (\text{mod } p)$$

Suppose (7) is true, using (2) we know (5) is truse for all $1\leqslant m\leqslant n-1$, therefore n is a prime number □

Corollary 2 A positive integer n is prime number if and only if

$$S(n+2,4)\equiv S(n+2,5)\equiv\cdots\equiv S(n+2,n)\equiv 0 \qquad (\text{mod } p) \qquad (8)$$

$$S(n+2,1)\equiv S(n+2,3)\equiv S(n+2,n+1)\equiv S(n+2,n+2)\equiv 1 \qquad (\text{mod } p)$$

$$S(n+2,2)\equiv 3 \qquad (\text{mod } p) \qquad \square$$

Corollary 3 Let $p(\geqslant 3)$ be a prime number, then

$$\begin{aligned}S(p+1,3)&\equiv S(p+1,4)\equiv\cdots\equiv S(p+1,p) && (\text{mod } p)\\ &\equiv S(p+2,4)\equiv\cdots\equiv S(p+2,p) && (\text{mod } p) \qquad (9)\\ &\equiv\cdots\equiv S(2p-2,p) && (\text{mod } p)\\ &\equiv 0 && (\text{mod } p). \qquad \square\end{aligned}$$

3. Modular properties of S(n, m) for m.

For fixed m we discuss modular properties of $S(n,m)$, we have

Theorem 2 A positive integer m is prime number if and only if

$$S(m+1,m)\equiv S(m+2,m)\equiv\cdots\equiv S(2m-1,m) \qquad (\text{mod } m)$$

$$\equiv S(2m-2,m)\equiv 0 \qquad (\text{mod } m) \quad (10)$$

Proof The Theorem 2 follows from Theorem 1 and Corrllary 3 □

Lemma 2 Let $p(\geqslant 3)$ be a prime number and $n\geqslant p$ be a positive integer, then

$$S(n,p)\equiv S(n+p-1,p) \qquad (\text{mod } p) \quad (11)$$

Proof When $0\leqslant k\leqslant p-1$, we have $(k,p)=1$. Using Fermat's little theorem we obtain

$$\begin{aligned} S(n+p-1,p) &= \{\sum_{0\leqslant k\leqslant p-1}(-1)^k\tbinom{m}{k}(m-k)^{n+p-1}\}/m! \\ &= \{\sum_{0\leqslant k\leqslant p-1}(-1)^k\tbinom{m}{k}(m-k)^n(m-k)^{p-1}\}/m! \\ &\equiv \{\sum_{0\leqslant k\leqslant p-1}(-1)^k\tbinom{m}{k}(m-k)^n\}/m! \equiv S(n,p) \qquad (\text{mod } p) \end{aligned}$$ □

Theorem 3 Let $p(\geqslant 3)$ be a prime number and $n>p$ be a natural number, if $n\equiv 1$ $(\text{mod } p-1)$, then we have

$$S(n,p)\equiv 1 \qquad (\text{mod } p), \quad (12)$$

otherwise

$$S(n,p)\equiv 0 \qquad (\text{mod } p). \quad (13)$$

Proof Since $S(p,p)\equiv 1$ $(\text{mod } p)$, by Lemma 2 the congruence (12) is true. Appling Lemma 2 and Corollary 2 we know the cogruence (13) is true □

Corollary 4 Let $p(\geqslant 3)$ be a prime number and $n>p$ be a natural number, if $n\equiv 1$ $(\text{mod } p-1)$, then we have

$$S(n,p-1)\equiv 1 \qquad (\text{mod } p), \quad (14)$$

otherwise

$$S(n,p-1)\equiv 0 \qquad (\text{mod } p). \quad (15)$$ □

Theorem 4 Let $p(\geqslant 3)$ be a prime number and $n>p$ be a natural number, if $n\equiv 1$ $(\text{mod } p-1)$, then we have

$$\begin{aligned} S(n,2)\equiv S(n,3) &\equiv\cdots\equiv S(n,p) && (\text{mod } p) \\ &\equiv S(n+1,3)\equiv\cdots\equiv S(n+1,p) && (\text{mod } p) \quad (16) \\ &\equiv\cdots\equiv S(n+p-1,p) && (\text{mod } p) \\ &\equiv 0 && (\text{mod } p). \end{aligned}$$ □

Corollary 5 Let $p(\geqslant 3)$ be a prime number and $n>p$ be a natural number, if $n=tp+1$ and $t\equiv k$ $(\text{mod } p)$, then we have

$$S(n,p+1)\equiv S(n+1,p+1)\equiv\cdots\equiv S(n+p-1,p+1)\equiv k \qquad (\text{mod } p)$$ □

References

[1] Aigner, M., Combinatorial Theory, Springer-Verlag, 1979.

[2] Gould, H. W., Combinatorial Identities, Morgantown, W. Va. 1972.

Reduction Techniques for Supereulerian Graphs and Related Topics — A Survey

Zhi-Hong Chen*
Butler University, Indianapolis, IN 46208

Hong-Jian Lai†
West Virginia University, Morgantown, WV 26506

October 14, 1994

Abstract

In [J. Graph Theory 12(1988) 29-44], Catlin developed a powerful reduction technique to study the existence of spanning and dominating eulerian subgraphs. This paper is intended as a survey article to cover recent developments of this reduction technique and its applications. It also covers some related topics such as supereulerian graphs, integer flows, cycle covers and hamiltonian line graphs.

I. Introduction

We use [8] for terminology and notation not defined here and consider only loopless finite graphs. A trail is a finite sequence $T = u_0e_1u_1e_2u_2\cdots e_ru_r$, whose terms are alternately vertices and edges, with $e_i = u_{i-1}u_i$ $(1 \leq i \leq r)$, where the edges are distinct. A trail T is a closed trail if $u_0 = u_r$ and is called a (u, v)-trail if $u_0 = u$ and $u_r = v$. A trail T is called a spanning trail if $V(T) = V(G)$ and is called a dominating trail if $E(G - V(T)) = \emptyset$. A graph G is supereulerian if G has a spanning closed trail. A cycle of order n is denoted by C_n.

---

*Partially supported by Butler University Academic Grant (1994)
†Partially supported by National Security Agency under Grant MDA904-94-H-2012.

A <u>nowhere zero k-flow</u> of G is an assignment of edge directions and integer weights in $\{1, 2, \cdots, k-1\}$ to edges of $E(G)$ such that at every vertex v, the total number of weights of the edges directed into v is equal to the total number of weights of edges directed out from v. The collection of all graphs admiting a nowhere-zero k-flow is denoted by F_k. By definition, F_2 consists of even graphs (graphs each of whose vertices has even degree). The graph K_1 is regarded as having a nowhere-zero k-flow for any $k \geq 1$. It has been noted that supereulerian graphs are all in F_4 [17].

Let G be a graph. The line graph of G, written $L(G)$, has $E(G)$ as its vertex set, where two vertices are adjacent in $L(G)$ if and only if the corresponding edges are adjacent in G. For a connected graph G, the <u>$i-th$ iterated line graph</u> $L^i(G)$ is defined recursively by $L^0(G) = G$, $L^i(G) = L(L^{i-1}(G))$. The minimum i such that $L^i(G)$ is hamiltonian is called the <u>hamiltonian index</u> of G ([25], [26]). Chartrand ([25]) showed that if a connected graph G is not a path, then the hamiltonian index of G exists. As shown by Harary and Nash-Williams [42], there is a close relationship between dominating eulerian subgraphs and Hamilton cycles in $L(G)$.

Theorem 1.1 ([42]). The line graph $L(G)$ of a connected graph G is hamiltonian if and only if G has a dominating eulerian subgraph and $G \notin \{K_1, K_2, K_{1,2}\}$. □

In [18], Catlin gave a survey of the research on supereulerian graphs, the reduction method, and its applications. In this paper, we shall focus on the technical aspects of the reduction method and cover some new development since then.

II. Reduction Methods

We start with Catlin's reduction method.

2.1 Catlin's reduction method

Let H be a connected subgraph of G. The <u>contraction</u> G/H is the graph obtained from G by contracting all edges of H and deleting any resulting loops, *i.e.*, replacing H by a new vertex v_H such that the number of edges in G/H joining any $v \in G - V(H)$ to v_H in G/H equals the number of edges joining v in G to H. A graph G is <u>contractible</u> to a graph G' if G contains pairwise vertex-disjoint connected subgraph $H_1, \cdots, H_c$ with $\bigcup_{i=1}^{c} V(H_i) = V(G)$ such that G' is obtained from G by successively contracting each H_i $(1 \leq i \leq c)$. Each subgraph $H \in \{H_1, \cdots, H_c\}$ is called the <u>preimage</u> of the vertex v_H of G'. A vertex v_H in G' is <u>nontrivial</u> if v_H is the contraction image of a nontrivial connected subgraph H of G.

For a graph G, let $O(G)$ denote the set of all odd-degree vertices of G. A graph G is <u>collapsible</u> ([14]) if for every subset $X \subseteq V(G)$ with $|X|$ even, there is a spanning connected subgraph H_X of G, such that $O(H_X) = X$. We use $\mathcal{CL}$ and $\mathcal{SL}$ to denote the families of collapsible graphs and supereulerian graphs, respectively. The trivial graph K_1, and cycle C_2, and C_3 are both supereulerian and collapsible, but $C_4 \in \mathcal{SL} - \mathcal{CL}$. Note that being collapsible is stronger than being supereulerian. In fact, for a collapsible graph G with $u, v \in V(G)$, let $R = \{u, v\}$ if $u \neq v$, and $R = \emptyset$ if $u = v$. Then $|R|$ is even. By the definition of collapsible graphs, G has a spanning subgraph H_R such that $O(H_R) = R$. Thus H_R is a spanning (u, v)-trail in G. When $u = v$, H_R

is a spanning eulerian subgraph of G, and so $\mathcal{CL} \subset \mathcal{SL}$.

In [14], Catlin showed that every graph G has a unique collection of pairwise vertex-disjoint maximal collapsible subgraphs $H_1, \cdots, H_c$ such that $\bigcup_{i=1}^{c} V(H_i) = V(G)$. The reduction of G, denoted by G', is the graph obtained from G by contracting each maximal collapsible subgraph H_i, $(1 \le i \le c)$, into a single vertex v_i. A graph G is reduced if $G = G'$.

What makes the reduction method and the collapsible graphs so useful is the following theorem:

Theorem 2.1 ([14]). Let G be a connected graph and G' the reduction of G. Then each of the following holds.

(a) $G \in \mathcal{CL} \iff G' = K_1$.

(b) $G \in \mathcal{SL} \iff G' \in \mathcal{SL}$.

(c) G has a dominating eulerian subgraph if and only if G' has a dominating eulerian subgraph containing all nontrivial vertices of G'. □

Jaeger [44] showed that a graph with two edge-disjoint spanning trees is supereulerian. Catlin [14] proved that

Theorem 2.2 ([14]). If G is at most one edge short of having two edge-disjoint spanning trees, then either $G \in \mathcal{CL}$ or G has a single cut-edge. □

2.2 A refinement of Catlin's reduction technique

In [76], Veldman gave a refinement of Catlin's reduction method that was used to solve a conjecture in [4].

Let G be a simple graph and define $D(G) = \{v \in V(G) \mid d(v) \in \{1,2\}\}$. For an independent subset X of $D(G)$, define $I_X(G)$ as the graph obtained from G by deleting the vertices in X of degree 1 and by replacing each path of length 2 whose internal vertex is a vertex in X of degree 2 by an edge. Note that $I_X(G)$ may not be simple. A graph G is called $X-$collapsible if $I_X(G)$ is collapsible. A subgraph H of G is an X-subgraph of G if $d_H(x) = d_G(x)$ for all $x \in X \cap V(H)$. An $X-$subgraph H of G is called $X-$collapsible if H is $(X \cap V(H))-$collapsible. Let $R(X)$ be the set of vertices in X that are not contained in an $X-$collapsible $X-$subgraph of G. Since $I_X(G)$ has a unique collection of vertex-disjoint maximal collapsible subgraphs $W_1, \cdots, W_k$ such that $\bigcup_{i=1}^{k} V(W_i) = V(I_X(G))$, the graph G has a unique collection of pairwise vertex-disjoint maximal $X-$collapsible $X-$subgraphs $H_1, \cdots, H_k$ such that $\bigcup_{i=1}^{k} V(H_i) \cup R(X) = V(G)$. The $X-$reduction of G, denoted by G'_X, is the graph obtained from G by contracting $H_1, \cdots, H_k$. The graph G is $X-$reduced if there exists a graph G_1 and an independent subset X_1 of $D(G_1)$ such that $X = R(X_1)$ and G is the X_1-reduction of G_1. An $X-$subgraph H of G is called $X-$reduced if H is $(X \cap V(H))-$reduced.

Note that if $X = \emptyset$, the refinement method is just the original method. A graph G is $\emptyset-$collapsible if and only if G is collapsible. Theorem 2.1(c) is a special case of Theorem 2.3 below.

Theorem 2.3 ([76]). Let G be a connected simple graph, X an independent subset of $D(G)$, and

G'_X the X−reduction of G. Then G has a dominating eulerian subgraph if and only if G'_X has a dominating eulerian subgraph containing all nontrivial vertices of G'_X. □

III. Reductions of 4-cycles

Suppose that a graph G contains an induced 4-cycle H. Since H is not collapsible, Theorem 2.1 cannot apply in this case. However, there are two ways which provide the extensions of the reduction method to subgraphs that are 4-cycles.

- Catlin's π−collapsibility reduction method.

Catlin [19] introduces π-collapsible graphs to deal with 4-cycles. Let H be a graph and let $\pi =< V_1, V_2 >$ be a partition of $V(H)$. Then H is called π^--collapsible (resp. π^+-collapsible) if for every subset $R \subseteq V(H)$ with $|R|$ even, the following holds:

(i) if $|R \cap V_1|$ is odd (reps., is even), then H has an R-subgraph;

(ii) if $|R \cap V_1|$ is even (reps., is odd), then $H + e$ has an R-subgraph $\Gamma_e \subseteq H$, for any added edge $w = w_1w_2$ with $w_i \in V_i$, $(1 \le i \le 2)$.

If H is either π^--collapsible or π^+-collapsible, then H is π-collapsible.

Here we describe a special case of this more general π−collapsibility reduction method. Let G be a graph containing a 4-cycle $xyzwx$, and define $E = \{xy, yz, zw, wx\}$. Define G/π to be the graph obtained from $G - E$ by identifying x and z to form a vertex v_1, by identifying w and y to form a vertex v_2, and by adding a new edge v_1v_2.

Theorem 3.1 ([19]). For the graphs G and G/π defined above, each of the following holds:
(a) If $G/\pi \in \mathcal{CL}$, then $G \in \mathcal{CL}$;
(b) If $G/\pi \in \mathcal{SL}$, then $G \in \mathcal{SL}$. □

- Lai's reduction method for m-cycles.

Let $C_m = x_1x_2 \cdots x_mx_1$ be an m−cycle of G with $m \ge 4$. Let e and f be two non-adjacent edges of C_m. Without loss of generality, we may assume that $e = x_1x_2$ and $f = x_ix_{i+1}$, where $m > i > 2$. Define

$$G^* = (G - [E(C_m) - \{e, f\}])/\{e, f\}.$$

Let w and z denote the two vertices of G^* to which the edges e and f are contracted, respectively. For convenience, we regard

$$V(G^*) = (V(G) - \{x_1, x_2, x_i, x_{i+1}\}) \bigcup \{w, z\}, \quad E(G^*) = E(G) - E(C_m).$$

Theorem 3.2 ([54]). For the graphs G and G^* defined above, each of the following holds:
(a) If $G^* \in \mathcal{CL}$, then $G \in \mathcal{CL}$;
(b) If $G^* \in \mathcal{SL}$, then $G \in \mathcal{SL}$. □

These techniques have been used to study the structures of small order reduced graphs ([35], [30]). Using these two theorems with Catlin's reduction technique, Lai ([51], [53]) proved two conjectures of Paulraja ([68], [69]) and a conjecture of Catlin [19]. One of the theorems in [51] is the following: If a 2-connected graph G has $\delta(G) \ge 3$ and each edge lies in a cycle of length at most 4, then $G \in \mathcal{CL}$. Since a spanning eulerian subgraph in a cubic graph is a Hamilton cycle, a corollary of the result above (in [51]) is that every cubic graph whose edges are lying in a cycle of length at most 4 is hamiltonian. (Proof: such graphs are 2-connected).

There are a few other results related to Catlin's π-collapsible graphs.

Theorem 3.3 ([54]). Let G' denote the reduction of G. Then G is π-collapsible for some partition π of $V(G)$ if and only if G' is π'-collapsible for some partition π' of $V(G')$. □

The converse of Theorem 3.1 is not true, as showed by Example 2 in [61]. However, the following gives a conditional converse of Theorem 3.1.

Theorem 3.4 ([61]). Let G be a graph of diameter at most 2 and let C_4, the 4-cycle, be a nonspanning subgraph of G. Let π denote the bipartition of C_4. Then G is collapsible if and only if one of the following holds:

(i) G/π is collapsible.

(ii) G is spanned by a subgraph $H \cong K_{2,n-2}$ such that there are two vertices of degree 2 adjacent to each other in G. □

IV. Reduced Graphs

Besides the theorems mentioned above, a number of useful results on reduced graphs are summarized in this section.

Theorem 4.1 ([14]). Let G be a connected graph. Then each of the following holds.

(a) G is reduced if and only if G has no nontrivial collapsible subgraphs.

(b) Any subgraph of a reduced graph is reduced.

(c) If G is reduced, then G is K_3-free with $\delta(G) \le 3$.

(d) If G is reduced and $G \notin \{K_1, K_2\}$, then $|E(G)| \le 2|V(G)| - 4$. □

For $X-$reduced graph, Veldman [76] proved the following:

Theorem 4.2 ([76]). Let G be a connected simple graph and X an independent subset of $D(G)$. Then each of the following holds.

(a) G is $X-$reduced if and only if $I_X(G)$ is reduced.

(b) G is $X-$reduced if and only if G contains no nontrivial $X-$collapsible $X-$subgraphs.

(c) If G is $X-$reduced, then every $X-$subgraph of G is $X-$reduced.

(d) If G is $X-$reduced, then $d(x) = 2$ for all $x \in X$ and exactly one of the following holds.

(d1) $G \in \{K_1, K_2\}$ and $X = \emptyset$.

(d2) $G = P_3$ and $|X| = 1$.

(d3) $|E(G)| \le 2|V(G)| - |X| - 4$. □

For the size of a maximum matching in a reduced graph, we have the following:

Theorem 4.3 ([34] [27]). Let $M(G)$ be the size of a maximum matching of a connected reduced graph G of order n. Let $D_2(G) = \{v \in V(G) \,|d(v) = 2\}$. If $\delta(G) \ge 2$ and $|D_2(G)| = l$, then

$$M(G) \ge \min\left\{\frac{n-1}{2}, \frac{n+4-l}{3}\right\}. \square$$

The reduced graphs of diameter two were characterized in [52]. For the independence number

$\alpha(G)$ of a reduced graph G, we have $(\delta(G)\alpha(G)+4)/2 \le n \le 4\alpha(G)-5$ if $\alpha(G)\ge 4$ ([28]).

By Theorem 2.1, the problem of determining whether a graph $G \in \mathcal{CL}$ (or $\mathcal{SL}$) can often be reduced to determining whether a much smaller graph (the reduction of G) is in $\mathcal{CL}$ (or $\mathcal{SL}$). Thus, knowing the structure of reduced graphs is very important even for small reduced graphs, when applying the reduction method. The smallest 2-edge-connected nonsupereulerian reduced graph is $K_{2,3}$, and the smallest 3-edge-connected nonsupereulerian reduced graph is the Petersen graph. In the following, we let P denote the Petersen graph.

Theorem 4.4 ([35], [31], [30], [48]). Let G be a graph of order n and G' the reduction of G.
(a) If $n \le 11$ and $\delta(G)\ge 3$ then either $G\in\mathcal{CL}$ or $G'\in\{K_2,P\}$.
(b) If $n\le 13$ and $\delta(G)\ge 3$ then either $G\in\mathcal{SL}$ or $G'\in\{K_2,K_{1,2},P\}$.
(c) If $n\le 17$ and $\kappa'(G)\ge 2$, then either $G\in F_4$ or G can be contracted to the Petersen graph. □

Let $P-v$ be the Petersen graph minus a vertex v. Let G be the graph on 14 vertices obtained from $P-v$ and $K_{2,3}$ by joining those vertices of degree 2 from $P-v$ to $K_{2,3}$ such that G is 3-edge-connected. Then G is a nonsupereulerian reduced graph. This shows that Theorem 4.4(b) is best possible. Catlin [16] conjectured that any 3-edge-connected simple graph of order at most 17 is either supereulerian, or it is contractible to the Petersen graph. Note that the contraction here is not necessary to be a contraction of collapsible subgraphs. The two Blanuša snarks (see [79]) of 18 vertices show that "17" is best possible in the conjecture and in Theorem 4.4(c).

V. Supereulerian Graphs.

There are many results on various sufficient conditions for the existence of spanning eulerian subgraphs (see [4], [9]-[12], [14]-[22], [27]-[35], [38], [49]-[53], [65], [66], [76]-[78], [80]). Several conjectures and open problems ([4], [2], [3], [9], [68], [69]) have been solved by using the reduction method ([14],[15], [22], [29], [49], [51], [76]). For sufficient condition in terms of a lower bound on the number of edges, Catlin and Chen proved the following:

Theorem 5.1 ([22], [33]). Let n, m and p be natural numbers, $m,p\ge 2$. Let G be a 2-edge-connected simple graph on $n>p+6$ vertices containing no K_{m+1}. If

$$|E(G)| \ge \binom{n-p+1-k}{2} + (m-1)\binom{k+1}{2} + 2p-4, \tag{1}$$

where $k=\lfloor\frac{n-p+1}{m}\rfloor$, then either G is supereulerian, or G can be contracted to a nonsupereulerian graph of order less than p, or equality holds in (1) and G can be contracted to $K_{2,p-2}$ (p is odd) by contracting a complete m-partite graph $T_{m,n-p+1}$ of order $n-p+1$ in G. □

The case $p=5$ and $m\ge n-4$ of Theorem 5.1 was proved by Cai in [9]. The case $p=10$ and $m\ge n-9$ for 3-edge-connected graphs [22] solved a conjecture of Cai [9]. For sufficient conditions in terms of lower bounds on degrees of nonadjacent vertices, we have the following:

Theorem 5.2 ([15],[29]). Let G be a k-edge-connected graph of order n with girth g, where

$k \in \{2,3\}$ and $g \in \{3,4\}$. Let G' be the reduction of G. If n is sufficiently large and if

$$d(u) + d(v) \geq \frac{2}{g-2}\left(\frac{n}{(k-1)5} - 4 + g\right),$$

whenever $uv \notin E(G)$, then exactly one of the following holds:

(a) $G \in \mathcal{CL}$;

(b) $k = 2$ and $n = (g-2)5s$ $(s \geq 20)$, and $G' = K_{2,3}$, and either

(b1) $g = 3$, the preimage of each vertex of $K_{2,3}$ is K_s or $K_s - e$, or

(b2) $g = 4$, the preimage of each vertex of $K_{2,3}$ is $K_{s,s}$ or $K_{s,s} - e$.

(c) $k = 3$ and $n = (g-2)10s$ $(s \geq 40)$, and $G' = P$, and either

(c1) $g = 3$, the preimage of each vertex of P is K_s or $K_s - e$; or

(c2) $g = 4$, the preimage of each vertex of P is $K_{s,s}$ or $K_{s,s} - e$. □

For degree conditions of adjacent vertices, we have the following:

Theorem 5.3 ([34], [27]). Let G be a k-edge-connected simple graph of order n, where $k \in \{2,3\}$. Let G' be the reduction of G and $l = |D_2(G)|$. If for any $uv \in E(G)$,

$$d(u) + d(v) \geq \frac{2n}{(k-1)5} - 2,$$

then exactly one of the following holds:

(a) $G \in \mathcal{CL}$;

(b) $k = 2$ and $l < n/5 - 19$, and either

(b1) $G' = K_{2,c-2}$, where $c \leq \max\{5, 3+l\}$, or

(b2) $n = 5s$ $(s > 19)$, $G' = C_5$, and the preimage of each vertex of $C_5 \in \{K_s, K_s - e\}$.

(c) $k = 3$, $n = 10s$ $(s > 24)$, $G' = P$, and the preimage of each vertex of $P \in \{K_s, K_s - e\}$. □

A graph G is almost bridgeless if every cut edge of G is incident with a vertex of degree 1. Veldman [76] proved the following theorem that solved a conjecture in [4].

Theorem 5.4 ([76]). Let G be a connected almost bridgeless simple graph of order n. If n is sufficiently large and for any $uv \in E(G)$,

$$d(u) + d(v) \geq \frac{2n}{5} - 2,$$

then $L(G)$ is hamiltonian. □

There are other degree condition results obtained by using the reduction technique ([28], [32], [58]) which improved some previous sufficient conditions for a graph to have a spanning or dominating eulerian subgraph. The general idea in using the reduction method to prove degree condition problems for the existence of spanning eulerian subgraphs or dominating eulerian subgraphs is the following: First, prove that the order of the reduction of a graph G with given degree conditions is small, or is independent on the order of G, or prove that the number of nontrivial vertices in the reduction graph of G is small. To do so, one may need to make use of some properties of reduced graphs, such as the size of matchings or independent numbers, together with the given degree condition. Then show that the order of the reduction graph is not more than 5 (or 13 if

G is 3-edge-connected). The former may require some structure results of reduced graphs. The latter may require some results on small reduced graphs. This part may be routine but it may also be tedious. Some other type of degree conditions (e.g., [82]) for spanning eulerian graphs may be improved by using the reduction method. However, it has been noted that the degree conditions for spanning eulerian subgraphs are somewhat restrictive. One may hope to study more structural conditions instead of degree conditions for the existence of spanning eulerian subgraphs.

Other types of sufficient conditions for supereulerian graphs have been investigated, as showed by the following theorems.

Theorem 5.5 ([59]). Let G be a simple graph with at least 61 vertices, and let G^c denote the complement of G. One of the following holds.

(a) G is supereulerian.

(b) G^c is supereulerian.

(c) Both G and G^c has a vertex of degree 1.

(d) One of G or G^c is contractible to a $K_{2,t}$ for some odd integer $t \geq 3$, and the other has either one or two vertices of degree 1.

(e) One of G and G^c is contractible to $K_{1,p}$ for some integer $p \geq 1$, and the other has exactly one isolated vertex. □

A <u>wheel</u> W_n is the graph obtained from the n-cycle $C_n = v_1v_2 \cdots v_nv_1$, where $n \geq 2$, by adding an extra vertex v and new edges $\{vv_i : 1 \leq i \leq n\}$. Define the <u>subdivided wheel</u> W_n^* to be the graph obtained from W_n by replacing each edge v_iv_{i+1}, $(1 \leq i \leq n, (\text{mod } n))$ by a path of length 2, $v_iv_i'v_{i+1}$ (say), where $\{v_1', \cdots, v_n'\} \cap V(W_n) = \emptyset$. Note that $W_2^* \cong K_{2,3}$. Let $\mathcal{F} = \{W_n^* : n \geq 2\}$.

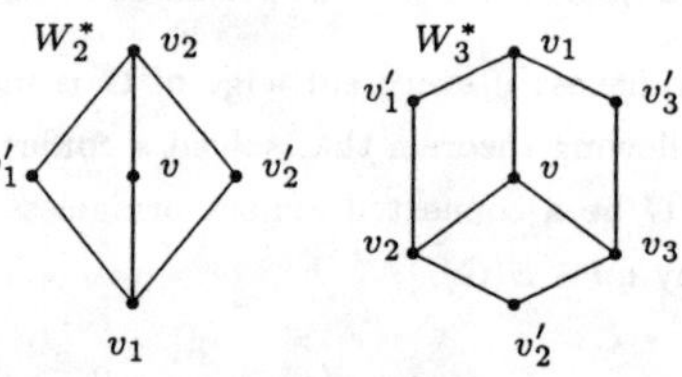

Figure 1: the graphs W_2^* and W_3^*

A graph H is a <u>minor</u> of G if H is isomorphic to the contraction image of a subgraph of G. We call H an <u>induced minor</u> of G if H is isomorphic to the contraction image of an induced subgraph of G.

Theorem 5.6 ([60]). Let G be a 2-edge-connected graph. If G has no induced minor isomorphic to a member in $\mathcal{F}$, then G is supereulerian. □

VI. Integer Flows and Cycle Covers

Flows have been discussed by many. See [44] for a comprehensive survey by Jaeger. The idea of reduction had also been applied to the problems involving integer flows and cycle covers. Tutte has three facinating conjectures on flows.

3-Flow Conjecture ([44]). Every 4-edge-connected graph is in F_3.

4-Flow Conjecture ([44]). Every 2-edge-connected graph either is in F_4, or has a subgraph contractible to the Petersen graphs.

5-Flow Conjecture ([44]). Every 2-edge-connected graph is in F_5.

It has been proved by Seymour [73] that every 2-edge-connected graph is in F_6. Fix an integer $k \geq 3$. Let F_k^o denote the collection of connected graphs H such that for any graph G that has H as a subgraph,

$$G \in F_k \iff G/H \in F_k.$$

Catlin [16] considers F_4^o, and he proves the following.

Theorem 6.1 ([16]). All collapsible graphs and the 4-cycle are in F_4^o. □

Catlin used this theorem ([16], [17]) to show that a 3-edge-connected graph with at most 10 edge-cuts of size 3 is either in F_4, or contractible to the Petersen graph. Theorem 6.1 and Theorem 4.4(c) were used ([48]) to show that if a 2-edge-connected simple graph G with $n \geq 18$ vertices satisfies

$$|E(G)| \geq \binom{n-17}{2} + 34,$$

then either $G \in F_4$, or G is contractible to the Petersen graph. This bound is asymptotically best possible.

For F_3^o, we have the following result.

Theorem 6.2 ([55]). Every complete graph K_m ($m \geq 5$) are in F_3^o. □

It is well known that K_4 is the smallest 2-edge-connected graph that is not in F_3. Among other applications, Theorem 6.2 is used to show that if a 2-edge-connected simple graph G with $n \geq 7$ vertices satisfies

$$|E(G)| \geq \binom{n-5}{2} + 49,$$

then either $G \in F_3$, or G is contractible to K_4. This bound is also asymptotically best possible.

A collection $\mathcal{C}$ of cycles of G is call a <u>cycle cover</u> (CC) if every edge of G lies in a cycle of $\mathcal{C}$. If every edge of G is in exactly k members $\mathcal{C}$, then $\mathcal{C}$ is called a <u>cycle k-cover</u> of G. Bermond, Jackson and Jaeger [5] have shown that every 2-edge-connected graph has a cycle 4-cover. A cycle 2-cover is also called a cycle double cover (CDC).

Cycle 2-Cover Conjecture ([72], [74]). Every 2-edge-connected graph has a cycle 2-cover.

For cycle 2-covers, see [45] for a survey. For a 2-edge-connected graph G, define

$$cc(G) = \min\{|\mathcal{C}| : \mathcal{C} \text{ is a CC of } G\},$$

and

$$sc(G) = \min\{|\mathcal{C}| : \mathcal{C} \text{ is a CDC of } G\}.$$

Bondy ([6], [7]) conjectures the following.

Conjecture SCDC ([6], [7]). If G is a 2-edge-connected simple graph with n vertices, then

$$sc(G) \leq n - 1.$$

Conjecture SCC ([6], [7]). If G is a 2-edge-connected simple graph with n vertices, then

$$cc(G) \leq \frac{2n-1}{3}.$$

No much has been done towards these two conjectures. Bondy and Seyffarth [7] proved Conjecture SCDC for planar triangulations. It was proved in [62] that Conjecture SDCD holds for graphs that do not have a subgraph contractible to K_4. Let $\mathcal{F}_{cc}$ denote the collection of graphs satisfying Conjecture SCC, and $\mathcal{F}_{sc}$ the collection of graphs satisfying Conjecture SCDC. It is still open whether all planar 2-edge-connected simple graphs are in $\mathcal{F}_{sc}$.

There are some contraction ideas used to attack Conjecture SCC. It is noticed ([63],[64]) that a certain configurations H satisfies the following: for any planar graphs that contain H as a subgraph,

$$G \in \mathcal{F}_{cc} \Longleftrightarrow G/H \in \mathcal{F}_{cc}.$$

The following lemma is a typical example.

Lemma 6.3 ([64]). Let G be a graph and $H = \Gamma_1$ (see Figure 2) be a subgraph of G such that the vertices of attachment of H in G are lying in $\{v_1, v_2, v_3\}$. Let $e \notin E(G)$ be an edge parallel to v_2v_3 and $G' = (G - V_H) + e_2$. Then

$$cc(G) \leq cc(G') + 1. \ \square$$

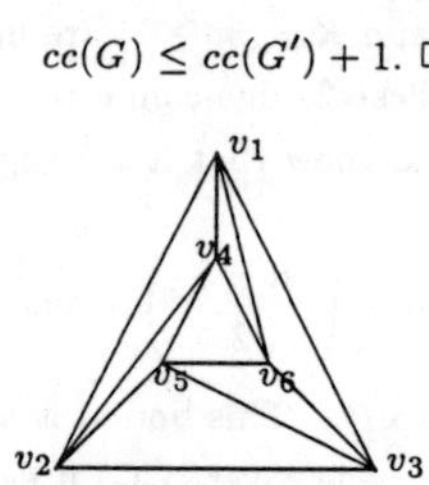

Γ_1

Figure 2: The graph Γ_1

Other similar reduction lemmas can be found in [64]. These "reducible" graphs seem to be very sporadic, and are described in [64]. These reduction ideas were used ([63], [64]) to show that for a 2-edge-connected simple graph G with n vertices, if G does not have a subgraph contractible to K_4, then

$$cc(G) \leq \frac{2n-2}{3},$$

and if G is a planar triangulation, then

$$cc(G) \leq \frac{2n-3}{3}.$$

Both bounds are best possible.

VII. Hamiltonian Line Graphs

A graph is essentially k-edge-connected if for any $X \subseteq E(G)$ with $|X| \leq k-1$, either $G - X$ is connected or at most one component of $G - X$ has an edge. It is easy to see that the line graph is k-connected if and only if G is essentially k-edge-connected. Theorem 1.1 gives a criterion for a line graph to be hamiltonian. It is known that any 4-edge-connected graph has a spanning eulerian subgraph [46]. Therefore, the line graph of any 4-edge-connected graph is hamiltonian. Thomassen [75] posed the following conjecture:

Conjecture ([75]). All 4-connected line graphs are hamiltonian.

Conjectures which are related to Thomassen Conjecture were discussed in [39]. In [1], Jackson made the following conjecture:

Conjecture ([1]). Let G be a 2-edge-connected graph, then G has an eulerian subgraph H with $|V(H)| \geq 2$ such that for each component F of $G - V(H)$, there are at most three edges between F and H.

It was proved in [37] that Jackson's conjecture implies Thomassen's conjecture. Several authors have proved Thomassen's conjecture for some special classes of graphs. Based on a similar idea of [56], Chen and Lai proved the following:

Theorem 7.1 ([36]). A 4-connected line graph of a projective planar graph is hamiltonian. □

Zhan [81] and Jackson [43] independently proved that:

Theorem 7.2 ([81]). If $L(G)$ is 7-connected, then $L(G)$ is hamiltonian. □

Here we present a proof of Theorem 7.2 that is based on a unpublished manuscript of Lai [57]. We prove the following theorem first.

Theorem 7.3. If G is 3-edge-connected and essentially 7-edge-connected, then G is collapsible.

Proof. Let G be a smallest counterexample. Since contraction preserves edge-connectivity, by the minimality of G, G must be reduced. For a vertex $v \in V(G)$, define $N(v) = \{u \in V(G) \mid vu \in E(G)\}$. Let $D_i = \{v \in V(G) \mid d(v) = i\}$ and $D_i^* = \{v \in V(G) \mid d(v) \geq i\}$. Since G is essentially 7-edge-connected, $N(v) \subseteq D_6^*$ if $v \in D_3$. Let $E^* = \{uv \in E(G) \mid u \in D_3 \text{ and } v \in D_6^*\}$. Let $H = G[E^*]$. Then by Theorem 4.1, H is a reduced bipartite graph with bipartition $V(H) = D_3 \cup D_6^*$. By Theorem 4.1(d), $|E(H)| \leq 2|V(H)| - 4$. Hence,

$$\begin{aligned} 3|D_3| &\leq |E(H)| \leq 2(|D_3| + |D_6^*|) - 4, \\ |D_3| &\leq 2|D_6^*| - 4. \end{aligned} \tag{2}$$

Since G is 3-edge-connected, $D_1 = D_2 = \emptyset$. By $\sum_{v \in V(G)} d(v) = 2|E(G)|$ and Theorem 4.1(d) again, we have

$$3|D_3| + 4|D_4| + 5|D_5| + 6|D_6^*| \leq 2|E(G)| \leq 4(|D_3| + |D_4| + |D_5| + |D_6^*|) - 8. \tag{3}$$

By (2) and (3), we have

$$|D_5| + 2|D_6^*| \leq |D_3| - 8 \leq (2|D_6^*| - 4) - 8.$$

Thus, $|D_5| \leq -12$, a contradiction. □

Proof of Theorem 7.2. Since $L(G)$ is 7-connected, G is essentially 7-edge-connected graph. Let $X = D_1 \cup D_2$. Then X is an independent set of G. Let $I_X(G)$ be the graph obtained from G as defined in section 2.2. Since G is essentially 7-edge-connected, $I_X(G)$ must be 3-edge-connected and essentially 7-edge-connected. By Theorem 7.3, $I_X(G)$ is collapsible. By Theorem 2.1 and then by Theorem 2.3, G has a dominating eulerian subgraph. Therefore, by Theorem 1.1, $L(G)$ is hamiltonian. □

The following result is best possible in the sense that there are graphs G such that $L^2(G)$ is 3-connected and $L^2(G)$ is not hamiltonian.

Theorem 7.4 ([37]). If $L^2(G)$ is 4-connected, then $L^2(G)$ is hamiltonian. □

The reduction method had also been applied in the study of the hamiltonian index ([23],[50]). A recent article [71] corrected an error of Theorem 20 in [50].

VIII. Final Remark and Open Problems

In this paper, we primarily focus on Catlin's reduction method, its applications and some recent developments. We choose our topics on how the reduction technique has been applied. It is not our intention to cover all the development of each of our topics. This article suppliments, not replaces, Catlin's survey ([18]). We present the following open problems and conjectures to conclude this article.

Conjecture 8.1 ([19]). Let H be a graph. If H is not collapsible, then H has a supergraph G such that this equivalence is false:

$$G \in \mathcal{SL} \Longleftrightarrow G/H \in \mathcal{SL}.$$

Conjecture 8.2 ([19]). Let $G(e)$ denote the graph obtained from G with an edge $e \in E(G)$ replaced by a simple path of length 2. If $G \in \mathcal{CL}$, then either

$$|E(G)| = \frac{3}{2}(|V(G)| - 1),$$

or for some edge $e \in E(G)$, $G(e)$ is collapsible.

Conjecture 8.3 ([19]). Let G be a graph. If H is a K_2 in G such that G/H is collapsible, then G is π-collapsible, for some partition π of $V(G)$.

Conjecture 8.4 ([19]). If a collapsible graph G has a vertex v of degree 2, then $G - v$ is π-collapsible, for some partition π of $V(G - v)$.

Conjecture 8.5 ([35]). Let G be a simple 3-edge-connected graph with n vertices. If for every edge $e = uv \in E(G)$,

$$d(v) + d(u) > \frac{n}{9} - 2,$$

then when n is large, either G is in $\mathcal{SL}$, or G is contractible to the Petersen graph.

Conjecture 8.6. Every 3-edge-connected, essentially 6-edge-connected graph is collapsible.

Conjecture 8.7. Every 3-edge-connected, essentially 5-edge-connected graph is supereulerian.

Problem 8.8. Determine

$$\min_{G \in \mathcal{SL}} \max \left\{ \frac{|E(H)|}{|E(G)|} : H \text{ is a spanning eulerian subgraph of } G \right\}.$$

It is known [31] that for the independence number $\alpha(G)$ of a reduced graph G of order n with $\delta(G) \geq 3$, $\frac{3\alpha(G)+4}{2} \leq n \leq 4\alpha(G) - 5$, and so $1/4 \leq \underline{\lim}_{n\to\infty} \frac{\alpha(G)}{n} \leq \overline{\lim}_{n\to\infty} \frac{\alpha(G)}{n} \leq 2/3$. However, there are no examples to show that those bounds are best possible.

Problem 8.9. Let G be a connected reduced graph of order n with $\delta(G) \geq 3$. What is the value of $\underline{\lim}_{n\to\infty} \frac{\alpha(G)}{n}$? What is the value of $\overline{\lim}_{n\to\infty} \frac{\alpha(G)}{n}$?

For the chromatic number of a reduced graph, Catlin [24] conjectured that

Conjecture 8.10 ([24]). Any nontrivial connected reduced graph has chromatic number 2 or 3.

By Theorem 4.1(c) and Grötzsch's Theorem on three-coloring K_3-free planar graph [40], one can see that this conjecture is true for reduced planar graphs. This conjecture implies that for a reduced graph G of order n, $n \leq 3\alpha(G)$, and so $\underline{\lim}_{n\to\infty} \frac{\alpha(G)}{n} \geq 1/3$.

Let G be a connected reduced graph of order n with $\delta(G) \geq 3$, and $M(G)$ the size of maximum matching of G. Then by Theorem 4.3, $1/2 \geq \underline{\lim}_{n\to\infty} \frac{M(G)}{n} \geq 1/3$. Note that all the snarks are reduced and have a perfect matching (e.g., the family of flower snarks [79]), and so $\overline{\lim}_{n\to\infty} \frac{M(G)}{n} = 1/2$. However, the following is unknown.

Problem 8.11. Let G be a connected reduced graph of order n with $\delta(G) \geq 3$. Determine

$$\underline{\lim}_{n\to\infty} \frac{M(G)}{n}?$$

References

[1] J. Bang-Jensen and B. Toft, Unsolved problems presented at the Julius Petersen Graph Theory Conference. Discrete Math. 101(1992) 351-360.

[2] D. Bauer, A note on degree conditions for hamiltonian cycles in line graphs. Congres. Numer. 49(1985) 11-18.

[3] D. Bauer, On hamiltonian cycles in line graphs. Stevens Research Report PAM No. 8501, Stevens Institute of Technology, Hoboken, NJ.

[4] A. Benhocine, L. Clark, N. Köhler and H.J. Veldman, On circuits and pancyclic line graphs. J. Graph Theory 10(1986) 411-425.

[5] J. C. Bermond, B. Jackson and F. Jaeger, Shortest coverings of graphs with cycles. J. Combin. Theory (B), 35(1983) 297-308.

[6] J. A. Bondy, Trighaphs. Discrete Math. 75(1989) 69-79.

[7] J. A. Bondy, Small double cycle cover of graphs. "Cycles and rays", eds. by G. Hahn, G. Sabidussi, and R. Woodrow. Kluwer Academic Publications, Dordrencht, (1990) 21-40.

[8] J. A. Bondy and U. S. R. Murty, "Graph Theory with Applications". American Elsevier, New York (1976).

[9] X. T. Cai, Connected eulerian spanning subgraphs. Chinese Quaterly J. of Mathematics, Vol. 5(1990) 78-84.

[10] X. T. Cai, S-circuits and hamiltonianity of line graphs. Kexue Tongbao 33(1988) 76-77.

[11] X. T. Cai, A sufficient condition of minimum degree for a graph to have an S-circuit. Chin. Ann. Math. 4A(1989) 117-122.

[12] X. T. Cai, On S-circuits and D-circuits of graphs. (Submitted).

[13] P. A. Catlin, Spanning Trails. J. Graph Theory 11(1987) 161-167.

[14] P. A. Catlin, A reduction method to find spanning eulerian subgraphs. J. Graph Theory 12(1988) 29-44.

[15] P. A. Catlin, Contractions of graphs with no spanning eulerian subgraphs. Comtinatorica 8(1988) 313-321.

[16] P. A. Catlin, Double cycle covers and the Petersen graph. J. Graph Theory 13(1989) 465-483.

[17] P. A. Catlin, Double cycle covers and the Petersen graph, II. Congressus Numerantium 76(1990) 173-181.

[18] P. A. Catlin, Supereulerian Graphs: A Survey. J. Graph Theory 16(1992) 177-196.

[19] P. A. Catlin, Supereulerian graphs, collapsible graphs, and four-cycles. Congressus Numerantium 58(1988) 233-246.

[20] P. A. Catlin, Spanning eulerian subgraph and matchings. Discrete Math. 76(1989)95-116.

[21] P. A. Catlin, A reduction method for graphs. Congres. Numer. 65(1988) 159-170.

[22] P. A. Catlin and Z.-H. Chen, Nonsupereulerian graphs with large size. "Graph Theory, Combinatorics, Algorithms, & Applications", eds. by Y. Alavi, F. R. K. Chung, R. L. Graham, and D. F. Hsu, SIAM (1991) 83-95

[23] P. A. Catlin, Iqblunnisa, T.N. Janakiraman and N. Srinivasan, Hamilton Cycles and Closed Trails in Iterated Line Graphs. J. Graph Theory 14(1990) 347-364.

[24] P. A. Catlin, personal communication (1990).

[25] G. Chartrand, On Hamiltonian line graphs. Trans. Amer. Math. Soc. 134, 3(1968) 559-566.

[26] G. Chartrand and C.E. Wall, On the Hamiltonian index of a graph. Studia. Sci. Math. Hung. (1973) 43-48.

[27] Z.-H. Chen, Spanning closed trails in graphs. Discrete Math. 117(1993) 57-71.

[28] Z.-H. Chen, Supereulerian Graphs, Independent Sets, and Degree-Sum Conditions. (Submitted).

[29] Z.-H. Chen, A degree Condition for Spanning Eulerian Subgraphs. J. Graph Theory 17(1993) 5-21.

[30] Z.-H. Chen, Supereulerian graphs and the Petersen graph. J. of Comb. Math. and Comb. Computing 9(1991) 70-89.

[31] Z.-H. Chen, Reduction of Graphs and Spanning Eulerian Subgraphs. Ph. D. Thesis, Wayne State University, (1991).

[32] Z.-H. Chen, The Fan-type conditions for collapsible graphs. (Submitted).

[33] Z.-H. Chen, On extremal nonsupereulerian graphs with clique number m. Ars Combinatoria 36(1993) 161-169.

[34] Z.-H. Chen and H.-J. Lai, Collapsible graphs and matchings. J. Graph Theory 17(1993) 597-605.

[35] Z.-H. Chen and H.-J. Lai, Supereulerian graphs and the Petersen graph, II. (Submitted).

[36] Z.-H. Chen and H.-J. Lai, Cycles in line graphs. Congressus Numerantium, accepted.

[37] Z.-H. Chen, H.Y. Lai, H.-J. Lai and G.Q. Weng, Jackson's conjecture on eulerian subgraphs. Proceeding of the 3rd China-USA Graph Theory Conference, (eds. by Alavi et al), accepted.

[38] Z.-Q. Chen and Y.-F. Xue, On spanning trails and dominating trails. Preprint.

[39] H. Fleischner and B. Jackson, A note concerning some conjectures on cyclically 4-edge-connected 3-regular graphs. Ann. Discrete Math. 41(1989) 171-178.

[40] H. Grötzsch, Ein Dreifarbensatz fur Dreikreisfreie Netze auf der Kugel, Wiss, Z. Martin-Luther Univ. Halle-Wittenberg Math. Naturwiss. Reihe, vol. 8(1958) 109-120.

[41] D. Gusfield, Connectivity and edge-disjoint spanning trees. Inform. Process. Lett. 16(1983) 87-89.

[42] F. Harary and C.S.J.A. Nash-Williams, On eulerian and hamiltonian graphs and line graphs. Canad. Math. Bull. 8(1965) 701-710.

[43] B. Jackson, personal communication.

[44] F. Jaeger, Nowhere-zero flow problems. "Topics in Graph Theory, 3", ed. by L. W. Beineke and R. J. Wilson. Academic Press, London (1988) 70-95.

[45] F. Jaeger, A survey of the cycle double cover conjecture. Ann. Discrete Math. 27(1985) 1-12.

[46] F. Jaeger, A note on subeulerian graphs. J. Graph Theory 3(1979) 91-93.

[47] S. Kundu, Bounds on the number of disjoint spanning trees. J. Combinat. Theory (B) 17(1974) 199-203.

[48] H.-J. Lai, The size of graphs without nowhere-zero 4-flows. J. Graph Theory, accepted.

[49] H.-J. Lai, Contractions and hamiltonian line graphs. J. Graph Theory 12(1988) 11-15.

[50] H.-J. Lai, On the hamiltonian index. Discrete Math. 69(1988) 43-53.

[51] H.-J. Lai, Graphs whose edges are in small cycles. Discrete Math. 94(1991) 11-22.

[52] H.-J. Lai, Reduced graphs of diameter 2. J. Graph Theory 14(1990) 77-87.

[53] H.-J. Lai, Contractions and eulerian subgraphs. Ph. D. thesis, Wayne State University (1988).

[54] H.-J. Lai, Reduction towards collapsiblity. "Graph Theory, Combinatorics, and Algorithms" eds. Y. Alavi and A. Schwenk, John Wiley and Sons, New York, (1994), accepted.

[55] H.-J. Lai, Graphs with nowhere-zero flows, in preparation.

[56] H.-J. Lai, Every 4-connected line graph of a planar graph is hamiltonian. Graphs and Combinatorics, accepted.

[57] H.-J. Lai, A note on supereulerian graphs.(1991) (unpublished)

[58] H.-J. Lai, Eulerian subgraphs containing given vertices and hamiltonian line graph. (Submitted).

[59] H.-J. Lai, Complementary supereulerian graphs. J. Graph Theory 17(1993) 263-273.

[60] H.-J. Lai, Supereulerian graphs and excluded induced minors. Discrete Math., accepted.

[61] H.-J. Lai, Graphs of diameter at most two. Ars Combinatoria 34(1992) 331-343.

[62] H.-J. Lai and H. Y. Lai, Graphs without K_4-minors. Ars Combinatoria 36(1993) 327-334.

[63] H.-J. Lai and H. Y. Lai, Cycle covers in graphs without subdivisions of K_4. J. of Combin. Math. and Combin. Computing 9(1991) 201-207.

[64] H.-J. Lai and H. Y. Lai, Cycle covers of planar triangulations. J. Combin. Math. and Combin. Computing 10(1991) 3-21.

[65] L. Lesniak-Foster, On n-hamiltonian line graphs. J. Combinat. Theory (B) 22(1977) 263-273.

[66] L. Lesniak-Foster and J. E. Williamson, On spanning and dominating circuits in graphs. Can. Math. Bull. 20(1977) 215-220.

[67] P. Paulraja, On graphs admitting spanning eulerian subgraphs. Ars Combinatoria 24(1987) 57-65.

[68] P. Paulraja, Research problem. Discrete Math. 64(1987) 109.

[69] P. Paulraja, Research problem. Discrete Math. 64(1987) 317.

[70] V. P. Polesskii, A lower bound for the reliability of information networks. Prob. Inf. Transmission 7(1971) 165-171.

[71] M. L. Saražin, On the hamiltonian index of a graph. Discrete Math. 122(1993) 373-376.

[72] P. D. Seymour, Sums of circuits. "Graph Theory and related topics", eds. J. A. Bondy and U. S. R. Murty, Academic Press, New York, San Francisco, London, (1979) 341-355.

[73] P. D. Seymour, Nowhere-zero 6-flows. J. Combinatorial Theory (B) 30(1981) 130-135.

[74] G. Szekeres, Polyhedral decomposition of cubic graphs. Bull. Austral. Math. Soc. 8(1973) 367-387.

[75] C. Thomassen, Reflections on graph theory. J. Graph Theory 10(1986) 309-324.

[76] H. J. Veldman, On dominating and spanning circuits in graphs. Discrete Math. 124(1994) 229-239.

[77] H. J. Veldman, Existence of spanning and dominating trails and circuits. J. Graph Theory 10(1986) 23-31.

[78] H. J. Veldman, A result on hamiltonian line graphs involving restrictions on induced subgraphs. J. Graph Theory 12(1988) 413-420.

[79] J.J. Watkins and R.J. Wilson, A survey of snarks. "Graph Theory, Combinatories, and Applications" (vol. 2), eds, by Y. Alavi, G. Chartrand, O. Oellermann and A. Schwenk, Kalamazoo (1991) 1129-1144.

[80] Y.-F. Xue and Z.-Q. Chen, Dominating trails and pancyclicity of line graphs. J. Graph Theory. To appear.

[81] S. M. Zhan, On hamiltonian line graphs and connectivity. Discrete Math. 89(1991) 89-95.

[82] L.C. Zhao and C.F. Liu, A sufficient condition for hamiltonian line graphs. Math. Applicata Vol.3 No.1 (1990) 22-26.

The Combinatorial Structure of Small Cut and Metric Polytopes

Antoine DEZA* Michel DEZA*

(* Tokyo Institute of Technology, Dept. of Math. and Computing Sciences, Japan)

(* CNRS, Ecole Normale Supérieure, Dépt. Math. et Informatique, Paris, France)

Abstract. We study the combinatorial structure of the cut and metric polytopes on n nodes for $n \leq 5$. Those two polytopes have a complicated geometrical structure, but using their large symmetry group, we can completely describe their face lattices. We present, for any n, some orbits of faces and give new result on the tightness of the wrapping of the cut polytope by the metric polytope, disproving a conjecture of [14] on their lattices.

Key words: metric polytope, complete bipartite subgraphs polytope, face lattice.

1 Introduction

We first recall the definition of the *metric polytope* m_n, the *cut polytope* c_n, and their relatives, the *metric cone* and the *cut cone*. Then we present some applications to well known optimization problems and some combinatorial and geometric properties of those polyhedra. The general references are [4, 24] for polytopes and [5] for graphs. For a complete study of the applications and the combinatorial optimization aspects of those polyhedra, we refer, respectively, to the surveys [13] and [21].

For all 3-sets $\{i, j, k\} \subset \{1, \ldots, n\}$, we consider the following inequalities:

$$x_{ij} - x_{ik} - x_{jk} \leq 0 \qquad (1)$$

$$x_{ij} + x_{ik} + x_{jk} \leq 2. \qquad (2)$$

The inequalities (1) define the metric cone and the metric polytope m_n is obtained by bounding the latter by the inequalities (2). The $3\binom{n}{3}$ facets defined by the inequalities (1),

This work was partially done while the second author visited SFB 343 Diskrete Strukturen in der Mathematik in the framework of European network DIMANET

which can be seen as triangle inequalities for distance x_{ij} on $\{1,2,\ldots,n\}$, are called *homogeneous triangle facets* and are denoted by $Tr_{ij,k}$. The $\binom{n}{3}$ facets defined by the inequalities (2) are called *non-homogeneous triangle facets* and are denoted by Tr_{ijk}.

Given a subset S of $V_n = \{1,2,\ldots,n\}$, the *cut* determined by S consists of the pairs (i,j) of elements of V_n such that exactly one of i, j is in S. $\delta(S)$ denotes both the cut and its incidence vector in $I\!R^{\binom{n}{2}}$, that is, $\delta(S)_{ij} = 1$ if exactly one of i, j is in S and 0 otherwise for $1 \leq i < j \leq n$. By abuse of language, we use the term cut for both the cut itself and its incidence vector, so $\delta(S)_{ij}$ are considered as coordinates of a point in $I\!R^{\binom{n}{2}}$. The cut polytope of the complete graph c_n, which is also called the *complete bipartite subgraphs polytope*, is the convex hull of all 2^{n-1} cuts, and the cut cone is the conic hull of all $2^{n-1}-1$ nonzero cuts. Those polyhedra were considered by many authors, see for instance [2, 3, 9, 11, 12, 13, 14, 15, 17, 18] and references therein. One of the motivations for the study of these polyhedra comes from their applications in combinatorial optimization, the most important being the max-cut and multicommodity flow problems.

Given a graph $G = (V_n, E)$ and nonnegative weights w_e, $e \in E$, assigned to its edges, the *max-cut* problem consists in finding a cut $\delta(S)$ whose weight $\sum_{e\in\delta(S)} w_e$ is as large as possible. It is a well-known NP-complete problem. By setting $w_e = 0$ if e is not an edge of G, we can consider without loss of generality the complete graph on V_n. Then the max-cut problem can be stated as a linear programming problem over the cut polytope c_n as follows:

$$\begin{cases} \max & w^T \cdot x \\ x \in c_n. \end{cases}$$

Since the metric polytope is a relaxation of the cut polytope, optimizing $w^T \cdot x$ over c_n instead of m_n provides an upper bound for the max-cut problem [3].

With E the set of edges of the complete graph on V_n, an instance of the *multicommodity flow problem* is given by two nonnegative vectors indexed by E: a capacity $c(e)$ and a requirement $r(e)$ for each $e \in E$. Let $U = \{e \in E : r(e) > 0\}$. If T denotes the subset of V_n spanned by the edges in U, then we say that the graph $G = (T, U)$ denotes the *support* of r. For each edge $e = (s,t)$ in the support of r, we seek a flow of $r(e)$ units between s and t in the complete graph. The sum of all flows along any edge $e' \in E$ must not exceed $c(e')$. If such a flow exists, we call c, r *feasible*. A necessary and sufficient condition for feasibility is given by the Japanese theorem [16]: a pair c, r is feasible if and only if $(c-r)^T x \geq 0$ is valid over the metric cone. For example, $Tr_{ij,k}$ can be seen as an elementary solvable flow problem with $c(ij) = r(ik) = r(jk) = 1$ and $c(e) = r(e) = 0$ otherwise, so the inequalities (1) correspond to $(c-r)^T x \geq 0$ for x in the metric cone. Therefore, the metric cone is the dual cone to the cone of feasible multicommodity flow problems.

2 Combinatorial and geometric properties of the cut and metric polytopes

The polytope c_n is a $\binom{n}{2}$ dimensional $0-1$ polyhedron with 2^{n-1} vertices and m_n is a polytope of same dimension with $4\binom{n}{3}$ facets inscribed in the cube $[0,1]^{\binom{n}{2}}$. We have $c_n \subseteq m_n$ with equality only for $n \leq 4$. It is easy to see that the point $\omega_n = (\frac{1}{2}, \frac{1}{2}, \ldots, \frac{1}{2})$ is the center of gravity of both c_n and m_n and is also the center of the sphere of radius $r = \frac{1}{2}\sqrt{n(n-1)}$ where all the cuts lie. Another two geometric characteristics of the cut polytope c_n are its *width* and *geometric diameter*. We recall that while the width of a polytope P is equal to the minimum distance between a pair of parallel hyperplanes containing P in the slice between them, the geometric diameter of P is the maximum distance between a pair of supporting hyperplanes. The width of c_n is 1 ([22]) and its geometric diameter is $\frac{n}{2}$ for n even and $\frac{1}{2}\sqrt{n^2-1)}$ for n odd, see [21]. Any facet, respectively ridge (that is, a face of codimension 2), of the metric polytope contains a facet, respectively a ridge, of the cut polytope and the vertices of the cut polytope are vertices of the metric polytope, in fact the cuts are precisely the integral vertices of the metric polytope. Actually the metric polytope m_n wraps the cut polytope c_n very tightly since, in addition to the vertices, all edges and 2-faces of c_n are also faces of m_n ([14]). In other words, c_n is a *segment of order* 2 of m_n and its dual, m_n^*, is a segment of order 1 of c_n^* in terms of [19]: a polytope P is a segment of order s of a polytope Q if they have the same dimension and if every i-face of P is a face of Q for $0 \leq i \leq s$. The polytope c_n is 3-neighbourly ([14]). Any two cuts are adjacent both on c_n ([3]) and on m_n ([20]); in other words m_n is *quasi-integral* in terms of [23], that is, the skeleton of the convex hull of its integral vertices, i.e. the skeleton of the c_n, is an induced subgraph of the edge graph of the metric polytope itself. While the diameter of m_n^* is 2 ([6]), the diameters of c_n^* and m_n are respectively conjectured to be 4 and 3 ([18, 6]). For a detailed study of the combinatorial and geometric properties of c_n and m_n, we refer to [8].

The metric polytope and the cut polytope share the same symmetry group, that is, the group of isometries preserving a polytope. This group is isomorphic to the automorphism group of the *folded n-cube*, that is, $Is(m_p) = Is(c_p) \approx Aut(\square_n)$, see [11, 17]. We recall that the folded n-cube is the graph whose vertices are the partitions of $V_n = \{1, \ldots, n\}$ into two subsets, two partitions being adjacent when their common refinement contains a set of size one, see [5]. More precisely, for $n \geq 5$, $Is(m_n) = Is(c_n)$ is induced by permutations on $V_n = \{1, \ldots, n\}$ and *switching reflections by a cut.* Given a cut $\delta(S)$, the switching reflection $r_{\delta(S)}$ is defined by $y = r_{\delta(S)}(x)$ where $y_{ij} = 1 - x_{ij}$ if $(i,j) \in \delta(S)$ and $y_{ij} = x_{ij}$ otherwise. These symmetries preserve the adjacency relations and the linear independency. For the study of their face lattices, we frequently use the fact that the faces of m_n and c_n are partitioned into orbits of their symmetry group.

We finally mention the following link with metrics: there is an evident $1-1$ correspondence between the elements of the metric cone and all the semi-metrics on n points, and the elements of the cut cone correspond precisely to the semi-metrics on n points that are isometrically embeddable into some l_1^m, see [1], it is easy to check that $m \leq \binom{n}{2}$.

3 Face lattices of small cut and metric polytopes

3.1 Face lattice of the $c_n = m_n$ for $n \leq 4$

For $n \leq 4$, we have $c_n = m_n$, moreover c_3 and c_4 are both well-known polytopes. While c_3 is the regular tetrahedron of edge length $\sqrt{2}$ and volume $v_3 = \frac{1}{3}$, c_4 is combinatorially equivalent to the 6-dimensional cyclic polytope with 8 vertices and its volume is $v_4 = \frac{2}{45}$. The f-vector of c_4 is obviously $f(c_4) = (8, 28, 56, 68, 48, 16)$; more precisely all proper faces of c_4 are partitioned into the following orbits of the symmetry group $Is(c_4) \approx Aut(\square_4)$:

- the 8 vertices of c_4 form the orbit O_0^1,
- the 2 orbits O_1^1 and O_1^2 of edges $\{\delta(S), \delta(S')\}$ are respectively formed by the 16 edges with $|S \Delta S'|$ odd and the 12 ones with $|S \Delta S'|$ even, (that is respectively represented by $\{\delta(\emptyset), \delta(1)\}$ and $\{\delta(\emptyset), \delta(1,2)\}$),
- the 2 orbits of 2-faces are: O_2^1 of size 48 which is represented by $\{\delta(\emptyset), \delta(2), \delta(1,2)\}$, and O_2^2 of size 8 which is represented by $\{\delta(1), \delta(2), \delta(3)\}$,
- the 3 orbits of 3-faces are: O_3^1 of size 12 which is represented by $\{\delta(\emptyset), \delta(1), \delta(2), \delta(1,2)\}$, O_3^2 of size 24 which is represented by $\{\delta(\emptyset), \delta(1), \delta(2), \delta(1,3)\}$, and O_3^3 of size 32 and represented by $\{\delta(\emptyset), \delta(1), \delta(2), \delta(3)\}$,
- the 48 ridges form the orbit O_4^1; they are the cofaces (that is the convex hull of the vertices not belonging to a face) corresponding to the 2-faces from the orbit O_2^1.
- the 16 facets form the orbit O_5^1; they are the cofaces corresponding to the edges from the orbit O_1^1.

Remark 3.1
(i) The skeleton of c_4^* *is the* (4×4)*-grid, which is also the line graph of* $K_{4,4} = \square_4$.
(ii) A set of vertices is not a face of c_4^* *if and only if it contains one of the following* 2 *sets of* 4 *vertices:* $\{\delta(1), \delta(2), \delta(3), \delta(4)\}$ *and* $\{\delta(\emptyset), \delta(1,2), \delta(1,3), \delta(1,4)\}$.

3.2 Face lattices of c_5 and m_5

The face lattices of c_5 and m_5 were obtained in the following way. We first got all the non-simplex faces by systematically checking all possible pairwise, 3-wise etc. intersections of non-simplex facets. Then, considering all 2, 3 and 4-sets of vertices and the remaining possible pairwise, 3-wise etc. intersections of facets, we obtained all i-faces for $i = 0, 1, 2, 3, 7, 8, 9$. Finally, noticing that few i-faces contains the complete lower part of the lattice (for example any 7-face of m_5 is a facet of a face belonging to a single orbit of 8-face), we found by a case by case analysis all the remaining simplex 4, 5 and 6-faces of m_5 and c_5. The dimensions of the faces were computed using the list of all affine dependencies of m_5 and c_5 given below.

Using [10] one can that check all affine dependencies on the vertices of m_5 and c_5, that is equations $\sum \lambda_i x_i = 0$ with $\sum \lambda_i = 0$, are, up to permutations, switchings and the bijection $\delta(S) \leftrightarrow \hat{\delta}(S)$ (which clearly preserves affine dependencies):

- $\sum_{S \subset \{1,2,3,4\}} (-1)^{|S|} \delta(S) = 0,$
- $\sum_{S \subset \{1,2,3,4,5\},\, |\{4,5\} \cap S| = 1} (-1)^{|S|} \delta(S) = 0,$
- $2\delta(\emptyset) - 2\delta(1) + \sum_{i=2}^{5} (\delta(1,i) - \delta(i)) = 0,$
- $3\hat{\delta}(\emptyset) + \delta(1) - \sum_{i=2}^{5} \delta(1,i) = 0.$

The restriction of the face lattices of c_5 and m_5 to their non-simplex faces are respectively given in Figures 3.1 and 3.8. All simplicial and all maximal (under inclusion) simplex faces are given in Proposition 3.2 and their complete face lattices are presented in detail in Section 3.2.1 and Section 3.2.2. The f-vectors of c_5 and m_5 are respectively:

- $f(c_5) = (16, 120, 560, 1780, \ldots, 3080, 640, 56),$
- $f(m_5) = (32, 280, 1280, 3620, \ldots, 2840, 480, 40).$

By f_j^i and g_j^i we respectively denote a representative j-face of the ith orbit C_j^i, respectively M_j^i, of c_5 and m_5.

3.2.1 Face lattice of c_5

In Figure 3.1 all the 4 orbits of proper non-simplex faces of m_5 are given. Each orbit is represented by the set of vertices belonging to a representative face of the orbit. A cut $\delta(i)$, respectively $\delta(ij)$, is denoted by a circled point i and respectively by an edge $\{i,j\}$.

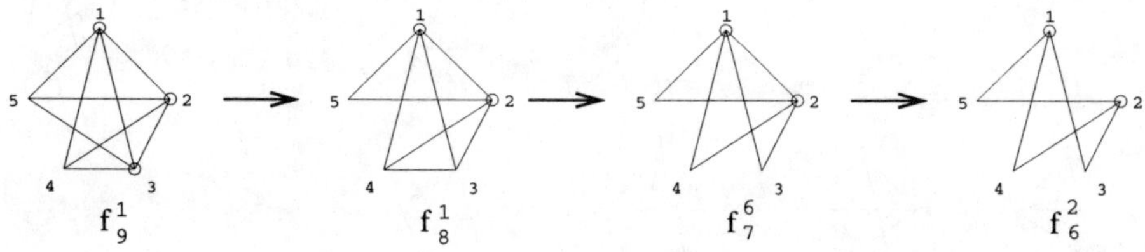

Figure 3.1: Non-simplices of the face lattice of c_5

The face lattice of c_5 is partitioned into the following orbits of $Is(c_5)$:

- the 56 facets are partitioned into the 2 orbits C_9^1 and C_9^2 respectively formed by the 40 triangle facets represented by $f_9^1 = Tr_{123}$ and the 16 switchings of the *equicut facet* f_9^2 which is defined by the inequality: $\sum_{1\le i<j\le 5} x_{ij} \le 6$,

- the 640 ridges are partitioned into the 3 orbits C_8^1, C_8^2 and C_8^3 of size 240, 240 and 160 and respectively represented by $f_8^1 = Tr_{123} \cap Tr_{124}$, $f_8^2 = Tr_{123} \cap Tr_{145}$ and $f_8^3 = Tr_{123} \cap f_9^2$,

- the 3080 7-faces are partitioned into the 7 orbits $C_7^1, C_7^2 \dots C_7^7$ of size 120, 960, 480, 240, 160, 160 and 960 respectively represented by the graphs given in Figure 3.2. We have: $f_7^1 = f_8^1 \cap Tr_{24,3} \cap Tr_{13,4}$, $f_7^2 = f_8^2 \cap Tr_{23,4}$, $f_7^3 = f_8^3 \cap Tr_{125}$, $f_7^4 = f_8^2 \cap f_9^2$ $(= f_8^3 \cap Tr_{145})$, $f_7^5 = f_8^1 \cap Tr_{134}$, $f_7^6 = f_8^1 \cap Tr_{125}$ and $f_7^7 = f_8^1 \cap Tr_{135}$,

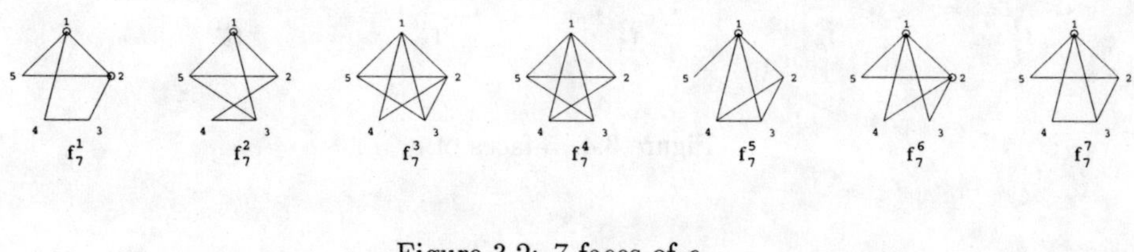

Figure 3.2: 7-faces of c_5

- the 6-faces are partitioned into the 10 orbits $C_6^1, \ldots, C_6^{10}$ respectively represented by the graphs given in Figure 3.3,

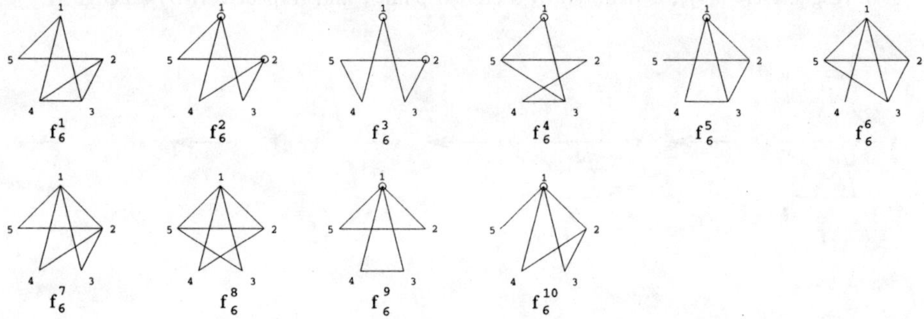

Figure 3.3: 6-faces of c_5

- the 5-faces are partitioned into the 11 orbits $C_5^1, \ldots, C_5^{11}$ respectively represented by the graphs given in Figure 3.4,

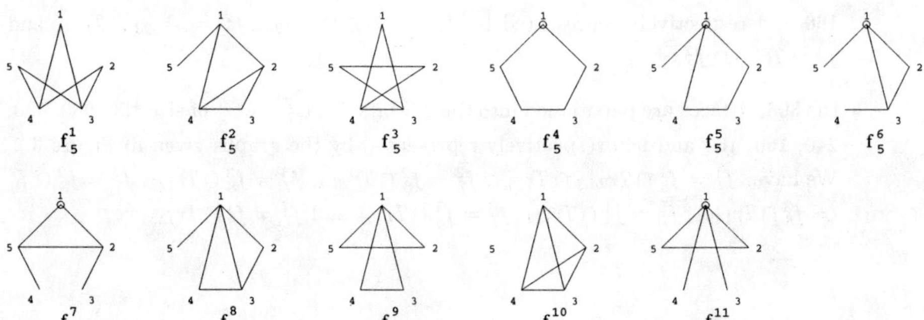

Figure 3.4: 5-faces of c_5

- the 4-faces are partitioned into the 8 orbits $C_4^1, \ldots, C_4^8$ respectively represented by the graphs given in Figure 3.5,

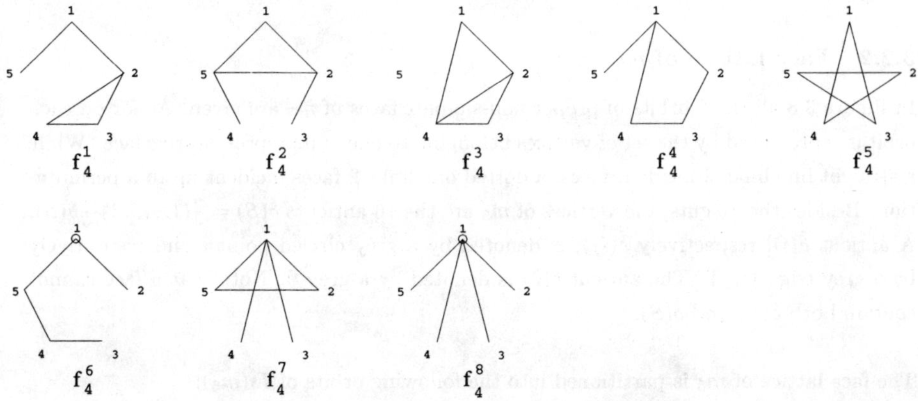

Figure 3.5: 4-faces of c_5

- the 1780 3-faces are partitioned into the 7 orbits $C_3^1, \ldots, C_3^7$ respectively represented by the graphs given in Figure 3.6. Actually, the only sets of 4 cuts which are not 3-faces of c_5 are the 40 members of the orbit of $\{\delta(\emptyset), \delta(1,2), \delta(1,3), \delta(2,3)\}$,

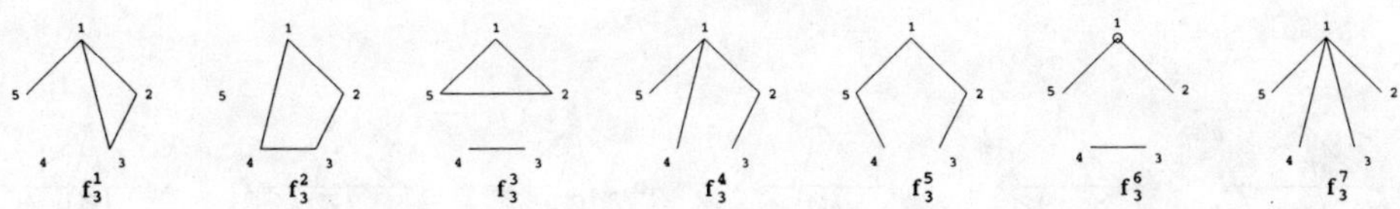

Figure 3.6: 3-faces of c_5

- the 560 2-faces are partitioned into the 3 orbits C_2^1, C_2^2 and C_2^3 of size 160, 160 and 240, and respectively represented by $f_2^1 = \{\delta(\emptyset), \delta(1), \delta(2)\}$, $f_2^2 = \{\delta(\emptyset), \delta(1,2), \delta(1,3)\}$ and $f_2^3 = \{\delta(\emptyset), \delta(1), \delta(2,3)\}$,

- the 120 edges are partitioned into the 2 orbits C_1^1 and C_1^2 respectively formed by the 40 edges $\{\delta(S), \delta(S')\}$ with $|S \Delta S'| = 1$ or 4 and the 80 ones with $|S \Delta S'| = 2$ or 3 (that is respectively represented by $f_1^1 = \{\delta(\emptyset), \delta(1)\}$ and $f_1^2 = \{\delta(\emptyset), \delta(1,2)\}$),
- the 16 vertices form the orbit C_0^1.

3.2.2 Face lattice of m_5

In Figure 3.8 all the 8 orbits of proper non-simplex faces of m_5 are given. As for c_5, each orbit is represented by the set of vertices belonging to one of its representative face. While a straight line links 2 incident faces, a dotted one links 2 faces incident up to a permutation. Besides the 16 cuts, the vertices of m_5 are the 16 anticuts $\hat{\delta}(S) = \frac{2}{3}(1, \ldots, 1)$-$\frac{1}{3}\delta(S)$. A anticut $\hat{\delta}(i)$, respectively $\hat{\delta}(ij)$, is denoted by a grey circled point i and respectively by a grey edge $\{i, j\}$. The anticut $\hat{\delta}(\emptyset)$ is denoted by a grey $\emptyset$. Note that a face cannot contain both $\delta(S)$ and $\hat{\delta}(S)$.

The face lattice of m_5 is partitioned into the following orbits of $Is(m_5)$:

- the 40 triangle facets form the orbit M_9^1 represented by $g_9^1 = Tr_{123}$,
- the 480 ridges are partitioned into the 2 orbits M_8^1 and M_8^2, both of size 240 and respectively represented by $g_8^1 = Tr_{123} \cap Tr_{124}$ and $g_8^2 = Tr_{123} \cap Tr_{145}$,
- the 1880 7-faces are partitioned into the 6 orbits $M_7^1, M_7^2, \ldots, M_7^6$ of size 120, 960, 480, 160, 160 and 960 respectively represented by the graphs given in Figure 3.7. We have: $g_7^1 = g_8^1 \cap Tr_{24,3} \cap Tr_{13,4}$, $g_7^2 = g_8^1 \cap Tr_{35,4}$, $g_7^3 = g_8^1 \cap Tr_{345}$, $g_7^4 = g_8^1 \cap Tr_{134}$, $g_7^5 = g_8^1 \cap Tr_{125}$ and $g_7^6 = g_8^1 \cap Tr_{135}$,

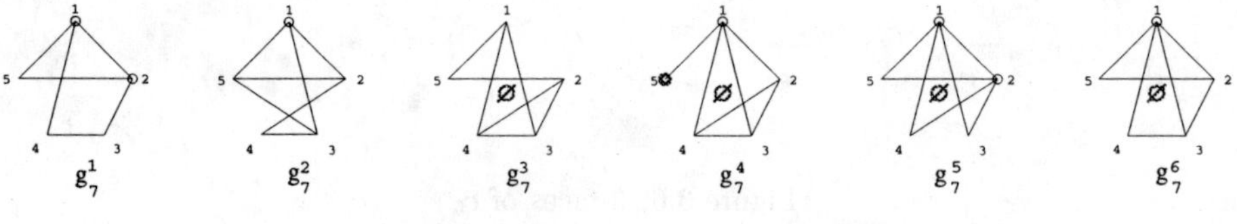

Figure 3.7: 7-faces of m_5

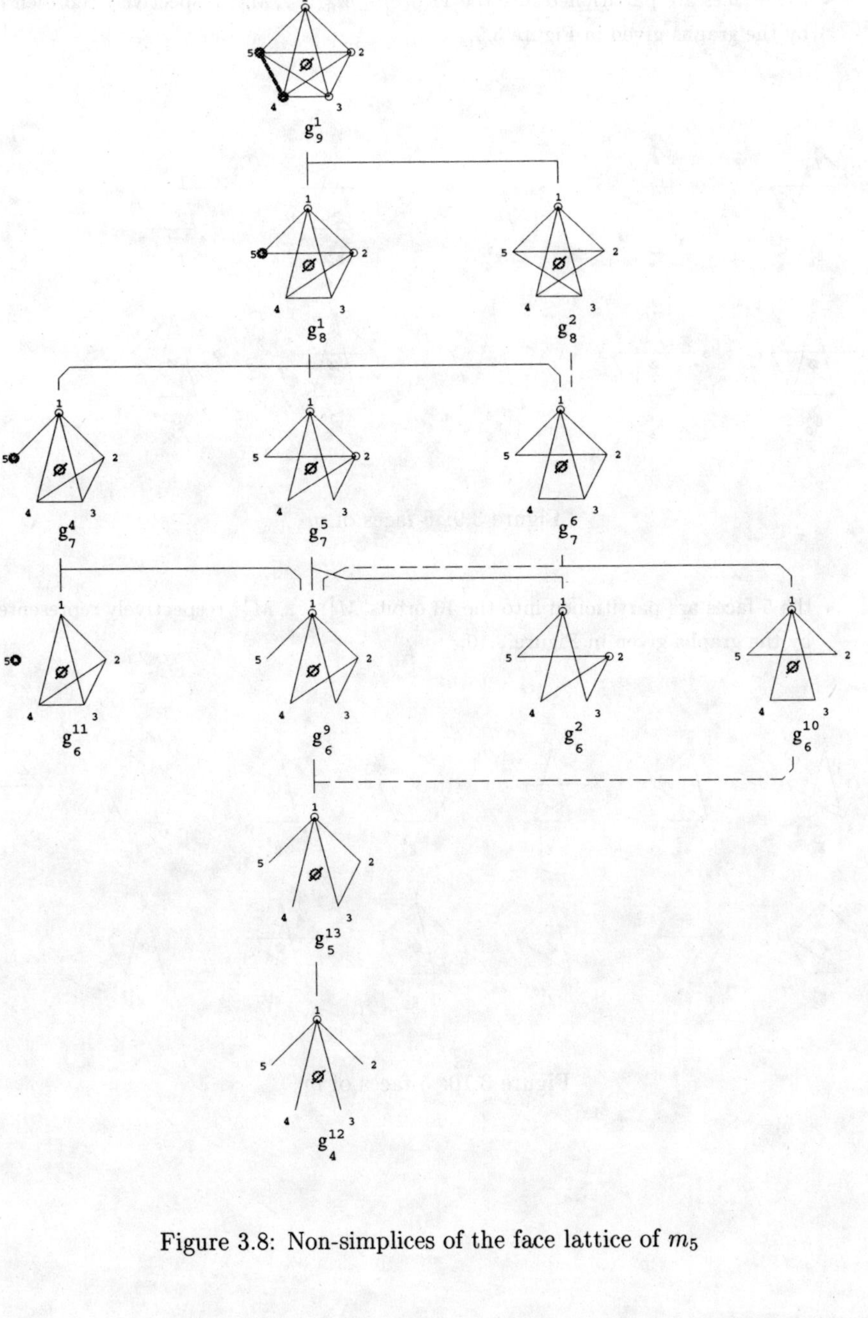

Figure 3.8: Non-simplices of the face lattice of m_5

- the 6-faces are partitioned into the 11 orbits $M_6^1, \ldots, M_6^{11}$ respectively represented by the graphs given in Figure 3.9,

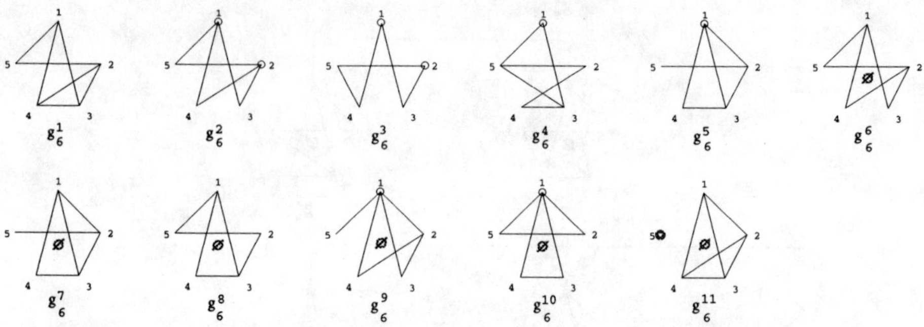

Figure 3.9: 6-faces of m_5

- the 5-faces are partitioned into the 13 orbits $M_5^1, \ldots, M_5^{13}$ respectively represented by the graphs given in Figure 3.10,

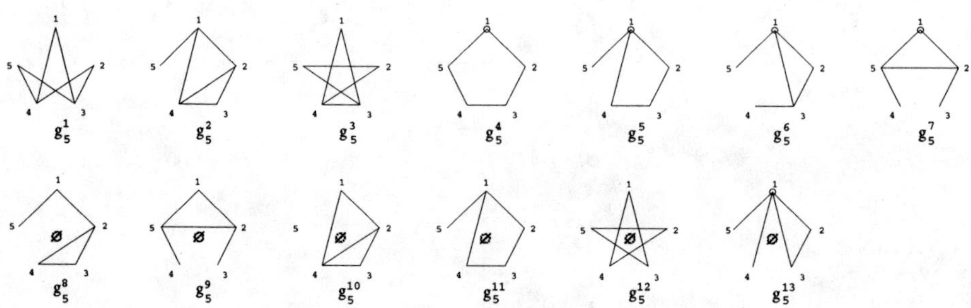

Figure 3.10: 5-faces of m_5

- the 4-faces are partitioned into the 12 orbits $M_4^1, \ldots, M_4^{12}$ respectively represented by the graphs given in Figure 3.11,

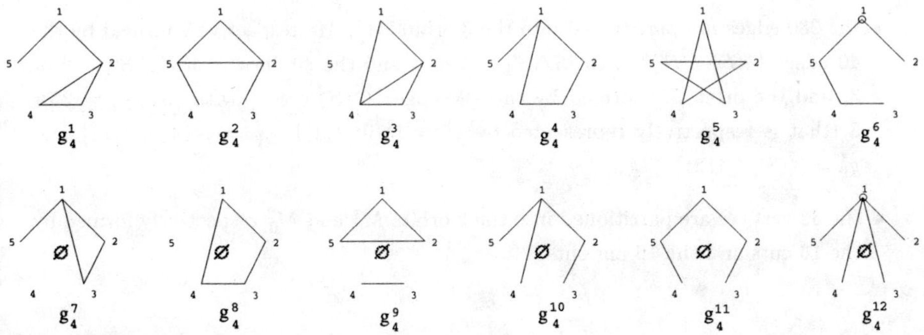

Figure 3.11: 4-faces of m_5

- the 3620 3-faces are partitioned into the 10 orbits $M_3^1, \ldots, M_3^{10}$ respectively represented by the graphs given in Figure 3.12,

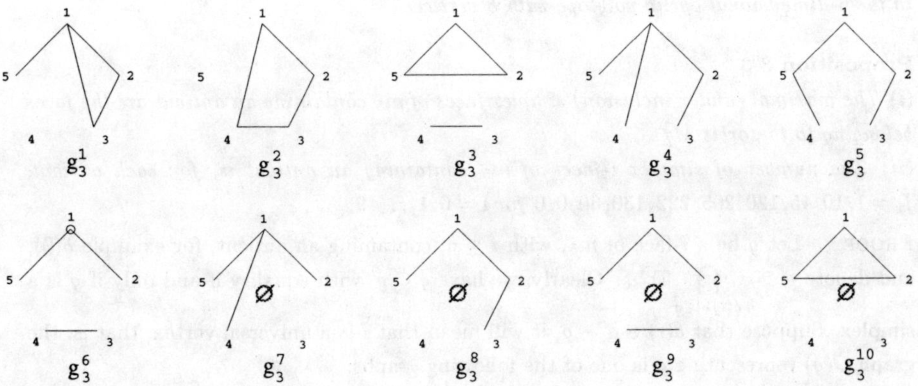

Figure 3.12: 3-faces of m_5

- the 1280 2-faces are partitioned into the 5 orbits M_2^1, $M_2^2, \ldots, M_2^5$ of size 160, 160, 240, 480 and 240, and respectively represented by $g_2^1 = \{\delta(\emptyset), \delta(1), \delta(2)\}$, $g_2^2 = \{\delta(\emptyset), \delta(1,2), \delta(1,3)\}$, $g_2^3 = \{\delta(\emptyset), \delta(1), \delta(2,3)\}$, $g_2^4 = \{\hat{\delta}(\emptyset), \delta(1,2), \delta(1,3)\}$ and $g_2^5 = \{\hat{\delta}(\emptyset), \delta(1,2), \delta(3,4)\}$,

- the 280 edges are partitioned into the 3 orbits M_1^1, M_1^2 respectively formed by the 40 edges $\{\delta(S), \delta(S')\}$ with $|S \triangle S'| = 1$ or 4 and the 80 ones with $|S \triangle S'| = 2$ or 3, and the orbit M_1^3 formed by the 160 edges $\{\hat{\delta}(S), \delta(S')\}$ with $|S \triangle S'| = 2$ or 3 (that is respectively represented by $g_1^1 = \{\delta(\emptyset), \delta(1)\}$, $g_1^2 = \{\delta(\emptyset), \delta(1,2)\}$ and $g_1^3 = \{\hat{\delta}(\emptyset), \delta(1,2)\}$).

- the 32 vertices are partitioned into the 2 orbits M_0^1 and M_0^2 respectively formed by the 16 cuts and the 16 anticuts.

Proposition 3.2
(i) The maximal (under inclusion) simplex faces of c_5 *are the faces belonging to the orbits* C_9^2, C_8^2, C_7^1 *and* C_7^5.
(ii) The maximal (under inclusion) simplex faces of m_5 *are the faces belonging to the orbits* M_7^1, M_7^2 *and* M_7^3.
(iii) The simplicial faces of c_5 *and* m_5 *are respectively the faces belonging to the orbit* C_6^2 *and* M_6^{11}, M_6^2 *and* M_4^{12}.
(iv) The faces belonging to C_6^2 *and* M_6^2 *are combinatorially equivalent to* $m_4 = c_4$ *and so to the 6-dimensional cyclic polytope with 8 vertices.*

Proposition 3.3
(i) The maximal (under inclusion) simplex faces of m_5 *containing an anticut are the faces belonging to the orbit* M_7^3.
(ii) The number of simplex i-faces of m_5 *containing an anticut is, for each anticut,* $\hat{f}_i = 1, 10, 45, 120, 205, 222, 130, 30, 0, 0$ *for* $i = 0, 1, \ldots, 9$.

PROOF. Let g be a i-face of m_5, with $i \leq 6$, containing an anticut, for example $\hat{\delta}(\emptyset)$, and denote $g' := \bigcap_{\delta(ijk) \notin g} Tr_{ijk}$. Clearly, we have $g \subseteq g'$ with equality if and only if g is a simplex. Suppose that $\delta(i) \in g' - g$, it will mean that i is a universal vertex, that is, the graph $G(g)$ representing g is one of the following graphs:

It turns out that those 4 graphs are respectively subgraphs (we require only inclusion of the edge-set) of the graphs representing the non-simplex faces: g_6^9, $g_6^{10}4\, g_5^{13}$, g_4^{12} (see Figure 3.8). Now, since g' is the intersection of homogeneous triangle facets, $\hat{\delta}(ij) \in g'$ implies that $\hat{\delta}(i), \hat{\delta}(j) \in g'$. Then $\hat{\delta}(i) \in g' - g$ if and only if $G(g)$ contains the clique K_4 which turns out to be a subgraph of the non-simplex face: g_6^{11}. □

3.3 Wrapping of c_n by m_n

Let call *extra i-face* of c_n (respectively m_n) a i-face of c_n (resp. m_n) which is not a i-face of m_n (resp. c_n). We recall that all i-faces of c_n are also i-faces of m_n for $i = 0, 1$ and 2; moreover, it was conjectured in [14] that for n large enough ($n > 2^i$) all i-faces of c_n are also i-faces of m_n. We disprove this conjecture by exhibiting an extra 3-face of c_n. For $n \geq 5$, let consider the following face of c_n:

$$f_3 = \{\delta(12), \delta(13), \delta(14), \delta(15)\}.$$

Proposition 3.4 *For $n \geq 5$, the face f_3 is an extra 3-face of c_n.*

PROOF. For $n = 5$, the face f_3 is the 3-face f_3^7 of c_5 which is itself a 10-face of c_n ($c_5 = \bigcap_{i=6}^{n}(Tr_{12,i} \cap Tr_{2i,1})$), that is, f_3 is a 3-face of c_n. Now let suppose that f_3 is also a face of m_n, it would implies that f_3 is a face of the following face of m_n:

$$g = (\bigcap_{2 \leq i < j \leq 5} Tr_{1ij}) \cap (\bigcap_{i=6}^{n}(Tr_{12,i} \cap Tr_{2i,1})),$$

where the triangle facets are seen as facets of m_n. For $n = 5$, we have $g = g_4^{12}$. One can easily check that g contains, besides the 5 cuts $\delta(1), \delta(12), \delta(13), \delta(14)$ and $\delta(15)$, the vertex x which coordinates $x_{ij} : 1 \leq i \leq j \leq n$ are: $x_{ij} = 0$ for $i \neq 2, 3, 4, 5 < j$ and $\frac{2}{3}$ otherwise. Then, to remove x we should intersect g with some $Tr_{ij,k}$ with $1 \leq i, j, k \leq 5$, but doing so will also eliminate one of the 4 cuts $\delta(12), \delta(13), \delta(14)$ or $\delta(15)$ of f_5 as well, which implies that f_3 cannot be a face of m_n. □

Corollary 3.5 *For $n \geq 5$, all i-faces of c_n are also i-faces of m_n for exactly $i = 0, 1$ and 2.*

PROOF. Let suppose that all i_0-faces of c_n are also i_0-faces of m_n with $i_0 \geq 3$. Then, the face g_i of m_n equal to a face f_i of c_n containing f_3 will contain a face g_3 equal to f_3, which contradicts Proposition 3.4. □

Proposition 3.6
(i) The representative extra i-faces of c_5 *belong to* $F = \{f_9^2, f_8^3, f_7^3, f_7^4, f_6^6, f_6^7, f_6^8, f_5^8, f_5^9, f_5^{10}, f_4^7, f_3^7\}$ *(that is, all extra faces belonging to* f_9^2*) and to* $F' = \{c_5 = f_{10}^1, f_9^1, f_8^1, f_8^2, f_7^5, f_7^6, f_7^7, f_6^9, f_6^{10}, f_5^{11}, f_4^8\}$.
(ii) The representative extra i-faces of m_5 *belong to* $G = \{g_7^3, g_6^6, g_6^7, g_6^8, g_5^8, g_5^9, g_5^{10}, g_5^{11}, g_5^{12}, g_4^7, g_4^8, g_4^9, g_4^{10}, g_4^{11}, g_3^7, g_3^8, g_3^9, g_3^{10}, g_2^4, g_2^5, g_1^3, g_0^2\}$ *(that is, all simplex extra faces, i.e. all extra faces belonging to* g_7^3*) and* $G' = \{m_5 = g_{10}^1, g_9^1, g_8^1, g_8^2, g_7^4, g_7^5, g_7^6, g_6^9, g_6^{10}, g_6^{11}, g_5^{13}, g_4^{12}\}$ *(that is, all faces given in Figure 3.8 except* $g_6^2 = f_6^2$ *which is the unique non-simplex common face of* c_5 *and* m_5*).*
(iii) All remaining faces, besides the ones given in (i) and (ii) are common faces of c_5 *and* m_5*, we ordered them so that* $f_j^i = g_j^i$*. Their representative i-faces belong to* $H = \{f_6^1, f_5^1, f_5^2, f_5^3, f_4^1, f_4^2, f_4^3, f_4^4, f_4^5, f_3^1, f_3^2, f_3^3, f_3^4, f_3^5, f_2^1, f_2^2, f_2^3, f_1^1, f_1^2, f_0^1, f_{-1}^1 = \emptyset\}$ *(that is, all common faces belonging to* f_6^1*) and to* $H' = \{f_7^1, f_7^2, f_6^2, f_6^3, f_6^4, f_6^5, f_5^4, f_5^5, f_5^6, f_5^7, f_4^6, f_3^6\}$.

Remark 3.7 *We observed that* $F = \{g \cap f_9^2 : g \in G'\}$, $F' = \{g \cap c_5 : g \in G' - \{g_6^{11}\}\}$, $H = \{g \cap c_5 : g \in G\}$ *and* $F - \{f_5^{10}\} = \{f \cap f_9^2 : f \in F'\}$ *and that the dimension of those intersections is one less than the one of the corresponding face. The four above equalities,* $g_3^7 \cap c_5 \simeq g_3^{10} \cap c_5$ *and* $g_6^{11} \cap c_5 = f_5^{10}$ *give the following four bijections:* $F \leftrightarrow G'$, $F' \leftrightarrow G' - \{g_6^{11}\}$, $H \leftrightarrow G - \{g_3^7\}$ *and* $F' \leftrightarrow F - \{f_5^{10}\}$.

dimension	-1	0	1	2	3	4	5	6	7	8	9	10
total # of orbits in c_5	1	1	2	3	7	8	11	10	7	3	2	1
# of orbits in F					1	1	3	3	2	1	1	
# of orbits in F'						1	1	2	3	2	1	1
# of orbits in H	1	1	2	3	5	5	3	1				
# of orbits in H'					1	1	4	4	2			
# of orbits in G		1	1	2	4	5	5	3	1			
# of orbits in G'						1	1	3	3	2	1	1
total # of orbits in m_5	1	2	3	5	10	12	13	11	6	2	1	1

Figure 3.13: Number of orbits of i-faces

Remark 3.8
(i) All minimal (by inclusion) faces from F *belong to the orbits of* f_3^7 *or* $f_5^{10} = f_9^2 - f_3^7$.
(ii) All minimal (by inclusion) faces from H' *belong to the orbits of* f_3^6 *or* f_5^7.
(iii) Besides f_3^7 *and the cofacet* $f_3^6 = Tr_{134} \cap Tr_{13,4} \cap Tr_{14,3} = \bar{Tr}_{34,1}$*, any of the* 5 *remaining representative 3-faces of* c_5 *belongs to exactly either 4 or 5 triangle facets.*

3.4 Some orbits of faces and nofaces of c_n and m_n

In Figure 3.14, we present 5 orbits of c_n and m_n. For each orbit a representative facet, the codimension, the size and the number of cuts (and anticuts for orbits of m_n) belonging to a face of the orbit are given. The orbit O_2^5 (respectively O_3^5, O_4^5 and O_5^5) corresponds to C_8^1 and M_8^1 (respectively C_8^2 and M_8^2, C_7^1 and M_7^1, and C_6^2 and M_6^2). In m_n, the faces belonging to O_5^n are combinatorially equivalent to m_{n-1}.

orbit	representative	codimension	size	# cuts	# anticuts
O_1^n	$Tr_{135} \cap Tr_{246}$	2	$160\binom{n}{6}$	$9 \times 2^{n-5}$	2^{n-5}
O_2^n	$Tr_{123} \cap Tr_{124}$	2	$48\binom{n}{4}$	$5 \times 2^{n-4}$	2^{n-4}
O_3^n	$Tr_{123} \cap Tr_{145}$	2	$240\binom{n}{5}$	$9 \times 2^{n-5}$	2^{n-5}
O_4^n	$Tr_{135} \cap Tr_{45,3}$	3	$24\binom{n}{4}$	2^{n-2}	0
O_5^n	$Tr_{135} \cap Tr_{15,3}$	$n-1$	$2\binom{n}{2}$	2^{n-2}	0

Figure 3.14: All pairwise intersections of facets of m_n

We then consider the following set of cuts: $A_n = \{\delta(i) : 1 \le i \le n\}$. It was remarked in [14] that no triangle facet contains A_n. Moreover, we have:

Proposition 3.9

(i) No proper face of c_n contains A_n.

(ii) For $n \ge 5$, any $(n-1)$-subset of A_n forms an extra $(n-2)$-face of c_n.

(iii) The size of the orbit represented by A_n is $|O(A_n)| = 1, 2, 2^{n-1}$ for $n = 3, 4, \ge 5$.

PROOF. Let F be a facet induced by the inequality: $\sum_{1\le i<j\le n} f_{ij}x_{ij} \le a$ and $F(S)$ the value of the left hand side of the inequality on the cut $\delta(S)$. Since we have $F(\delta(ij)) = F(\delta(i)) + F(\delta(j)) - 2f_{ij} \le a$, A_n belongs to F will implies that $f_{ij} \ge \frac{a}{2}$. So, we have $a \ge F(\delta(S)) \ge \frac{a|S|(n-|S|)}{2}$, this implies $a = 0$ and therefore, for $n \ge 4$ all $f_{ij} \ge \frac{a}{2} = 0$ which contradicts $F(\delta(1)) \le a = 0$. To prove *(ii)*, remark that $A_n - \{\delta(1)\}$ is the following face of c_n: $A_n - \{\delta(1)\} = E_{\delta(1)} \cap (\bigcap_{2\le i<j\le n} Tr_{ij,1})$ where $E_{\delta(1)}$ is the switching by the cut $\delta(1)$ of the face defined by the inequality: $\sum_{1\le i<j\le n} x_{ij} \le \lceil\frac{n}{2}\rceil \cdot \lfloor\frac{n}{2}\rfloor$. On the other hand, $\bigcap_{2\le i<j\le n} Tr_{ij,1}$ contains $\delta(\emptyset)$ which can be removed only by intersecting $A_n - \{\delta(1)\}$ with a non-homogeneous triangle facet, but this will also eliminate some cuts $\delta(i)$, that is, $A_n - \{\delta(1)\}$ cannot be a face of m_n. Since for $n \ge 5$ all switchings of A_n are different and any permutation amounts to a switching, $|O(A_n)| = 2^{n-1}$; cases $n = 3$ and 4 are clear. □

Remark 3.10 *Proposition 3.8 means that A_n is a minimal (by inclusion) blocker, that is $A_n \cap \bar{f} \neq \emptyset$, for the clutter $T_c := \{\bar{f} : f$ is a facet of $c_n\}$ of complements of facets of c_n and, a fortiori, for the clutter $T_m := \{\bar{g} : g$ is a facet of $m_n\}$. Perhaps, A_n is also a minimum, that is of minimal cardinality, blocker for T_c. Remark that minimum blockers for T_m are exactly the pairs $\{\delta(S), \hat{\delta}(S)\}$.*

We call *noface* of c_n a set of cuts which does not form a face of c_n.

Proposition 3.11
(i) For $n \geq 4$, any set containing member of $O(A_n)$ is a noface (see Proposition 3.8).
(ii) For $n = 4$, any noface contains a member of $O(A_4)$; the only nofaces which are cofaces belong to the orbits represented by $\bar{f}_1^2$ and $\bar{f}_2^2$.
(iii) For $n = 5$, there are exactly $1, 2, 3, 8, 13$ orbits of i-sets of cuts in c_5 for $i = 1, 2, 3, 4, 5$ and among there is exactly 6 orbits of nofaces represented by the following graphs:

One can check that the following holds for any $n \geq 3$:

- the 2^{n-1} vertices of c_n form the orbit represented by $\{\delta(\emptyset)\}$,
- the $\binom{2^{n-1}}{2}$ edges of c_n are partitioned into the $\lfloor \frac{n}{2} \rfloor$ orbits respectively represented by $f_1^i = \{\delta(\emptyset), \delta(1, \ldots, i)\}$ for $i = 1, \ldots, \lfloor \frac{n}{2} \rfloor$ with $|O_1^i| = 2^{n-2}\binom{n}{i}$ for $1 \leq i < \lfloor \frac{n}{2} \rfloor$ and $|O_1^i| = 2^{n-3}\binom{n}{\frac{n}{2}}$ for n even,
- the $\binom{2^{n-1}}{3}$ 2-faces of c_n are partitioned into the orbits respectively represented by $f_2^{r,s,t} = \{\delta(\emptyset), \delta(1, \ldots, r+s), \delta(r+1, \ldots, r+s+t)\}$ for all triplets of integers $\{r, s, t\}$ such that $1 \leq r \leq \lfloor \frac{n}{3} \rfloor$, $0 \leq s \leq r$, $r \leq t \leq min(\lfloor \frac{n-r}{2} \rfloor, \lfloor \frac{n}{2} \rfloor - s, n - 2r - s)$,
- for $n \geq 6$, the $\frac{16(n^2-7)\binom{n}{4}}{3}$ ridges, that is $(\binom{n}{2} - 2)$-face of m_n, are partitioned into the 3 orbits O_1^n, O_2^n and O_3^n given in Figure 3.14 (see [6]),
- the $4\binom{n}{3}$ facets of m_n form the orbit represented by Tr_{123}.

References

[1] P. Assouad and M. Deza, "Metric subspaces of L^1", *Publications mathématiques d'Orsay* **3**, (1982).

[2] D. Avis , "On the extreme rays of the metric cone", *Canadian Journal of Mathematics XXXII* **1**, (1980) 126-144.

[3] F. Barahona and R. Mahjoub, "On the cut polytope", *Mathematical Programming* **36** (1986) 157-173.

[4] M. Bayer and C. Lee, "Combinatorial aspects of convex polytopes ", in P. Gruber and J. Wills (eds.) *Handbook on Convex Geometry*, North Holland (1994) 485-534.

[5] A. Brouwer, A. Cohen and A. Neumaier, *Distance-Regular Graphs*, Springer-Verlag, Berlin, 1989.

[6] A. Deza and M. Deza, "The ridge graph of the metric polytope and some relatives", in T. Bisztriczky, P. McMullen, R. Schneider & A. Ivic Weiss (eds.) *Polytopes: Abstract, Convex and Computational* (1994) 359-372.

[7] A. Deza and M. Deza, "On the skeleton of the dual cut polytope", to appear in *AMS Series Contemporary Mathematics*, H. Barcelo & G. Kalai (eds.), Proceedings of *Jerusalem Combinatorics 93.*

[8] A. Deza and M. Deza, "Geometry of cut and metric polytopes", *in preparation.*

[9] M. Deza, "Matrices de formes quadratiques non négatives pour des arguments binaires", *Comptes rendus de l'Académie des Sciences de Paris* **277**, (1973) 873-875.

[10] M. Deza, "Isometries of hypergraphs", *Proceedings of the International Conference on the Theory of Graphs*, in A. Rao (ed.), McMillan, Calcutta (1979) 174-189.

[11] M. Deza, V. Grishukhin and M. Laurent, "The symmetries of the cut polytope and of some relatives", in P. Gritzmann and B. Sturmfels (eds.) *Applied Geometry and Discrete Mathematics, the "Victor Klee Festschrift" DIMACS Series in Discrete Mathematics and Theoretical Computer Science* **4** (1991) 205-220.

[12] M. Deza and M. Laurent, "Facets for the cut cone I", *Mathematical Programming* **56 (2)**, (1992) 121-160.

[13] M. Deza and M. Laurent, "Applications of cut polyhedra" Rapport de Recherche du LIENS 92-18, Ecole Normale Supérieure, Paris, to appear in *Journal of Computational and Applied Mathematics*, (1992).

[14] M. Deza, M. Laurent and S. Poljak, "The cut cone III: on the role of triangle facets", *Graphs and Combinatorics* **8**, (1992) 125-142.

[15] V. P. Grishukhin, "Computing extreme rays of the metric cone for seven points", *European Journal of Combinatorics* **13**, (1992) 153-165.

[16] M. Iri, "On an extension of maximum-flow minimum-cut theorem to multicommodity flows", *Journal of the Operational Society of Japan* **13**, (1970-1971) 129-135.

[17] M. Laurent, "Graphic vertices of the metric polytope", Research report No. 91737-OR, Institut für Diskrete Mathematik, Universität Bonn, (1991).

[18] M. Laurent and S. Poljak, "The metric polytope", in E. Balas, G. Cornuejols & R. Kannan (eds.), *Integer Programming and Combinatorial Optimization*, Carnegie Mellon University, GSIA, Pittsburgh, (1992) 274-286.

[19] K. G. Murty and S. J. Chung, "Segments in enumerating faces", to appear in *Mathematical Programming.*

[20] M. Padberg, "The boolean quadric polytope: some characteristics, facets and relatives", *Mathematical Programming* **45**, (1989) 139-172.

[21] S. Poljak and Z. Tuza, "Maximum Cuts and Large Bipartite Subgraphs", in W. Cook, L. Lovasz and P. D. Seymour (eds.) DIMACS volume, *in print.*

[22] G. Rote, personal communication.

[23] V. Trubin, "On a method of solution of integer linear problems of a special kind", *Soviet Mathematics Doklady* **10**, (1969) 1544-1546.

[24] G. M. Ziegler, *Lectures on Polytopes*, Graduate Texts in Mathematics **152**, Springer-Verlag, New York, Berlin, Heidelberg, 1995.

ANTOINE DEZA

Tokyo Institute of Technology, department of mathematical and computing sciences, Meguro-ku, Ookayama, Tokyo, Japan. E-mail: deza@is.titech.ac.jp

MICHEL DEZA

CNRS, Ecole Normale Supérieure, département mathématiques et informatique, 45 rue d'Ulm, Paris, France. E-mail: deza@dmi.ens.fr

On the Problem of Optimization of Sociogram

Jiyu Ding *
(Qiqihar Teachers' College, China)

Guodong Song
(Qiqihar Light Industry Institute, China)

Abstract This paper puts forword the problem of studying optimization in sociogram and its mathematical models with the method of graph theory. Some of these models have been summed up as a typical problem in graph theory and have got the algorithms, and this paper provides algorithms for others.

Key words sociogram; optimization; human relations; mathematical model.

1. Introduction

The relationship between men and men is one of the most complicated relations in society. The words "men" and " relationship" dealt with in this paper are in their broad sense. " men" can be "countries", "organizations" or "things", the relations between men and men may refer to the relations among groups of people, countries or circles, they may also refer to the relations among different elements within a system. The group of "people" with such relations can be represented with a graph. Early in 1934 Mereno had represented the relations between men and men with a graph and called it as Sociogram. [1]

Suppose $V = \{v_1, v_2, \cdots, v_n\}$ denote a set of n people, to construct a digraph $G = G(V, E)$: its vertex set is V. The arc set is constructed with the following method: If v_i has a certain relation H to v_j, the intensity of the relation is represented with real number a_{ij}, then in E there is an arc (v_i, v_j) from v_i to v_j whose weight is a_{ij}. Thus we get a weighted digraph $G(V, E)$ which is called Human relation graph or Sociogram and whose weight matrix is $A(G) = [a_{ij}]_{n\times n}$. The relation H may be one-way or two-way relations of control, trust, coorperation, alliance, influence, affiliation, connection, etc.

In the field of studying the classifications of sociogram, Hubbell had studied the problem of a_{ij} which can be both positive and negative and which is called "sentiment". Guodong Song takes $0 \leqslant a_{ij} \leqslant 1$, sets a_{ij} as the probability value of relation (v_i, v_j) appearing in G [3]; Based on sign(a_{ij}), Harary provided some theorems for the measurement of the balance of sociogram G; $G. D.$ Song did some study of the classification of sociogram with graph invariant such as Weiner's index number, non-adjoining number etc.

In the field of studying local circumstances and the position of individuals in groups, Davis had studied the clusterability[6] in sociogram; Holland and others had studied the grade

* This research was supported by the Natural Science Foundation of Heilongjiang Education Committee No 024.

system of cluster of sociogram; G. D. Song had described the clusterability with graph invariant; French and Harary had studied the influence of an individual on the rest of the individuals, *i. e.* the position of an individual in a group and provided the definition of index number and its calculation formulas of individual position for graphs without circles[8]. G. D. Song developde the above-mentioned work and provided eigenvector index definition of individual, calculation formulae and algorithm.

From the brief retrospect mentioned above we can see that soiogram as a graph theory model for the description of Human relation has been systematically studied, but a great number of problems about the optimization in this model have not been brought forword and studied. This paper raise these problems of optimization and studied them with some of the graph theory models.

2. Members with the highest position in the sociogram

Suppose that $G=G(V,E)$ is a sociogram, $A(G)=[a_{ij}]_{n\times n}$ is matrix of weight of G, where a_{ij} is intensity of relation H from vertex v_i to vertex v_j. Let $L_{ij}^{(k)}$ be a path from v_i to v_j with length k (it contain k arcs) and product of the intensities of k arcs is called intensity of the path $L_{ij}^{(k)}$. In sociogram G, sum of intensities of all paths with length k from v_i is called k degree intensity of v_i, and it is denoted by $d_k(v_i), k=1,2,\cdots$;

$$\sigma_k(v_i) = \sum_{m=1}^{k} d_m(v_i) \tag{1}$$

is called k-degree complete intensity of v_i.

Definition 1 In sociogram G, $v_i \in V$. If the limit

$$d(v_i) = \lim_{k\to\infty}[d_k(v_i)/\sum_{i=1}^{n} d_k(v_i)] \tag{2}$$

exists, then $d(v_i)$ is called position index of v_i. $D=(d(v_1), d(v_2), \cdots, d(v_n))^T$ is called position index vector.

G. D. Song proved: If G is biconnected and $n \geqslant 4$, then position index vector is

$$D = \lim_{k\to\infty}(A^k I/I^T A I) \tag{3}$$

where $A=A(G)$, $I=(1,1,\cdots,1)^T$. This limit exactly exist and equals dominant eigenvector of A, and it is obtained by Power method.

Definition 2 If the limit

$$\sigma(v_i) = \lim_{m\to\infty}[\sum_{k=1}^{m} d_k(v_i)/\sum_{i=1}^{n}\sum_{k=1}^{m} d_k(v_i)] \tag{4}$$

exists , $\sigma(v_i)$ is called position complete index of v_i, and vector $\sum = (\sigma(v_1), \sigma(v_2), \cdots, \sigma(v_n))^T$ is called position complete index vector.

Jiyu Ding and G. D. Song proved: If G is biconnected and $n \geqslant 4$, $a_{ij} \geqslant 0, i, j = 1, 2, \cdots, n$. Then $\sum = D$[10]. Based on this result we can provide the optimization problem: to find $v_i \in V$ so that

$$\sigma(v_i) = \max_{v_k \in V} \{\sigma(v_k)\} \tag{5}$$

We know that v_i is the very member with highest position in the sociogram.

3. The most reliable relation between two members in sociogram

A certain relation between members is represented by H, and intensity of this ralation between v_i and v_j is represented by a_{ij}, and $0 \leqslant a_{ij} \leqslant 1$. If $a_{ij} = 0$, then there is not any edge (v_i, v_j) in G; If there is a path L_{ij} from v_i to v_j and its intensity $d(L_{ij}) \neq 0$, then we have

$$d(L_{ij}^*) = \max_{v_i, v_j \in V} \{d(L_{ij})\} \tag{6}$$

The optimization problems can be given as following:

① How do we find the path L_{ij} for given v_i and v_j?

② How do we find v_i for given v_j so that $\max\{d(L_{ij})\} = d(L_{ij}^*)$.

We can reduce the problem ② to the problem ①, and reduce problem ① to the model of finding the shortest path between two given points.

Suppose $d(L_{ij}^*) = p_1 p_2 \cdots p_k$, where $p_1, p_2, \cdots, p_k$ are intensities of k arcs on the path L_{ij}^* respectively, $0 < p_i \leqslant 1, i = 1, 2, \cdots, k$, ,and p_i is exactly the element of $A = [a_{ij}]_{n \times n}$.

Because that

$$\ln d(L_{ij}^*) = \sum_{i=1}^{k} \ln p_i \geqslant \ln d(L_{ij})$$
$$|\ln d(L_{ij}^*)| = \sum_{i=1}^{k} |\ln p_i| \leqslant |\ln d(L_{ij})| \tag{7}$$

A new weighted digraph G' is constructed by $|\ln a_{ij}|$ as weight of arc (v_i, v_j) in $G(V, E)$, then the shortest path L_{ij}^* in G' is the solution of problem ① and the solution for v_j

$$\max_{v_i \in V} \{|\ln d(L_{ij}^*)|\} \tag{8}$$

is the solution of problem ②.

Algorithm finding the strongest relation from point s to point t in G(V,E)

Step 1: Take absolute value of logarithm for every element a_{ij} of A

$$w(v_i, v_j) = |ln\ a_{ij}|,\ i,j = 1,2,\cdots,n.$$

Step 2: Give initial mark values

```
while (x ∈ V){
        label(x)=∞;
        final(x)=false;
    }
label(s)=0;
final(x)=true;
y=s;
```

Step 3: Modify marks by Ford method:

```
while (! final(t)){
    while (x ∈ V){
        if(! final(x) && lable(x)>label(y)+w(y,x){
            lable(x)=lable(y)+w(y,x);
            pred(x)=y;
        }
    }
    let k be any vertex with minimum temporary label;
    final(k)=y
    y=k;
}
```

Step 4: Produce results $d(H(s,t))=e^{label(t)}$.

4. The strongest supporting relation

If there is only one path L between points x and y in sociogram, then the relation from x to y is $H(x,y)=H(L)$. Suppose that the direct relation linking arc e (from beginning point of the arc e to terminal point of e) is denoted by $H(e)$, then

$$H(x,y) = \bigwedge_{e \in L} H(e) \tag{9}$$

If there are more than one paths from x to y, and $P(x,y)$ denote the set of these paths, now we have

$$H(x,y) = \bigvee_{L \in P(x,y)} H(L) \tag{10}$$

Suppose T is an arbitrary supporting tree, obviously, the path between arbitrary two points in T is uniqe, so

$$H(T) = \bigwedge_{e \in T} H(e) \tag{11}$$

Let $p(e)$ denote intensity of $H(e)$ and $d(T)$ denote intensity of $H(T)$, we have

$$d(T) = \prod_{e \in T} p(e) \tag{12}$$

To find supporting tree T^* so that

$$d(T^*) = \max_{T}\{d(T)\} \tag{13}$$

where T^* is called the strongest supporting tree and $H(T^*)$ is called the strongest supporting relation.

From above mentioned two formulae, we have

$$d(T^*) = \prod_{e \in T^*} p(e) \geqslant d(T)$$
$$ln[d(T^*)] = \sum_{e \in T^*} ln\, p(e) \geqslant ln[d(T)]$$

Because $ln\, p(e) \leqslant 0$, for arbitrary $e \in T^*$, so

$$|ln[d(T^*)]| = \sum_{e \in T^*} |ln\, p(e)| \leqslant |ln[d(T)]| \tag{14}$$

The formula (14) means: the problem of finding the strongest supporting relation can be turned into the problem of finding minimum tree with weight $|ln\, p(e)|$ of edge e in G.

The minimum supporting tree in set E_T (set of edges of supporting tree T) is constructed by the method of adding shortest edge. When the vertexes of V_T (set of vertexes of T) increases from $V_T = \varnothing$ to $V_T = V$, the algorithm is completed.

Algorithm finding the strongest supporting relation:

Step 1: Take absolute value of logarithm of elements of A.

$$w(v_i, v_j) = |ln\, a_{ij}|, i, j = 1, 2, \cdots, n.$$

Step 2: Arbitrarily take a point $x_0 \in V$

Step 3: Give initial value:

$\forall x \in V$ $dist(x) = w(x, x_0)$, $near(x) = x_0$;

$V_T = \{x_0\}$;

$E_T = \varnothing$;

$w = 0$;

Step 4: Iterat calculating by adding the shortest edge method.

```
while(VT! =V){
    x=vertex in V-VT with the smallest value of dist(x);
    VT=VT∪{x};
    ET=E∪{(x,near(x))};
    w+=dist(x);
    while(y ∈V-VT){
    if (dist(y)>w(x,y)){
    dist(y)=w(x,y);
    near(y)=x;
        }
    }
}
```

Step 5: Result: $T^* = \bigcup_{e \in E_T} e$, $d(T^*)=e^w$

5. The strongest complete relation between two members

Let $P(x,y)$ denote the set of all paths from point x to point y and $d[x,y]$ denote the intensity of complete relation between x and y. By (9) and (10), we have

$$H[x,y] = H(x,y) \vee H(y,x) \tag{15}$$

$$H(x,y) = \vee_{k=1}^{n-1}\left(\bigvee_{L_k \in P(x,y)} H(L_k)\right) \tag{16}$$

so

$$\begin{aligned} d[x,y] &= d(x,y) + d(y,x) \\ &= \sum_{k=1}^{n-1}\sum_{L_k \in P(x,y)} d(L_k) + \sum_{k=1}^{n-1}\sum_{L'_k \in P(y,x)} d(L'_k) \\ &= \sum_{k=1}^{n-1}\left(\sum_{L_k \in P(x,y)} d(L_k) + \sum_{L'_k \in P(y,x)} d(L'_k)\right) \end{aligned} \tag{17}$$

When $k=1$, put $v_i=x, v_j=y, L_1=(x,y), L'_1=(y,x)$ and we have $d(L_1)=a_{ij}$, $d(L'_1)=a_{ji}$. When $k=2$, $A^k=[a_{ij}^{(k)}]_{n\times n}$, we have that

$$a_{ij}^{(2)} = a_{i1}a_{1j} + a_{i2}a_{2j} + \cdots + a_{in}a_{nj}$$

Obviously, $a_{it}a_{tj}$ is the intensity of path $v_i\, v_t\, v_j$, $t=1,2,\cdots,n$, that is $a_{ij}^{(2)}$ is sum of all intensities of paths with length 2 from v_i to v_j.

By (9) and (10), we have

$$\sum_{L_k \in P(x,y)} d(L_k) = a_{ij}^{(k)} \tag{18}$$

$$\sum_{L'_k \in P(y,x)} d(L'_k) = a_{ji}^{(k)} \tag{19}$$

By induction we can easily prove (18) and (19) are true for arbitrary k. The algorithm is obtained by(17),(18) and (19).

Algorithm finding the strongest complete relation between two members in sociogram

Step 1: Give initial value:

$A^1 = A(G) = [a_{ij}]_{n \times n}$

Step 2: Calculate matrix:

$A^k = [a_{ij}^{(k)}]_{n \times n}$;

$B = \sum_{k=1}^{n-1} A^k$;

$B = [b_{ij}]$;

Step 3: Find $\max_{x,y \in V} \{d[x,y]\}$

```
x=1;
y=2;
maxd=0;
for(i=1;i<=n;i++){
   for (j=i+1;j<=n;j++){
      if(maxd<b_ij+b_ji){
         maxd =b_ij+b_ji;
           x=i;
           y=j;
      }
   }
}
```

Step 4: Produce results

$\max\{d[x,y]\} = \text{maxd}$;

$x = v_x$;

$y = v_y$;

Many optimization models in graph theory such as the matching and covering problems, the problem of decomposition of a single factor and so on can be made as models of optimization for sociogram.

Refernces

[1] J. L. Moreno, who shall survive? Nervous and Mental Disease Publ. Co. Washington, D. C. 1934.

〔2〕 C. H. Hubbell, E. C. Johnsen and M. Marcus, Structural Balance in Group Networks, Handbook of Social Science Methods, Irvington Publ. Helted Press, New York, 1978.

〔3〕 G. D. Song, Fixed quantity method of graph theory for individual position in sociogram, O. R. and Management. Vol 3—4(1993). 24—28.

〔4〕 F. Harary, On the Notion of Balance of a Signed graph, Michigan Math. J. 1954; (2): 143—146.

〔5〕 G. D. Song, Quantization graph theory method for sociogram, mathematical application in sociology, 1989, 12—15.

〔6〕 J. A. Davis, Clusering and Structure Balance in Graph, Humen relations, 1967: (20)181—187.

〔7〕 P. W. Holland and S. Leinhardt, Transitivity in Strncture Model of Small Group, comparetive Group Studies, 1971; (2): 107—124.

〔8〕 F. Harary, A Criterion for Unanimity in French's Theory of Social Power, Studies in Social Power, Inst. Soc. Res., Ann Arbor, Michigan, 1959, 168—182.

〔9〕 G. D. Song, Characteristic vector index for individual position in sociogram, Management science and AHP, 1991, 89—99.

〔10〕 Jiyu Ding and G. D. Song, The relations between Individual Status Index and Complete Status Index in Sociogram, Northeast Operations Resesrch 8(1994): 174—176.

HEILBRONN PROBLEM FOR SIX POINTS IN A PLANAR CONVEX BODY

DRESS, A.W.M. YANG Lu* ZENG Zhenbing

Fakultät für Mathematik, Universität Bielefeld
D-4800 Bielefeld 1, Germany

Abstract For any six points in a planar convex body K there must be at least one triangle, formed by three of these points, with area not greater than 1/6 of the area of K. This upper bound 1/6 is best possible.

1 Introduction

Let K be a planar convex body (that means a compact convex set with non-empty interior), $|K|$ the area of K; for any triangle $r_1r_2r_3$, by $(r_1r_2r_3)$ denote its area; and let

$$(r_1r_2\cdots r_n) := \min\{(r_ir_jr_k)\,|1 \le i < j < k \le n\};$$

$$H_n(K) := \frac{1}{|K|}\sup\{(r_1r_2\cdots r_n)\,|r_i \in K, i = 1,\cdots,n\}.$$

The values $H_n(K), n = 3,4,\cdots$ defined as above are called *Heilbronn numbers*. Obviously, Heilbronn numbers do not change under affine transformations. If $\{r_1,r_2,\cdots,r_n\}$ is a subset of K such that

$$\frac{1}{|K|}(r_1r_2\cdots r_n) = H_n(K),$$

we say that $\{r_1,r_2,\cdots,r_n\}$ or $r_1r_2\cdots r_n$ is a *Heilbronn arrangement* of n points in K, simply, an *H-arrangement* in K.

Usually, we drop the K in $H_n(K)$ and write H_n to denote the Heilbronn numbers for K, when K is a square or parallelogram. There has been a lot of work concerning these numbers, see [2-11]. Even if n is a small integer, it is not easy to compute the exact value of H_n.

---

*Permanent address: Institute of Mathematical Sciences, Academia Sinica, 610015 Chengdu, People's Republic of China

In his paper [1], M. Goldberg considered exact values of the first several Heilbronn numbers. Besides the trivial cases, $H_3 = H_4 = 1/2$, he asserted that, for $n < 8$, H_n can be reached by some affine regular n-gon contained in the square, i.e. there must be an affine regular n-gon $r_1r_2\cdots r_n$ in K such that $\frac{1}{|K|}(r_1r_2\cdots r_n) = H_n$. And he listed these values as follows:

$$H_5 = \frac{3-\sqrt{5}}{4} = 0.1909\cdots,$$

$$H_6 = \frac{1}{8} = 0.125,$$

$$H_7 = 0.0794\cdots.$$

But he didn't give any proof.

The above-mentioned assertions were examined in [12-16] where it was shown that only one of the three is true, that is $H_6 = 1/8$. By a careful and detailed analysis, Yang, Zhang and Zeng proved

$$H_5 = \frac{\sqrt{3}}{9} = 0.1924\cdots;$$

$$H_6 = \frac{1}{8} = 0.125.$$

The first disproves Goldberg's conjecture for $n = 5$, the latter confirms it for $n = 6$. In addition, they showed by a simple example that

$$H_7 \geq \frac{1}{12} = 0.08333\cdots > 0.0794\cdots,$$

disproving Goldberg's conjecture for $n = 7$. From these discussions we know that, in general, Heilbronn arrangements in a square are not necessarily affine regular n-gons even if n is small. As to the problem for a triangular region, please refer to [17] where it was proved that

$$H_5(\triangle) = 3 - 2\sqrt{2}, \quad H_6(\triangle) = \frac{1}{8}.$$

So far as we know the above results concerning H_5, $H_5(\triangle)$, H_6, and $H_6(\triangle)$ give the first exact values of Heilbronn numbers. No further results appear to be known, in particular none concerning general planar convex bodies.

In this paper, we prove the following

Theorem 1 *For any six points in a planar convex body K there must be at least one triangle, formed by three of these points, with area not greater than $1/6$ of $|K|$. This upper bound $1/6$ is best possible.*

To the corresponding problem for seven points, we conjecture that for any planar convex body K, it holds

$$H_7(K) \leq \frac{1}{9},$$

and the upper bound 1/9 is best possible.

One could easily find examples to check that, in general, $H_n(K)$ are not necessarily bounded by $H_n(K_n)$ for $n \geq 7$, where by K_n denote the regular n-gons.

2 Prerequisites

The lemmata stated in this section, except Lemma 3, all are known and can be found in [16]. We will give the arguments here, because [16] has not been published formally.

Given n points $r_1, \cdots, r_n$ in the plane, we define a triangle $r_ir_jr_k$, formed by three of these points, to be *tight* relative to $\{r_1, \cdots, r_n\}$ if $(r_ir_jr_k) = (r_1 \cdots r_n)$, and we define it to be *loose* if $(r_ir_jr_k) > (r_1 \cdots r_n)$. Given a convex polygon with vertices $r_1, r_2, \cdots, r_n$, a triangle $r_ir_jr_k$ is said to be *peripheral* if r_i, r_j, r_k are three consecutive adjacent vertices in the polygon.

Lemma 1 *Given a convex polygon $r_1r_2 \cdots r_n$, the triangle $r_ir_jr_k$ must be peripheral if it is tight relative to $\{r_1, r_2, \cdots, r_n\}$.*

This is easily established. □

Lemma 2 *Given a convex n-gon $r_1r_2 \cdots r_n$ with two loose peripheral triangles sharing an edge and a positive number δ, there is another convex n-gon $r'_1 \cdots r'_n$, contained in the former, such that*

$$(r'_1 \cdots r'_n) > (r_1 \cdots r_n)$$

and $|r'_i - r_i| < \delta$ for $i = 1, \cdots, n$.

Proof. Let $r_1r_2r_3$ and $r_2r_3r_4$ be two loose peripheral triangles sharing the edge r_2r_3. Put $r_3' := (1-\varepsilon)r_3 + \varepsilon r_2$, (as shown in Fig. 1), where $\varepsilon > 0$ small enough, which keeps

$$(r_1r_2r_3') > (r_1r_2\cdots r_n), \quad (r_2r_3'r_4) > (r_1r_2\cdots r_n)$$

and makes $(r_3'r_4r_5) > (r_1r_2\cdots r_n)$ because

$$(r_3'r_4r_5) = (1-\varepsilon)(r_3r_4r_5) + \varepsilon(r_2r_4r_5)$$

and $r_2r_4r_5$ is not tight. Replacing r_3 by r_3', we have a new n-gon with 3 peripheral triangles of area greater than $(r_1r_2\cdots r_n)$. Repeating this procedure, one finally gets an n-gon which is contained in the first one and every of its peripheral triangles has an area greater than $(r_1r_2\cdots r_n)$. □

Lemma 3 *Given points $B, C \in \mathbf{R}^2$ with coordinates $B = (1,0)$, $C = (0,1)$, let D, E be the intersections of the line connecting points $(\xi, 0)$ and $(0,\eta)$ with the lines $x = \rho$, $y = \tau$, respectively, where $0 < \rho < \xi < 1$, $0 < \tau < \eta < 1$. Then*

$$(BDE) = \frac{1}{2}\eta(1-\xi)(1-\frac{\rho}{\xi}-\frac{\tau}{\eta}),$$

$$(CDE) = \frac{1}{2}\xi(1-\eta)(1-\frac{\rho}{\xi}-\frac{\tau}{\eta}). \tag{1}$$

Proof. A direct computation gives

$$D = (\rho,\ \eta(1-\frac{\rho}{\xi})), \quad E = (\xi(1-\frac{\tau}{\eta}),\ \tau)$$

as shown in Fig. 2, hence,

$$(BDE) = \frac{1}{2}\begin{vmatrix} 1 & 0 & 1 \\ \rho & \eta(1-\frac{\rho}{\xi}) & 1 \\ \xi(1-\frac{\tau}{\eta}) & \tau & 1 \end{vmatrix} = \frac{1}{2}\eta(1-\xi)(1-\frac{\rho}{\xi}-\frac{\tau}{\eta}),$$

$$(CDE) = \frac{1}{2}\begin{vmatrix} 0 & 1 & 1 \\ \rho & \eta(1-\frac{\rho}{\xi}) & 1 \\ \xi(1-\frac{\tau}{\eta}) & \tau & 1 \end{vmatrix} = \frac{1}{2}\xi(1-\eta)(1-\frac{\rho}{\xi}-\frac{\tau}{\eta}).$$

□

Remark 1 Obviously, (1) remains true at least up to sign, if the conditions $0 < \rho < \xi < 1$, $0 < \tau < \eta < 1$ are not fulfilled.

Corollary 1 *For $\rho \in (0,1)$ let $X(\rho)$ denote the set of all pairs of distinct points $D, E \in \mathbf{R}^2$ whose coordinates are at least as large as ρ and for which the line connecting D and E intersects x-axis and y-axis at points $(\xi, 0)$ and $(0, \eta)$, respectively, with $\xi, \eta \in (0,1)$. Then*

$$\sup_{(D,E)\in X(\rho)} \min\{(BDE),(CDE)\} = \frac{1}{2}(\frac{1}{2}-\rho)^2, \tag{2}$$

where $B = (1,0)$, $C = (0,1)$.

Proof. With ξ and η as above and

$$D_0 := (\rho, \quad \eta(1-\frac{\rho}{\xi})), \quad E_0 := (\xi(1-\frac{\eta}{\rho}), \quad \rho)$$

one has

$$(BDE) \leq (BD_0E_0) = \frac{1}{2}\eta(1-\xi)(1-\frac{\rho}{\xi}-\frac{\rho}{\eta}),$$

$$(CDE) \leq (CD_0E_0) = \frac{1}{2}\xi(1-\eta)(1-\frac{\rho}{\xi}-\frac{\rho}{\eta}),$$

for all $(D,E) \in X(\rho)$. Without loss of generality one can assume that $\eta \leq \xi$, hence

$$\begin{aligned}\sup_{(D,E)\in X(\rho)} \min\{(BDE),(CDE)\} &\leq (BD_0E_0) \\ &\leq \frac{1}{2}(\xi-2\rho)(1-\xi) \leq \frac{1}{2}(\frac{1}{2}-\rho)^2.\end{aligned}$$

□

Corollary 2 *Given points r_4, r_5 in a triangle $r_1r_2r_3$, one has*

$$(r_1r_2\cdots r_5) \leq \frac{1}{4+2\sqrt{3}}(r_1r_2r_3), \tag{3}$$

where the coefficient $1/(4+2\sqrt{3})$ is best possible.

Proof. Assume that the line through r_4, r_5 intersects the edges r_1r_2 and r_1r_3 of the triangle $r_1r_2r_3$. Choose an affine coordinate system such that

$$r_1 = (0,0), \quad r_2 = (1,0), \quad r_3 = (0,1)$$

as shown in Fig. 3. Let $\rho_0 := \frac{1}{4+2\sqrt{3}}$. If the conclusion does not hold, that is, if

$$(r_i r_j r_k) > \frac{1}{4+2\sqrt{3}}(r_1 r_2 r_3), \text{ whenever } 1 \leq i < j < k \leq 5,$$

then $(r_4, r_5) \in X(\rho_0)$. In view of Corollary 1 we have

$$\min\{(r_2 r_4 r_5), (r_3 r_4 r_5)\} \leq \frac{1}{2}(\frac{1}{2} - \rho_0)^2.$$

On the other hand,

$$(r_2 r_4 r_5) > \frac{1}{4+2\sqrt{3}}(r_1 r_2 r_3), \quad (r_3 r_4 r_5) > \frac{1}{4+2\sqrt{3}}(r_1 r_2 r_3),$$

hence

$$\frac{1}{4+2\sqrt{3}} < 2\min\{(r_2 r_4 r_5), (r_3 r_4 r_5)\} \leq (\frac{1}{2} - \rho_0)^2 = \frac{1}{4+2\sqrt{3}},$$

a contradiction! □

Lemma 4 *Given a convex quadrilateral $r_1 r_2 r_3 r_4$ with*

$$(r_2 r_3 r_4) = (r_1 r_2 r_3 r_4)$$

and a point $r_5 \in \Delta r_2 r_3 r_4$, one has

$$(r_1 r_2 \cdots r_5) \leq \frac{1}{8}|r_1 r_2 r_3 r_4|,$$

where equality holds if and only if $r_1 r_2 r_3 r_4$ is a parallelogram and either

$$(r_2 r_3 r_5) = (r_3 r_1 r_5) = (r_4 r_2 r_5) = \frac{1}{8}|r_1 r_2 r_3 r_4|$$

or

$$(r_3 r_4 r_5) = (r_1 r_3 r_5) = (r_4 r_2 r_5) = \frac{1}{8}|r_1 r_2 r_3 r_4|.$$

Proof. We may assume that

$$(r_1 r_2 \cdots r_5) \geq \frac{1}{8}|r_1 r_2 r_3 r_4|$$

and $r_5 \in \Delta r_1 r_2 r_3$. Choose an affine coordinate system such that

$$r_1 = (0,0), \quad r_2 = (1,0), \quad r_4 = (0,1)$$

and let

$$r_3 = (a, b), \quad r_5 = (x, y),$$

as shown in Fig. 4. $(r_2r_3r_4) = (r_1r_2r_3r_4)$ obviously implies $a \leq 1$ and $b \leq 1$, while $r_3 \notin \Delta r_1r_2r_4$ implies $a + b \geq 1$.

Since $(r_3r_1r_5), (r_4r_2r_5)$ and $(r_2r_3r_5)$ are not less than $\frac{1}{8}|r_1r_2r_3r_4| = \frac{1}{8}(a+b)$, we have

$$\begin{cases} bx - ay \geq \frac{1}{8}(a+b), \\ x + y - 1 \geq \frac{1}{8}(a+b), \\ -bx + (a-1)y + b \geq \frac{1}{8}(a+b). \end{cases} \tag{4}$$

If this system of linear inequalities for x and y has a solution at all, these must form a triangle, the intersection (x_0, y_0) of the lines

$$bx - ay = \frac{1}{8}(a+b), \quad x + y - 1 = \frac{1}{8}(a+b),$$

being one of its vertices, that is, we must have

$$-bx_0 + (a-1)y_0 + b \geq \frac{1}{8}(a+b). \tag{5}$$

Substituting the values of x_0 and y_0, we get

$$-2a^2 + 3ab + 5b^2 + a - 7b \geq 0. \tag{6}$$

Put $g(a,b) := -2a^2 + 3ab + 5b^2 + a - 7b$, which has a unique critical point $(\frac{31}{49}, \frac{25}{49})$ with the critical value

$$g(\frac{31}{49}, \frac{25}{49}) = -\frac{72}{49} < 0.$$

Thus, what we only need to compute the maximum on the boundary of the compact region:

$$0 \leq a \leq 1, \quad 0 \leq b \leq 1, \quad a + b \geq 1.$$

Now

$$\begin{array}{ll} 1) & g(1,b) = (5b+1)(b-1); \\ 2) & g(a,1) = -2(1-a)^2; \\ 3) & g(1-b,b) = -1-b. \end{array}$$

Therefore, $g(a,b) \le 0$ for all admissable choices of a and b, and $g(a,b) = 0$ if and only if $a = b = 1$ and in which case $r_5 = (3/4, 1/2)$. That means, $r_1r_2r_3r_4$ is a parallelogram and

$$(r_2r_3r_5) = (r_3r_1r_5) = (r_4r_2r_5) = \frac{1}{8}|r_1r_2r_3r_4|.$$

Analogously, when $r_5 \in \Delta r_1r_3r_4$, that $(r_1 \cdots r_5) = \frac{1}{8}|r_1r_2r_3r_4|$ if and only if $r_1r_2r_3r_4$ is a parallelogram and

$$(r_3r_4r_5) = (r_1r_3r_5) = (r_4r_2r_5) = \frac{1}{8}|r_1r_2r_3r_4|.$$

□

Lemma 5 *If the convex hull of six points $r_1r_2\cdots r_6$, which belong to a convex region K, is a quadrilateral, then*

$$(r_1r_2\cdots r_6) \le \frac{1}{8}|K|.$$

Proof. Let $r_1r_2r_3r_4$ be the convex hull of $r_1r_2\cdots r_6$ and and assume that $(r_2r_3r_4) = (r_1r_2r_3r_4)$. If either r_5 or r_6 belongs to $\Delta r_2r_3r_4$, the conclusion holds in view of Lemma 4. If r_5 and r_6 both belong to $\Delta r_1r_2r_4$, by Corollary 2, we have

$$(r_1r_2r_4r_5r_6) \le \frac{1}{4+2\sqrt{3}}(r_1r_2r_4) < \frac{1}{7}(r_1r_2r_4),$$

hence

$$|K| \ge |r_1r_2r_3r_4| = (r_1r_2r_4) + (r_2r_3r_4) > 8(r_1r_2r_3r_4r_5r_6).$$

□

3 Proof of the Main Theorem

In order to prove Theorem 1, we must consider the three distinct cases in which the convex hull of the six points is a 4-gon, a 5-gon, or a 6-gon, separately. The case of a 4-gon has been settled already in Lemma 5.

As we shall see, however, if the convex hull is a pentagon, things are much more complicated. Several further definitions, theorems and lemmata are needed.

Definition 1 *Given* $S = \{r_1, \cdots, r_n\} \subset \mathbf{R}^2$, *let*

$$h(S) := \frac{(r_1 \cdots r_n)}{|\mathrm{conv}(S)|}$$

where $|\mathrm{conv}(S)|$ *is the area of the convex hull of* S.

Definition 2 *Given* $S = \{r_1, \cdots, r_6\} \subset \mathbf{R}^2$, *we say* S *belongs to class* Σ_k $(k = 0, 1, 2)$ *if*

i. *the convex hull of* S *is a pentagon, say,* $r_1 \cdots r_5$;

ii. r_6 *is at neither the edges nor the diagonals of* $r_1 \cdots r_5$;

iii. r_6 *belongs to the intersection of* k *peripheral triangles of* $r_1 \cdots r_5$, *exactly.*

Lemma 6 *Given* $\{r_1, \cdots, r_6\} \in \Sigma_2$ *with* $r_6 \in \Delta r_3 r_4 r_5 \cap \Delta r_4 r_5 r_1$, *and*

$$h(\{r_1, \cdots, r_6\}) = \sup_{S \in \Sigma_2} h(S),$$

then $r_1 r_2 r_3$ *is tight relative to* $\{r_1, \cdots, r_6\}$.

Proof. Otherwise, let $r_2' := (1 - \varepsilon) r_2 + \varepsilon r_6$ where $\varepsilon > 0$ small enough, that keeps the area of every triangle formed by three of $r_1, r_2', r_3, r_4, r_5, r_6$ not less than $(r_1 \cdots r_6)$ and makes

$$(r_1 r_2' r_3) = (1 - \varepsilon)(r_1 r_2 r_3) - \varepsilon (r_1 r_3 r_6) < (r_1 r_2 r_3),$$

hence

$$(r_1 r_2' r_3 r_4 r_5 r_6) \geq (r_1 \cdots r_6), \quad |r_1 r_2' r_3 r_4 r_5| < |r_1 \cdots r_5|.$$

This leads to

$$h(\{r_1, r_2', r_3, r_4, r_5, r_6\}) > h(\{r_1, \cdots, r_6\}),$$

a contradiction! □

Lemma 7 *With the notations and assumptions as in the last lemma, there is a point* r_2' *such that when replacing* r_2 *by* r_2' *the following holds:*

i. $r_1 r_2' r_3 r_4 r_5$ *is still a convex pentagon;*

ii. $h(\{r_1, r_2', r_3, r_4, r_5, r_6\}) = \sup_{S\in\Sigma_2} h(S)$;

iii. *besides* $r_1r_2'r_3$, *at least one of the two triangles* $r_5r_1r_2'$ *and* $r_2'r_3r_4$ *is tight relative to* $\{r_1, r_2', r_3, r_4, r_5, r_6\}$.

Proof. Let

$$r_2' := r_2 + \frac{(r_2r_3r_4) - (r_1r_2r_3)}{(r_1r_3r_4)}(r_3 - r_1). \tag{7}$$

Then, we have

$$(r_1r_2'r_3) = (r_1r_2r_3), \quad (r_2'r_3r_4) = (r_1r_2r_3), \quad (r_5r_1r_2') \geq (r_5r_1r_2)$$

and $(r_2'r_jr_k) \geq (r_2r_jr_k)$ for j,k else, hence

$$(r_1r_2'r_3r_4r_5r_6) = (r_1\cdots r_6),$$
$$|r_1r_2'r_3r_4r_5| = |r_1\cdots r_6|.$$

That means

$$h(\{r_1, r_2', r_3, r_4, r_5, r_6\}) = h(\{r_1,\cdots, r_6\}) = \sup_{S\in\Sigma_2} h(S).$$

□

Theorem 2 *Given a convex pentagon* $r_1\cdots r_5$, *if a point* r_6 *belongs to the intersection of two peripheral triangles of the pentagon, then*

$$(r_1\cdots r_6) \leq \mu_2|r_1\cdots r_5| \tag{8}$$

where $\mu_2 = 0.1397365\cdots$ *is the smallest root of the equation*

$$25\mu^3 - 31\mu^2 + 11\mu - 1 = 0; \tag{9}$$

and μ_2 *is best possible, that is, there exist* $r_1,\cdots, r_5, r_6$ *as above with* $(r_1\cdots r_6) = \mu_2|r_1\cdots r_5|$.

Proof. By Lemmata 6-7, we may assume that both $r_1r_2r_3$ and $r_2r_3r_4$ are tight relative to $\{r_1,\cdots, r_6\}$. Choose an affine coordinate system such that

$$r_1 = (0,1), \quad r_2 = (0,0), \quad r_3 = (1,0),$$

and let

$$r_4 = (a, 1), \quad r_5 = (u, b), \quad r_6 = (x, y),$$

as shown in Fig. 5. In this case,

$$|r_1 \cdots r_5| = \frac{1}{2}(ab + 1),$$

hence

$$h(\{r_1, \cdots, r_6\}) = \frac{1}{ab + 1}.$$

And $(r_1r_2r_4) \geq (r_1 \cdots r_6) = (r_1r_2r_3) = 1/2, (r_2r_3r_5) \geq (r_1 \cdots r_6) = (r_1r_2r_3) = 1/2$ obviously implies $a \geq 1, b \geq 1$.

Since $(r_6r_1r_4), (r_6r_5r_3)$ and $(r_6r_4r_5)$ are not less than $(r_1r_2r_3)$, we have

$$\begin{cases} a(y-1) \geq 1, \\ bx - (u-1)y - b \geq 1, \\ (1-b)x + (u-a)y + ab - u \geq 1. \end{cases} \tag{10}$$

Similarly as above, the consistency of the system of inequalities leads to

$$a^2b^2 - ab(a+b) - ab + b - 1 \geq (ab - a - 1)u. \tag{11}$$

And $(r_5r_1r_2) \geq (r_1r_2r_3)$ implies $u \geq 1$. Noting $b \geq y \geq (a+1)/a$, hence $ab - a - 1 \geq 0$, by (11) we have

$$a^2b^2 - ab(a+b+2) + (a+b) \geq 0. \tag{12}$$

Observing the symmetry in a and b, put $p := ab$ and $s := a + b$. It follows that

$$p^2 - s(p-1) - 2p \geq 0. \tag{13}$$

Set $f(p,s) := p^2 - s(p-1) - 2p$. Since $2\sqrt{p} \leq s$ and $f(p,s)$ is monotone increasing with s while $ab \geq 1$, we have

$$f(p, 2\sqrt{p}) \geq f(p,s) \geq 0,$$

that is

$$p^2 - 2\sqrt{p}(p-1) - 2p \geq 0. \tag{14}$$

Let $\mu := h(\{r_1 \cdots, r_6\}) = \dfrac{1}{p+1}$, then (14) leads to

$$25\mu^3 - 31\mu^2 + 11\mu - 1 \leq 0 \tag{15}$$

under which and a trivial bound for μ,

$$\mu = h(\{r_1 \cdots, r_6\}) \le \frac{1}{5},$$

the maximum of μ is

$$\mu_2 = 0.1397365\cdots,$$

the smallest positive root of the equation

$$25\mu^3 - 31\mu^2 + 11\mu - 1 = 0.$$

Furthermore, μ_2 is best possible because the equality holds when

$$x = y = 1.403031\cdots, \quad u = 1, \quad a = b = 2.481194\cdots.$$

□

Lemma 8 *Given* $\{r_1, \cdots, r_6\} \in \Sigma_1$ *with* $r_6 \in \Delta r_4 r_5 r_1$, *if*

$$h(\{r_1, \cdots, r_6\}) = \sup_{S \in \Sigma_1} h(S),$$

then $r_6 r_1 r_4$ *is tight relative to* $\{r_1, \cdots, r_6\}$.

Proof In the case $\angle r_1 r_2 r_6 + \angle r_2 r_6 r_5 > \pi$ or $\angle r_5 r_6 r_3 + \angle r_6 r_3 r_4 > \pi$, say, the former holds, the triangles $r_1 r_2 r_5$ and $r_1 r_6 r_5$ are loose. If $r_1 r_4 r_6$ is also loose, let $r_1' := (1 - \varepsilon) r_1 + \varepsilon r_5$ where $\varepsilon > 0$ small enough, that keeps $(r_1' r_2 r_5), (r_1' r_6 r_5), (r_1' r_4 r_6)$ still greater than $(r_1 \cdots r_6)$ and

$$(r_1' r_2 r_3 r_4 r_5 r_6) \ge (r_1 \cdots r_6),$$

while

$$|r_1' r_2 r_3 r_4 r_5| = |r_1 \cdots r_5| - \varepsilon (r_1 r_2 r_5) < |r_1 \cdots r_5|.$$

Thus,

$$h(\{r_1, \cdots, r_6\}) < h(\{r_1', r_2, r_3, r_4, r_5, r_6\}) \le \sup_{S \in \Sigma_1} h(S)$$

which contradicts the hypothesis.

In the case $\angle r_1 r_2 r_6 + \angle r_2 r_6 r_5 \le \pi$ and $\angle r_5 r_6 r_3 + \angle r_6 r_3 r_4 \le \pi$, we have

$$(r_6 r_1 r_2) \ge (r_5 r_1 r_2), \quad (r_6 r_3 r_4) \ge (r_5 r_3 r_4).$$

If $(r_6 r_1 r_4)$ is loose, let

$$r'_5 := r_5 + \varepsilon(r_6 - r_5), \quad r'_6 := r_6 + \varepsilon(r_6 - r_5)$$

with $\varepsilon > 0$ small enough, that keeps $(r'_6 r_1 r_4) > (r_1 \cdots r_6)$ and the area of every triangle formed by three of $r_1, r_2, r_3, r_4, r'_5, r'_6$ not less than $(r_1 \cdots r_6)$ and makes $(r'_5 r_1 r_4) < (r_5 r_1 r_4)$. Thus,

$$(r_1 r_2 r_3 r_4 r'_5 r'_6) \geq (r_1 \cdots r_6), \quad |r_1 r_2 r_3 r_4 r'_5| < |r_1 \cdots r_5|,$$

and therefore,

$$h(\{r_1, \cdots, r_6\}) < h(\{r_1, r_2, r_3, r_4, r'_5, r'_6\}) \leq \sup_{S \in \Sigma_1} h(S)$$

which contradicts the hypothesis. □

Theorem 3 *Given a convex pentagon $r_1 \cdots r_5$, if a point r_6 belongs to one and only one peripheral triangle of the pentagon, then*

$$(r_1 \cdots r_6) \leq \mu_1 |r_1 \cdots r_5| \tag{16}$$

where $\mu_1 = 0.14860979 \cdots$ is the unique real root of the equation

$$11\mu^3 + 10\mu^2 + 5\mu - 1 = 0; \tag{17}$$

and μ_1 is best possible.

Proof. By Lemma 8, we may assume $r_6 r_1 r_4$ is tight relative to $\{r_1, \cdots, r_6\}$. Choose an affine coordinate system such that

$$r_6 = (0,0), \quad r_1 = (1,0), \quad r_4 = (0,1),$$

and let

$$r_2 = (a, v), \quad r_3 = (u, b), \quad r_5 = (-x, -y),$$

as shown in Fig. 6. It is easy to find the area of the pentagon:

$$|r_1 \cdots r_5| = \frac{1}{2}(ab + 1 + x + y - (u-1)(v-1)). \tag{18}$$

Since $(r_6 r_3 r_4), (r_6 r_1 r_2), (r_6 r_3 r_5)$ and $(r_6 r_5 r_2)$ are not less than $(r_6 r_1 r_4)$, we have

$$\begin{cases} u \geq 1, \\ v \geq 1, \\ bx - uy \geq 1, \\ ay - vx \geq 1. \end{cases} \tag{19}$$

The system leads to

$$\begin{cases} u-1 \le \frac{1}{y}(bx-y-1), \\ v-1 \le \frac{1}{x}(ay-x-1), \\ bx \ge y+1, \\ ay \ge x+1. \end{cases} \tag{20}$$

Substituting the former two of (20) into (18), we have

$$|r_1 \cdots r_5| \ge \frac{1}{2xy}(x^2y + xy^2 + bx(1+x) + ay(1+y) - x - y - 1); \tag{21}$$

and then, substituting the latter two of (20) into (21), we obtain

$$|r_1 \cdots r_5| \ge \frac{1}{2xy}(x^2y + xy^2 + 2xy + x + y + 1). \tag{22}$$

Set

$$g(x,y) := \frac{xy}{x^2y + xy^2 + 2xy + x + y + 1}, \tag{23}$$

then

$$\frac{(r_1 \cdots r_6)}{|r_1 \cdots r_5|} \le g(x,y). \tag{24}$$

To find the critical values of $g(x,y)$, we need solve the following system:

$$\begin{cases} \mu(x^2y + xy^2 + 2xy + x + y + 1) - xy = 0, \\ x^2y - y - 1 = 0, \\ xy^2 - x - 1 = 0. \end{cases} \tag{25}$$

We employ an efficient algorithm, "successive resultant computation"[18], which is supported by current softwares for computer algebra such as MAPLE, MACSYMA, REDUCE or MATHEMATICA. The following program was written in REDUCE:

$g_0 := \mu(x^2y + xy^2 + 2xy + x + y + 1) - xy\$$
$g_1 := x^2y - y - 1\$$
$g_2 := xy^2 - x - 1\$$
$R_2 :=$ resultant$(g_1, g_2, y)\$$
$R_1 :=$ resultant$(g_0, g_1, y)\$$
$R_0 :=$ resultant$(R_1, R_2, x)\$$
$R^* :=$ factorize(R_0);
;end;

The result for R^* is the polynomial

$$(\mu - 1)^2(11\mu^3 + 10\mu^2 + 5\mu - 1) \tag{26}$$

whose roots consist of all the real and complex critical values of $g(x, y)$.

Now let us go to find the global maximum of $g(x, y)$ in region $\{x \geq 0, y \geq 0\}$. When $x \geq 6$ or $y \geq 6$, clearly $g(x, y) < 1/8 = 0.125$. Then we consider the compact region $D : \{0 \leq x \leq 6, 0 \leq y \leq 6\}$. Since $g(x, y) < 1/8$ when $x = 6$ or $y = 6$ and $g(x, y) = 0$ when $x = 0$ or $y = 0$, we have $g(x, y) < 1/8$ over the boundary of D. On the other hand, if $g(x, y)$ takes its maximum at some point belonging to the interior of D, that maximum has to be a real root of (26). Notting $g(x, y) < 1$ whenever $x > 0$ and $y > 0$, we can assert that the unique critical value of $g(x, y)$ over region D is the unique real root of the equation

$$11\mu^3 + 10\mu^2 + 5\mu - 1 = 0,$$

$$\mu_1 = 0.14860949 \cdots,$$

so is the global maximum of $g(x, y)$ over region $\{x \geq 0, y \geq 0\}$. Then it follows (24) that

$$(r_1 \cdots r_6) \leq \mu_1 |r_1 \cdots r_5|.$$

And μ_1 is best possible because the equality holds when

$$x = y = 1.3247 \cdots, \quad u = v = 1, \quad a = b = 1.7548 \cdots.$$

□

Lemma 9 *Given* $\{r_1, \cdots, r_6\} \in \Sigma_0$ *with* r_6 *in the convex pentagon* $r_1 \cdots r_5$, *if*

$$h(\{r_1, \cdots, r_6\}) = \sup_{S \in \Sigma_0} h(S),$$

then there is at least one diagonal of the pentagon, its end points with r_6 *form a triangle tight relative to* $\{r_1, \cdots, r_6\}$.

Proof. Otherwise, we can prove that at most one of triangles $r_6 r_{i-1} r_i$ $(i = 1, \cdots, 5; r_0 := r_5)$ is tight relative to $\{r_1, \cdots, r_6\}$. If there are two, they have either one or no common edge. In the former case, say,

$$(r_6 r_1 r_2) = (r_6 r_2 r_3) = (r_1 \cdots r_6),$$

the triangle $r_6r_1r_3$ has to be tight because

$$(r_6r_1r_3) + (r_1r_2r_3) = 2(r_1 \cdots r_6).$$

In the latter case, say,

$$(r_6r_1r_2) = (r_6r_3r_4) = (r_1 \cdots r_6),$$

then

$$\angle r_6r_1r_2 + \angle r_1r_2r_3 \geq \pi, \quad \angle r_2r_3r_4 + \angle r_3r_4r_6 \geq \pi,$$

which makes the sum of the interior angles of quadrilateral $r_1r_2r_3r_4$ greater than 2π, impossible!

Thus, we can assume that triangles $r_6r_{i-1}r_i$ $(i = 2, 3, 4, 5)$ are loose relative to $\{r_1, \cdots, r_6\}$. Let $r_6' := r_6 + \varepsilon(r_3 - r_6)$ where $\varepsilon > 0$ small enough such that

$$(r_6'r_ir_j) \geq (r_1 \cdots r_6), \qquad \text{for } 1 \leq i < j \leq 5.$$

On the other hand, at most two peripheral triangles are tight relative to $\{r_1, \cdots, r_5, r_6'\}$. Otherwise, there must be two of them with a common edge, say, $r_1r_2r_3$ and $r_2r_3r_4$ are tight. In this case,

$$(r_6'r_2r_3) < \max\{(r_1r_2r_3), (r_2r_3r_4)\} = (r_1 \cdots r_5r_6'),$$

that is impossible. And therefore, there are two loose peripheral triangles with a common edge, that leads to a contradiction in view of Lemma 2.

□

Theorem 4 *Given a convex pentagon $r_1 \cdots r_5$, if a point r_6 in the pentagon belongs to none of the peripheral triangles, then*

$$(r_1 \cdots r_6) \leq \mu_0 |r_1 \cdots r_5| \tag{27}$$

where $\mu_0 = (\sqrt{5} - 1)/10 = 0.1236 \cdots$ is best possible.

Proof. By Lemma 9, we may assume $r_6r_4r_1$ is tight relative to $\{r_1, \cdots, r_6\}$. Choose an affine coordinate system such that

$$r_6 = (0, 0), \quad r_4 = (1, 0), \quad r_1 = (0, 1),$$

and let

$$r_2 = (-a, -v), \quad r_3 = (-u, -b), \quad r_5 = (x, y),$$

as shown in Fig. 7. It is easy to find the area of the pentagon:

$$|r_1\cdots r_5| = \frac{1}{2}(ab + a + b + x + y - uv). \tag{28}$$

Since $(r_6r_1r_3), (r_6r_2r_4), (r_6r_3r_5)$ and $(r_6r_5r_2)$ are not less than $(r_6r_4r_1)$, we have

$$\begin{cases} u \geq 1, \\ v \geq 1, \\ bx - uy \geq 1, \\ ay - vx \geq 1. \end{cases} \tag{29}$$

The latter two of the system leads to

$$\begin{cases} a \geq \dfrac{1}{y}(vx + 1), \\ b \geq \dfrac{1}{x}(uy + 1). \end{cases} \tag{30}$$

Substituting the former two of (30) into (28), we have

$$|r_1\cdots r_5| \geq \frac{1}{2xy}(x^2y + xy^2 + vx^2 + uy^2 + (v+1)x + (u+1)y + 1); \tag{31}$$

and then, substituting the former two of (29) into (31), we obtain

$$|r_1\cdots r_5| \geq \frac{1}{2xy}(x^2y + xy^2 + x^2 + y^2 + 2x + 2y + 1). \tag{32}$$

Set

$$g(x,y) := \frac{xy}{x^2y + xy^2 + +x^2 + y^2 + 2x + 2y + 1}, \tag{33}$$

then

$$\frac{(r_1\cdots r_6)}{|r_1\cdots r_5|} \leq g(x,y). \tag{34}$$

To find the critical values of $g(x,y)$, we need solve the following system:

$$\begin{cases} \mu(x^2y + xy^2 + x^2 + y^2 + 2x + 2y + 1) - xy = 0, \\ y(y+1)(1 + y - x^2) = 0, \\ x(x+1)(1 + x - y^2) = 0. \end{cases} \tag{35}$$

Solving it we obtain 8 critical points of $g(x,y)$ but only one of them is in the first quadrant, that is, $(\frac{1}{2}(\sqrt{5}+1), \frac{1}{2}(\sqrt{5}+1))$.

Now let us go to find the global maximum of $g(x,y)$ in region $\{x \geq 0, y \geq 0\}$. When $x \geq 7$ or $y \geq 7$, clearly $g(x,y) < 1/9 = 0.1111\cdots$. Then we consider the compact region $D: \{0 \leq x \leq 7, 0 \leq y \leq 7\}$. Since $g(x,y) < 1/9$ when $x = 7$ or $y = 7$ and $g(x,y) = 0$ when $x = 0$ or $y = 0$, we have $g(x,y) < 1/9$ over the boundary of D. On the other hand, if $g(x,y)$ takes its maximum at some point belonging to the interior of D, that maximum has to be a critical value, i.e.

$$\mu_0 := g(\frac{1}{2}(\sqrt{5}+1), \frac{1}{2}(\sqrt{5}+1)) = \frac{\sqrt{5}-1}{10} = 0.1236\cdots,$$

so is the global maximum of $g(x,y)$ over region $\{x \geq 0, y \geq 0\}$. Then it follows (34) that

$$(r_1 \cdots r_6) \leq \mu_0 |r_1 \cdots r_5|.$$

And μ_0 is best possible because the equality holds when

$$a = b = x = y = \frac{1}{2}(\sqrt{5}+1), \quad u = v = 1.$$

□

Now we are ready to complete the proof of Theorem 1.

Proof of Theorem 1. We have pointed out that the conclusion holds when the convex hull of the six points is a quadrilateral. Also, by Theorems 2-4, the conclusion holds when the convex hull is a pentagon because of

$$(r_1 \cdots r_6) \leq \mu_1 |r_1 \cdots r_5| < \frac{1}{6}|K|.$$

It remains that points $r_1, \cdots, r_6$ form a convex hexagon. Let

$$p_1 := r_1r_4 \cap r_3r_6, \quad p_2 := r_2r_5 \cap r_3r_6, \quad p_3 := r_1r_4 \cap r_2r_5,$$

as shown in Fig. 8. Then it is easy to see that each of the six triangles $r_1r_2p_1$, $r_2r_3p_1, r_3r_4p_2, r_4r_5p_2, r_5r_6p_3, r_6r_1p_3$ is of area not less than $(r_1 \cdots r_6)$, e.g.

$$(r_1r_2p_1) \geq \min\{(r_1r_2r_6), (r_1r_2r_3)\} \geq (r_1 \cdots r_6).$$

And therefore,

$$\begin{aligned} |r_1 \cdots r_6| &= |r_1r_2p_1| + |r_2r_3p_1| + |r_3r_4p_2| \\ &\quad + |r_4r_5p_2| + |r_5r_6p_3| + |r_6r_1p_3| + |p_1p_2p_3| \\ &\geq 6(r_1 \cdots r_6), \end{aligned}$$

hence,

$$(r_1 \cdots r_6) \le \frac{1}{6}|r_1 \cdots r_6| \le \frac{1}{6}|K|.$$

And the constant $\frac{1}{6}$ is best possible because the equality holds when K is an affine regular hexagon and $r_1, \cdots, r_6$ are those vertices. □

References

[1] Goldberg, M., Maximizing the smallest triangle made by points in a square, *Math. Mag.*, **45**(1972), 135-144.

[2] Komlos, J. et al., On Heilbronn's problem, *J. London Math. Soc.* (2), **24**(1981), 385-396.

[3] Komlos, J. et al., A lower bound for Heilbronn's problem, *J. London Math. Soc.* (2), **25**(1982), 13-14.

[4] Moser, W., Problems on extremal properties of a finite set of points, in "Discrete Geometry and Convexity", *Ann. New York Acad. Sci.*, **440**(1985), 52-64.

[5] Moser, W. & Pach,J., "100 Research Problem in Discrete Geometry", MacGill University, Montreal, Que. 1986.

[6] Roth, K.F., On a problem of Heilbronn, *J. London Math. Soc.*, **26**(1951), 198-204.

[7] Roth, K.F., On a problem of Heilbronn II, *Proc. London Math. Soc.*, **25**(1972), 193-212.

[8] Roth, K.F., On a problem of Heilbronn III, *Proc. London Math. Soc.*, **25**(1974), 543-549.

[9] Roth, K.F., Estimation of the area of the smallest triangle obtained by selecting three out of n points in a disc of unit area, *Amer. Math. Soc. Proc. Symp. Pure Math.*, **24**: 251-262, 1973.

[10] Roth, K.F., Developments in Heilbronn's triangle problem, *Advances in Math.*, **22**(1976), 364-385.

[11] Schmidt, W.M., On a problem of Heilbronn, *J. London Math. Soc.* (2), **4**(1971), 545-550.

[12] Yang Lu & Zhang Jingzhong, The problem of 6 points in a square, in "Lectures in Math. (2)", 151-175, Sichuan People's Publishing House 1980, pp. 151-175. (in Chinese).

[13] Yang Lu & Zhang Jingzhong, A conjecture concerning six points in a square, in "Mathematical Olympiad in China", Hunan Education Publishing House 1990.

[14] Yang Lu, Zhang Jingzhong & Zeng Zhenbing, On the conjecture and computing for exact values of the first several Heilbronn numbers, *Chin. Ann. Math.*, to appear. (in Chinese).

[15] Yang Lu, Zhang Jingzhong & Zeng Zhenbing, On the first several Heilbronn numbers of a triangle, *Acta Math. Sinica*, to appear. (in Chinese).

[16] Yang Lu, Zhang Jingzhong & Zeng Zhenbing, Heilbronn problem for five points, Preprint, International Centre for Theoretical Physics, 1991, IC/91/252.

[17] Yang Lu, Zhang Jingzhong & Zeng Zhenbing, On Goldberg's conjecture: computing the first several Heilbronn numbers, Preprint, Universität Bielefeld, 1991, ZiF-Nr.91/29, SFB-Nr.91/074.

[18] Yang Lu, Zhang Jingzhong & Zeng Zhenbing, Searching dependency between algebraic equations: an algorithm applied to automated reasoning, Preprint, International Centre for Theoretical Physics, 1991, IC/91/6.

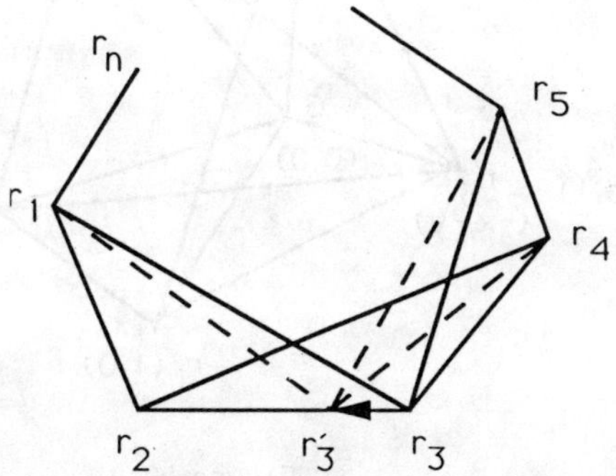

Figure 1

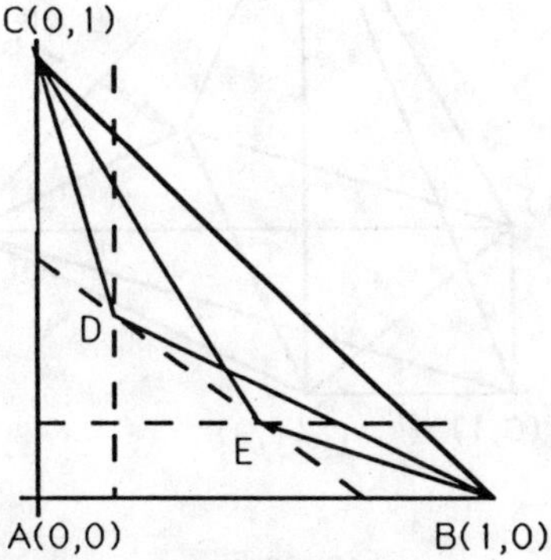

Figure 2

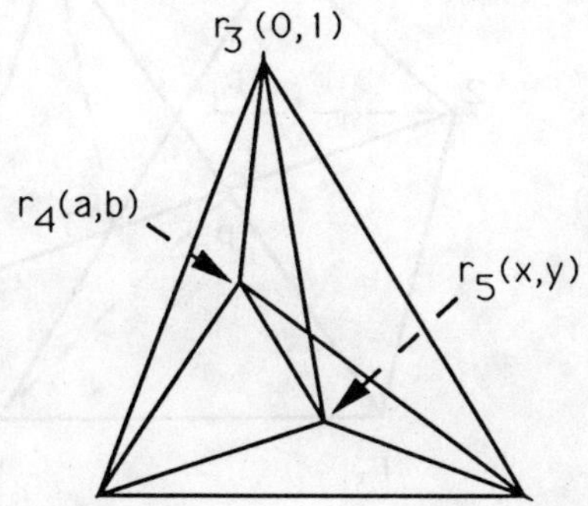

Figure 3

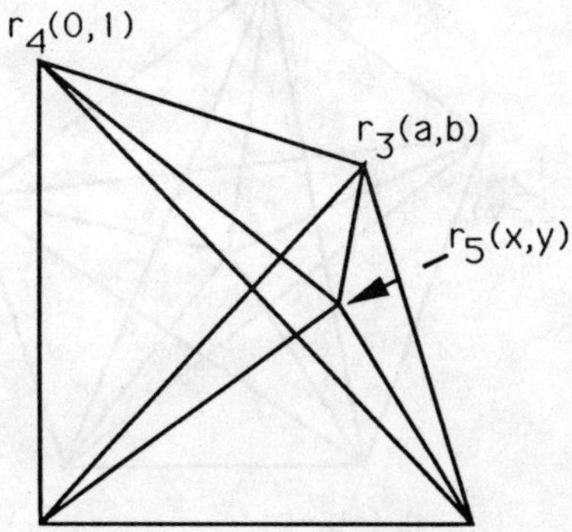

Figure 4

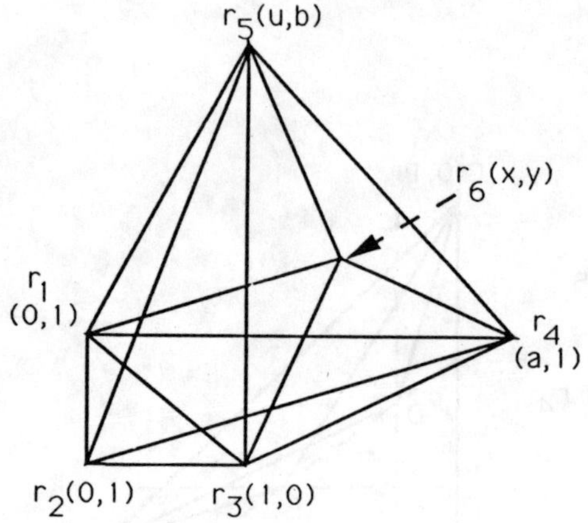

Figure 5

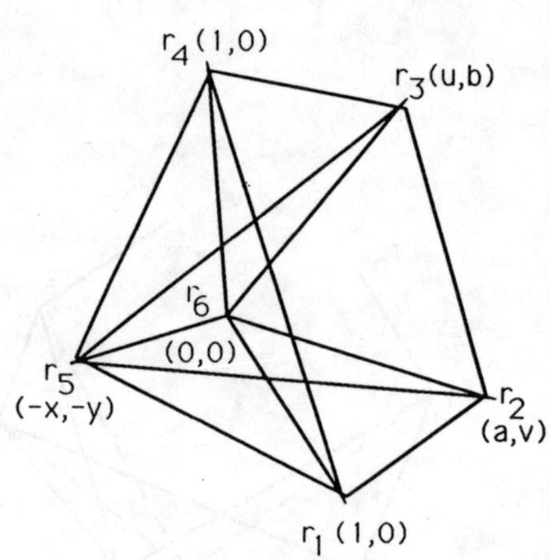

Figure 6

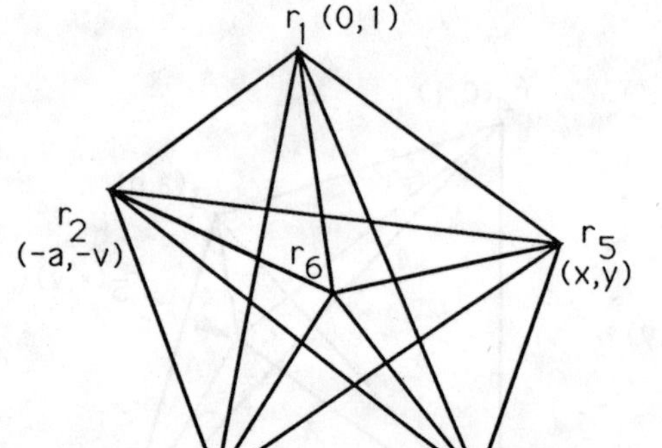

Figure 7

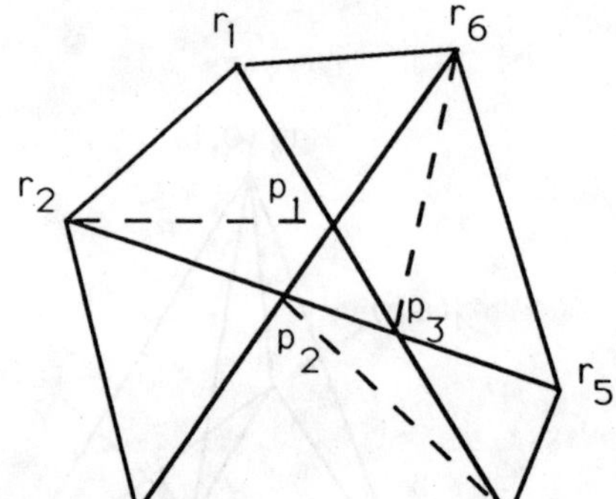

Figure 8

A characterization of infinite, bipartite Toeplitz graphs

by

R.Euler* , H.Le Verge** and T.Zamfirescu***

September 1994

Abstract

Toeplitz graphs are graphs whose adjacency matrix is Toeplitz. Infinite Toeplitz graphs having no loops and no multiple edges can thus be defined by an infinite 0-1 sequence. We characterize bipartite such Toeplitz graphs by introducing the concept of odd T-cycles and by using the base-circuit language of matroid theory.

---

* : Laboratoire d'Informatique de Brest (LIBr), Université de Bretagne Occidentale, 6, Avenue Le Gorgeu, B.P. 809, F-29285 Brest Cedex, France.

**: IRISA, Campus de Beaulieu, F-35042 Rennes Cedex, France.

***: Fachbereich Mathematik, Universität Dortmund, D-44221 Dortmund, Germany.

---

1) Introduction

An n×n matrix $A=(a_{ij})$ is called Toeplitz, if $a_{ij}=a_{(i+1)(j+1)}$ for i,j=1,...,n. Since any diagonal of such a matrix contains identical elements it is uniquely determined by its first row and column and therefore easy to memorize. Beyond this straightforward advantage there are other particular applications of these matrices in computer science (cf. Aho et al. [1, p.249], JaJa [7]) and also in mathematics or physics.

A Toeplitz graph is a graph whose adjacency matrix is Toeplitz. Not much is known to date about structural and algorithmic properties of these graphs. Connectivity properties have been studied by Garfinkel et al. [6] , Burkard and Sandholzer [3], van Doorn [5], and hamiltonian properties by Medova and Dempster [8], van der Veen et al. [9] and van Dal et al. [4].

The Toeplitz graphs we consider in this paper are infinite, symmetric ,without loops and multiple edges; they will be denoted by $T=(\mathbb{N}, E)$. Consequently, any such graph is uniquely defined by a sequence from $\{0,1\}^{\mathbb{N}}$, whose first element is 0 (throughout the paper we assume that this is the case for the sequences we study.). Given two such sequences A and B we say that A dominates B (or $A \geq B$) if $A_i \geq B_i$ for all $i \in \mathbb{N}$. We will make heavy use of the concept of an odd T-cycle, i.e. a sequence from $\{0,1\}^{\mathbb{N}}$, which induces ≥1 odd cycles in its associated Toeplitz graph and whose number of 1-entries is minimal with respect to this property. Our main result consists in characterizing the bipartite graphs among the Toeplitz graphs under study by the non-domination of odd T-cycles. We also fully describe the maximal bipartite such graphs. We think that our structural results might help to also obtain algorithmic results, for instance in relation with the max-cut problem over such graphs.

It will become clear from the structure of odd T-cycles that the finite case cannot be treated the same way but instead requires some special attention. This situation and conditions for planarity of Toeplitz-graphs are the subject of a companion paper.

2) Bipartite Toeplitz graphs

For $\alpha \in \mathbb{N}$ let $\underline{B}^{\alpha}$ denote the $\{0,1\}^{\mathbb{N}}$-sequence (consisting of α 0's first, then of a 1 etc.)

$$(\underbrace{0\dots0}_{\alpha}\ 1\ \underbrace{0\dots0}_{\alpha-1}\ \underbrace{0\dots0}_{\alpha}\ 1\ \underbrace{0\dots0}_{\alpha-1}\ \underbrace{0\dots0}_{\alpha}\ 1\ \underbrace{0\dots0}_{\alpha-1}\ \dots),$$

and let $\underline{T}^{\alpha}$ denote the associated Toeplitz graph. As illustrated in Figure 1, T^{α} is the disjoint union of complete, bipartite subgraphs. The nodes $\mathbb{N}$ of T^{α} can be classified according to the remainder r mod 2α, i.e.

i) $0 \le r < \alpha$ and ii) $\alpha \le r < 2\alpha$.

More precisely, a node $i \in \mathbb{N}$ is either

i') an even multiple of α plus the remainder r with $0 \le r < \alpha$,

or

ii') an odd multiple of α plus the remainder r with $0 \le r < \alpha$.

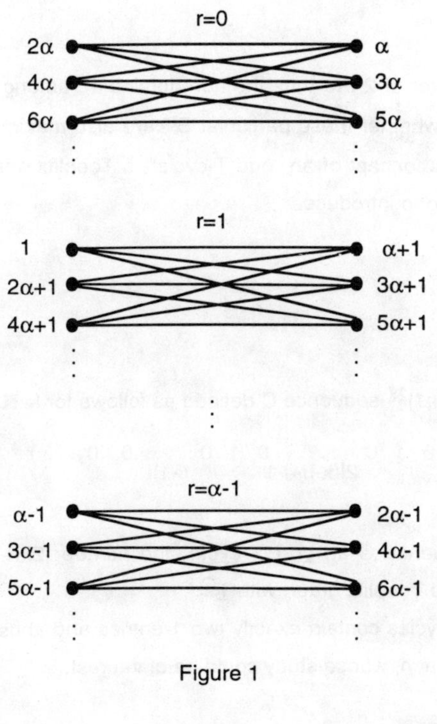

Figure 1

Observe that if $\alpha=1$ the graph T^α reduces to a (single) complete, bipartite graph over the nodes $2\mathbb{N}$ and $2\mathbb{N}-1$.

Proposition 1

If $\alpha\equiv\beta \bmod 2\beta$ then $B^\alpha \leq B^\beta$.

Proof:

Consider B^α and subdivide it according to the condition given. ■

Proposition 2

Let $\beta\in\mathbb{N}$, $\beta\neq 2^m$ for all $m\in\{0,1,2,...\}$. Then there is an α with $\alpha=2^{m'}$ for some $m'\in\{0,1,2,...\}$ such that $B^\beta \leq B^\alpha$.

Proof:

Obvious from the the fact that β contains an odd divisor and from the condition given in Proposition 1. ■

The B^α with α a power of 2 are thus the non-dominated among all these sequences. We may further ask whether these particular B^α are also maximal w.r.t. bipartiteness. For this we need the concept of an "odd T-cycle", a Toeplitz-analogon to odd cycles, that we are now going to introduce.

3) Odd T-cycles

Let us consider the $\{0,1\}^{\mathbb{N}}$-sequence C defined as follows for $l\in\mathbb{N}$, $k\in\mathbb{N}$ and $i\in\{1,...,k\}$:

$$C = (\underbrace{0...0}_{il}\ 1\ \underbrace{0........0}_{2l(k-i)+l-1}\ 1\ \underbrace{0......0}_{(i-1)l}\ 0\ ...)\ .$$

We call such a sequence C an <u>odd T-cycle</u>, the number $2kl+1$ its <u>length</u>, and we denote the associated Toeplitz graph with $\underline{T^C}$.

Observe that odd T-cycles contain exactly two 1-entries and ,thus, give rise to another (infinite) symmetric graph, whose study could be of interest.

Lemma 1

T^C contains an odd cycle.

Proof:

Let us first illustrate two simple instances:

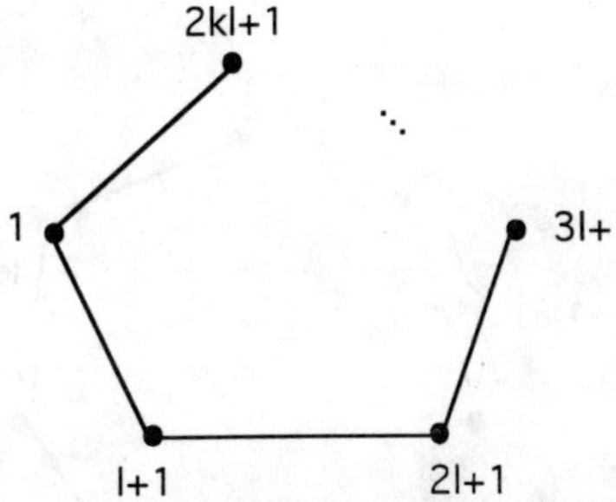

a) the case i=1

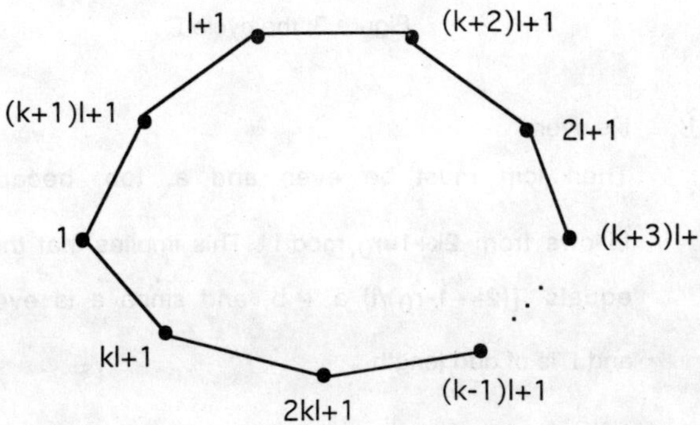

b) the case i=k

Figure 2

c) the general case:

We suppose that $2k+1 \equiv r_0 \bmod i$.

c1) If $r_0=0$, the there is a cycle in T^C of total length $(2k+1)/i$, which is odd.

c2) For $r_0>0$ let <u>lcm</u> denote the least common multiple of r_0 and i and let $lcm=ar_0=bi$ with a and b prime to each other. Then we can identify within T^C a cycle Γ of total length $[(2k+1-r_0)/i]\ lcm/r_0 + lcm/i$, as shown in Fig.3:

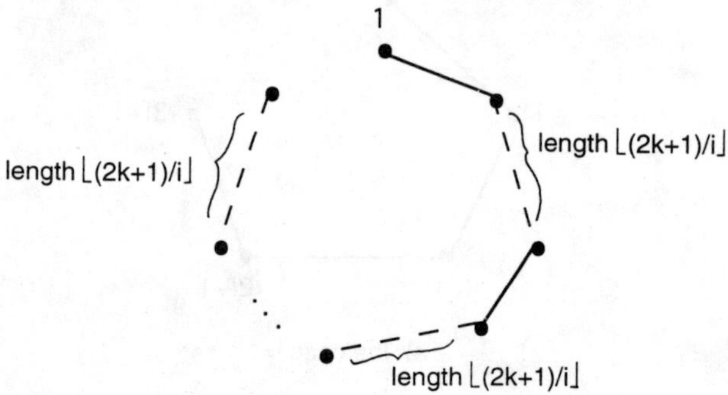

Figure 3: the cycle Γ

<u>Case 1</u>: i is even:

Then lcm must be even and a, too, because r_0 is odd as follows from $2k+1 \equiv r_0 \bmod i$. This implies that the total length of Γ equals $[(2k+1-r_0)/i]\ a + b$, and since a is even b must be odd and Γ is of odd length.

<u>Case 2</u>: i is odd:

Subcase 2.1: lcm is also odd. Then b must be odd and so are a and r_0. The latter implies $(2k+1-r_0)/i$ to be even, and therefore the total length of Γ is odd.

Subcase 2.2: lcm is even. This implies b to be even and a, r_0 to be odd, even, respectively. Hence $(2k+1-r_0)/i$ is odd and, altogether, Γ is shown to be of odd length. ■

Observe that changing one of the two 1's to 0 within such an odd T-cycle C leads to a sequence inducing a bipartite Toeplitz graph. Therefore, C (i.e. its number of 1-entries) is minimal w.r.t. inducing an odd cycle in T^C. Also note that more than one such odd cycle can be induced in T^C. Finally, observe that to fully induce these odd cycles it is (in general) necessary that T^C contains at least 2kl+1 nodes. This is always the case for the node set $\mathbb{N}$. However, if the node set is finite, the situation needs special attention. A companion paper is devoted to the study of bipartiteness (and planarity) of finite Toeplitz graphs.

We will now ask whether the number of 1-entries of a sequence B^α ,with $\alpha=2^m$ and $m\in\{0,1,2,...\}$, is maximal w.r.t. inducing a bipartite graph (we know already that B^α is not properly dominated by any of the B^β, $\beta\in\mathbb{N}-\{1,2,4,8,...\}$).

For this let $\underline{B_j^\alpha}$ denote B^α with position j ($j\neq\alpha+1, 3\alpha+1, 5\alpha+1,...$) changed from 0 to 1. We distinguish 4 main cases:

<u>Case 1</u>: $\alpha=1$ i.e. $B_j^\alpha = (0\,1\,0\,1\,0\,1 \ldots 0\,1\,\overset{j}{1}\,1\,0\,1\,0 \ldots)$:

with l=1, k=i=j-2 there is an odd T-cycle dominated by B_j^α.

<u>Case 2</u>: $j\equiv 2p \bmod 2\alpha$, $0 \le p \le \alpha-1$, $j \ge \alpha+2$:

with l=1, i=α and k=$(j+\alpha-2)/2$ there is an odd T-cycle dominated by B_j^α.

<u>Case 3</u>: $j\equiv 2p+1 \bmod 2\alpha$, $0 \le p \le \alpha-1$, $j \ge \alpha+2$ and $p\neq\alpha/2$:

subcase 3.1: p=0 (observe that this case also covers the situation $\alpha=2$):

with l=α, i=1 and k=$(j-1)/2\alpha$ there is an odd T-cycle dominated by B_j^α.

subcase 3.2: $1 \le p \le \alpha-1$. Let $p=2^r t$ with t odd. Then with $l=2^{r+1}$, $i=2^{m-(r+1)}$ and $k=(t-1)/2+2^{m-r-2}(2\beta+1)$, where β comes from $j=(2p+1)+\beta\, 2\alpha$, there is an odd T-cycle dominated by B_j^α.

Case 4: $2 \leq j \leq \alpha$:

subcase 4.1: α-j = 2p , $p \in \{0,1,2,...\}$:

with l=1, i=j-1 and k=(α+j-2)/2 there is an odd T-cycle dominated by B_j^α.

subcase 4.2: α-j = 2p+1 , $p \in \{0,1,2,...\}$:

- if p is even then with l=2, i=(j-1)/2 and k=(α-p)/2 we are done;
- if p is odd let $p+1=2^r t$, t odd. With $l=2^{r+1}$, $i=2^{m-(r+1)}$ and $k=(t-1)/2+2^{m-r-1}$ there is an odd T-cycle dominated by B_j^α.

This completes the proof of

Lemma 2

A sequence B^α with $\alpha \in \{1,2,4,...\}$ is maximal w.r.t. inducing a bipartite Toeplitz graph.

4) The main result

We will now proceed to show that the B^α, α=1,2,4,8,... , are the *only* 0-1 sequences that induce bipartite Toeplitz graphs and that are not properly dominated by any other such sequence. For this let us consider a sequence $I \in \{0,1\}^{\mathbb{N}}$ which we suppose to induce a bipartite Toeplitz graph. By Lemma 1, I cannot dominate an odd T-cycle.

Suppose that I is not dominated by any of the B^α, α=1,2,4,8,... . To obtain a contradiction we will exhibit an odd T-cycle dominated by I. Let j_1 and j_2 be the indices of the first and second 1-entry in I, respectively, and let $\gamma=j_1-1$, $\delta=j_2-j_1-1$, as indicated in Figure 4:

$$I = (\underbrace{0........0}_{\gamma}\ \overset{j_1}{1}\ \underbrace{0........0}_{\delta}\ \overset{j_2}{1}\ \ ...\ ) .$$

Figure 4

Proposition 3

δ must be odd.

Proof:

Suppose $\delta=2p$, $p\in\{0,1,2,...\}$. Then with $k=i+p$, $l=1$ and $i=\gamma$ we can exhibit an odd T-cycle dominated by I. ■

Consequently, for some $\alpha\in\mathbb{N}$, I can be written as

$$I = (\underbrace{0\ldots\ldots0}_{\gamma}\ \overset{j_1}{1}\ \underbrace{0\ldots\ldots0}_{\alpha-1}\ \underbrace{0\ldots\ldots0}_{\alpha}\ \overset{j_2}{1}\ \ldots\ )\ .$$

Figure 5

a) We will first show that for our I there exists β, $\beta=2^m$ with $m\in\{0,1,2,...\}$ such that

$\alpha\equiv0 \bmod \beta$ and $\gamma\equiv\beta \bmod 2\beta$; in other words I has the following form:

$$I = (\underbrace{0.0}_{\beta}\ \underbrace{0..0}_{2\beta}\ \ldots\ \underbrace{0..0}_{2\beta}\ \overset{j_1}{1}\underbrace{0.0}_{\beta}\ \underbrace{0.0}_{\beta}\ \ldots\ \underbrace{0.0}_{\beta}\ \overset{j_2}{1}\ \ldots\ )\ ,$$

Figure 6

So let us suppose that I is not of this form.

Proposition 4

If $\alpha=2^r t$ with t odd and if $\gamma\not\equiv\beta \bmod 2\beta$ for any divisor β of α then $\gamma\equiv0 \bmod 2^{r+1}$.

Proof:

t is odd and so are all of its divisors. It is therefore sufficient to assume that $\gamma \not\equiv \beta \bmod 2\beta$ for $\beta=1,2,\ldots,2^r$. But then the only remaining case is $\gamma \equiv 0 \bmod 2^{r+1}$. ■

It is for this particular case that we are now going to exhibit an odd T-cycle dominated by l. Let $\gamma=2^{r+1}p$ for some $p\in\mathbb{N}$. Then we choose $l=2^{r+1}$, $i=p$ and $k=p+(t-1)/2$ to obtain

$$2l(k-i)+l-1 = 2^{r+1}(2(k-p)+1)-1 = 2^{r+1}t-1 = 2\alpha-1, \text{ as required.}$$

We can thus complete part a) with the conclusion that if l induces a bipartite Toeplitz graph then there is a β with $\beta \mid \alpha$ and $\gamma\equiv\beta \bmod 2\beta$. Moreover, β can be chosen to be a power of 2.

b) So let $\beta=2^m$ for some $m\in\{0,1,2,\ldots\}$, let j_3 be the index of a third 1-entry in our sequence l; we set $\varepsilon=j_3-j_1-1$.

We assume that $\varepsilon\not\equiv 2^{m+1}-1 \bmod 2^{m+1}$ and that $\varepsilon\geq 2^{m+1}$. Then either $\varepsilon\equiv 0 \bmod 2$ or $\varepsilon\equiv 1 \bmod 4$ or ... or $\varepsilon\equiv 2^{m-1} \bmod 2^{m+1}$. For any of these cases we will now exhibit an odd T-cycle which is dominated by l, the desired contradiction.

So let $\varepsilon=(2^q-1)+p2^{q+1}$ for $p\in\mathbb{N}$ and $q\in\{0,1,\ldots,m\}$. We choose $l=2^q$, $i=2^{m-q}$ and $k=i+p$. Then we have

$$il = 2^m \quad\text{and}\quad 2l(k-i)+l-1 = 2^{q+1}p+2^q-1 = \varepsilon,$$

which yields an odd T-cycle dominated by l.

Altogether we have shown:

Theorem

An infinite 0-1 sequence I induces a bipartite Toeplitz graph

if and only if

I does not dominate an odd T-cycle

if and only if

I is dominated by one of the sequences B^{α}, where $\alpha \in \{1,2,4,8,...\}$.

5) Conclusion and final remarks

As already mentioned in the introduction, we feel that studying (infinite or finite) Toeplitz graphs might not only lead to structural results: several types of optimization problems such as the stable set problem, the matching problem or the max-cut problem might be faster solvable for these graphs.It could also be helpful within this context to study the facial structure of related polyhedra.

As for bipartiteness in finite Toeplitz graphs and results on planarity we refer the reader to our companion paper currently being written.

References

[1] Aho, A., J.E.Hopcroft and J.D.Ullman (1974) , "The design and analysis of computer algorithms", Addison-Wesley, Reading.

[2] Boesch, F. and R.Tindell (1984), "Circulants and their connectivities", *Journal of Graph Theory*, **8**, 487-499.

[3] Burkard, R.E. and W.Sandholzer (1991), "Efficiently solvable special cases of bottleneck travelling salesman problems", *Discrete Applied Mathematics*, **32**, 61-67.

[4] Dal, R. van, G.Tijssen, J.A.A. van der Veen, Ch.Zamfirescu and T.Zamfirescu (1993), "Hamiltonian properties of Toeplitz graphs", Technical Report, Fachbereich Mathematik, Universität Dortmund.

[5] Doorn, E.A. van (1986), "Connectivity of circulant digraphs", *Journal of Graph Theory*, **10**, 9-14.

[6] Garfinkel, R.S. (1977), "Minimizing wallpaper waste, part 1: A class of traveling salesman problems", *Operations Research*, **25**(5), 741-751.

[7] JaJa, J. (1992), "An introduction to parallel algorithms", Addison-Wesley, Reading.

[8] Medova-Dempster, E.A. (1988), "The circulant traveling salesman problem", Technical report, Department of Mathematics, University of Pisa.

[9] Veen, J.A.A. van der, R. van Dal and G.Sierksma (1991), "The symmetric circulant traveling salesman problem", Research Memorandum N° 429, Insitute of Economic Research, University of Groningen.

On Sub-bandwidth of Graph[①]

Fan Guimei

(Qingdao Institute of Architecture and Engineering, Qingdao, China)

Ku Tung—Hsin

(Hefei Branch Research Centre of Combinatorial Mathematics and Computer Science, Academia Sinica, Hefei, China)

Yao Bing

(Northwest Normal University, Lanzhou, China)

Wang Jianfang

(Institute of Applied Mathematics, Academia Sinica, Beijing, China)

Abstract The sub-bandwidth of a graph G, denoted by S(G), is defined with

$\min\{\max S_f(G)-\min S_f(G) \mid f \text{ is a numbering of } G\}$, where $S_f(G)=\{f(u)+f(v) \mid (u,v)\in E(G)\}$

for a numbering f of G. We have proved that

(1) $q(G)\leqslant\alpha'(G)[S(G)+1]$;

(2) $2q(G)\leqslant p(G)[S(G)+1]$;

(3) $x'(G)-1\leqslant S(G), 2(\delta(G)-1)\leqslant S(G)$;

(4) $B(G)\leqslant S(G), 4B(G)\leqslant 2p(G)+S(G)$.

and so on, where B(G), α′(G) and x′(G) are the bandwidth, the edge—independent number and edge—chromatic number of G.

Key words bandwidth, sub—bandwidth, graph.

§ 1. Definitions and Properties

The order of a graph G is denoted by p(G), the size |E(G)| of G by q(G); Δ(G) and δ(G) stand for the maximum degree and minimum degree of G respectively; $\overline{G}$ is the complement of G. A numbering f of a graph G is one to one mapping from V(G) to {1,2,⋯,p(G)}

$$f=V(G)\rightarrow\{1,2,\cdots,p(G)\}.$$

① Supported by National Natural Science Foundation of China

$|x|$ denotes the greatest integer number which is not greater than x.

In this paper, all graphs are finite and simple graphs with at least tow edges.

Definition Let G be a graph of order p and F denote the set of all numberings of G. The sub—bandwidth of G is the integer

$$S(G)=\min\{\max S_f(G)-\min S_f(G) \mid f\in F\}$$

where

$$S_f(G)=\{f(u)+f(v) \mid (u,v)\in E(G)\}.$$

It is easy, by the definition, to see that for a numbering f of G with $S(G)=\max S_f(G)-\min S_f(G)$, the all non—zero elements of the adjacent A(G) of G corresponding f lie in a banded field which contains the sub—diagonal line of A(G). So we try to provide a way of studying bandwidth of graphs.

Theorem 1. Let G be a graph with at least 2 edges, then

$$0\leqslant S(G)\leqslant 2p(G)-4$$

and the bound are sharp.

Proof immediately follows the definition. The uper—bound is arrived by K_p. The lower—bound is reached by a graph of order p which maximum component is K_2.

Now we define two classes of sub—bandwidth graphs.

A sub—bandwidth graph of type 1 of order n, write H(n), is defined as following.

$$H(1)=K_1$$

For $n\geqslant 2$, let $V_1=\{v_1,v_2,\cdots v_n\}, V_2=\{u_1,u_2,\cdots,u_{n-h}\}$

where $h=[\frac{n}{2}]$.

$$V(H(n))=V_1\cup V_2$$

$$E(H(n))=\{(u_i,u_j) \mid i,j\in\{1,2,\cdots,n-h\}, i\neq j\}$$
$$\cup (V_i,u_{n-h-j+1}) \mid i=1,2,\cdots,h, j=1,2,\cdots,i\}$$

Set $V_{h+i}=u_i$ for $i=1,2,\cdots,n-h$. Then

$$V(H(n))=\{v_1,v_2,\cdots,v_n\}$$

It is easily seen that $d(V_i)=i$ and

$$N(V_i)=\{V_n,V_{n-1},\cdots,V_{n-i+1}\} \quad \text{for} \quad 1\leqslant i\leqslant h, \text{ and}$$

$$d(V_{h+i})=h+i-1 \text{ and } N(V_{h+i})=\{V_n,\cdots,V_{n+i+1},V_{n+i-1},\cdots,V_{n-h},V_{n-h-1},\cdots,V_{n-h-i+1}\}$$

for $1\leqslant i\leqslant n-h$.

Given two sub—bandwidth graphs of type 1;

H(n) and H(m).

Let $V(H(n))=\{V_1,\cdots,V_n\}$. $V(H(m))=\{u_1,u_2,\cdots,u_m\}$.

Assume $1\leqslant r\leqslant \min\{m-1,n-1\}$ and

$$V_i=u_{r-i+1} \quad \text{for} \quad 1\leqslant i\leqslant r.$$

The sub-bandwidth of type 1 of order $m+n-r$, write $H(r;m+n)$, is defined as

$V(H(r;m,n))=\{v_1,v_2,\cdots,v_1,u_{r+1},u_{r+2},\cdots,u_m\}$

$=\{v_n,v_{n-1},\cdots,v_{r+1},u_1,\cdots,u_m\}$

$E(H(r;m+n))=E(H(n))\bigcup E(H(m))$

It is easy to check $E(H(n))\bigcap E(H(m))=\varnothing$

If $r\leqslant 0$. Then $V(H(n))\bigcap V(H(m))=\varnothing$

$H(r;m,n)$ is defined as

$V(H(r;m+n))=V(H(n))\bigcup V(H(m))\bigcup\{w_1,w_2,\cdots,w_{-r}\}$ and

$E(H(r;m+n))=E(H(n))\bigcup E(H(m))$

It is not difficult to show that any G contains a sub-bandwidth graph $H(n)$ of type 1. We stipulate that a sub-bandwidth graph $H(r;m+n)$ of type 2 is a subgraph of G if and only if $H(r;m+n)$ is a spanning subgraph of G and $r=m+n-p(G)$. So every graph G contains a sub-bandwidth graph $H(r;m+n)$ of type 2 for $2-p(G)\leqslant r\leqslant p(G)-2$.

A subgraph $H(n)$ of G is called the greatest in G if $m\leqslant n$ for any $H(m)\subseteq G$. Similarly, $H(r;m+n)\subseteq G$ is the greatest in G if $r'\leqslant r$ for any $H(r',m',n')\subseteq G$.

Lemma 1. For any $H(r;m+n)\subseteq\overline{G}$, there exists a numbering f of the graph G such that

$$n+2\leqslant \min S_f(G)\leqslant \max S_f(G)\leqslant 2p(G)-m.$$

Proof. Let f be a numbering of $\overline{G}$ satisfying

(a) $f(v_i)=i,\ 1\leqslant i\leqslant n$

(b) $f(u_i)=p(G)-i+1\ \ 1\leqslant i\leqslant m$

(c) If $V(H(n))\bigcup V(H(m))\neq V(G)$. then

$f(x)\in\{n+1,\ n+2,\cdots\cdots P(G)-m\}$ for $x\in V(\overline{G})\backslash V(H(n))\bigcup V(H(m))$.

It is that from the construction of subgraphs $H(n)$ and $H(m)$ of $H(r;\ m+n)$ that $(x,y)\in E(m)$ if $f(x)+f(y)\geqslant 2p(G)-m+1$; $(x,y)\in E(n)$ if $f(x)+f(y)\leqslant n+1$. So

$$n+2\leqslant f(x)+f(y)\leqslant 2p(G)-m$$

for any $(x.y)\in E(G)$, moreover

$$n+2\leqslant \min S_f(G)\leqslant \max S_f(G)\leqslant 2p(G)-m$$

Lemma 2. For any numbering f of the graph G, there exists $H(r;m+n)\subseteq\overline{G}$ such that

$$n+2=\min S_f(G)\leqslant \max S_f(G)=2p(G)-m.$$

Proof Let $P(G)=p$. and for any numbering f of G, Let

$$V(G)=\{w_1,w_2\cdots\cdots,w_p\}$$

such that $f(w_i)=i$ for $1\leqslant i\leqslant p$.

Let $M=\max S_f(G)$, $m=2p-m$,

$V'=\{w_p,w_{p-1},\cdots\cdots,w_{M+1-p}\}$

$E'=\{(w_{p-k+1},\ w_{p-k+1-j})\,|\,k=1,2\cdots\cdots[(2p-m+1)/2],$

$j=1,2\cdots 2(p-k)-M+1\}$

When $M\leqslant p$ $V'=\{W_p\}$, $E'=\varnothing$

Let $A=(V',E')$ then $H(m)\cong A\subseteq\overline{G}$

On the other hand, let $N=\min S_f(G)$. there is $H(N-2)\subseteq\overline{G}$ which consists of the vertices

$w_1, w_2 \cdots, w_{N-2}$ and the edges $(w_t, w_{N-t-1-s})$ satisfying inequalities

$N-1 \geqslant f(W_t)+f(W_{N-t-1-s})=N-1-s$. $0 \leqslant S \leqslant N-2t-2$. for $t=1,2,\cdots[\frac{N-1}{2}]$. Write $n=N-2$. then $\min S_f(G)=n+2$. It is obvious that $E(H(n)) \cap E(H(m))=\varnothing$. If $N-2<M+1-P$. We get a sub-bandwidth graph $H(r;m,n)$ of type 2. Where $-r=M-N+2-P$. If $N-2 \geqslant M+1-P$. then $H(r;m+n)$ is in $\overline{G}$ where $r=p-(M-N+2)$. The proof is over. □

Corollary 1. For any $H(r;m.n) \subseteq \overline{G}$

$S(G) \leqslant 2p(G)-m-n-2$ or $S(G) \leqslant p(G)-r-2$

It follows from **Lemma 1**.

Basic Theorem. Let G be a simple and finite graph having at least two edges, then $H(r;m+n)$ is a greatest sub-bandwidth of $\overline{G}$ if and only if

$$S(G)=p(G)-r-2=2p(G)-m-n-2$$

Proof. Necessity, Assume $H(r;m+n)$ is the greatest. By **Lemma 1**. there is a numbering f of G such that

$$n+2 \leqslant \min S_f(G) \leqslant \max S_f(G) \leqslant 2p(G)-m.$$

Hence

$$S(G) \leqslant \max S_f(G)-\min S_f(G) \leqslant 2p(G)-m-n-2.$$

If $S(G)<2p(G)-m-n-2$, then there exists a numbering h of G such that

$$S(G)=\max S_h(G)-\min S_h(G).$$

According to lemma 2, there is $H(r';m',n') \subseteq \overline{G}$ such that

$$m'=2p(G)-\max S_h(G) \text{ and } n'=\min S_h(G)-2.$$

Thus

$$2p(G)-m'-n'-2<2p(G)-m-n-2.$$

$$m+n<m'+n', r<r'$$

a evident contradiction.

Sufficienty. Assume $H(r;m+n) \subseteq \overline{G}$ such that

$$S(G)=p(G)-r-2=2p-m-n-2$$

Suppose that there exists $H(r'',m''+n'') \subseteq \overline{G}$ such That $r<r''$. By **lemma 1**. there is a numbering g of G such that

$$S(G) \leqslant \max S_g(G)-\min S_g(G) \leqslant 2p(G)-m''-n'-2.$$

Thus

$$S(G) \leqslant p(G)-r''-2<p(G)-r-2=S(G).$$

It is obviously imposible. The proof is completed.

Corollary 2. If $K_n \subseteq G$ then $S(G) \geqslant 2(n-1)$

Corollary 3. Let $\alpha(G)$ denote the independent number of the graph G and V_1 be the maximum vertex-independent set of G. then

$$2p(G)-4\alpha(G) \leqslant S(G) \leqslant 2p(G)-\alpha(G)-\alpha(G-V_1)-2$$

Proof From Basic Theorem, $S(G)=2p(G)-m-n-2$ for a largest $H(r;m+n)$ of $\overline{G}$. Assume $m \geqslant n$. Let A be the complement of $H(r;m+n)$, then A includs G. Since both H

(m) and H(n) are subgraphs of H(r; m+n), we have that

$$4\alpha(G)\geqslant 4\alpha(A)\geqslant 2(m+1)\geqslant m+n+2$$

Hence $S(G)=2p(G)-m-n-2\geqslant 2p(G)-4\alpha(A)\geqslant 2p(G)-4\alpha(G)$

On ther hand, $\overline{G}$ has a subgraph $H(r_1;\alpha(G),\alpha(G-V_1))$ where $r_1=P(G)-\alpha(G)-\alpha(G-V_1)$

By Lemma 1.

$$S(G)\leqslant 2p(G)-\alpha(G)-\alpha(G-V_1)-2$$

Corollary 4. $S(G)\geqslant 2(\delta(G)-1)$

Proof. Since $n-1$, $m-1\leqslant\Delta(\overline{G})$ for $H(n),H(m)$ of $\overline{G}$. It follows Basic Theorem.

Corollary 5. Let $p=p(G)$ $q=q(G)$, then

$$S(G)\geqslant 2\{p-1-\sqrt{p(p-1)-2q+1}\}$$

Proof. It is easy to calculate that

$$|E(H(2n))|=n^2 \text{ and } |E(H(2n-1))|=n(n-1)=\frac{(2n-1)^2-1}{4}$$

Hence

$$\frac{m^2+n^2-2}{4}\leqslant|E(H(m))|+|E(H(n))|\leqslant\frac{p(p-1)}{2}-q$$

for any $H(r;m,n)$ in $\overline{G}$ and moreover

$$2p(p-1)-4q+2\geqslant m^2+n^2=(m+n)^2-2mn\geqslant(m+n)^2-2\left(\frac{m+n}{2}\right)^2=\frac{(m+n)^2}{2}$$

or $m+n\leqslant 2\sqrt{p(p-1)-2q+1}$. The conclution of **corollary** 5 by Basic Theorem, can be obtained at once. □

Let $\alpha'(G)$ and $X'(G)$ denote The edge—independent number and the edge—chromatic of the graph G respectively.

Corollary 6. (1) $X'(G)\leqslant S(G)+1$

(2) $2q(G)\leqslant p(G)(S(G)+1)$

(3) $q(G)\leqslant\alpha'(G)(S(G)+1)$

(4) If $\Delta(G)\delta(G)\geqslant 4[p(G)-k]$, then

$[S(G)+4][p(G)-k]\leqslant\Delta(G)q(G)$.

Proof. Let f be a numbering of G such that

$S(G)=\max S_f(G)-\min S_f(G)$

Let $M=\max S_f(G)$ an $N=\min S_f(G)$. There is a postition of E(G) as following

$$E(G)=E_N\cup E_{N+1}\cup\cdots\cup E_m$$

where $E_j=\{(u,v)\,|\,(u,v)\in E(G) \text{ and } f(u)+f(v)=j\}$ for $N\leqslant j\leqslant M$. It is clear that every E_j is an edge—independent set of G. Let $A^*(G)$ stand for the average

$$\sum_{j=N}^{M}\frac{|E_j|}{S(G)+1}=\frac{q(G)}{S(G)+1}$$

(1) Since $S(G)+1\geqslant$the number of all non—empty E_j for $N\leqslant i\leqslant M$,

$$S(G)+1\geqslant X'(G)$$

(2) It is clear that $A^*(G)\leqslant\frac{P(G)}{2}$ for $|E_j|\leqslant\frac{P(G)}{2}$ for $N\leqslant j\leqslant M$.

$$2q=2A^*(S(G)+1)\leqslant p(G)(S(G)+1).$$

(3). It is clear that $\alpha'(G)\geqslant A^*(G)$. So $q(G)\leqslant\alpha'(G)(S(G)+1)$

(4) Let $\Delta(G)=\Delta$, $\delta(G)=\delta$, $S(G)=S$, $q(G)=q$ and $p(G)=p$ then by **Theorem** 1

$$\frac{q}{(S+4)}=\frac{1}{2(S+4)}\sum_{i=1}^{p}d_G(V_i)\geqslant\frac{p\delta}{2(s+4)}\geqslant\frac{\delta(s+4)}{4(s+4)}\geqslant\frac{\delta}{4}\geqslant\frac{p-K}{\Delta}.\ \square$$

§ 2. The Sub-bandwidth Between Graphs

Theorem 2. Assume $|E(G)|\geqslant 2$ and $|E(\overline{G})|\geqslant 2$ then

$S(G)+S(\overline{G})\geqslant 2p(G)-5-\min\{\Delta(G),\Delta(\overline{G})\}$

Proof. By **Corollary** 4 and 6(1), we have that

$$S(G)+S(\overline{G})\geqslant\Delta(G)-1+2(\delta(\overline{G})-1)$$
$$=\Delta(G)-1+2\{P(G)-\Delta(G)-2\}=2p(G)-\Delta(G)-5$$

and

$$S(G)+S(\overline{G})\geqslant 2\{\delta(G)-1\}+\Delta(\overline{G})-1=2p(G)-\Delta(\overline{G})-5$$

so

$$S(G)+S(\overline{G})\geqslant 2p(G)-5-\min\{\Delta(G),\Delta(\overline{G})\},$$

Corollary 7. If G is a regular graph, then

$$S(G)+S(\overline{G})\geqslant 2\{P(G)-3\}$$

On two vertices u and v in $V(G)$, the merge of a graph G, denoted by $G|_{u,v}$, can be obtained from G by identifying u and v into one vertex, and deleting the possible resulting loop and multi-edges from G. If $e=(u,v)\in E(G)$, $G|_{u,v}=G|_e$ is called the contraction of G along edge e.

Theorem 3. $S(G)\leqslant S(G|_{u,v})+2$

Proof. Let f be a numbering of $G|u,v$ such that

$$S(G)=\max S_f(G|_{u,v})-\min S_f(G|_{u,v})$$

It yield one vertex u,v of $G|_{u,v}$ to identify u and v of G.

Case1. $f(u,v)>f(x)$ for any $x\in N(u,v)$

Take $x_0\in N(u,v)$ such that $f(x_0)>f(x)$ for any $x\in N(u,v)\backslash\{X_0\}$

We make a numbering g of G as following:

(a) $g(y)=f(y)$ if $f(y)\leqslant f(x_0)$

(b) $g(y)=f(y)+1$ if $f(y)>f(x_0)$ and $y\neq u\cdot v$

(c) $g(y)=f(u.v)$ if $y=u$

(d) $g(y)=f(x_0)+1$ if $y=v$

Estimate $g(x)+g(y)$ for every edge (x,y) in $E(G)$. If $(u,v)\in E(G)$, we have

$$\min S_f(G|_{u,v})+1\leqslant g(u)+g(v)=f(u.v)+f(x_0)+1\leqslant\max S_f(G|_{u,v})+1$$

For $(v,y)\in E(G)$, namely, $(v,y)=(u,v,y)\in E(G|_{u,v})$

then

$$\min S_f(G|_{u,v})+1\leqslant g(v)+g(y)=f(x_0)+1+g(y)$$
$$\leqslant f(x_0)+f(y)+2\leqslant f(x_0)+f(u,v)+2\leqslant \max S_f(G|_{u,v})+2$$

For other $(x,y)\in E(G)$ being different from (u,v) and (v,y) obviously

$$\min S_f(G|_{u,v})\leqslant g(x)+g(y)\leqslant \max S_f(G|_{u,v})+2$$

Hence

$$S(G)\leqslant \max S_g(G)-\min S_g(G)\leqslant \max S_f(G|_{u,v})+2-\min S_f(G|_{u,v})=S(G|_{u,v})+2$$

Case2. $f(y_1)<f(u,v)<f(y_0)$ for some two ventices y_0, $y_1\in N(u,v)$

We can obtain a numbering h of G by changing f as following.

(a′) $h(x)=f(x)$ if $f(x)<f(u,v)$;

(b′) $h(x)=f(x)+1$ if $f(x)>f(u,v)$;

(c′) $h(x)=f(u,v)$ if $x=u$

(d′) $h(x)=f(u,v)+1$ if $x=V$

If $(u,v)\in E(G)$

$$\min S_f(G|_{u,v})+1\leqslant f(y_1)+f(u,v)+1\leqslant 2f(u,v)+1$$
$$=h(u)+h(v)\leqslant f(y_0)+f(u,v)+1\leqslant \max S_f(G|_{u,v})+1$$

For $(v,x)\in E(G)$.

$$\min S_f(G|_{u,v})+2\leqslant f(y_1)+f(u,v)+2\leqslant f(u,v)+2+f(v)$$
$$=h(x)+h(v)\leqslant f(y_0)+f(u,v)+2\leqslant \max S_f(G|_{u,v})+2$$

For other edge $(x,y)\in E(G)$,

$$\min S_f(G|_{u,v})\leqslant h(x)+h(y)\leqslant \max S_f(G|_{u,v})+2$$

Thus $S(G)\leqslant \max S_h(G)-\min S_h(G)\leqslant S(G|_{u,v})+2$

Case 3. $f(u,v)<f(x)$ for any $x\in N(u,v)$

Using the numbering g satisfied

$$g(x)=p(G|_{u,v})+1-f(x) \text{ for } x\in V(G|_{u,v})$$

and

$$\max S_g(G|_{u,v})-\min S_g(G|_{u,v})$$
$$=\max S_f(G|_{u,v})-\min S_f(G|_{u,v})$$

then $g(u,v)>g(x)$ for $x\in N(u,v)$. The proof is the same as case 1

Lemma 3. For any $v\in V(G)$, $S(G-V)\leqslant S(G)+1$.

Proof. Take arbitrarity a numbering f of G, and there is a numbering g of $G-V$ as below. For any $x\in V(G-V)$.

(i) $g(x)=f(x)$ if $f(x)<f(v)$;

(ii) $g(x)=f(x)-1$ if $f(x)>f(v)$.

Let

$M_g=\max S_g(G-V)$, $N_g=\min S_g(G-V)$

$M_f=\max S_f(G)$, and $N_f=\min S_f(G)$

The possible occuring situations are in the following

(a) $M_g=M_f$, $N_g=N_f$ or $N_g=N_f-1$

(b) $M_g=M_f-1$, $N_g=N_f-2$ or $N_g=N_f-1$ or $N_g=N_f$.

(c) $M_g=M_f-2$, $N_g=N_f-2$ or $N_g=N_f-1$ or $N_g=N_f$

Thus

$S(G-V)\leqslant M_g-N_g\leqslant M_f-N_f+1.$

Lemma 4. (1). For any $v\in V(H(n))$. $H(n-2)\subseteq H(n)-V$.

(2) For any $e\in E(H(n))$, $H(n-2)\subseteq H(n)-e$.

Proof. It is clear that $H(n-1)\subseteq H(n)-v$ for any $v\in V(H(n))$ when $2\leqslant n\leqslant 3$ Now assume $n\geqslant 4$,

Case 1. $n\equiv 0 \pmod 2$

Let $h=\frac{n}{2}$ and v be a vertex of $H(n)$. By the definition of $H(n)$. we have that:

Sub-case1-1. If $v=v_{h-t}$ for $0\leqslant t\leqslant h-1$. we need only to remove edges $(V_{h+t+1-k}, V_{h-t+k})$ for $k=0,1,2,\cdots,t$ away from $H(n)-v$, and then obtain the graph $H(n-1)$

sub-case1-2

If $v=v_{n-s}$ for $0\leqslant S\leqslant h-1$,

$H(n-2)=H(n)-\{v_{n-s}, v_{s+1}\}$. It is easy to notice that the case $H(n-1)\subseteq H(n)-v$ does not exist always.

Case 2. $n\equiv 1 \pmod 2$

The proof can be omitted for it is much the same

Case 1.

Let $e=(u,v)\in E(H(n))$, the case on $H(n)-e$ follows above. The proof is over

Lemma 5 Given $H=H(r;m+n)$ such that

$r=m+n-P(G)\geqslant 1$. If $r\leqslant\min\{[\frac{m}{2}].[\frac{n}{2}]\}$ then for any $v\in v(H)$

$H(r;m_1+n_1)\subseteq H-V$ or $H(r_2;m_2+n_2)\subseteq H-V$

where $m_1+n_1-r_1=P(G)-1$ or $m_2+n_2-r_2=P(G)-2$.

proof: We use the notations in the definition of $H(r;m,n)$ in the following discussion.

Case 1. $v=v_{h-t}$ for $0\leqslant t\leqslant h-1$ where $h=[\frac{n}{2}]$

If $h-t>r$, by the proof of lemma 4. $H(n-1)$ can be obtained by removing appropriately some edge from $H(n)-v$, and those edges are not in $H(m)$.

Therefore

$H(r;m,n-1)\subseteq H(r;m,n)$ and $m+(n-1)-r=p(G)-1$

If $h-t\leqslant r$ then $v_{h-t}=U_{r-h+t+1}$, namely

$r-h+t+1\leqslant r\leqslant[\frac{m}{2}]$ form the condition. By the proof of **Lemma** 4.

$H(n-1)\subseteq H(n)-v$ and $H(m-1)\subseteq H(m)-v$

It shows that $H(r-1;(m-1)+(n-1))$ can be obtained by deleting suitably some edges from $H(r;m+n)-v$, and

$$(m-1)+(n-1)-(r-1)=p(G)-1$$

Case 2. $v=v_{n-s}$ for $0\leqslant s\leqslant h-1$ where $h=[\frac{n}{2}]$

In this case, by the proof of **Lemma** 4 we get

$$H(n-2)=H(n)-\{v_{n-s},\ v_{s+1}\}$$

If $s+1>r$ then

$H(r;(n-2)+m)\subseteq H(r;m+n)$, so

$(n-2)+m-r=p(G)-2$

Lemma 6. Let G be a graph having a isolated vertex v, then $S(G-v)\leqslant S(G)$

Proof Suppose that G be a graph having a isolated vertex v such that $S(G)+1\leqslant S(G-v)$. By **Lemma** 3, $S(G)+1=S(G-v)$, and there is the greatest $H(r;m,n)\subseteq\overline{G}(m+n-r=p(G))$ and $H(r';m'+n')\subseteq\overline{(G-r)}(m'+n'-r'=p(G-v))$ holding

$$2p(G)-m-n-1=p(G)-r-2+1=S(G)+1=S(G-v)$$
$$=p(G-v)-r'-2=2p(G-v)-m'-n'-2$$

in the light of Basic Theorem. We get

$$r'=r-2 \text{ and } m'+n'=m+n-3$$

case 1. If $r\leqslant 0$ then $r'\leqslant -2$

By **lemma** 4. assume v, without loss of generality, is a vertex of H(n), then $H(n-2)\subseteq H(n)$. This shows that $H(r-2;m+n-2)\subseteq\overline{(G-V)}$ at least. It derivers by **Corollary** 1 that

$$S(G-V)\leqslant 2p(G-v)-(m+n-2)-2$$
$$=2p(G)-m-n-2=S(G)$$

a clear contradiction.

Case 2 If $r\geqslant 1$ Then $r'\geqslant -1$

Suppose v belong to H(n). Let H be a graph whose complement is $H(r;m,n)+A$ where

$$A=w\backslash[w\cap E(H(r;m+n))] \quad W=\{(u,v)\mid u\in N(v)\}$$

Then the vertex v is isolated in H. Therefore

$S(H)=S(G)$ and $S(H-v)\geqslant S(G-v)$

Sub-case 2-1. If $v=v_{n-j}, 0\leqslant j\leqslant[\frac{n}{2}]=h$

After removing the vertex v, by the proof of **Lemma** 4. there is no $H(n-1)$ as a subgraph of H(n) and only there is $H(n-2)=H(n)-\{v_{n-j},v_{j+1}\}$ in $H(n)-v$. it shows that the graph $\overline{(H-v)}$ has no $H(r',m',n')$ such that

$m'+n'-r'=p(H-v)=P(G)-1$ and $r'\geqslant -1$

and $\overline{(H-v)}$ contains at most a certain $H(r'';m'',n'')$ in which $r''\leqslant -2$

Sub-case 2-2(1). If $v=v_t$ for $t>r$, $0\leqslant t\leqslant h$, i,e. the vertex V_t is not adjacent with any vertex of subgraph H(m) of H(r;m,n)

From the proof of sub-case 1.1 in **Lemma** 4, it is true that $H(n-1)\subseteq H(n)-v$, It shows $H(r;m+n-1)\subseteq H(r;m+n)-v$, Thus, by corollary 1.

$$S(G-v)\leqslant S(H-v)\leqslant p(H-v)-r-2$$
$$=p(G)-1-r-2=S(G-1)$$

This inequality just contradicts with $S(G-v)=S(G)+1$

In this case, $v=v_t=u_{r-t+1}\in V(H(m))$ and $r-t+1\leqslant[\frac{m}{2}]$. We have, by the sub-case 1.1 in the proof of **Lemma 4**, that $H(n-1)\subseteq H(n)-v$ and $H(m-1)\subseteq H(m)-v$ So $H(r-1;m+n-1)\subseteq H(r;m+n)$, and by corollary 1.

$$S(G-v)\leqslant S(H-v)\leqslant p(G-v)-(r-1)-2$$
$$=p(G)-1-(r-1)-2=S(G)$$

It yields a contradiction.

The proof is completed

Theorem 4. For any subgraph H of G with $|E(H)|\geqslant 2$, $S(H)\leqslant S(G)$

Proof. If $V(H)=V(G)$, it is evident. If $V(H)\neq V(G)$, we can take a subgraph H' of G which satisfies

$$V(H')=V(G) \text{ and } E(H')=E(H)$$

Let H' has isolated vertices $w_1, w_2\cdots, w_t$, then

$$V(H)=V(H')\backslash\{w_1,w_2\cdots,w_t\}$$

By **Lemma 6**.

$$S(H)\leqslant S(H'-\{w_1,w_2,\cdots,w_t\})\leqslant S(H'-\{w_1,\cdots,w_{t-1}\})$$
$$\leqslant\cdots\leqslant S(H'-w_1)\leqslant S(G)$$

Corollary 8. If $\overline{G}$ has no subgraph $H(r;m+n)$ for any integer $r\geqslant 1$, then for any vertex $v\in V(G)$.

$$S(G+E(v))\leqslant S(G)+2$$

where $E(v)=\{(u,v)\mid(u,v)\in(\overline{G})\}$

Proof Since $S(G)\geqslant p(G)-2$ follows that $\overline{G}$ has no subgraph $H(r;m,n)$ for only integer $r\geqslant 1$, then

$$S(G)=2p(G)-m'-n'-2$$

for a greatest $H(r';m+n)\subseteq\overline{G}$ with $r'\leqslant 0$. Therefore, one of

$$H(n'-2)\subseteq H(n')-v \text{ and } H(m'-2)\subseteq H(m')-v$$

is in $\overline{(G+E(v))}$ by Lemma 4, so $H(r'-2,m'+n'-2)\subseteq\overline{(G+E(v))}$ Hence, by corollary 1.

$$S(G+E(v))\leqslant 2p(G)-(m'+n'-2)-2=S(G)-2$$

Lemma 7. For any $u,v\in V(H(m))$, $H(n-1)\subseteq H(n)|_{u,v}$

Proof. Using the symbole of $H(n)$. Let $u=v_s$ and $v=v_t$ such that $s>t$ and $h=[\frac{n}{2}]$. If $h+1\leqslant t<s$ then $H(n-1)$ can be obtained from $H(n)|_{u,v}$ by deleting the loop which a edge in $H(n)$, If $t<s<h$ the rest is just $H(n-1)$ after removing the edges $(v_{t+k}, v_{n-t-k+1})$ from $H(n)|_{u,v}$ for $1\leqslant k<h-t$. Finally, $H(n-1)=H(n)|_{u,v}-(v_h,v_{n-h+1})$ when $t=h$ and $s\geqslant h+1$.

For $1<h<s$, $H(n-1)$ can be made well by removing the loop and the edges $(v_{t+k}, v_{n-t-k+1})$for $1\leqslant k\leqslant h-t$ from $H(n)|_{u,v}$

Corollary 9. If $\overline{G}$ has a subgraph $H(r;m+n)$ for any integer $r\leqslant 0$, then $S(G|_{u,v})\leqslant S(G)$

§ 3. The Sub-bandwidth of Several Particular Classes of Graphs

It is necessary to know the sub-bandwidth values about some graphs. For examples, $S(K_p)=2p-4$, $S(P_k)=1$, $S(C_p)=2$ and $S(K_{m,n})=m+n-2$.

Theorem 5. If $\overline{G}$ has no triangle, then

$$2p(G)-8\leqslant S(G).$$

Theorem 6. If $\overline{G}$ is a planar, then $2p(G)-16\leqslant S(G)$

A tree is a caterpillar if $T-V_e(T)$ is a path where

$$V_e(T)=\{v|v\in V(T) \text{ and } d_T(v)=1\}$$

A caterpillar T is k-regular if $d_T(v)=k$ for every $v\in V(T)\backslash V_e(T)$.

Lemma 8. If T is a k-regular caterpillar, then

$$S(T)=K$$

Proof. Let the path $P_n=T-V_e(T)=u_1e_1u_2\cdots u_{n-1}e_{n-1}u_n$,
then

$d(u_1)=d(u_2)=\cdots=d(u_n)=k$

$N_1=N(u_1)\backslash\{u_2\}=\{v_{1,1},\cdots v_{1,k-1}\}$

$N_n=N(u_n)\backslash\{u_{n-1}\}=\{v_{n,1},\cdots,v_{n,k-1}\}$

$N_t=N(u_t)\backslash\{u_{t-1},u_{t+1}\}=\{v_{t,1},\cdots,v_{t,k-2}\}$, $2\leqslant t\leqslant n-1$

Thus $V_e(T)=N_1\cup N_2\cup\cdots\cup N_n$

Let $p(T)=p$, Then $p=n+n(k-2)+2=n(k-1)+2$

Let f be a numbering such that

1. $f(u_1)=1$ $f(v_{1,t})=p-t+1$ for $1\leqslant t\leqslant k-1$,
2. $f(2)=f(v_{1,k-1})-1$, $f(v_{2,t})=f(u_1)+t$,
3. $f(u_i)=f(v_{i-1,k-2})+(-1)^{i-1}$, $f(v_{i,t})=f(u_{i-1})+t(-1)^i$

for $1\leqslant t\leqslant k-2$ and $3\leqslant i\leqslant n-1$,

4. $f(u_n)=f(v_{n-1,k-2})+(-1)^{n-1}$

   $f(v_{n,t})=f(v_{n-1})+t(-1)^n$ for $1\leqslant t\leqslant k-1$

Obviously

$p-k+2=f(u_1)+f(u_2)\leqslant f(x)+f(y)\leqslant f(u_1)+f(v_{1,1})=p+1$

So $S(T)=k-1=\Delta(T)-1$ by Corollary 6(1)

Theorem 7. For every caterpillar T, $S(T)=\Delta(T)-1$

Proof Since T is a subgraph of a regular caterpillar T^* having maximum degree $\Delta(T^*)=\Delta(T)$, it follows that corollary 6(1). Theorem 4 and Lemma 8. □

§ 4. On S(G) and B(G)

The bandwidth of a graph G, denoted by B(G), is defined as follows

$$\min_{f\in F}\max_{(u,v)\in E(G)}\{|f(u)-f(v)|\}$$

where F is the set of all numberings of G (to see [3])

Theorem 8. For any numbering f of the graph G.

$2B(G)-1\leqslant 2p(G)-\min S_f(G)$ and $2B(G)+1\leqslant \max S_f(G)$

Using **Lemma** 9 below it is not difficult to get the proof of Theorem 8.

Lemma 9. For any $H(n)\subseteq\overline{G}$, $B(G)\leqslant p(G)-[\frac{n}{2}]-1$

Proof Let $n=2k$ and

$$V(H(2k))=\{v_{2k},v_{2k-1},\cdots v_1\}$$

We can set up an one to one mapping

$g:V(\overline{G})\to\{1,2,\cdots,p(G)\}$

as following:

(a) $g(v_{2t})=p-k+t$, $g(V_{2t-1})=k-t+1$ for $1\leqslant t\leqslant k$,

(b) let $V(\overline{G})\backslash V(H(2k))=\{w_1,w_2,\cdots w_{p-2k}\}$

where $p=p(\overline{G})$, then $g(w_j)=k+j$ for $1\leqslant j\leqslant p-2k$.

Obviously

$$|g(x)-g(y)|\geqslant p-k$$

if and only if $(x,y)\in E(H(2k))$, so

$$B(G)\leqslant B_g(G)\leqslant p-k-1$$

If $n=2k+1$, then $H(2k)=H(2k+1)-V_{k+1}$, where $d(V_{k+1})=k+1$ in $H(2k+1)$. This ends the proof.

Theorem 9. Let G be a connected graph, then

$$B(G)\leqslant S(G)$$

Proof. It is clear when G is a complete graph. Let G be not a complete graph. If for any $H(r;m+n)$ in $\overline{G}$, $r\leqslant 0$. Then $S(G)\geqslant p(G)-2$, by Basic Theorem. This shows that $B(G)\leqslant S(G)$. It is only need to discuss the case that $S(G)=P(G)-r-2$ for the greatest $H(r;m+n)$ in $\overline{G}$ with $r\geqslant 1$. We take such a graph H so that the complement graph $\overline{H}$ of H is isomoripc to $H(r;m+n)$. By the definition of $H(r;m+n)$, $H(m)$, $H(n)\subseteq H(r;m+n)$. Use entirly the notation of $H(r;m+n)$ here, rewrite every vertex of H as

$w_i=u_{m-i+1}$ when $1\leqslant i\leqslant m$, $w_j=v_{r+j-m}$ when $m+1\leqslant j\leqslant p$

and labele w_i by an one to on mapping $f:V(\overline{H})\to\{1,2,\cdots,p\}$ such that $f(w_i)=i$ for $1\leqslant i\leqslant p$. Since f is a numbering, for $(x,y)\in E(H)$.

$$\min S_f(H)=f(w_1)+f(w_{m+1})\leqslant f(x)+f(y)$$
$$\leqslant f(w_p)+f(w_{m-r})=\max S_f(H)$$

or

$$m+2\leqslant f(x)+f(y)\leqslant p+m-r,$$

namely, $S(H)=p-r-2=S(G)$. Let $N=m-r$, we construct a new graph G_0 from H by adding vertices $w_{p+1},w_{p+2},\cdots,w_{p+N-1}$ to H such that $N(w_j)=\{w_{p+1},w_{p+2},\cdots,w_{p+n-j}\}$ for $1\leqslant j$

$\leqslant N-1$. Give a numbering g of G as $g(w_i)=i$ for $1\leqslant i\leqslant p+N-1$, and then

$$\min S_g(G_0)=m+2,\ \max S_g(G_0)=p+N=p+m-r.$$

farther

$$S(G_0)\leqslant \max S_g(G_0)-\min S_g(G_0)=p-r-2=S(G)$$

Notice that the degree of w_1 in G_0 is $p-r-1$, so that

$$S(G_0)\geqslant p-r-2$$

by corollary 6(1). then we get $S(G_0)=S(G)$. Let $S=S(G_0)$, then for $w_i\in V(G_0)$

$$N(w_i)=\{w_{p+N-i},w_{p+n-i-1},\cdots,w_{a(i)}\} \qquad (*)$$

where $a(i)=\max\{1,p+N-s-i\}$ for $1\leqslant i\leqslant p+N-1$, and

$$|N(w_i)|\leqslant(p+N-i)-(p+N-S-i)+1=S+1$$

We give a new numbering h of G_0 as following:

(1)$h(w_1)=2$, $h(w_{p+N-1})=1$

(2)$h(w_{i+1})=(t+1)S+2+i$, $h(w_{p+N-i-1})=tS+2+i$ for $tS+1\leqslant i\leqslant(t+1)S$ and $t=0,1,2,\cdots,[\frac{p+N-1}{2s}]-1$

Now, we inspect $|h(x)-h(y)|$ for $(x,y)\in E(G_0)$ For $w\in N(w_i)$ when $t=0$

$$|h(w)-h(w_i)|\leqslant h(w_{p+n-s-1})-h(w_1)=S+2-2=S$$

For $w\in N(w_{i+1})$ when $tS+1\leqslant i\leqslant(t+1)S$ and $t\geqslant 1$,

$$h(w_{i+1})=(t+1)S+2+i \text{ and } (t+1)S+1\leqslant i+S\leqslant(t+2)S.$$

By (*) and the definition of h,

$$h(w_{p+N-(i+1)})\leqslant h(w)\leqslant h(w_{p+N-s-(i+1)}) \quad\text{or}\quad h(w_{p+N-i-1})\leqslant h(w)\leqslant h(w_{p+N-(i+S)-1})$$

Namely,

$$tS+2+i\leqslant h(w)\leqslant(t+1)S+2+(i+S)=(t+2)S+2+i$$

That is

$$|h(w)-h(w_{i+1})|\leqslant|h(w_{i+1})-h(w_{p+N-i-1})|=S$$

or

$$|h(w)-h(w_{i+1})|\leqslant|h(w_{i+1})-h(w_{p+N-(i+S)-1})|=S$$

This shows that

$$B(G_0)\leqslant B_h(G_0)\leqslant S$$

So

$$B(G)\leqslant B(G_0)\leqslant S=S(G)$$

because $G\subseteq H\subseteq G_0$

Lemma 10. If $\overline{G}$ is not connected then $2B(G)\geqslant p(G)-1$

Proof Since G contains a certain k-partite graph for $k\geqslant 2$ as its subgraph, if follows that the Theorem 4.1.2 in [3] which deals with the bandwidth about k-partite graph

Theorem 10. Let G be a graph of order not less than 5 and $E(G)\geqslant 2$. If $\overline{G}$ has at least two complements,

then

$$S(G)\leqslant p(G)-7+2B(G)$$

Proof By Basic Theorem. $S(G)=2p(G)-(m+n)-2$ for the greatest $H(r;m+n)$ of $\overline{G}$ naturally, $m,n\geqslant 1$ and $r\leqslant 0$ Let $m\geqslant n$

If $m+n=2$ then $G=K_n$

If $m+n=3$, then there are $H(2)$ and $H(1)$, it is clear that $B(G)=p(G)-2$

$S(G)\leqslant 2p(G)-5\leqslant 2B(G)-1$

If $m+n=4$, then $H(m)=k_2=H(n)$ or $H(m)=p_3$ and $H(n)=k_1$, so $B(G)\geqslant p(G)-3$

$S(G)\leqslant 2p(G)-6\leqslant 2B(G)$

If $m+n=5$ then $H(m)=p_3$ and $H(n)=k_2$, or there are $H(4)$ and $H(1)$. So $B(G)\geqslant p(G)-3$ and $S(G)\leqslant 2B(G)-1$

When $m+n\geqslant 6$, by **corollary** 1 and **Lemma** 10,

$$S(G)\leqslant 2p(G)-8\leqslant p(G)-7+2B(G).$$

Corollary 10. $4B(G)\leqslant 2p(G)+S(G)$

Corollary 11. $2[B(G)-B(\overline{G})-2]\leqslant S(G)$

Proof It follows that $B(G)+B(\overline{G})\geqslant p(G)-2$ (to see [3]) and **corollary** 10

Corollary 12. For any $H(r;m+n)$ of $\overline{G}$

$$B(G)\leqslant p(G)-\frac{m+n+2}{4}.$$

Corollary 13.

(1) If $\overline{G}$ has no triangle, then $2B(G)\geqslant P(G)-2$ so that $S(G)\leqslant 4B(G)$.

(2) If $\overline{G}$ is a planar , then $B(G)\geqslant p(G)-5$, so

$$S(G)\leqslant 2B(G)+6.$$

References

[1] J. A. Bondy and U. S. R. Murty, Graph Theory with Applications. The Macmiland Press LTD, 1976

[2] F. Harary, Graph Theory, Addison—Wesley, Reading, Mass, 1969

[3] P. Z. Chinn, J. Chvatolova, A. K Dewdney, and N. E. Gibbs, Bandwidth problem for graphs and matrices—A Survey, J. Graph Theory, Vol. 6(1982)223—254.

On Generalization of the Edmonds-Karp Maximum Flow Algorithm in the Setting of Quasi Pure Networks①

Wei He② and Zhemin Wang③

(*Dept. of Statistics & Operations Research, Fudan University, Shanghai* 200433, *P. R. China*)

Abstract This paper makes a substantial generalization of the Edmonds-Karp maximum flow algorithm in the context of quasi pure networks. The computational complexity of our scheme remains basically the same as theirs.

Key words pure network, quasi pure network, maximum flow

1. Introduction

Combinatorial algorithms for generalized network flow problems are among the major concerns in the modern studies on combinatorial optimization [2][5]. In this paper, we are concerned with making generalization of the Edmonds-Karp maximum flow algorithm [3] in the context of quasi pure networks.

A maximum flow problem can be formulated as:

$$\text{maximize}\{x_{e_0} \mid Ax=0 \text{ and } a\leqslant x\leqslant b\},$$

where $A=[a_{ve}]\in R^{V\times E}$ denotes the node-edge incidence matrix of a certain pure network $N=(V, E)$, V and E are respectively the node and edge sets, $e_0\in E$, $a\in(R\cup\{-\infty\})^E$, $b\in(R\cup\{+\infty\})^E$ and $a\leqslant b$. By means of adding properly into N a certain number of parallel edges and/or splitting some of the edges into pairs of adjacent ones, the maximum flow problem above can be reformulated into a form as:

$$\text{maximize}\{x_{e_0} \mid Ax=0 \text{ and } x\geqslant a\},$$

where $a\in R^E$. In the following discussion, we assume, without loss of generality, the maximum flow problem is set in the latter form. In this paper, we are concerned with the maxi-

① This research was supported by the National Natural Science Foundation of China under grant No. 79370020.

② Assistant professor, Dept. of Statistics & Operations Research, School of Management, Fudan University, Shanghai 200433, P. R. China

③ Professor, Dept. of statistics & Operations Research, School of Management, Fudan University, Shanghai 200433, P. R. China

mum flow problem where the network concerned is a so-called *quasi pure network* i. e. if two nonzero entries happen to be in the same column vector of A, they are not necessarily oppositely signed as is the case of pure network (in [6], a quasi pure network is referred to as a bidirected graph). In this paper, an edge e if only incident to a single node, is called specifically a *source edge*, and an edge e is said to be *positively* (*negatively*) incident to a node v if $a_{ve}=1(-1)$. Two edges e_1 and e_2 are said to be *complementary* if $a_{ve_1}=-a_{ve_2}$ for $v\in V$. For an ordinary maximum flow problem i. e. of which the N is a pure network, and the underlying graph of N is a connected simple graph; the e_0 there is not, by itself, a source edge; however, it is not difficult to verify that if e_0 is splitted into two source edges that they are incident respectively to the nodes which e_0 is incident to, then it is easy to see that the maximum flow problem can be regarded equivalently as one where the maximum is taken over one of these source edges. However, this is not true for N that is generally supposed to be of a quasi pure network.

It is known that the Edmonds-Karp method with respect to a pure network is based on making flow augmentations along shortest augmenting paths in the related residual graph, and this entails that their algorithm is of a strongly polynomial character i. e. the total number of augmentations is at most $O(|V||E|)$ (which is independent of variable bounds), and the running time for each augmentation is bounded by $O(|V|^2)$.

For N that is generally supposed to be of a quasi pure network, no definite results concerning this kind of maximum flow problem have been reported so far, and it is our purpose to have this problem fully dealt with in this paper. First, we study such problem for e_0 that is supposed to be a source edge; basing on this result, we then set to design a primal-dual scheme for the case where e_0 is not a source edge. Our approach is characterized in that the flow augmentation is carried out along so-called restricted augmenting walks in the related residual network, instead of shortest augmenting paths as proposed by Edmonds and Karp. In order to find out such walks (or deny their existence), a valid scheme based on the Breadth First Search (BFS) principle is developed in this paper, of which the running time is bounded by $O(k|V|^2)$, where k denotes the maximum number of parallel edges between pairs of adjacent nodes in the network related. The major result of this paper is that with our scheme the total number of augmentations involved is proved to be bounded by $O(|V||E|)$, which is the same as that given by Edmonds and Karp.

2. Preliminaries

By the general principle of complementary slackness the optimality condition for the maximum flow problem:

$$\text{maximize}\{x_{e_0} \mid Ax=0 \text{ and } x\geqslant a\},$$

with respect to a quasi pure network $N=(V,E)$, can be stated as: A feasible flow x of this problem is a maximum flow if and only if there exists a vector $\pi\in R^V$ such that for $y=A^T\pi$, (1) $y\geqslant 0$ and $y_{e_0}>0$, (2) for $e\neq e_0$, $y_e>0$ implies $x_e=a_e$. We call such π an *optimal node potential vector*. (Indeed, the "if" part of the optimal condition can be easily proved because of the fact that every feasible flow of the problem must be orthogonal to y; and the "only if" part can be proved constructively with the major results obtained in the last two sections of this paper.)

For a feasible flow x of the problem, a *residual network* (related to x) is defined to be a quasi pure network $N(x)$ that: (1) N forms a spanning subnetwork of $N(x)$, (2) the edge set of $N(x)$ is $E\cup\{e^*\,|\,e\in E, x_e>a_e\}$, where e^* and e are complementary.

In the following discussion of this section, e_0 is assumed specifically to be a source edge. An *augmenting* e_0-*walk* W in $N(x)$ is meant for an alternating sequence of edges and nodes of $N(x)$:

$$e_0v_0e_1v_1\cdots e_{i-1}v_{i-1}e_iv_i\cdots e_{t-1}v_{t-1}e_t,$$

where (1) e_0 and e_t are source edges which are incident respectively to v_0 and v_{t-1}, (2) for $i\in\{1,2,\cdots,t-1\}$, e_i is incident to v_{i-1} and v_i, and $v_{i-1}\neq v_i$, (3) *for* $i\in\{0,1,\cdots,t-1\}$, e_i is incident to v_i positively if and only if e_{i+1} is incident to v_i negatively.

It should be noted that nodes, and edges as well, in W are allowed to occur repeatedly. A walk becomes a path, if none of the nodes and edges appear in it repeatedly.

An augmenting e_0-walk is called *closed* if the terminal source edge e_t is e_0 itself.

The following properties can be readily proved by negation.

Property 1 If N is a pure network, then a shortest augmenting e_0-walk in $N(x)$ must be a path.

Property 2 In a shortest augmenting e_0-walk in $N(x)$, any node, and edge as well, can only appear at most twice.

Let W be an augmenting e_0- walk in $N(x)$, a vector $\lambda\in R^E$ is called to be *induced by* W if satisfying: for $e\in E$, $\lambda_e=m_1-m_2$, where m_1 and m_2 denote respectively the times that e and e^* appear in W.

It is not difficult to verify that such λ can act as an augmenting vector of x; if $\lambda\geqslant 0$, then λ is an infinite augmenting vector. For λ induced by W that is not an infinite augmenting one, to augment x along W is meant for to obtain a feasible flow $x'=x+k\lambda$, where k is specifically chosen to be $\min\{\frac{x_e-a_e}{-\lambda_e}\,|\,\lambda_e<0\}$.

From Property 2, follows immediately the property below.

Property 3 Let W be a shortest augmenting e_0-walk among all augmenting e_0-walks which terminate at a given source edge (it could be e_0 itself), then the λ induced by W satisfies:

(1) for $e\in E$, $\lambda_e=0$, ± 1 or ± 2,

(2) $|\{e|\lambda_e \neq 0\}| \leqslant 2|V|$.

For any given source edge e_t, one would like to know whether there is an augmenting e_0-walk that terminates at e_t or not, if such walks do exist, how to produce a shortest walk of such kind. The following Scan-and-Label Rule which works on BFS principle is devised specifically for this purpose.

A *label string* is defined to be a 4-tuple (k,s,v,e), where k, called level-tag, denotes a certain nonnegative integer; s, called sign-tag, represents a certain sign "+" or "−"; v, called node-tag, denotes a certain node; and e, called edge-tag, denotes a certain edge.

With this Scan-and-Label Rule, the operation is carried out, level to level, by attaching to nodes with new label strings through "scanning" label strings unscanned (until all label strings produced are scanned). It should be noted that with this rule label strings attached to a node are not necessarily unique, and if this is so, then there could only be two and their sign-tags are different. In the following, we call a node positively (negatively) labeled, if it is attached with a label string, where the sign-tag is a "+"("−").

The Scan-and-Label Rule is as follows:

Step 0: Let e_0 be incident to, say, v_0 positively (negatively). Attach to v_0 with a label string where the level, sign, and edge-tags are respectively 0, + (−) and e_0(for initiation step the node-tag is left out).

Step 1: Among all unscanned label strings (if existing), let (k,s,v,e) be the one with the smallest level-tag. Suppose this label string is attached to node v', then "*scan*" this label string as follows:

For every nonsource edge e' satisfying:

(1) e' is incident positively to v' if s is a "−" or e' is incident negatively to v' if s is a "+",

(2) e' is incident positively (negatively) to $v''(\neq v')$ that is not yet positively (negatively) labeled,

attach to v'' a new label string $(k+1, +(-), v', e')$.

Go back to Step 1.

The Proposition 2 given below will show that this rule is "valid" in the sense that: For a given source edge e_t that is incident positively (negatively) to v_{t-1} if v_{t-1} is not negatively (positively) labeled after applying the Scan-and-Label Rule, then there happens no augmenting e_0-walk that terminates at e_t; otherwise, a label string with a "−"("+") sign-tag must have been attached to v_{t-1}. Now, by means of backtracing starting from this label string, a shortest augmenting e_0-walk, among all augmenting e_0-walks terminating at e_t, of length $k+2$, where k is the level-tag of this label string, can be finally found out with the help of the knowledge regarding the node and edge-tags in the successive label strings backtraced.

From the Scan-and-Label Rule above, it can be seen that a node could be labeled both positively and negatively; however, there could be no more than one label string with the same kind of sign-tag attached to a single node.

Proposition 1 The runing time of the rule above is bounded by $O(k|V|^2)$, where k denotes the maximum number of parallel edges between pairs of adjacent nodes in the underlying graph of the network considered.

Proof Since the number of label strings produced by this rule is at most $2|V|$, and the running time required for scanning each label string is at most $k|V|$, so the total running time is $O(k|V|^2)$. ■

Proposition 2 The rule above is valid.

Proof It is easy to see that we need only to prove that: for every shortest augmenting e_0-walk among all augmenting e_0-walks terminating at e_t: $e_0v_0e_1v_1 \cdot\cdot\cdot e_{i-1}v_{i-1}e_iv_i \cdot\cdot\cdot e_{t-1} v_{t-1}e_t$, if e_i is, say, positively (negatively) incident to v_i, then a label string, of which the sign-tag is a "+"("−") and the level-tag is i, must have been attached to v_i, for $i=0,1,\cdot\cdot\cdot,h,h+1,\cdot\cdot\cdot,t-1$. This is obvious for $i=0$. Suppose the argument is true for $i=h$. Now, if e_{i+1} is incident to v_{i+1}, say, positively (negatively), then there must be a label string that has been attached to v_{i+1}. This label string should be such one, of which the sign-tag is a "+"("−") and the level-tag is less or equal to $i+1$. However, if the level-tag is less than $i+1$, then by backtracing from this label string, a shorter walk between e_0 and e_t can be found, this of course is contradictory. The proof is completed. ■

Proposition 3 A closed augmenting e_0-walk exists if and only if that: after the Scan-and-Label Rule is applied, there exists at least one node that has been labeled both positively and negatively.

Proof If a closed augmenting e_0-walk exists, then for every shortest closed augmenting e_0-walk there must be a node in it which appears twice. It can then be verified (referring to the proof of Proposition 2) that this node must be labeled both positively and negatively. If a node is attached with two label strings with different kinds of sign-tags, then by backtracing from them respectively, a closed augmenting e_0-walk could be obtained accordingly. This completes the proof. ■

Proposition 4 There exists no augmenting e_0-walk if and only if that: after the Scan-and-Label Rule is applied, every node can only be attached with at most one label string, and if it is positively (negatively) labeled, then there can be no source edge that is incident to this node negatively (positively).

Proof From Proposition 2 and 3, this can be readily proved. ■

Proposition 5 If there exists no augmenting e_0-walk in $N(x)$, then an optimal node potential vector π can be found by means of applying the Scan-and-Label Rule.

Proof According to Proposition 4 that once the Scan-and-Label Rule is applied, every node of N must be attached with at most one label string. Now, let $\pi\in R^V$ be defined as such:

$$\pi_v = \begin{cases} 0, \text{ if } v \text{ is not attached with a label string,} \\ +1, \text{ if } v \text{ is positively labeled,} \\ -1, \text{ if } v \text{ is negatively labeled.} \end{cases}$$

It can be specified that this π is an optimal node potential vector. ■

Form the proposition above and the optimality condition, the corollary below follows:

Corollary If there exists no augmenting e_0-walks in $N(x)$, then x is a maximum flow over e_0.

3. Maximum flows over source edges

In this section, the maximum flow problem is studied under supposition that e_0 is a source edge. In the following discussion, we call an augmenting e_0-walk *restricted* if it is either a shortest closed augmenting e_0-walk, or a shortest augmenting e_0-walk in case that closed augmenting e_0-walk is nonexistent.

Suppose a feasible flow $x=x^0$ is to be augmented successively along a sequence of augmenting e_0-walks: $W_0, W_1, \cdots, W_p$, and $x^1, x^2, \cdots, x^{p+1}$ are respectively the feasible flows thus entail. An edge of N is called an *off-coupled edge* of $N(x^i)$ $(i\neq 0)$ if both e and e^* are members of $N(x^{i-1})$, and among e and e^* only e is a member of $N(x^i)$. A e^* is called an *in-coupled edge* of $N(x^i)$ $(i\neq 0)$ if both e and e^* are members of $N(x^i)$, and among e and e^* only e is a member of $N(x^{i-1})$. Obviously, any edge e of N is an edge of $N(x^i)$. The augmenting e_0-walk sequence is called *independent* if W_k passes no off and in-coupled edge of $N(x^i)$, where $i\leqslant k$.

Property 4 Suppose the augmenting e_0-walk sequence is independent and e^* happens to be an in-coupled edge of $N(x^i)$, then e^* must be an edge of $N(x^j)$ for $j\geqslant i$.

Proof The argument fails only if there is a W_k, $i\leqslant k<j$, which passes e^* first; however this is contradictory to that the sequence concerned is independent. ■

Property 5 Suppose the augmenting e_0-walk sequence is independent, then W_k must be an augmenting e_0-walk in $N(x^j)$ for $j\leqslant k$.

Proof The argument fails only if W_k passes an in-coupled edge of $N(x^i)$ where $j<i\leqslant k$, this of course is contradictory to the supposition that the sequence concerned is independent. ■

The lemmas below are prepared for the proving of the following theorem that plays a key role in this paper.

Lemma 1 Suppose the augmenting e_0-walk sequence is independent, then it follows that $p+1\leqslant |E|$.

Proof It is clear that each augmentation of x^k by W_k brings to $N(x^{k+1})$ at least one off-coupled edge of $N(x^{k+1})$. On the other hand, since the augmenting walk sequence is supposed to be independent; therefore, once an edge becomes an off-coupled edge of a certain $N(x^i)$, it can no longer be an off-coupled edge of another $N(x^j)$. This implies that the total number of distinct edges, of which each acts as an off-coupled edge of a certain $N(x^i)$,, is at least $p+1$. Since the number of distinct edges must not be greater than $|E|$, hence follows $p+1 \leqslant |E|$. ■

Lemma 2 Suppose the augmenting e_0-walk sequence is independent, and is composed of restricted augmenting e_0-walks. We have:

(1) For $i \leqslant j$, if W_i and W_j are both closed or nonclosed augmenting e_0-walks, then the length of W_j must not be less than that of W_i.

(2) For $i \leqslant j$, if W_i is a nonclosed augmenting e_0-walk, then W_j must also be a nonclosed augmenting e_0-walk.

Proof The argument follows immediately from Property 5 and the supposition that the sequence concerned is composed of restricted augmenting e_0-walks. ■

Lemma 3 Let a feasible flow be augmented successively along a sequence of restricted augmenting e_0-walks: $W_0, W_1, \cdots, W_p, W_{p+1}$, where the subsequence $W_0, W_1, \cdots, W_p$ is assumed to be independent. Supposed that $W_0, W_1, \cdots, W_p, W_{p+1}$ is not independent and both W_0 and W_{p+1} are closed or nonclosed augmenting e_0-walks, then the length of W_{p+1} must be greater than that of W_0.

Proof Since $W_0, W_1, \cdots, W_p$ is independent, and $W_0, W_1, \cdots, W_p, W_{p+1}$ is not, hence W_{p+1} must contain at least one off or in-coupled edge of a certain $N(x^i)$. Let $i_0 = \max\{i \mid \text{for } N(x^i) \text{ where } W_{p+1} \text{ contains an off or in-coupled edge of } N(x^i),\ l \leqslant i \leqslant p\}$, let W_{p+1} be: $e_0 v_0 \varepsilon_1^{p+1} v_1^{p+1} \cdots v_{i_1-1}^{p+1} \varepsilon_{i_1}^{p+1} v_{i_1}^{p+1} \cdots \varepsilon_{i_k}^{p+1} v_{i_k}^{p+1} \cdots \varepsilon_t^{p+1}$, where ε_j^{p+1} represents either a certain edge e or e^* that locates in W_{p+1}; and $\varepsilon_{i_k}^{p+1}$ denotes the kth off or in-coupled edge of $N(x^{i_0})$ which W_{p+1} passes in order. It is clear that if $\varepsilon_{i_k}^{p+1}$ is an e (e^*) then W_{i_0-1} must passes e^* (e). Let W_{i_0-1} be: $e_0 v_0 \varepsilon_1^{i_0-1} v_1^{i_0-1} \cdots v_{j_1-1}^{i_0-1} \varepsilon_{j_1}^{i_0-1} v_{j_1}^{i_0-1} \cdots \varepsilon_{s-1}^{i_0-1} v_{s-1}^{i_0-1} \varepsilon_s^{i_0-1}$, where $\varepsilon_{j_1}^{i_0-1}$ is an e^* (e) if $\varepsilon_{i_1}^{p+1}$ is e (e^*), and (1) $v_{j_1-1}^{i_0-1} = v_{i_1}^{p+1}$ and $v_{j_1}^{i_0-1} = v_{i_1-1}^{p+1}$ or (2) $v_{j_1-1}^{i_0-1} = v_{i_1-1}^{p+1}$ and $v_{j_1}^{i_0-1} = v_{i_1}^{p+1}$.

It is not difficult to vindicate that all $\varepsilon_1^{p+1}, \varepsilon_2^{p+1}, \cdots, \varepsilon_{i_1-1}^{p+1}$ are edges of $N(x^{i_0-1})$ (referred to Property 4). Now, suppose that (1) happens. Since W_{i_0-1} is a shortest augmenting e_0-walk among all augmenting e_0-walks terminating at $\varepsilon_s^{i_0-1}$, therefore it can be verified that the number of $\varepsilon_1^{i_0-1}, \varepsilon_2^{i_0-1}, \cdots, \varepsilon_{j_1}^{i_0-1}$ is not greater than that of $\varepsilon_1^{p+1}, \varepsilon_2^{p+1}, \cdots, \varepsilon_{i_1-1}^{p+1}$, hence the number of $\varepsilon_1^{i_0-1}, \varepsilon_2^{i_0-1}, \cdots, \varepsilon_{j_1-1}^{i_0-1}$ is less than that of $\varepsilon_1^{p+1}, \varepsilon_2^{p+1}, \cdots, \varepsilon_{i_1-1}^{p+1}, \varepsilon_{i_1}^{p+1}$; this implies that the following augmenting e_0-walk $W_{p+1}^{(1)}$:

$$e_0 v_0 \varepsilon_1^{i_0-1} v_1^{i_0-1} \cdots v_{j_1-1}^{i_0-1} \varepsilon_{i_1+1}^{p+1} v_{i_1+1}^{p+1} \cdots \varepsilon_{i_2-1}^{p+1} v_{i_2-1}^{p+1} \varepsilon_{i_2}^{p+1} \cdots \varepsilon_t^{p+1}$$

is shorter than W_{p+1}. (Note: All edges which precede $\varepsilon_{i_2}^{p+1}$ in $W_{p+1}^{(1)}$ are edges of $N(x^{i_0-1})$). Now suppose that (2) happens. It can be seen that $e_0 v_0 \varepsilon_1^{i_0-1} v_1^{i_0-1} \cdots v_{j_1-1}^{i_0-1} \varepsilon_{i_1-1}^{p+1} v_{i_1-2}^{p+1} \cdots v_1^{p+1} \varepsilon_1^{p+1} v_0 e_0$ forms a closed augmenting e_0-walk in $N(x^{i_0-1})$. This implies that W_{i_0-1} must be a closed augmenting e_0-walk in $N(x^{i_0-1})$, therefore it can be verified that the number of $\varepsilon_{j_1}^{i_0-1}$, $\varepsilon_{j_1+1}^{i_0-1}$, $\cdots$, $\varepsilon_s^{i_0-1}(=e_0)$ is not greater than that of ε_1^{p+1}, ε_2^{p+1}, $\cdots$, $\varepsilon_{i_1-1}^{p+1}$, hence the number of $\varepsilon_{j_1+1}^{i_0-1}$, $\varepsilon_{j_1+2}^{i_0-1}$, $\cdots$, $\varepsilon_s^{i_0-1}(=e_0)$ is less than that of ε_1^{p+1}, ε_2^{p+1}, $\cdots$, $\varepsilon_{i_1-1}^{p+1}$, $\varepsilon_{i_1}^{p+1}$; this implies that the following augmenting e_0-walk $W_{p+1}^{(1)}$:

$$e_0 v_0 \varepsilon_{s-1}^{i_0-1} \cdots v_{j_1}^{i_0-1} \varepsilon_{i_1+1}^{p+1} v_{i_1+1}^{p+1} \cdots \varepsilon_{i_2-1}^{p+1} v_{i_2-1}^{p+1} \varepsilon_{i_2}^{p+1} \cdots \varepsilon_t^{p+1}$$

is shorter than W_{p+1}. (Note: All edges which precede $\varepsilon_{i_2}^{p+1}$ in $W_{p+1}^{(1)}$ are edges of $N(x^{i_0-1})$).

Apply the above policy repeatedly, we get finally a sequence of augmenting e_0-walks (all of them terminate at ε_t^{p+1}) W_{p+1}, $W_{p+1}^{(1)}$, $\cdot\cdot\cdot$, $W_{p+1}^{(r)}$ that each of them is shorter than the one preceding, and the last augmenting e_0-walk $W_{p+1}^{(r)}$ happens to be an augmenting e_0-walk in $N(x^{i_0-1})$. On the other hand, since it is supposed that both W_0 and W_{p+1} are closed or non-closed augmenting e_0-walks, and each W_k is a restricted augmenting e_0-walk, therefore from the conclusion (2) of Lemma 2, it follows that the length of W_{i_0-1} is not greater than that of $W_{p+1}^{(r)}$, hence is less than that of W_{p+1}. Since, according to the conclusion (1) of lemma 2, the length of W_0 is not greater than that of W_{i_0-1}, this ensures that the length of W_{p+1} is greager than that of W_0. ■

From the lemmas and Property 2 above, follows immediately the important theorem of this paper.

Theorem 1 If feasible flows are kept to be augmented along restricted augmenting e_0-walks, then either an infinite augmenting vector or a maximum flow over e_0 can be obtained in at most $O(|V||E|)$ times of such augmentations.

Remark 1 Suppose that the maximum flow problem over a source edge e_0 is generally formulated in the form of:

$$\text{maximize}\{x_{e_0} \mid Ax=0 \text{ and } a\leqslant x\leqslant b\},$$

where A is the incidence matrix of N, $a\in(R\cup\{-\infty\})^E$, $b\in(R\cup\{+\infty\})^E$ and $a\leqslant b$. Now, according to the general principle of complementary slackness, the optimality condition for the problem above can now be stated as: a feasible flow x is a maximum flow if and only if there exists a vector $\pi\in R^V$ such that for $y=A^T\pi$, (1) $y_{e_0}>0$, (2) for $e\neq e_0, y_e>0$ implies $x_e=a_e$, and (3) $y_e<0$ implies $x_e=b_e$. We call such π an *optimal node potential vector*. In order to obtain a maximum flow x over e_0 together with an optimal node potential vector, we can first transform N (by adding into it a certain number of parallel edges and/or splitting a certain number of the edges into pairs of adjacent ones) into another quasi generalized network $N^1=(V^1,E^1)$ where V^1 contains V, so that the above problem is transformed equivalently to a maximum flow problem that assumes the following standard form:

$$\text{maximize}\{x_{e_0} \mid A^1x=0 \text{ and } x\geqslant a^1\},$$

where $a^1\in R^{E^1}$, and A^1 is the incidence matrix of N^1. Now, apply the method suggested in this section to the solving of this problem. If this results in obtaining an optimal solution together with an optimal node potential vector π^1 relating to N^1, then the optimal solution can be transformed back into a maximum flow of the original problem, and πb can be readily proved to be an optimal node potential vector π relating to N.

Remark 2 It is known that in order to find out a feasible flow (or deny its existence) of a certain maximum flow problem, the Phase I approach of the general linear programming can be adopted; and it is not difficult to see that this doing can be interpreted as solving a maximum flow problem over a source edge e_0, where the network concerned is enlarged, and e_0 together with a certain number of edges and a node v_0 are newly-added-in entities. For this problem a feasible flow is readily in possession.

4. Maximum flows over nonsource edges

In this section, the maximum flow problem is studied under supposition that e_0 is a nonsource edge.

Without loss of generality, we assume that a feasible flow x^0 is already in possession (see to Remark 2). Our scheme is basically based on a primal-dual approach and runs in two stages as follows:

The Initial Stage:

Let e_0 be incident to nodes v' and v'' respectively. First, we "split" e_0 into two source edges e_0' and e_0'' so that they are incident respectively to v' and v'' in the same manner as e_0. Accordingly, we "split" the $x^0_{e_0}$ into two components $x^0_{e'_0}$ and $x^0_{e''_0}$ such that $x^0_{e'_0}=x^0_{e''_0}=x^0_{e_0}$. It is easy to see that the new x^0 remains a feasible flow with respect to this "recasted" N.

With this x^0 and N at hand, the following steps are to be executed.

Step 1: Let x^0 act as the current feasible flow, and apply the method suggested in section 3 above to this "recasted" N for a maximum flow over e_0'. If this ends up in obtaining a maximum flow, then turn to the next stage below; otherwise, an infinite augmenting vector λ' with respect to e_0' is obtained. We assume that this λ' has been so normalized that $\lambda'_{e'_0}=1$. If $\lambda'_{e''_0}=0$, then turn to Step 2; otherwise, it can be verified that $\lambda'_{e''_0}$ must be equal to $\lambda'_{e'_0}$. Now, recast λ' by merging $\lambda'_{e'_0}$ and $\lambda'_{e''_0}$ into a single component λ'_{e_0}, and restore N to its original form (i. e. replace e_0' and e_0'' by e_0), then it is easy to see that this recasted λ' turns out an infinite augmenting vector with respect to e_0; thus the maximum flow problem with respect to e_0 is unbounded.

Step 2: With x^0 as the current feasible flow, apply the method suggested in section 3 above to this "recasted" N for a maximum flow over e_0''. Suppose this ends up in obtaining a maximum flow, then turn to the next stage below; otherwise, an infinite augmenting vector λ'' with respect to e_0'' is obtained. We assume that this λ'' has been so normalized that $\lambda''_{e''_0}=1$.

2.1: If $\lambda''_{e'_0}\neq 0$, then it can be verified that $\lambda''_{e'_0}$ must be equal to $\lambda''_{e''_0}$. Now, recast λ'' by merging $\lambda''_{e'_0}$ and $\lambda''_{e''_0}$ into a single component λ''_{e_0}, and restore N to its original form (i. e. replace e_0' and e_0'' by e_0), then it is easy to see that this recasted λ'' turns out an infinite augmenting vector with respect to e_0; thus the maximum flow problem with respect to e_0 is unbounded.

2.2: If $\lambda''_{e'_0}=0$, then let $\lambda=\lambda'+\lambda''$. Now, recast λ by merging $\lambda_{e'_0}$ and $\lambda_{e''_0}$ into a single component λ_{e_0}, and restore N to its original form (i. e. replace e_0' and e_0'' by e_0), then it is easy to see that this recasted λ turns out an infinite augmenting vector with respect to e_0; thus the maximum flow problem with respect to e_0 is unbounded.

The Primal-Dual Stage:

Without loss of generality, we assume that the operation is switched over from Step 1 of the above stage. First, it should be noted that for the current feasible flow x, x'_{e_0} must be equal or greater than $x_{e''_0}$, this is because that the augmentations of the feasible flows are supposed to be carried out along restricted augmenting e_0'-walks. Let π denote an optimal node potential vector which is obtained as an outcome brought about by Step 1 of the stage above. Now, add into the current network an additional source edge e_0^1 so that e_0^1 and e_0' are complementary, and restore e_0 by removing e_0' and e_0'' from the network concerned. Accordingly, recast the x obtained in Step 1 of the stage above by adding into it a new component $x_{e_0^1}$ which attains currently the value of $-(x_{e'_0}-x_{e''_0})$ (Note: $x_{e_0^1}$ is now nonpositive), and merging $x_{e'_0}$ and $x_{e''_0}$ into a single component x_{e_0} which attains currently the value of $x_{e''_0}$. Now, let P_1 denote the following maximum flow problem over e_0.

$$
\begin{array}{lll}
P_1 & \text{maximize} & x_{e_0} \\
 & \text{subject to} & A^1x=0, \\
 & & x_E\geqslant a, \\
 & & x_{e_0^1} \text{ is parametrically fixed,}
\end{array}
$$

where A^1 is the incidence matrix of N^1 which differs from A in that it possesses an additional column vector which corresponds to e_0^1. Obviously, the current x can be regarded as a feasible flow of P_1; besides, since now $y=A^{1T}\pi$ and x satisfy the optimality condition, therefore x can also be regarded as a maximum flow of P_1. Clearly, if $x_{e_0^1}=0$, then x induces a maximum flow over e_0 of the original problem. Now, with x, y and P_1 at hand, the following two steps are to be executed.

Step 1: Let $E_1=\{e|e(\neq e_0, e_0^1)\in E, y_e=0\}$,

$E_2=\{e|e(\neq e_0, e_0^1)\in E, y_e>0\}$,

and let d be a vector of R^{E_2}, of which d_e attains the current value of x_e. Now, let P_2 denote the following maximum flow problem over e_0^1.

$$\begin{array}{lll} P_2 & \text{maximize} & x_{e_0^1} \\ & \text{subject to} & A^1x=0, \\ & & x_{E_1}\geqslant a_{E_1}, \\ & & x_{E_2}=d, \\ & & x_{e_0^1}\leqslant 0. \end{array}$$

Obviously, the current x can be regarded as a feasible flow of P_2. Solve P_2 by means of the method suggested in section 3 above (refer to Remark 1 given in section 3). It should be noted that every updated feasible flow occurred during this process must be a maximum flow of P_1, because it can be specified that such feasible flow and the current y satisfy invariably the optimality condition. Now, if the optimal value of $x_{e_0^1}$ reaches at zero, then the current x induces a maximum flow over e_0 of the original problem; otherwise, let the correspondingly produced optimal node potential vector be π^1, and turn to Step 2 below.

Step 2: Let $y^1=A^{1T}\pi^1$ and $E_3=\{e|e(\neq e_0, e_0^1)\in E, y_e^1<0\}$. Since the original problem is supposed to be feasible, therefore, it can be verified that E_3 must be nonempty. Reset y: $=y+k_0y^1$ where $k_0=\min\{\frac{y_e}{-y_e^1}|e\in E_3\}$; (Note: It can be specified that the updated y and the current x still staisfy the optimality condition.) and turn back to Step 1.

Theorem 2 With the scheme above the total number of augmentations of feasible flows is at most $O(|V||E|)$, while the total times of iterations of y is bounded by $O(|V|)$.

Proof It can be specified that with the three arguments given below, the former part of this theorem can be derived from the first two, while the latter from the third.

1. For each y occurred, y_{e_0} remains constant, and $y_{e_0^1}$ increases for each iteration of y.
2. For each augmenting vector λ induced by the augmenting e_0^1-walk when maximizing $x_{e_0^1}$ (where λ is assumed to have been normalized that $\lambda_{e_0^1}=1$),
   (1) $\lambda_{e_0}y_{e_0}+\lambda_{e_0^1}y_{e_0^1}=0$ holds invariably for every y occurred.
   (2) The value of λ_{e_0} can only be 0, $\frac{1}{2}$, or 1 (due to Property 3).
3. If the value of $x_{e_0^1}$ stays unimproved for two consecutive iterations of y, the $|\{v|\pi_v^1\neq 0$ for $y^1=A^{1T}\pi^1\}|$ must increase by one at least.

Of these arguments, the first two are easy to be verified, while the third can be justified by investigating into the intrinsic features of the scheme proposed. ■

Remark 3 Based on the results gained in this paper, a primal-dual plot can be developed [1][4] for the solving of the optimal circulation problem: maximize $\{c^T x \mid Ax=0 \text{ and } a \leq x \leq b\}$, where A is the incidence matrix of a quasi pure network N, and $c \in R^E$. By means of scaling techniques [3][7], this plot can be improved into a polynomial scheme; and, if the Tardos' approach [8] is properly exercised furthermore, a strongly polynomial scheme, which can be regarded as a substantial extension of the well-known Tardos' algorithm for optimal circulations in pure networks, can be finally obtained.

References

[1] Bland, R. G. and Jensen, D. L. On the Computational Behavior of a Polynomial-Time Network Flow Algorithm. Math. Prog. 54, 1 1-39. 1992.

[2] Cohen, E. and Megiddo, New Algorithms for Generalized Network Flows. Math. Prog. 64, 3 325-336. 1994.

[3] Edmonds, J. and Karp, R. M. Theoretical Improvements in Algorithmic Efficiency for Network Flow Problems. J. Assoc. Comput. Math. 19 248-264. 1972.

[4] Fulkerson, D. R. An Out-of-Kilter Algorithm for Minimal Cost Flow Problems. J. SIAM. 9 18-27. 1961.

[5] Goldberg, A. V., Plotkin, S. A. and Tardos, E. Combinatorial Algorithms for the Generalized Circulation Problem. Math. Oper. Res. 16 351-381. 1991.

[6] Lawler, E. L. Combinatorial Optimization: Networks and Matroids. Holt, Reinhart, and Winston, New York. 1976.

[7] Röck, H. Scaling Techniques for Minimal Cost Network Flows. In: V. Page, ed., Discrete Structures and Algorithms (Carl Hansen, Munich). 1980.

[8] Tardos, E. A Strongly Polynomial Minimum Cost Circulation Algorithm. Combinatorica 5, 3 247-255. 1985.

REPRESENTATION OF VARIOUS SPECIAL POLYNOMIALS VIA THE CYCLE INDICATOR OF SYMMETRIC GROUP

L.C. Hsu
(Dalian Univ. of Tech., China)

Peter J.S. Shiue
(Univ. of Nevada LasVegas, U.S.A.)

Abstract Here investigated is a kind of general procedures that can be used to express various classical special polynomials in the form of the cycle indicator of symmetric group.

Key words Faa di Bruno's formula, set of partitions, cycle indicator, C_n-type formula.

1. Intorudction

The cycle indicator C_n of the symmetric permutation group of n letters, is known as a useful tool in Enumerative Combinatorics, which may be written in the form

$$C_n(t_1, t_2, \cdots, t_n) = \sum_{\sigma(n)} \frac{n!}{k_1!k_2!\cdots k_n!} \left(\frac{t_1}{1}\right)^{k_1} \left(\frac{t_2}{2}\right)^{k_2} \cdots \left(\frac{t_n}{n}\right)^{k_n}, \tag{1}$$

where the sum is over all non-negative integral solutions of k_i's for the equation $k_1+2k_2+\cdots+nk_n=n$, or what is the same thing, over the set $\sigma(n)$ of all partitions of n. Usually, $\sigma(n)$ may be denoted by

$$\sigma(n) = \{1^{k_1}2^{k_2}\cdots n^{k_n}\}$$

in which $k = k_1 + k_2 + \cdots + k_n$ denotes the number of parts of the partition $1^{k_1}2^{k_2}\cdots n^{k_n}$ of n.

Clearly, the famous identity due to Cauchy

$$\sum_{\sigma(n)} \frac{1}{k_1!k_2!\cdots k_n!1^{k_1}2^{k_2}\cdots n^{k_n}} = 1 \tag{2}$$

just corresponds to the case $n! = C_n(1, 1, \cdots, 1)$. Also, a well-known identity for the Bell numbers if given by

$$B_n = C_n\left(\frac{1}{0!}, \frac{1}{1!}, \cdots, \frac{1}{(n-1)!}\right), \tag{3}$$

where the numbers B_n are generated by the function

$$\exp(e^t - 1) = \sum_{n\geq 0} B_n \frac{t^n}{n!}. \tag{4}$$

The object of this note is to show that the cycle indicator C_n can be employed in a unified manner to express various special polynomials, including the Laguerre polynomials $L_n^{(\alpha)}(z)$, the Gegenbauer polynomials $C_n^{(\lambda)}(z)$, Horadam-Mahon's $m-$th convolution of Pell polynomials $P_{n+1}^{(m)}(z)$, and the Jacobi polynomials $F(-n, \beta, \gamma, z)$, etc. Several cases will be mentioned as examples.

2. A Basic Lemma

We will present a lemma in terms of the theory of formal power series. In our statement of the lemma, the formal logarithmic function $f(x) = \log x$ is defined by the formal expansion of $\log[1+(x-1)]$ in powers of $(x-1)$, so that the formal exponential e^x and $\log x$ are compositional inverse of each other, as may be verified by formal substitutions.

Lemma *Let $\phi(t)$ be a formal power series over the complex field* $\mathbf{C}$ *with $\phi(0) > 0$, and let is be written in the form*

$$\phi(t) = \sum_{n\geq 0} \left[{\phi \atop n}\right] t^n. \tag{5}$$

Suppose that $\psi(t)$ is defined by the formal logarithm $\psi(t) = \log \phi(t)$ and that it may be expanded in the form

$$\psi(t) = \sum_{n\geq 0} \left[{\psi \atop n}\right] t^n. \tag{6}$$

Then there holds the C_n−type formula

$$\left[{\phi \atop n}\right] = \frac{\phi(0)}{n!} C_n\left(1\left[{\psi \atop 1}\right], 2\left[{\psi \atop 2}\right], \cdots, n\left[{\psi \atop n}\right]\right). \tag{7}$$

Proof Denote by $f(x) = \log x$ and $f^{-1}(x) = e^x$. Cleraly $\log \phi(t) = \log \phi(0) + \log(\phi(t)/\phi(0))$, so that a formal expansion of $\psi(t)$ like (6) can be obtained. Now, applying Faa di Bruno's formula (cf. [1]) to the composite function $\phi(t) = f^{-1} \circ (f \circ \phi)(t)$ with $D_x = d/dx$ denoting the formal differential operator, we find

$$\begin{aligned}
\left[{\phi \atop n}\right] &= \sum_{\sigma(n)} \left[D_x^k f^{-1}(x)\right]_{x=\psi(0)} \prod_{i=1}^{n} \frac{1}{k_i!} \left[{\psi \atop i}\right]^{k_i} \\
&= \sum_{\sigma(n)} [e^x]_{x=\log\phi(0)} \frac{\left[{\psi \atop 1}\right]^{k_1} \left[{\psi \atop 2}\right]^{k_2} \cdots \left[{\psi \atop n}\right]^{k_n}}{k_1! k_2! \cdots k_n!} \\
&= \frac{\phi(0)}{n!} C_n\left(1\left[{\psi \atop 1}\right], 2\left[{\psi \atop 2}\right], \cdots, n\left[{\psi \atop n}\right]\right). \qquad \square
\end{aligned}$$

Evidently, (2) and (3) follow easily from (7) by taking $\phi(t) = 1/(1-t)$ and $\phi(t) = \exp(e^t - 1)$ respectively.

Note that if $\phi(t)$ is a function of C^∞ analytic in $t = 0$ with $\phi(0) > 0$, then both the coefficients $\left[{\phi \atop n}\right]$ and $\left[{\psi \atop n}\right]$ contained in (5) and (6) respectively may also be determined via Taylor's series expansion. Certainly, the C_n−type formula (7) can be applied, in principle, to any analytic function or any formal power series $\phi(t)$ provided that the coefficients $\left[{\psi \atop n}\right]$

of the function $\psi(t) = \log \phi(t)$ can be computed explicitly. Actually this is the case when $\phi(t)$ is taken to be the generating functions of those special polynomials as mentioned at the end of §1.

3. Various Expressions

Recall that the polynomials $L_n^{(\alpha)}(z), C_n^{(\lambda)}(z), P_{n+1}^{(m)}(z)$ and $F(-n,\beta,\gamma,z)$ are given respectively by the following generating functions

$$(1-t)^{-(\alpha+1)}\exp(\frac{zt}{t-1}) = \sum_{n\geq 0} L_n^{(\alpha)}(z)t^n, \tag{8}$$

$$(1-2zt+t^2)^{-\lambda} = \sum_{n\geq 0} C_n^{(\lambda)}(z)t^n, \tag{9}$$

$$(1-2zt-t^2)^{-(m+1)} = \sum_{n\geq 0} P_{n+1}^{(m)}(z)t^n, \tag{10}$$

$$(1-t)^{\beta-\gamma}[1-(1-z)t]^{-\beta} = \sum_{n\geq 0} \frac{(\gamma)_n}{n!}F(-n,\beta,\gamma,z)t^n, \tag{11}$$

where $\alpha > -1, \lambda \neq 0, m \in \mathbf{Z}_0$ (set of non-negative integers) and $(\gamma)_n = \gamma(\gamma+1)\cdots(\gamma+n-1)$, $\gamma \neq$ negative integer or zero.

Note that the particular cases of (9) with $\lambda = 1/2$ and $\lambda = 1$

$$C_n^{(\frac{1}{2})}(z) = \varphi_n(z), \quad C_n^{(1)}(z) = U_n(z)$$

give the familiar Legendre polynomial and Chebyshev polynomial respectively. Also, $P_{n+1}^{(m)}(\frac{1}{2}) = F_{n+1}^{(m)}$ just represents Hoggatt's $m-$th convolution sequence of Fibonacci numbers. In particular, $F_{n+1}^{(0)} = F_{n+1}$ is the $(n+1)$th Fibonacci number which may be generated by the function

$$\frac{1}{1-t-t^2} = \sum_{n\geq 0} F_{n+1}t^n. \tag{12}$$

What we wish to present and prove are the following four propositions.

Proposition 1 *For any real number $\alpha > 0$ we have*

$$L_n^{(\alpha-1)}(z) = \frac{1}{n!}C_n(\alpha - z, \alpha - 2z, \cdots, \alpha - nz). \tag{13}$$

Proposition 2 *For any real number $\lambda \neq 0$ we have*

$$C_n^{(\lambda)}(z) = \frac{1}{n!}C_n(X_1, X_2, \cdots, X_n), \tag{14}$$

where $X_j = \lambda\left((z+\Delta)^j + (z-\Delta)^j\right)$ with $\Delta = \sqrt{z^2-1}$.

Proposition 3 *For any integer $m \geq 0$ we have*

$$P_{n+1}^{(m)}(z) = \frac{1}{n!} C_n(Y_1, Y_2, \cdots, Y_n), \tag{15}$$

where $Y_j = (m+1)\left((z+\Delta)^j + (z-\Delta)^j\right)$ *with* $\Delta = \sqrt{z^2+1}$.

Proposition 4 *For any real number $\beta \neq 0$ and $\gamma \neq 0$ or negative integer we have*

$$(\gamma)_n F(-n, \beta, \gamma, z) = C_n(Z_1, Z_2, \cdots, Z_n), \tag{16}$$

where $Z_j = \gamma - \beta + \beta(1-z)^j$.

These propositions may be proved using the basic lemma in a unified manner. The first step is to write down the following functions that can generate the special functions involved in (13), (14), (15) and (16) respectively,

(*i*) $$\phi(t) = (1-t)^{-\alpha} \exp(\frac{zt}{t-1}), \quad (\alpha > 0)$$

(*ii*) $$\phi(t) = (1 - 2zt + t^2)^{-\lambda},$$

(*iii*) $$\phi(t) = (1 - 2zt - t^2)^{-(m+1)},$$

(*iv*) $$\phi(t) = (1-t)^{\beta-\gamma}[1 - (1-z)t]^{-\beta}.$$

Then, in accordance with the C_n-type formula (7) we have to compute the coefficients $[{\psi \atop j}]$ of t^n involved in each of the power series expansion of $\psi(t) = \log \phi(t)$.

For the case (i) we have $\phi(0) = 1$ and $\psi(t) = -\alpha \log(1-t) + zt/(t-1)$. Consequently we find $[{\psi \atop j}] = \alpha/j - z$, $(j = 1, 2, \cdots, n)$. Hence (13) may be inferred from (7).

For the case (ii), note that

$$1 - 2zt + t^2 = (1 - (z+\Delta)t)(1 - (z-\Delta)t), \ \Delta = \sqrt{z^2 - 1}.$$

Thus it follows that

$$\begin{aligned} \psi(t) &= -\lambda[\log(1-(z+\Delta)t) + \log(1-(z-\Delta)t)] \\ &= \lambda \sum_{n\geq 1} \frac{(z+\Delta)^n + (z-\Delta)^n}{n} t^n, \end{aligned}$$

$$[{\psi \atop j}] = \frac{\lambda}{j}[(z+\Delta)^j + (z-\Delta)^j], \ (j = 1, 2, \cdots, n).$$

Hence (14) is implied by (7).

The verification of (15) is entirely similar to that of (14) and may be omitted.

Finally, let us consider the function displayed in (iv). Clearly we have

$$\begin{aligned}\psi(t) &= -(\gamma-\beta)\log(1-t)-\beta\log[(1-(1-z)t] \\ &= \sum_{n\geq 1}\frac{(\gamma-\beta)+\beta(1-z)^n}{n}t^n.\end{aligned}$$

Hence (16) is also implied by (7) in view of (11). □

Here let us mention in passing that an expression equivalent to (13) was obtained as an example in our recent paper [6]. Also it has been shown that the explicit formula given by (13) can be used to construct an asymptotic expansion of $L_n^{(\lambda\alpha)}(\mu z)$ for $\lambda\uparrow\infty,\mu\uparrow\infty$ and $n\uparrow\infty$ with $n=o(|\lambda+\mu|^{1/2})$, (cf. [6]). Most likely, the C_n-type formulas (14), (15) and (16) may be also applicable to finding some asymptotic expansions.

4. Some Special Examples

In what follows we will mention some special cases of the Propositions 2 and 3 as examples.

Example 1 As an immediate consequence of Proposition 2, we see that the Legendre polynomial $\varphi_n(x)$ and the Chebyshev polynomial $U_n(x)$ have the following representations

$$\varphi_n(x) = \frac{1}{n!}C_n(\frac{(z+\Delta)^1+(z-\Delta)^1}{2},\cdots,\frac{(z+\Delta)^n+(z-\Delta)^n}{2}), \tag{17}$$

$$U_n(x) = \frac{1}{n!}C_n\left((z+\Delta)^1+(z-\Delta)^1,\cdots,(z+\Delta)^n+(z-\Delta)^n\right), \tag{18}$$

respectively, where $\Delta=\sqrt{z^2-1}$.

Example 2 Recall that the Lucas polynomials $V_n(x)$ are usually defined by the second-order linear recurrence relations

$$V_n(x)=xV_{n-1}(x)+V_{n-2}(x),\ (V_0(x)=2,V_1(x)=x).$$

The Binet form of $V_n(x)$ is given by $V_n(x)=\alpha^n+\beta^n$, where

$$\alpha=\frac{1}{2}(x+\Delta^*),\ \beta=\frac{1}{2}(x-\Delta^*),\ \Delta^*=\sqrt{x^2+4}.$$

Let us take $z=x/2$ in (15). Then we may infer from Proposition 3 the following

$$P_{n+1}^{(m)}(\frac{x}{2})=\frac{1}{n!}C_n\left((m+1)V_1(x),\cdots,(m+1)V_n(x)\right). \tag{19}$$

In fact, Y_j contained in (15) may be written as

$$\begin{aligned}Y_j &= (m+1)[\left(\frac{x}{2}+\sqrt{(\frac{x}{2})^2+1}\right)^j+\left(\frac{x}{2}-\sqrt{(\frac{x}{2})^2+1}\right)^j] \\ &= (m+1)(\alpha^j+\beta^j)=(m+1)V_j(x).\end{aligned}$$

Hence (19) follows from (15) as a consequence.

Example 3 Taking $m = 0$ in (19) we obtain a relation connecting the Pell polynomial with the Lucas polynomials, viz.

$$P_{m+1}(\frac{x}{2}) = \frac{1}{n!}C_n(V_1(x), V_2(x), \cdots, V_n(x)). \tag{20}$$

Moreover, for $x = 1$ we obtain from (19) the following

$$F_{n+1}^{(m)} = P_{n+1}^{(m)}(\frac{1}{2}) = \frac{1}{n!}C_n\left((m+1)L_1, \cdots, (m+1)L_n\right), \tag{21}$$

where $L_j = V_j(1)$ just stands for the j−th Lucas number, and $F_{n+1}^{(m)}$ is Hoggatt's m−th convolution of Fibonacci number sequence.

Example 4 When $x = 1$ and $m = 0$ we see that (19) implies the following, by comparing (12) with (10),

$$F_{n+1} = \frac{1}{n!}C_n(L_1, L_2, \cdots, L_n). \tag{22}$$

This shows that the Fibonacci number F_{n+1} can be expressed in the form of the cycle indicator C_n using the Lucas numbers.

References

[1] L. Comtet, *Advanced Combinatorics*, Dordrecht, 1974.

[2] P. Filipponi & A.F. Horadam, *Derivative sequences of Fibonacci and Lucas polynomials*, in *Applications of Fibonacci Numbers*, Vol.4, pp.99–108, Edited by G.E. Bergum, A.N. Philippou and A.F. Horadam, Dordrecht: Kluwer, 1991.

[3] V.E. Hoggatt & M. Bicknell-Johnson, *Fibonacci convolution sequences*, The Fibonacci Quarterly, **15:2**(1977), 117–122.

[4] A.F. Horadam & J.M. Mahon, *Convolution for Pell polynomials*, in *Fibonacci Numbers and Their Applications*, pp.55–80, Edited by A.N. Philippou, G.E. Bergum and A.F. Horadam, Dordrecht, 1986.

[5] L.C. Hsu, *Finding some strange identities via Faa di Bruno's formula*, J. Math. Res. & Exposition, **13**(1993), 159–165.

[6] L.C. Hsu, *Certain asymptotic expansions for the Laguerre polynomials and Charlier polynomials*, to appear in Approximation Theory and Its Applications, 1995.

[7] W. Magnus, F. Oberhettinger & R.P. Soni, *Formulas and Theorems for the Special Functions of Mathematical Physics*, 3rd Edition, Springer-Verlag, Berlin, New York, 1966.

[8] J. Riordan, *An Introduction to Combinatorial Analysis*, Wiley, 1958.

AMS Classification Numbers: 33C25, 11B39, 11B83

LATTICES GENERATED BY ORBITS OF SUBSPACES UNDER FINITE CLASSICAL GROUPS

Yuan-ji Huo

Department of Mathematics, Zhangjiakou Teacher's College

Zhangjiakou, Hebei 075028, China

and

Zhe-xian Wan

Institute of Systems Science, Chinese Academy of Sciences, Beijing 100080, China

1. Introduction

Let $\mathbb{F}_q$ be the finite field with q elements, where q is a prime power,

$$\mathbb{F}_q^{(n)} = \{(x_1, x_2, \cdots, x_n) | x_i \in \mathbb{F}_q\}$$

be the n-dimensional row vector space over $\mathbb{F}_q$, and G_n be one of the classical groups of degree n over $\mathbb{F}_q$, i.e., $G_n = GL_n(\mathbb{F}_q), Sp_{2\nu}(\mathbb{F}_q)$ (where $n = 2\nu$), $U_n(\mathbb{F}_q)$(where q is a square),$O_{2\nu+\delta}(\mathbb{F}_q)$ (where $n = 2\nu + \delta$ and $\delta = 0$, 1 or 2), or $Ps_{2\nu+\delta}(\mathbb{F}_q)$ (where $q = 2^k$, $n = 2\nu + \delta$, and $\delta = 1, 2$). Then G_n acts on $\mathbb{F}_q^{(n)}$ as follows:

$$\begin{array}{ccc} \mathbb{F}_q^{(n)} \times G_n & \longrightarrow & \mathbb{F}_q^{(n)} \\ ((x_1, x_2, \cdots, x_n), T) & \longmapsto & (x_1, x_2, \cdots, x_n)T \end{array}$$

This action induces an action on the set of subspaces of $\mathbb{F}_q^{(n)}$, i.e., if P is an m-dimensional space of $\mathbb{F}_q^{(n)}$, then it is carried into the m-dimensional subspace PT under $T \in G_n$. The set of subspaces of $\mathbb{F}_q^{(n)}$ is partitioned into orbits under G_n.

Let $\mathcal{M}$ be any orbit of subspaces under G_n. Denote by $\mathcal{L}$ the set of subspaces which are intersections of subspaces of $\mathcal{M}$. We order $\mathcal{L}$ by reverse inclusion and agree that the intersection of the empty set of subspase of $\mathbb{F}_q^{(n)}$ is $\mathbb{F}_q^{(n)}$. Then $\mathcal{L}$ is a finite geometric lattice with $\mathbb{F}_q^{(n)}$ as its least element and $\mathcal{M}$ as its set of atoms. We define the characteristic polynomial of $\mathcal{L}$ by

$$\chi(\mathcal{L}, t) = \sum_{P \in \mathcal{L}} \mu(\mathbb{F}_q^{(n)}, P) t^{\dim P}$$

where μ is the Möbius function of $\mathcal{L}$.

It is natural to study the various geometric lattices $\mathcal{L}$ under a given classical group G_n. At first we study the inclusion relation between different lattices $\mathcal{L}$. Then we deduce a characterization of the subspaces contained in a given lattice $\mathcal{L}$. Finally we compute the characteristic polynomial of $\mathcal{L}$.

Furthermore, when $G_n = O_{2\nu+\delta}(\mathbb{F}_q)$ and q is odd, we choose a fixed non-square element $z \in \mathbb{F}_q$, and let

$$S_{2\nu+\delta,\Delta} = \begin{pmatrix} 0 & I^{(\nu)} & \\ I^{(\nu)} & 0 & \\ & & \Delta \end{pmatrix},$$

where

$$\Delta = \begin{cases} \phi & \text{if } \delta = 0, \\ (1) \text{ or } (z) & \text{if } \delta = 1, \\ \begin{pmatrix} 1 & \\ & -z \end{pmatrix} & \text{if } \delta = 2. \end{cases}$$

A subspace P of $\mathbb{F}_q^{(n)}$ is of rank r if $PS_{2\nu+\delta,\Delta}\,{}^tP$ is of rank r. If $PS_{2\nu+\delta,\Delta}\,{}^tP$ is nonsingluar, then P is called a non-isotropic subspace. Given m and r, where $0 \leq m \leq 2\nu+\delta$ and $0 \leq r \leq m$, let $\mathcal{M}_1$ be the set of subspaces of dimension m and rank r. Denote by $\mathcal{L}_1$ the set of subspaces which are intersections of subspaces of $\mathcal{M}_1$. We order $\mathcal{L}_1$ by reverse inclusion and agree that the intersection of the empty set of subspace of $\mathbb{F}_q^{(n)}$ is $\mathbb{F}_q^{(n)}$. Then $\mathcal{L}_1$ is also a finite geometric lattice. When $G_n = O_{2\nu+\delta}(\mathbb{F}_q)$ and q is even, and $G_n = Ps_{2\nu+\delta}(\mathbb{F}_q)$, we can define the lattices $\mathcal{L}_1$ in a similar way. The same problems as above for $\mathcal{L}$ can also be studied for $\mathcal{L}_1$.

In 1985 P. Orlik and L. Solomon[3] gave the results in the following cases

(i) $G_n = U_n(\mathbb{F}_q)(q = q_0^2)$, $\mathcal{M}$ is the orbit of $(n-1)$-dimensional nonisotropic subspaces.

(ii) $G_n = O_{2\nu+\delta,\delta}(\mathbb{F}_q)(n = 2\nu+\delta$, q is a power of odd prime), $\mathcal{M}_l$ is the set of $(n-1)$-dimensional nonisotropic subspaces.

In 1991-1992 D. Chen and Z. Wan[2,4] gave the results in the following cases

(iii) $G_n = O_{2\nu+\delta}(\mathbb{F}_q)(n = 2\nu+\delta$, q is a power of 2), $\mathcal{M}_l$ is the set of $(n-1)$-dimensional nonsingular subspaces.

(iv) $G_n = Sp_{2\nu}(\mathbb{F}_q)(n = 2\nu)$, $\mathcal{M}$ is the orbit of $2(\nu-1)$-dimensional nonisotropic subspaces.

In 1992- 1995 Y. Huo, Y. Liu, and Z. Wan[10-12] gave the results of the general $\mathcal{M}$, and Y. Huo and Z. Wan[13,14,17,18] gave the results of the general $\mathcal{M}_1$.

2. Main results

Take the case of $G_n = O_{2\nu+\delta}(\mathbb{F}_q)$ and q is odd as an example.

Let P be an m-dimensional subspace of $\mathbb{F}_q^{(n)}$. We use the same symbol P to denote any $m \times (2\nu+\delta)$ matrix whose rows form a basis of the subspace P. It is well-known that $PS_{2\nu+\delta,\Delta}\,{}^tP$ is cogredient to a symmetric matrix of the form

$$\begin{pmatrix} 0 & I^{(s)} & & \\ I^{(s)} & 0 & & \\ & & \Gamma & \\ & & & I^{(m-2s-\gamma)} \end{pmatrix},$$

where

$$\Gamma = \begin{cases} \phi, & \text{if } \gamma = 0, \\ (1) \text{ or } (z), & \text{if } \gamma = 1, \\ \begin{pmatrix} 1 & \\ & -z \end{pmatrix}, & \text{If } \gamma = 2. \end{cases}$$

Then we say that P is a subspace of type $(m,\ 2s+\gamma,\ s,\ \Gamma)$ with respect to $S_{2\nu+\delta,\Delta}$ in $\mathbb{F}_q^{(2\nu+\delta)}$. Denote the set of all subspaces of type $(m,\ 2s+\gamma,\ s,\ \Gamma)$ with respect to $S_{2\nu+\delta,\Delta}$ in $\mathbb{F}_q^{(2\nu+\delta)}$ by $\mathcal{M}(m,\ 2s+\gamma,\ s,\ \Gamma;\ 2\nu+\delta,\ \Delta)$, then $\mathcal{M}(m,\ 2s+\gamma,\ s,\ \Gamma;\ 2\nu+\delta,\ \Delta)$ is an orbit of subspaces under $O_{2\nu+\delta}(\mathbb{F}_q)$. Denote by $\mathcal{L}(m,\ 2s+\gamma,\ s,\ \Gamma;\ 2\nu+\delta,\ \Delta)$ the set of intersections of subspaces of $\mathcal{M}(m,\ 2s+\gamma,\ s,\ \Gamma;\ 2\nu+\delta,\ \Delta)$. We have

THEOREM 1. *Let $n = 2\nu + \delta \geq 1$. Assume that $(m, 2s+\gamma, s, \Gamma)$ satisfies*

$$2s + \gamma \leq m \leq \begin{cases} \nu + s + \min(\gamma, \delta), & \text{if } \gamma \neq \delta, \text{ or } \gamma = \delta \text{ and } \Gamma = \Delta, \\ \nu + s, & \text{if } \gamma = \delta = 1 \text{ and } \Gamma \neq \Delta, \end{cases} \tag{1}$$

TABLE 1

δ	γ	γ_1	Δ	Γ	Γ_1	$\mathbb{F}_q$	m_1	s_1	t
0	1	0	ϕ	1 or z	ϕ	$\mathbb{F}_3$	$m-1-t$	$s-t$	$0 \leq t \leq s$
0	0	1	ϕ	ϕ	1 or z	$\mathbb{F}_q$	$m-1-t$	$s-1-t$	$0 \leq t \leq s-1$
0	0	2	ϕ	ϕ	D	$\mathbb{F}_q$	$m-2-t$	$s-2-t$	$0 \leq t \leq s-2$
1	1	0	1 (or z)	1 (or z)	ϕ	$\mathbb{F}_q$	$m-1-t$	$s-t$	$0 \leq t \leq s$
1	2	1	1 (or z)	D	1 (or z)	$\mathbb{F}_3$	$m-1-t$	$s-t$	$0 \leq t \leq s$
1	0	1	1 (or z)	ϕ	1 (or z)	$\mathbb{F}_3$	$m-1-t$	$s-1-t$	$0 \leq t \leq s-1$
1	1	2	1 (or z)	1 (or z)	D	$\mathbb{F}_q$	$m-1-t$	$s-1-t$	$0 \leq t \leq s-1$
1	1	1	1 (or z)	1 (or z)	z (or 1)	$\mathbb{F}_q$	$m-2-t$	$s-1-t$	$0 \leq t \leq s-1$
2	2	1	D	D	1 or z	$\mathbb{F}_q$	$m-1-t$	$s-t$	$0 \leq t \leq s$
2	2	0	D	D	ϕ	$\mathbb{F}_q$	$m-2-t$	$s-t$	$0 \leq t \leq s$
2	1	2	D	1 or z	D	$\mathbb{F}_3$	$m-1-t$	$s-1-t$	$0 \leq t \leq s-1$

where $D = \begin{pmatrix} 1 & \\ & -z \end{pmatrix}$.

$m \neq n$, and that $(m_1, 2s_1+\gamma_1, s_1, \Gamma_1)$ satisfies

$$2m - 2m_1 \geq \begin{cases} (2s+\gamma) - (2s_1+\gamma_1) + |\gamma-\gamma_1| \geq 2|\gamma-\gamma_1|, \\ \qquad\qquad \text{if } \gamma_1 \neq \gamma, \text{ or } \gamma_1 = \gamma \text{ and } \Gamma_1 = \Gamma, \\ (2s+\gamma) - (2s_1+\gamma_1) + 2 \geq 4, \\ \qquad\qquad \text{if } \gamma_1 = \gamma = 1 \text{ and } \Gamma_1 \neq \Gamma. \end{cases} \tag{2}$$

Then

$$\mathcal{L}(m, 2s+\gamma, s, \Gamma; 2\nu+\delta, \Delta) \supset \mathcal{L}(m_1, 2s_1+\gamma_1, s_1, \Gamma_1; 2\nu+\delta, \Delta)$$

unless

$$2s+\gamma \leq m = \begin{cases} \nu+s+\min(\gamma,\delta), & \text{if } \gamma \neq \delta, \text{ or } \gamma = \delta \text{ and } \Gamma = \Delta, \\ \nu+s, & \text{if } \gamma = \delta = 1 \text{ and } \Gamma \neq \Delta \end{cases} \tag{3}$$

holds and one of the cases listed in Table 1 occurs.

THEOREM 2. *Let $n = 2\nu+\delta \geq 1$, $(m, 2s+\gamma, s, \Gamma)$ satisfies(1), and $m \neq n$.If*

$$2s+\gamma \leq m < \begin{cases} \nu+s+\min(\gamma,\delta), & \text{if } \gamma \neq \delta \text{ or } \gamma = \delta \text{ and } \Gamma = \Delta, \\ \nu+s, & \text{if } \gamma = \delta = 1 \text{ and } \Gamma \neq \Delta \end{cases} \tag{4}$$

holds, then $\mathcal{L}(m, 2s+\gamma, s, \Gamma; 2\nu+\delta, \Delta)$ consists of $\mathbb{F}_q^{(2\nu+\delta)}$ and all subspaces of type $(m_1, 2s_1+\gamma_1, s_1, \Gamma_1)$ which satisfies (2). If (3) holds, then $\mathcal{L}(m, 2s+\gamma, s, \Gamma; 2\nu+\delta, \Delta)$ consists of $\mathbb{F}_q^{(2\nu+\delta)}$ and all subspaces of type $(m_1, 2s_1+\gamma_1, s_1, \Gamma_1)$ which satisfies (2) and are not listed in Table 1.

THEOREM 3. *Let $n = 2\nu+\delta \geq 1$, $(m, 2s+\gamma, s, \Gamma)$ satisfies(1), and $m \neq n$.*

(a) If $\gamma = 0$, then

$$\begin{aligned} &\chi(\mathcal{L}(m, 2s, s; 2\nu+\delta, \Delta), t) \\ &= \Big(\sum_{\gamma_1=0,2} + \sum_{\gamma_1=1}\sum_{\Gamma_1=1,z}\Big)\Big(\sum_{s_1=s-\gamma_1+1}^{\nu}\ \sum_{m_1=2s_1+\gamma_1}^{l} + \sum_{s_1=0}^{s-\gamma_1}\ \sum_{m_1=m-s+s_1+1}^{l}\Big) \\ &\quad \cdot N(m_1, 2s_1+\gamma_1, s_1, \Gamma_1; 2\nu+\delta, \Delta)(t-1)(t-q)\cdots(t-q^{m_1-1}), \end{aligned}$$

where

$$l=\begin{cases}\nu+s_1+\min(\gamma_1,\delta), & \text{if } \gamma_1\neq\delta, \text{ or } \gamma_1=\delta \text{ and } \Gamma_1=\Delta,\\ \nu+s_1, & \text{if } \gamma_1=\delta=1 \text{ and } \Gamma_1\neq\Delta.\end{cases} \tag{5}$$

(b) If $\gamma=1$, then

$$\chi(\mathcal{L}(m,2s+1,s;\Gamma;2\nu+\delta,\Delta),t)$$

$$=\sum_{\gamma_1=0,1,2}\Big(\sum_{s_1=s-[\gamma_1/2]+1}^{\nu}\ \sum_{m_1=2s_1+\gamma_1}^{l}+\sum_{s_1=0}^{s-[\gamma_1/2]}\ \sum_{m_1=m-s+s_1-[(2-\gamma_1)/2]+1}^{l}\Big)$$

$$\cdot N(m_1,2s_1+\gamma_1,s_1,\Gamma_1;2\nu+\delta,\Delta)(t-1)(t-q)\cdots(t-q^{m_1-1})$$

$$+\Big(\sum_{s_1=s}^{\nu}\ \sum_{m_1=2s_1+1}^{l}+\sum_{s_1=0}^{s-1}\ \sum_{m_1=m-s+s_1}^{l}\Big)$$

$$\cdot N(m_1,2s_1+\gamma_1,s_1,\bar{\Gamma};2\nu+\delta,\Delta)(t-1)(t-q)\cdots(t-q^{m_1-1}),$$

where $\Gamma=\Gamma_1$ if $\gamma_1=1$, and $\bar{\Gamma}=(z)$ or (1) when $\Gamma=(1)$ or (z), respectively, and l is also defined by (5).

(c) If $\gamma=2$, then

$$\chi(\mathcal{L}(m,2s+2,s,\Gamma;2\nu+\delta,\Delta),t)$$

$$=\Big(\sum_{\gamma_1=0,2}+\sum_{\gamma_1=1}\sum_{\Gamma_1=1,z}\Big)\Big(\sum_{s_1=s+1}^{\nu}\ \sum_{m_1=2s_1+\gamma_1}^{l}+\sum_{s_1=0}^{s}\ \sum_{m_1=m-s+s_1+\gamma_1-1}^{l}\Big)$$

$$\cdot N(m_1,2s_1+\gamma_1,s_1,\Gamma_1;2\nu+\delta,\Delta)(t-1)(t-q)\cdots(t-q^{m_1-1}),$$

where l is also defined by (5)

Remark. A closed formula of $N(m_1,2s_1+\gamma_1,s_1,\Gamma_1;2\nu+\delta,\Delta)$ is obtained by Z. Dai and X. Feng[7].

Let P be an m-dimensional subspace of $\mathbb{F}_q^{(n)}$ which is of rank $2s+\tau$ where $\tau=0$ or 1, then we say that P is an $(m,2s+\tau)$-subspace in $\mathbb{F}_q^{(2\nu+\delta)}$ with respect to $S_{2\nu+\delta,\Delta}$. Denote the set of all $(m,2s+\tau)$-subspace in $\mathbb{F}_q^{(2\nu+\delta)}$ by $\mathcal{M}(m,2s+\tau;2\nu+\delta,\Delta)$ and the set of intersections of subspaces of $\mathcal{M}(m,2s+\tau;2\nu+\delta,\Delta)$ by $\mathcal{L}(m,2s+\tau;2\nu+\delta,\Delta)$. We have also

THEOREM 4. *Let $n=2\nu+\delta\geq1$. Assume that $(m,2s+\tau)$ satisfies*

$$2s+\tau\leq m\leq\begin{cases}\nu+s, & \text{if } \tau=0 \text{ and } s=0,\\ \nu+s+\max(0,\delta-1), & \text{if } \tau=0 \text{ and } s\geq1,\\ \nu+s+\min(1,\delta), & \text{if } \tau=1,\end{cases} \tag{6}$$

$m \neq n$, and $(m_1, 2s_1 + \tau_1)$ satisfies

$$2m - 2m_1 \geq (2s + \tau) - (2s_1 + \tau_1) \geq 0 \tag{7}$$

Then

$$\mathcal{L}(m, 2s + \tau; 2\nu + \delta, \Delta) \supset \mathcal{L}(m_1, 2s_1 + \tau_1; 2\nu + \delta, \Delta)$$

unless

$$2s + \tau \leq m = \begin{cases} \nu + s + \max(0, \delta - 1), & \text{if } \tau = 0 \text{ and } s \geq 1, \\ \nu + s + \min(1, \delta), & \text{if } \tau = 1 \end{cases} \tag{8}$$

holds and one of the cases listed in Table 2 occurs.

TABLE 2

δ	Δ	τ	τ_1	m_1	s_1	*
0	ϕ	0	0	$m - t - t'$	$s - t \geq 1$	$(m_1, 2(s_1 - 1) + 2, s_1 - 1)$
0	ϕ	0	1	$m - t - 1$	$s - t - 1 \geq 0$	$(m_1, 2s_1 + 1, s_1, 1)$or $(m_1, 2s_1 + 1, s_1, z)$
1	1 or z	1	0	$m - t - 1$	$s - t \geq 0$	$(m_1, 2s_1, s_1)$ or $(m_1, 2(s_1 - 1) + 2, s_1 - 1)$
1	1 or z	1	1	$m - t - t'$	$s - t \geq 0$	$(m_1, 2s_1 + 1, s_1, z\Delta)$
2	$\begin{pmatrix} 1 & \\ & -z \end{pmatrix}$	0	0	$m - t - t'$	$s - t \geq 1$	$(m_1, 2s_1, s_1)$
2	$\begin{pmatrix} 1 & \\ & -z \end{pmatrix}$	0	1	$m - t - 1$	$s - t - 1 \geq 0$	$(m_1, 2s_1 + 1, s_1, 1)$ or $(m_1, 2s_1 + 1, s_1, z)$

where * denotes type of $(m_1, 2s_1 + \tau_1)$-subspace.

THEOREM 5. *Let $n = 2\nu + \delta \geq 1$, $(m, 2s + \tau)$ satisfies (6), and $m \neq n$. If*

$$2s + \tau \leq m < \begin{cases} \nu + s + \max(0, \delta - 1), & \text{if } \tau = 0, \\ \nu + s + \min(1, \delta), & \text{if } \tau = 1 \end{cases}$$

holds then $\mathcal{L}(m, 2s + \tau; 2\nu + \delta, \Delta)$ consists of $\mathbb{F}_q^{(2\nu+\delta)}$ and all $(m_1, 2s_1 + \tau_1)$-subspaces, where $(m_1, 2s_1 + \tau_1)$ satisfies (7). However, if (8) holds then $\mathcal{L}(m, 2s + \tau, 2\nu + \delta, \Delta)$ consists of $\mathbb{F}_q^{(2\nu+\delta)}$ and all $(m_1, 2s_1 + \tau_1)$-subspaces, where $(m_1, 2s_1 + \tau_1)$ satisfies (7) and are not listed in Table 2.

THEOREM 6. *Let $n = 2\nu + \delta \geq 1$, $(m, 2s + \tau)$ satisfies (6), and $m \neq n$. When (8) holds, we assume that any of the three cases "$\delta = \tau = 0$", "$\delta = \tau = 1$", and "$\delta = 2, \tau = 0$" does not occur. Then*

$$\begin{aligned} &\chi(\mathcal{L}(m, 2s + \tau; 2\nu + \delta, \Delta), t) \\ &= \sum_{\gamma_1 = 0,1} \Big(\sum_{s_1 = (s+1) - (1-\tau)\tau_1}^{[n/2]} \sum_{m_1 = 2s_1 + \tau_1}^{l} + \sum_{s_1 = 0}^{s - (1-\tau)\tau_1} \sum_{m_1 = m - s + s_1 + \tau(\tau_1 - 1) + 1}^{l} \Big) \\ &\qquad \cdot N(m_1, 2s_1 + \tau_1; 2\nu + \delta, \Delta)(t-1)(t-q)\cdots(t - q^{m_1 - 1}) \end{aligned}$$

where

$$l = \begin{cases} \nu + s_1, & \text{if } \tau_1 = 0 \text{ and } s_1 = 0, \\ \nu + s_1 + \max(0, \delta - 1), & \text{if } \tau_1 = 0 \text{ and } s_1 \geq 1, \\ \nu + s_1 + \min(1, \delta), & \text{if } \tau_1 = 1. \end{cases}$$

Notice that

$$\begin{aligned} &N(m, 2s + \tau; 2\nu + \delta, \Delta) \\ &= \begin{cases} N(m, 2s, s; 2\nu + \delta, \Delta) + N(m, 2(s-1) + 2, s - 1; 2\nu + \delta, \Delta), \text{if } \tau = 0, \\ N(m, 2s + 1, s, 1; 2\nu + \delta, \Delta) + N(m, 2s + 1, s, z; 2\nu + \delta, \Delta), \text{if } \tau = 1, \end{cases} \end{aligned}$$

where

$$N(m, 2s + \gamma, s, \Gamma; 2\nu + \delta, \Delta) = |\mathcal{M}(m, 2s + \gamma, s, \Gamma; 2\nu + \delta, \Delta)|.$$

As a special case of Theorm 6, we have the following result of P. Orlik and L. Solomon

COROLLARY 7. *Let $n = 2\nu + \delta \geq 1$, Then*

$$\begin{aligned}
&\chi(\mathcal{L}(n-1, n-1; 2\nu+\delta, \Delta), t) \\
&= \sum_{s_1=0}^{\nu} N(\nu + s_1 + \delta, 2s_1 + \delta; 2\nu + \delta, \Delta)(t-1)(t-q)\cdots(t-q^{\nu+s_1+\delta-1}) \\
&= (t-1)(t-q)\cdots(t-q^{n-\nu-1})\gamma(t),
\end{aligned}$$

where $\gamma(t) \in \mathbb{Z}[t]$ is a monic polynomial of degreen ν

References

[1] M. Aigner, *Combinatorial Theory*, Springer-Verlay, Berlin, 1979.

[2] D. Chen and Z. Wan, The characteristic polynomial of a geometric lattice in symplectic geometry over finite fields, *Kexue Tongbao*,1990, No.**21**,1627-1630(in Chinese).

[3] P. Orlik and L. Solomon, Arrangements in unitary and orthogonal geometry over finite fields, *J. Comb. Theory, ser.* A, **38**(1985), 217-229.

[4] D. Chen and Z. Wan., An arrangement in orthogonal geometry over finite fields of char=2, *Acta Mathematica Sinica*, New Series,9(1993),39-47.

[5] L. E. Dickson, *Linear Groups*, Teubner, 1900.

[6] J. Dieudonné, *Sur les groupes classiques*, Hermann, Paris, 1948.

[7] Z. Dai and X. Feng, Studies in finite geometries and the construction of incomplete block designs IV, Some "Anzahl" theorems in orthogonal geometry over finite fields of characteristic$\neq$ 2, *Acta Mathematica Sinica* **15** (1965), 533-544 (In Chinese). English translation: Chinese Mathematics **7** (1965), 265-280.

[8] Z. Wan, Finite geometries and block designs, *Sankhya: The Indian Journal of Statistics, Series A*, **54**(1991).

[9] Z. Wan, Z. Dai, X. Feng, and B. Yang, *Studies on Finite Geometries and Incomplete Block Designs*, Science Press, Beijing, 1966(in Chinese).

[10] Y. Huo, Y. Liu, and Z. Wan, Lattices generated by transitive sets of subspaces under finite classical groups I, *Communications in Algebra* **20** (1992), 1123-1144.

[11] Y. Huo, Y. Liu, and Z. Wan, Lattices generated by transitive sets of subspaces under finite classical groups II, the orthogonal case of odd characteristic, *Communications in Algebra* **20**(1992), 2685-2727.

[12] Y. Huo, Y. Liu, and Z. Wan, Lattices generated by transitive sets of subspaces under finite classical groups III, the orthogonal case of even characteristic, *Communications in Algebra* **21**(1993), 2351-2393.

[13] Y. Huo and Z. Wan, Lattices generated by subspaces of the same dimension and rank in orthogonal geometry over finite fields of odd characteristic, *Communications in Algebra* **21**(1993), 4219-4252.

[14] Y. Huo and Z. Wan, Lattices generated by subspaces of the same dimension and rank in orthogonal geometry over finite fields of even characteristic, *Communications in Algebra* **22**(1994), 2015-2037.

[15] Y. Liu, and Z. Wan, Pseudo-symplectic geometries over finite fields of characteristic two, *Advances in Finite Geometries and Designs*, ed. by J. W. P. Hirschfeld et al., Oxford University Press, 1991, 265-288.

[16] Z. Wan, Singular pseudo-symplectic geometry over finite fields of characteristic two, *Northeastern Mathematical Journal* **8** (1992),391-416.

[17] Y. Huo and Z. Wan, Lattices generated by transitive sets of subspaces under finite pseudo-symplectic groups, preprint.

[18] Y. Huo and Z. Wan, Lattices generated by subspaces of the same dimension and rank in finite pseudo-symplectic space, preprint.

[19] Z. Wan, Studies in finite geometries and the construction of incomplete block designs I, Some "Anzahl" theorems in symplectic geometry over finite fields, *Acta Mathematica Sinica* **15** (1965), 354-361 (In Chinese). English translation: Chinese Mathematics **7** (1965), 55-62.

[20] Z. Wan, *Geometry of Classical Groups over Finite fields*, Studentlitteratur, Lund, 1993.

Generalized Kaplansky Line and Cycle Theorems with Multiple Constraints

Frank K. Hwang

and

Paul E. Wright

AT&T Bell Laboratories
Murray Hill, NJ 07974

ABSTRACT

In giving a solution to the problème des menages, Kaplansky enumerated the numbers of ways of selecting k objects, no two consecutive, from n objects arranged in a line and a cycle. These two results have been generalized in two directions. The first involves an extension of the requirement "no two consecutive" to "exactly l consecutive pairs." The second is an extension of the notion of "consecutiveness" to "separation by s objects." However, all these generalizations allow only one constraint, i.e., s-separation for a single s, and the fastest algorithm for these single-constraint problem requires $O(kn^2)$ time. In this paper we give a general method to handle multiple constraints, and an $O(n^c \log n)$ algorithm for c constraints.

Keywords. Problème des Ménages, Restricted Permutations, Kaplansky's Line and Circle Problems.

1. Introduction

In giving a solution to the Problème des Ménages, Kaplansky [7], studied two related problems and obtained the following results:

Theorem 1.1 (Kaplansky's Line Theorem) *The number of ways of selecting k objects, with no two consecutive, from n objects arranged in a line is*

$$\binom{n-k+1}{k}.$$

■

Theorem 1.2 (Kaplansky's Cycle Theorem) *The number of ways of selecting k objects, with no two consecutive, from n objects arranged in a cycle is*

$$\frac{n}{n-k}\binom{n-k}{k}.$$

■

These two theorems have been generalized in two directions. The first involves an extension of the requirement "no two consecutive", to "exactly l consecutive pairs". The second is an extension of the notion of "consecutiveness" to that of s-separation, i.e., the two objects in question are separated by exactly s objects between them. Kaplansky's original theorems then study the case $s = l = 0$.

Define $f(n,k,l,s)$ and $g(k,n,l,s)$ as the number of ways of selecting k objects, with exactly l s-separated pairs, from an n-line and an n-cycle, respectively. Viewing the objects selected as black beads and the objects not selected as white beads, it is easily verified that the number of consecutive pairs equals n minus the number of black runs. Therefore, from the theory of runs [2], we obtain

$$f(k,n,l,0) = \binom{k-1}{l}\binom{n-k+1}{k-l} \tag{1.1}$$

while the number for an n-cycle is,

$$g(k,n,l,0) = \frac{n}{n-k}\binom{k-1}{k-l-1}\binom{n-k}{k-l}. \tag{1.2}$$

Konvalina [9], first ventured beyond the case $s = 0$ and was able to show that

$$f(k,n,0,1) = \sum_{i\geq 0}\binom{n+1-k-2i}{k-2i}. \tag{1.3}$$

Meanwhile, Hwang [4] demonstrated a combinatorial method for deriving the generating function of Whitworth's run on many cycles. This method was later adopted by Hwang, Korner and Wei [6], and Hwang [5] to derive the generating functions of $f(k,n,0,s)$ and $g(k,n,l,s)$. The latter leads to the closed form solution:

$$g(k,n,0,s)=\sum_{j=0}^{\lfloor m/2\rfloor}\binom{m}{j}(-1)^{nj/m}\frac{n-2nj/m}{k-nj/m}\binom{n-k-nj/m-1}{k-nj/m-1}, \tag{1.4}$$

where $m=\gcd(n,s)$.

Kirschenhofer and Prodinger [8] used a more direct approach (see [3]) to derive the generating functions of $g(k,n,l,s)$ and $f(k,n,l,s)$, as well as $f(k,n,l,s)$ itself.

Two essential points must be noted with respect to the above solutions however. The first observation is that the expressions for the generating functions and $f(k,n,l,s)$ mentioned above, are in full summation form, requiring $O(kn^2)$ time to compute. Secondly, and more importantly, the existing solutions cover only the case of one constraint i.e. s-separatedness for a single s. This limits their utility somewhat, as in many applications, e.g. reliability theory, one is often confronted with several constraints. For this latter, important case, no published solutions appear to exist. Further aggravating the situation, is the fact that no effective computational procedures short of brute force search of the $O(2^n)$ elements of the state space is known.

Let $S=(s_1,\ \ldots,\ s_c)$ and $L=(l_1,\ \ldots,\ l_c)$ denote the joint constraints that the k-selected objects contain among them, exactly l_i pairs which are s_i-separated. This paper gives a general methodology of computing $f(k,n,L,S)$ and $g(k,n,L,S)$, which represents their generating functions as linear functionals (traces or inner products) of matrix polynomials.

For small s_i (e.g. constant or $o(\log n)$, say), the f- and g- values can be computed in $O\left(n^c \log n\right)$ time. The only extant case treated in the literature for $c > 1$, is of the sets $S = (1, 2, \ldots, c)$ and $L = (0, 0, \ldots, 0)$. This case is covered in a completely different context, as a special case of the extended birthday problem [10] which asks for the probability that every two birthdays (among k birthdays out of n distinct dates) are separated by at least c days, given that no two birthdays fall on the same day.

2. The Missing Guest Problem

As a means of illustrating the methodology, we consider the following specific problem:

> *There are n guests seated around a table. In how many ways may k of them be excused so that, (i) no two neighbors are removed, and (ii) of those remaining, there are exactly l pairs of neighbors, with each member of the pair having a missing neighbor?*

Note that condition (ii) can be restated as there are l pairs of missing guests distance-3 apart, since the two conditions together imply that the two guests sandwiched between a pair of missing guests distance-3 apart must both be present, the generating function method [3] works for this problem. Namely,

$$
\begin{aligned}
g(k, n, (0, l), (0, 2)) &= \frac{n}{k}[s^n t^l](s^2 + ts^3 + s^4 + \cdots)^k \\
&= \frac{n}{k}[s^{n-2k} t^l](1 + ts + s^2 + \cdots)^k ,
\end{aligned}
$$

where $[s^i t^j]$ means "coefficient" of $s^i t^j$. But the working of the generating function method in this instance is purely coincidental. Suppose that "distance-3" in condition (ii) is replaced by "distance d" for any $d > 3$. Then the generating function method does not work since it counts only those pairs of missing guests distance-d apart with no other missing guests in between. But for $d > 3$, there exist such pairs with a missing guest in between whose neighbors are both present (thus not violating condition (i)). Our method works for any d. However, we set $d = 3$ for easier illustration.

To avoid trivialities, we assume that $n \geq 4$. Define the bivariate generating function,

$$G_n(s,t) = \sum_{k=0}^{n}\sum_{l=0}^{n} g(k,n,(0,l),(0,2))s^k t^l. \tag{2.1}$$

Define also the matrix polynomial $V = V(s,t)$ via.,

$$V = \begin{bmatrix} 1 & 1 & 1 & 1 & 1 \\ s & st & s & 0 & 0 \\ s & s & st & s & s \\ s & s & s & st & st \\ s^2 & s^2 t & s^2 & 0 & 0 \end{bmatrix}, \tag{2.2}$$

as well as the matrix polynomials $W_i,\ i = 1,2$ via.,

$$W_1 = \begin{bmatrix} 1+s & 1+st & 1+s & 1 & 1 \\ s & s & st & s & s \\ s+s^2 & s+s^2 t & s+s^2 & st & st \\ s+s^2 t & s+s^2 t^2 & s+s^2 t & s & s \\ s^2 & s^2 & s^2 t & s^2 & s^2 \end{bmatrix}, \tag{2.3}$$

and

$$W_2 = \begin{bmatrix} 1+2s & 1+s+st & 1+s+st & 1+s & 1+s \\ s+s^2 & s+s^2t & s+s^2 & st & st \\ s+s^2+s^2t & s+s^2+s^2t^2 & s+2s^2t & s+s^2 & s+s^2 \\ s+s^2+s^2t & s+2s^2t & s+s^2+s^2t^2 & s+s^2t & s+s^2t \\ s^2+s^3 & s^2+s^3t & s^2+s^3 & s^2t & s^2t \end{bmatrix}. \tag{2.4}$$

We then have the following result:

Theorem 2.1.

$$G_n(s,t) = \begin{cases} Trace\, V^r, & if\ n = 3r, \\ Trace\, V^{r-1}W_1, & if\ n = 3r+1, \\ Trace\, V^{r-1}W_2, & if\ n = 3r+2. \end{cases} \tag{2.5}$$

Proof. Label the n objects by $0, 1, \ldots, n-1$. For $0 \le i \le n-1$, let

$$\omega_i = \begin{cases} 1, & \text{if ball } i \text{ is selected,} \\ 0, & \text{if not.} \end{cases} \tag{2.6}$$

The global state of the system is then specified by the word $\omega = \omega_0\omega_1 \ldots \omega_{n-1} \in \{0,1\}^n$.

$\underline{n = 3r}$. For the sake of simplicity, we begin with the case of n divisible by 3. Let $\sigma_i = \omega_{3i}\omega_{3i+1}\omega_{3i+2}$ denote the joint state of the nodes $3i, 3i+1, 3i+2$, $i = 0, 1, \ldots, r-1$. The non-consecutiveness constraint implies that σ_i may only assume one of 5 possible values: (a) 000, (b) 001, (c) 010, (d) 100, and (e) 101. Denote this set of allowed states by Ω. For $\sigma \in \Omega$, let λ_σ denote its weight, i.e. the number of ones. Thus for example, $\lambda_{101} = 2$. Furthermore, for $\eta \in \Omega$, define

$$\psi_{\sigma,\eta} = \#\{\text{distance-3, 1's in the sequence } \sigma\eta\}, \tag{2.7}$$

n	$G_n(s,t)$
4	$1+4\,s+2\,s^2$
5	$1+5\,s+5\,s^2t$
6	$1+6\,s+6\,s^2+3\,s^2t^2+2\,s^3$
7	$1+7\,s+7\,s^2+7\,s^2t+7\,s^3t$
8	$1+8\,s^2t+8\,s+12\,s^2+8\,s^3+8\,s^3t^2+2\,s^4$
9	$1+9\,s^2t+9\,s+18\,s^2+18\,s^3t+9\,s^3+3\,s^3t^3+9\,s^4t$
10	$1+10\,s^2t+10\,s+25\,s^2+20\,s^3t+20\,s^3+10\,s^3t^2+10\,s^4+15\,s^4t^2+$ $2\,s^5$
11	$1+11\,s^2t+11\,s+33\,s^2+33\,s^3t+33\,s^3+11\,s^3t^2+11\,s^4+33\,s^4t+$ $11\,s^4t^3+11\,s^5t$
12	$1+12\,s^2t+12\,s+42\,s^2+48\,s^3t+52\,s^3+12\,s^3t^2+30\,s^4+$ $36\,s^4t+36\,s^4t^2+12\,s^5+3\,s^4t^4+24\,s^5t^2+2\,s^6$

Table 1: $G_n(s,t)$ for $4 \le n \le 12$.

that is, using the lexicographic ordering of Ω above,

$$\psi = \begin{bmatrix} 0 & 0 & 0 & 0 & 0 \\ 0 & 1 & 0 & 0 & 1 \\ 0 & 0 & 1 & 0 & 0 \\ 0 & 0 & 0 & 1 & 1 \\ 0 & 1 & 0 & 1 & 1 \end{bmatrix}. \tag{2.8}$$

Call a pair $\sigma, \eta \in \Omega$ *taboo* if the concatenation $\sigma\eta$ contains two adjacent 1's, and *non-taboo* otherwise. Let

$$\chi_{\sigma,\eta} = \begin{cases} 1, & \text{if } (\sigma,\eta) \text{ is non-taboo,} \\ 0, & \text{otherwise.} \end{cases} \tag{2.9}$$

and define the 5×5 transfer matrix V (again indexed by elements of Ω) by

$$V_{\sigma,\eta} = s^{\lambda_\sigma}\, t^{\psi_{\sigma,\eta}}\, \chi_{\sigma,\eta}. \tag{2.10}$$

Using the same lexicographic ordering of Ω, V has the representation (2.2). The essential observation is that,

$$\begin{aligned} G_n(s,t) &= \sum_{k=0}^{n}\sum_{l=0}^{n} g(k,n,(0,l),(0,2))s^k t^l \\ &= \sum_{\sigma_0\in\Omega}\cdots\sum_{\sigma_{r-1}\in\Omega} s^{\lambda_{\sigma_1}+\ldots+\lambda_{\sigma_r}} t^{\psi_{\omega_0,\omega_1}+\ldots+\psi_{\omega_{r-2},\omega_{r-1}}+\psi_{\omega_{r-1},\omega_0}} \prod_{i=0}^{r-1}\chi_{\sigma_i,\sigma_{i+1}} \\ &= \sum_{\sigma_0\in\Omega}\cdots\sum_{\sigma_{r-1}\in\Omega} V_{\sigma_0,\sigma_1}V_{\sigma_1,\sigma_2}\ldots V_{\sigma_{r-2},\sigma_{r-1}}V_{\sigma_{r-1},\sigma_0} \\ &= \sum_{\sigma_0\in\Omega} V^r_{\sigma_0,\sigma_0} \\ &= \mathit{Trace}\, V^r. \end{aligned} \tag{2.11}$$

Recall that V is a polynomial in s and t, so that (2.11) expresses $G_n(s,t)$ as the trace of a 5×5 matrix polynomial.

$n = 3r+1$ and $3r+2$. Reasoning similar to that leading to (2.11), taking into account the states of objects $3r+1$ and $3r+2$ as the case may be yields the remaining cases. ■

Table 1 contains a list of $G_n(s,t)$ for $4 \le n \le 12$ objects while Table 2 contains entries for $13 \le n \le 21$. The claims advanced in the abstract may be verified with the help of an examination of $G_{21}(s,t)$. For example, the coefficient of s^7t is 1491 and the coefficient of s^5t^4 is 21, whence the assertions.

Computational Issues. The matrix computations involved in Theorem 2.1 may be executed symbolically in $O(5^{\log_2(n)}) = O(n^{\log_2 5})$ steps. In the most naive implementation, the number of terms needed for each entry increasing at most five-fold at each step. A better approach for large values of n goes via the Discrete Fast Fourier Transform (DFFT) [1]. Here one makes use of the standard fact that recovering the coefficients of a polynomial from its values

n	$G_n(s,t)$
13	$1 + 13\,s^2t + 13\,s + 52\,s^2 + 65\,s^3t + 78\,s^3 + 13\,s^3t^2 + 52\,s^4 + 78\,s^4t + 39\,s^4t^2 + 13\,s^4t^3 + 52\,s^5t + 13\,s^5 + 26\,s^5t^3 + 13\,s^6t$
14	$1 + 14\,s^2t + 14\,s + 63\,s^2 + 84\,s^3t + 112\,s^3 + 14\,s^3t^2 + 91\,s^4 + 126\,s^4t + 63\,s^4t^2 + 14\,s^4t^3 + 56\,s^5t + 42\,s^5 + 84\,s^5t^2 + 14\,s^6 + 14\,s^5t^4 + 35\,s^6t^2 + 2\,s^7$
15	$1 + 15\,s^2t + 15\,s + 75\,s^2 + 105\,s^3t + 155\,s^3 + 15\,s^3t^2 + 150\,s^4 + 195\,s^4t + 90\,s^4t^2 + 15\,s^4t^3 + 150\,s^5t + 75\,s^5 + 90\,s^5t^2 + 15\,s^6 + 60\,s^5t^3 + 75\,s^6t + 15\,s^7t + 50\,s^6t^3 + 3\,s^5t^5$
16	$1 + 16\,s^2t + 16\,s + 88\,s^2 + 128\,s^3t + 208\,s^3 + 16\,s^3t^2 + 236\,s^4 + 288\,s^4t + 120\,s^4t^2 + 16\,s^4t^3 + 256\,s^5t + 144\,s^5 + 192\,s^5t^2 + 56\,s^6 + 64\,s^5t^3 + 80\,s^6t + 16\,s^5t^4 + 160\,s^6t^2 + 16\,s^7 + 40\,s^6t^4 + 48\,s^7t^2 + 2\,s^8$
17	$1 + 17\,s^8t + 17\,s^2t + 17\,s + 102\,s^2 + 153\,s^3t + 272\,s^3 + 17\,s^3t^2 + 357\,s^4 + 408\,s^4t + 153\,s^4t^2 + 17\,s^4t^3 + 442\,s^5t + 255\,s^5 + 306\,s^5t^2 + 102\,s^6 + 102\,s^5t^3 + 255\,s^6t + 17\,s^5t^4 + 170\,s^6t^2 + 102\,s^7t + 170\,s^6t^3 + 17\,s^7 + 85\,s^7t^3 + 17\,s^6t^5$
18	$1 + 18\,s^2t + 18\,s + 117\,s^2 + 180\,s^3t + 348\,s^3 + 18\,s^3t^2 + 522\,s^4 + 558\,s^4t + 189\,s^4t^2 + 18\,s^4t^3 + 720\,s^5t + 432\,s^5 + 468\,s^5t^2 + 2\,s^9 + 213\,s^6 + 144\,s^5t^3 + 450\,s^6t + 18\,s^5t^4 + 450\,s^6t^2 + 108\,s^7t + 180\,s^6t^3 + 72\,s^7 + 90\,s^6t^4 + 270\,s^7t^2 + 63\,s^8t^2 + 90\,s^7t^4 + 3\,s^6t^6 + 18\,s^8$
19	$1 + 133\,s^8t + 19\,s^2t + 57\,s^7t^5 + 133\,s^8t^3 + 19\,s + 133\,s^2 + 209\,s^3t + 437\,s^3 + 19\,s^3t^2 + 741\,s^4 + 741\,s^4t + 228\,s^4t^2 + 19\,s^4t^3 + 1121\,s^5t + 703\,s^5 + 684\,s^5t^2 + 399\,s^6 + 190\,s^5t^3 + 855\,s^6t + 19\,s^5t^4 + 760\,s^6t^2 + 399\,s^7t + 380\,s^6t^3 + 133\,s^7 + 95\,s^6t^4 + 285\,s^7t^2 + 19\,s^9t + 380\,s^7t^3 + 19\,s^6t^5 + 19\,s^8$
20	$1 + 140\,s^8t + 20\,s^2t + 20\,s + 150\,s^2 + 240\,s^3t + 540\,s^3 + 20\,s^3t^2 + 1025\,s^4 + 960\,s^4t + 270\,s^4t^2 + 20\,s^4t^3 + 1680\,s^5t + 1104\,s^5 + 960\,s^5t^2 + 20\,s^9 + 720\,s^6 + 240\,s^5t^3 + 1500\,s^6t + 20\,s^5t^4 + 1300\,s^6t^2 + 2\,s^{10} + 720\,s^7t + 600\,s^6t^3 + 300\,s^7 + 80\,s^9t^2 + 175\,s^8t^4 + 20\,s^7t^6 + 150\,s^6t^4 + 900\,s^7t^2 + 420\,s^8t^2 + 300\,s^7t^4 + 400\,s^7t^3 + 20\,s^6t^5 + 90\,s^8$
21	$1 + 588\,s^8t + 21\,s^2t + 126\,s^7t^5 + 735\,s^8t^3 + 21\,s + 168\,s^2 + 273\,s^3t + 658\,s^3 + 21\,s^3t^2 + 1386\,s^4 + 1218\,s^4t + 315\,s^4t^2 + 21\,s^4t^3 + 2436\,s^5t + 1680\,s^5 + 1302\,s^5t^2 + 21\,s^9 + 1246\,s^6 + 294\,s^5t^3 + 2520\,s^6t + 21\,s^5t^4 + 2100\,s^6t^2 + 21\,s^{10}t + 147\,s^8t^5 + 196\,s^9t^3 + 3\,s^7t^7 + 1491\,s^7t + 910\,s^6t^3 + 588\,s^7 + 210\,s^6t^4 + 1575\,s^7t^2 + 168\,s^9t + 441\,s^8t^2 + 315\,s^7t^4 + 1050\,s^7t^3 + 21\,s^6t^5 + 168\,s^8$

Table 2: $G_n(s,t)$ for $13 \leq n \leq 21$.

at the roots of unity is equivalent, after an elementary transformation, to the FFT. Evaluation of $G_n(s,t)$ at the n^2 points in the complex plane, both of whose coordinates are n-th roots of unity, may be effected using (2.5) in $O(n^2 \log n)$ steps. The 2-dimensional FFT is equivalent to n 1-dimensional FFTs, each of which may be done in $O(n \log n)$ time. For details, the reader is referred to [1].

3. General Constraints

The extension to the general case is straightforward. Indeed, let $m = \max\{s_i \in S\}$. Consider the n-cycle. Then partition the state $\omega \in \{0,1\}$ into segments of length m (for simplicity, we treat first the case where m divides n).

Define the $(c+1)$-variate generating function,

$$G_n(t_0, t_1, \ldots, t_c) = \sum_{k=0}^{n} \sum_{l_1=0}^{n} \ldots \sum_{l_c=0}^{n} g(k, n, (l_1, \ldots, l_c)(s_1, \ldots s_c)) t_0^k t_1^{l_1 1} t_2^{l_2} \ldots t_c^{l_c}.$$

$$\psi_{\sigma,\eta}^{(i)} = \#\{s_i\text{-separated 1's in } \sigma\eta\}, \quad 1 \le i \le c.$$

Let Ω denote the subset of binary m-vectors. Define the $|\Omega| \times |\Omega|$ transfer matrix V (again indexed by elements of Ω) by

$$V_{\sigma,\eta} = t_0^{\lambda_\sigma} \prod_{i=1}^{c} t_i^{\psi_{\sigma,\eta}^{(i)}} \chi_{\sigma,\eta},$$

where λ_σ denotes the weight of σ as before. Then one obtains

$$G_n(t_0, t_1, \ldots, t_c) = Trace\, V^{n/m}.$$

A variation on the theme which reduces the size of Ω from 2^m, is the introduction of taboo states, as in section 2. Call a pair $\sigma, \eta \in \Omega$ *taboo* if the concatenation $\sigma\eta$ violates some fixed, easily verified subset of the constraints and *non-taboo* otherwise. Then let χ be as in (2.9), and define $\psi^{(i)}$ and V accordingly.

The procedure is similar for the cases where n is not congruent to m, the result then depending on the residue class of n modulo m. The linear case is treated in a similar fashion. Instead of a trace, one obtains a formula of the form $G_n(t_0, t_1, \ldots, t_c) = \mathbf{u}^t V^{n/m} \mathbf{w}$ for specific column $|\Omega|$-vectors $\mathbf{u}$ and $\mathbf{w}$. The details are left to the reader. In all cases, the computational complexity is easily seen to be $O(n^c \log n)$ time, as claimed, if advantage is taken of the taboo state technique. Finally, we note that for the case $c = 1$, i.e. case of a single constraint, our proposed method is already faster than existing methods.

Acknowledgements

We thank A. R. Calderbank for reading the manuscript.

References

[1] A. V. Aho, J. E. Hopcroft, and J. D. Ullman. *The Design and Analysis of Computer Algorithms.* Addison-Wesley, Reading, Mass., 1974.

[2] F. N. David and D. E. Barton. *Combinatorial Chance.* Hafner Publishing Co., New York, NY., 1962.

[3] I. P. Goulden and D. M. Jackson. *Combinatorial Enumeration.* Wiley, New York, NY., 1983.

[4] F. K. Hwang. Cycle polynomials. *Proc. Amer. Math. Soc.*, 83:215–220, 1981.

[5] F. K. Hwang. Selecting k objects from a cycle with p pairs of separation s. *Jour. Comb. Theory A*, 37:197–199, 1984.

[6] F. K. Hwang, J. Korner, and V. K. Wei. Selecting non-consecutive balls arranged in many lines. *Jour. Comb. Theory A*, 37:327–336, 1984.

[7] I. Kaplansky. Solution of the Problème des Ménages. *Bull. Amer. Math. Soc.*, 49:784–785, 1943.

[8] P. Kirschenhofer and H. Prodinger. Two selection problems revisited. *Jour. Comb. Theory A*, 42:311–316, 1981.

[9] J. Konvalina. Solution of the Problème des Ménages. *Jour. Comb. Theory A*, 31:101–107, 1981.

[10] J. I. Naus. An extension of the birthday problem. *Amer. Statistician*, 22:27–29, 1968.

Some methods of construction of rectangular PBIB designs

Sanpei KAGEYAMA and Ying MIAO
Department of Mathematics
Hiroshima University
Higashi-Hiroshima 739, Japan

Abstract. Several methods of construction of partially balanced incomplete block designs based on a rectangular association scheme are suggested with some examples. Series of such designs are also provided for illustrations.

Key words and phrases: rectangular PBIB design; rectangular association scheme; BIB design; Hadamard matrix; pairwise balanced design; orthogonal array.

1. Introduction

Rectangular partially balanced incomplete block (PBIB) designs, which have three associate classes, were introduced by Vartak [12] and subsequently studied by Shah [11], Vartak [13] and others. Such designs were also called two way group divisible (GD) designs by Ghosh and Das [7]. In addition to their statistical properties such as their balance for two way group comparison [7], these designs are also useful in the construction of combinatorial designs [2]. However, methods of constructing rectangular PBIB designs are not rich in the literature. Here, we shall present construction methods of such designs by several approaches. Some series of the design are also provided.

A rectangular PBIB design is an arrangement of v treatments into b blocks such that

(i) each treatment is replicated r times;

(ii) each block contains k $(< v)$ treatments;

(iii) there exists an association scheme between $v = mn$ treatments, arranged in a rectangle of m rows and n columns, such that

(a) any two treatments from the same row are the first associates;

(b) any two treatments from the same columns are the second associates; and

(c) any two treatments from different rows and different columns are the third associates;

(iv) any two treatments which are ith associates occur together in λ_i blocks, $i = 1, 2, 3$.

We shall denote the parameters of a rectangular PBIB design by $(v, b, r, k; \lambda_1, \lambda_2, \lambda_3; m, n)$.

Note that in the case of $m = n$, a rectangular PBIB design can reduce to an L_2 design (cf. [9]). Also, if $\lambda_1 = \lambda_3$ or $\lambda_2 = \lambda_3$, a rectangular PBIB design can reduce to a GD design by combining the first and the third, or the second and the third associate classes (cf. [9]).

We assume the reader is familiar with design theory. For the undefined terms used in this paper, we refer to [3, 9].

2. Method of differences

Most of direct constructions for designs are based on the method of differences, which can also be used to produce rectangular PBIB designs.

Theorem 2.1. Let p and q be primes such that $q \geq p$. Then there exists a rectangular PBIB design with parameters

$$\begin{aligned} &v = pq, b = gq(q-1) + fg, r = g(g-1) + f, k = p; \\ &\lambda_1 = 0, \lambda_2 = f, \lambda_3 = g; m = p, n = q \end{aligned} \tag{2.1}$$

for a non-negative integer f and a positive integer g.

Proof. Let the set of treatments be $Z_p \times Z_q$. Two distinct treatments (x, y) and (x', y') are first associates if $x = x'$, second associates if $y = y'$, and third associates otherwise. Assaf [1] showed that

$$\{(0,0),(1,\alpha),(2,\alpha),...,(p-1,(p-1)\alpha)\} \bmod (-,q), \alpha = 1,2,...,q-1,$$

form the blocks of a rectangular PBIB design with parameters $v = pq, b = q(q-1), r = q-1, k = p, \lambda_1 = \lambda_2 = 0, \lambda_3 = 1, m = p, n = q$. Then taking g copies of these blocks and adding f copies of the blocks $\{(0,x),(1,x),...,(p-1,x)\}, x = 0,1,...,q-1$, can provide the blocks of the required rectangular PBIB design. □

Theorem 2.2. Let p and q be primes such that $k|(p-1)$ and $k|(q-1)$. Then there exists a rectangular PBIB design with parameters

$$\begin{aligned} &v = pq, b = epq(p-1)(q-1)/k + gqp(p-1)/k + fpq(q-1)/k, \\ &r = e(p-1)(q-1) + g(p-1) + f(q-1), k; \lambda_1 = f(k-1), \\ &\lambda_2 = g(k-1), \lambda_3 = e(k-1); m = p, n = q \end{aligned}$$

for non-negative integers f and g, and a positive integer e.

Proof. For the set of treatments GF(p)× GF(q), two distinct treatments (x,y) and (x',y') are first associates if $x = x'$, second associates if $y = y'$, and third associates otherwise. Let $p = sk+1$ and $q = tk+1$. Further let α and β be the primitive roots of GF(p) and GF(q) respectively. Then it follows that

$$\begin{gathered} \{(\alpha^i,\beta^j),(\alpha^{i+e},\beta^{j+f}),...,(\alpha^{i+(k-1)e},\beta^{j+(k-1)f})\}, \\ \{(\alpha^i,\beta^{j+f}),(\alpha^{i+e},\beta^{j+2f}),...,(\alpha^{i+(k-1)e},\beta^{j})\}, \\ \vdots \\ \{(\alpha^i,\beta^{j+(k-1)f}),(\alpha^{i+e},\beta^{j}),...,(\alpha^{i+(k-1)e},\beta^{j+(k-2)j})\}, \\ i = 0,1,...,e-1;\ j = 0,1,...,f-1, \end{gathered} \tag{2.2}$$

give the base blocks of a rectangular PBIB design with parameters $v = pq, b = pq(p-1)(q-1)/k, r = (p-1)(q-1)/k, k; \lambda_1 = \lambda_2 = 0, \lambda_3 = k-1; m = p, n = q$. In this case, by taking e copies of the blocks generated by blocks (2.2), and adding g copies of the blocks generated by

$$\{(\alpha^i,y),(\alpha^{i+e},y),...,(\alpha^{i+(k-1)e},y)\},\ i = 0,1,...,e-1, \tag{2.3}$$

for all $y \in$ GF(q), and f copies of the blocks generated by

$$\{(x,\beta^j),(x,\beta^{j+f}),...,(x,\beta^{j+(k-1)f})\},\ j=0,1,...,f-1, \tag{2.4}$$

for all $x \in \mathrm{GF}(p)$, we can obtain the blocks of the required design. □

Remark 2.3. The rectangular PBIB design constructed in Theorem 2.2 has nice block structure. For example, the blocks generated by blocks (2.2) can be partitioned into some classes each of which partitions $(\mathrm{GF}(p)-\{x\})\times(\mathrm{GF}(q)-\{y\})$ for some $(x,y)\in\mathrm{GF}(p)\times\mathrm{GF}(q)$; the blocks generated by blocks (2.3) can be partitioned into some classes each of which partitions $(\mathrm{GF}(p)-\{x\})\times\mathrm{GF}(q)$ for some $x\in\mathrm{GF}(p)$; the blocks generated by blocks (2.4) can be partitioned into some classes each of which partitions $(\mathrm{GF}(p)\times(\mathrm{GF}(q)-\{y\})$ for some $y\in\mathrm{GF}(q)$. Furthermore, we can combine some of the blocks generated by (2.2), (2.3) and (2.4), respectively, to form the second and the third types of classes described above, and even to form some classes each of which partitions $\mathrm{GF}(p)\times\mathrm{GF}(q)-\{(x,y)\}$ for some $(x,y)\in\mathrm{GF}(p)\times\mathrm{GF}(q)$. Such properties are of interest in both combinatorial and statistical sense. For detailed information on these properties, the reader is referred to [6].

3. Patterned method

This method had been used extensively to construct various types of experimental designs (cf. [4,5]). Variations of some patterns used in the literature can also produce rectangular PBIB designs.

In what follows, N denotes the $v'\times b'$ incidence matrix of a balanced incomplete block (BIB) design with parameters v',b',r',k',λ', I_s is the identity matrix of order s, $O_{s\times t}$ is an $s\times t$ matrix with all elements zero, $J_{s\times t}$ is an $s\times t$ matrix with all elements unity, $\overline{N}=J_{v'\times b'}-N$, A^T is the transpose of a matrix A, $A\otimes B$ denotes the Kronecker product of matrices A and B.

Theorem 3.1. The existence of a BIB design with parameters v',b',r',k',λ' implies the existence of a rectangular PBIB design with parameters

$$v=tv',b=tb',r=b'+(t-2)r',k=v'+(t-2)k';$$
$$\lambda_1=b'-2r'+t\lambda',\lambda_2=(t-2)r',\lambda_3=\lambda'(t-4)+2r';m=t,n=v'$$

for a positive integer $t \geq 2$.

Proof. Let the mn treatments be arranged in a rectangle of m rows and n columns as (standard form of a rectangular association scheme, i.e. AS(m,n))

$$\begin{array}{cccc} 1 & m+1 & \cdots & (n-1)m+1 \\ 2 & m+2 & \cdots & (n-1)m+2 \\ \vdots & \vdots & & \vdots \\ m & 2m & \cdots & nm \end{array} . \tag{3.1}$$

Then the result follows from the fact that $I_t \otimes \overline{N} + (J_{t\times t} - I_t) \otimes N$ is the incidence matrix of the required design. □

Remark 3.2. The pattern used in Theorem 3.1 is a variation of one of the pattern used in [4, Theorem 3.2].

Theorem 3.3. The existence of a BIB design with parameters $v' = 2k', b', r', k', \lambda'$ implies the existence of a rectangular PBIB design with parameters

$$v = tv', b = (t+1)b', r = (t+1)r', k = tk'; \lambda_1 = (t+1)\lambda',$$
$$\lambda_2 = (t-1)r', \lambda_3 = (t-3)\lambda' + 2r'; m = t, n = v'$$

for a positive integer $t \geq 3$.

Proof. Let the mn treatments have AS(m,n) as in (3.1). Then the result follows from that $[J_{t\times 1} \otimes N : I_t \otimes \overline{N} + (J_{t\times t} - I_t) \otimes N]$ is the incidence matrix of the required design. □

Remark 3.4. The pattern described above is a variation of the pattern used in [4, Theorem 3.1].

As a matter of fact, most of the patterns used in the literature are still valid if we utilize N to denote the incidence matrix of a BIB design. For example, when $v' = 2k'$, patterns as in

$$\begin{bmatrix} N \\ \overline{N} \end{bmatrix}, \begin{bmatrix} N & N & O_{v'\times b'} & N & \overline{N} & O_{v'\times b'} \\ N & O_{v'\times b'} & N & \overline{N} & O_{v'\times b'} & N \\ O_{v'\times b'} & N & N & O_{v'\times b'} & N & \overline{N} \end{bmatrix}, \begin{bmatrix} N & N \\ N & \overline{N} \end{bmatrix},$$

$$\begin{bmatrix} N & N & J_{v'\times b'} & O_{v'\times b'} \\ N & \overline{N} & O_{v'\times b'} & J_{v'\times b'} \end{bmatrix}, \begin{bmatrix} N & N & N & O_{v'\times b'} \\ N & \overline{N} & O_{v'\times b'} & N \\ N & O_{v'\times b'} & \overline{N} & \overline{N} \\ O_{v'\times b'} & N & \overline{N} & N \end{bmatrix}, \text{ etc.,}$$

become incidence matrices of some rectangular PBIB designs with suitable parameters.

Another type of configuration which can be utilized to produce rectangular PBIB designs is a Hadamard matrix. A Hadamard matrix H of order s is a matrix with elements $+1$ and -1 such that $HH^T = sI_s$. In this case, $s = 2$ or $s \equiv 0 \bmod 4$. For the Hadamard matrix, the reader is referred to [10].

Theorem 3.5. The existence of a normalized Hadamard matrix of order s and a BIB design with $v' = 2k'$ implies the existence of a rectangular PBIB design with parameters

$$v = (s-i)v', b = sb', r = sr', k = (s-i)k'; \lambda_1 = s\lambda',$$
$$\lambda_2 = sr', \lambda_3 = sr'/2; m = s-i, n = v'$$

for a non-negative integer $i \leq s-2$, where $s = 2$ or $s \equiv 0 \bmod 4$.

Proof. Let the mn treatments have AS(m,n) as in (3.1). Let A denote the $(s-i)\times i$ matrix obtained by deleting the first and any other $i-1$ rows of the Hadamard matrix H for $i \leq s-2$. Then it can be shown that

$$\frac{1}{2}\{(J_{(s-i)\times s} + A)\otimes N + (J_{(s-i)\times s} - A)\otimes \overline{N}\}$$

is the incidence matrix of the required design. Note that when $i = 0$, no deletion of rows in H occur. □

Remark 3.6. The pattern used in Theorem 3.5 is just the same as in [5, Theorem 2.1] that constructed group divisible designs.

We conclude this section by pointing out that most of the patterns using the Hadamard matrix in the literature are still valid in the present case. The reader may check those patterns and give explicit statements if necessary.

4. Method of breaking up blocks

This method is combinatorial in some sense. Starting from a rectangular PBIB design, we can get another one by breaking up the original blocks using other types of designs. On the other hand, we can also start from other types of designs, and break up its blocks by rectangular PBIB designs to get new rectangular PBIB designs. For example, we have the following.

Theorem 4.1. The existence of a rectangular PBIB design with parameters $v', b', r', k'; \lambda_1', \lambda_2', \lambda_3'; m', n'$, and a BIB design with parameters

$$v'' = k', b'', r'', k'', \lambda'' \tag{4.1}$$

implies the existence of a rectangular PBIB design with parameters

$$v = v', b = b'b'', r = r'r'', k = k'k''; \lambda_1 = \lambda_1'\lambda'',$$
$$\lambda_2 = \lambda_2'\lambda'', \lambda_3 = \lambda_3'\lambda''; m = m', n = n'.$$

Proof. Let the rectangular association scheme be the same as the original one. We break up the blocks of the original rectangular PBIB design as follows. On each original block, we construct a BIB design with parameters (4.1). Then the union of the new blocks just forms the blocks of the required design. □

Let p and q be primes such that $q \geq p$, and k be a positive integer such that $k|(p-1)$. Let α be a primitive root of GF(p). Then base blocks $\{\alpha^i, \alpha^{i+d}, ..., \alpha^{i+(k-1)d}\}, i = 0, 1, ..., d-1, d = (p-1)/k$, provide a BIB design with parameters $v'' = p, b'' = pd, r'' = p-1, k'' = k, \lambda'' = k-1$. Applying Theorems 2.1 and 4.1, we can get a rectangular PBIB design with parameters

$$v = pq, b = pd\{gq(q-1) + fg\}, r = (p-1)\{g(q-1) + f\}, k;$$
$$\lambda_1 = 0, \lambda_2 = f(k-1), \lambda_3 = g(k-1); m = p, n = q$$

for a non-negative integer f and a positive integer g.

We can also utilize pairwise balanced designs (cf. Mullin and Stanton [8]).

Theorem 4.2. If there exists a regular pairwise balanced design $(\mathcal{X}, \mathcal{B})$ with parameters v', r', K', λ' and for each block $B \in \mathcal{B}$, there further exists a rectangular PBIB design with parameters

$$v_B'' = |B|n'', b_B'', r_B'', k''; \lambda_1'', \lambda_2'', \lambda_3''; m'' = |B|, n'', \tag{4.2}$$

then there exists a rectangular PBIB design with parameters

$$v = v'n'', b = \sum_{B\in\mathcal{B}} b''_B, r = bk/v, k = k'';$$
$$\lambda_1 = r'\lambda''_1, \lambda_2 = \lambda'\lambda''_2, \lambda_3 = \lambda'\lambda''_3; m = v', n = n''. \tag{4.3}$$

Proof. On each block $B \in \mathcal{B}$, we construct a rectangular PBIB design $(B\times\{1,2,...,n''\}, \mathcal{B}_B)$ with parameters (4.2), where the underlined association scheme on $B \times \{1,2,...,n''\}$ is defined as follows. Two distinct treatments (x,y) and (x',y') are first associates if $x = x'$, second associates if $y = y'$, and third associates otherwise. Now we will construct a rectangular PBIB design on $\mathcal{X} \times \{1,2,...,n''\}$ with parameters (4.3). Arrange the $v = v'n''$ treatments in a rectangle of v' rows (of form $\{x\}\times\{1,2,...,n''\}$) and n'' columns (of form $\mathcal{X} \times \{i\}$). Two treatments (x,y) and (x',y') are first associates if $x = x'$, second associates if $y = y'$, and third associates otherwise. Then $(\mathcal{X} \times \{1,2,...,n\}, \sum_{B\in\mathcal{B}} \mathcal{B}_B)$ provides the required design with the association scheme defined as above. □

When $\lambda''_1 = 0$, we can further have the following result, whose proof is similar to that of Theorem 4.2 and hence it is omitted.

Theorem 4.3. If there exists a pairwise balanced design with parameters v', K', λ', and for each block B there further exists a rectangular PBIB design with parameters

$$v''_B = |B|n'', b''_B, r''_B, k''; \lambda''_1 = 0, \lambda''_2, \lambda''_3; m'' = |B|, n'',$$

then there exists a rectangular PBIB design with parameters

$$v = v'n'', b = \sum_B b''_B, r = bk/v, k = k'';$$
$$\lambda_1 = 0, \lambda_2 = \lambda'\lambda''_2, \lambda_3 = \lambda'\lambda''_3; m = v', n = n''.$$

Let $\mathrm{B}(K) = \{v :$ there exists a pairwise balanced design with parameters $v, K, \lambda = 1\}$. If $K = \mathrm{B}(K)$, then K is called a PBD-closed set. Now Theorem 4.3 can be restated as

Theorem 4.4. $\mathrm{RPBIBD}(k, \lambda_2, \lambda_3, n) = \{m :$ there exists a rectangular PBIB design with parameters $v = mn, b, r, k; \lambda_1 = 0, \lambda_2, \lambda_3; m, n\}$ is a PBD-closed set.

This property is useful in constructing rectangular PBIB designs. For example, for each positive integer $n' \equiv 1$ or 3 mod 3, $n' \neq 6$, there exists a pairwise balanced design with parameters $v = n', K = \{3, 4\}, \lambda = 1$ (cf. [3, IX, Proposition 2.13]). Let $q = 3t + 1$ be a prime, then there exist a rectangular PBIB design with parameters $v = 3q, b = 2eq(q-1) + 2gq, r = 2e(q-1) + 2g, k = 3; \lambda_1 = 0, \lambda_2 = 2g, \lambda_3 = 2e; m = 3, n = q$ (by Theorem 2.1) and a rectangular PBIB design with parameters $v = 4q, b = 4eq(q-1)+4gq, r = 3e(q-1)+3g, k = 3; \lambda_1 = 0, \lambda_2 = 2g, \lambda_3 = 2e; m = 4, n = q$ (by Theorem 2.2) for a positive integer e and a non-negative integer g. Then there exists a rectangular PBIB design with parameters $v = n'q, b = vr/3, r = \sum_{i=1}^{3} n_i\lambda_i/2, k = 3; \lambda_1 = 0, \lambda_2 = 2g, \lambda_3 = 2e; m = n', n = q$ for $e \geq 1$ and $g \geq 0$, provided that $n' \equiv 1$ or 3 mod 3, $n \neq 6$ and $q = 3t+1$ is a prime, where $n_1 = q-1, n_2 = n'-1, n_3 = (q-1)(n'-1)$ (by Theorem 4.3 or Theorem 4.4).

5. Method of breaking up rows and/or columns

The method described in this section is also combinatorial in some sense. Starting from a rectangular PBIB design, similarly to the previous section, we can get another rectangular PBIB design by breaking up the rows and/or columns by other designs, for example, BIB designs.

Theorem 5.1. The existence of a rectangular PBIB design with parameters $v' = mn, b', r', k'; \lambda_1', \lambda_2', \lambda_3'; m, n$, and two BIB designs with respective parameters

$$v'' = m, b'', r'', k'' = k', \lambda'', \tag{5.1}$$

$$v''' = n, b''', r''', k''' = k', \lambda''' \tag{5.2}$$

implies the existence of a rectangular PBIB design with parameters

$$\begin{aligned} &v = mn, b = b' + nb'' + mb''', r = r' + r'' + r''', k = k'; \\ &\lambda_1 = \lambda_1' + \lambda''', \lambda_2 = \lambda_2' + \lambda'', \lambda_3 = \lambda_3'; m, n. \end{aligned} \tag{5.3}$$

Proof. Let the mn treatments be arranged in the rectangular array (3.1). We now construct a BIB design with parameters (5.1) on each of columns $\{jm+1, jm+2, ..., (j+1)m\}$, for $j = 0, 1, ..., n-1$, with blocks B_j'', and a BIB design with parameters (5.2) on each of rows $\{i, m+i, ..., (n-1)m+i\}$, for $i = 1, 2, ..., m$, with blocks B_i'''. Then $B' \cup (\cup_{0 \leq j \leq n-1} B_j'') \cup (\cup_{1 \leq i \leq m} B_i''')$

forms the blocks of the required design with parameters (5.3), where B' is the collection of the blocks of the starting rectangular PBIB design. □

For example, we have the following.

Theorem 5.2. Let p and q be primes such that $q \leq p$. Then the existence of a BIB design with parameters

$$v'' = q, b'' = \frac{eq(q-1)}{p(p-1)}, r'' = \frac{e(q-1)}{p-1}, k'' = p, \lambda'' = e \tag{5.3}$$

implies the existence of a rectangular PBIB design with parameters

$$v = pq, b = gq(q-1) + fg + eq(q-1)/\{p(p-1)\}, r = g(q-1) + f + e(q-1)/(p-1), k = p; \lambda_1 = e, \lambda_2 = f, \lambda_3 = g; m = p, n = q$$

for non-negative integers e and f and a positive integer g.

Proof. By Theorem 2.1 there exists a rectangular PBIB design with parameters (2.1). Now on each row of the association scheme of this design, construct a BIB design with parameters (5.3). Then the blocks of the required design is given by the union of the blocks of these p BIB designs and the blocks of the rectangular PBIB design with parameters (2.1). □

Applying Theorem 5.1 with the series of designs constructed at the end of Section 4, we have the following result, for which the proof is obvious.

Theorem 5.3. Let $n' \equiv 1$ or $3 \bmod 3$, $n' \neq 6, q = 3t+1$ be a prime. Then there exists a rectangular PBIB design with parameters

$$v = n'q, b = n'q\sum_{i=2}^{3} n_i\lambda_i/6 + fq(q-1)n'/3, r = \sum_{i=2}^{3} n_i\lambda_i/2 + f(q-1), k = 3; \lambda_1 = 2f, \lambda_2 = 2g, \lambda_3 = 2e; m = n', n = q$$

where $n_1 = q-1, n_2 = n'-1, n_3 = (q-1)(n'-1), e \geq 1, f, g \geq 0$.

Theorem 5.4. The existence of an orthogonal array of size $N = \lambda s^2$, k constraints, s levels, strength 2, and index λ, containing t (≥ 0) copies of columns $(x, x, ..., x)^T$ for all s entries implies a rectangular PBIB design with parameters

$$v = ks, b = \lambda s^2 + (f-t)s, r = \lambda s + f - t, k,$$
$$\lambda_1 = 0, \lambda_2 = \lambda + f - t, \lambda_3 = \lambda; m = k, n = s$$

for a non-negative integer f $(\geq t)$.

Proof. It is well known (cf. [3, VIII, Lemma and Definition 8.2]) that the orthogonal array is equivalent to a rectangular PBIB design with parameters $v' = ks, b' = \lambda s^2, r' = \lambda s, k; \lambda_1 = 0, \lambda_2 = \lambda_3 = \lambda; m = k, n = s$ where $(x, x, ..., x)^T$ in the theorem corresponds to the column of the rectangular association scheme AS(k, s) underlined. Now breaking up these columns by adding $(f - t)$ copies of the BIB design with parameters $v = k, b = r = 1, \lambda = 1$, we can get the required design. □

The next theorem is a variation of [5, Theorem 3.1].

Theorem 5.5. The existence of a rectangular PBIB design with parameters

$$v' = mn', b', r', k'; \lambda_1' = 0, \lambda_2', \lambda_3'; m, n' \tag{5.4}$$

and a BIB design with parameters

$$v'' = n, b'' = n', r'', k'', \lambda'' \tag{5.5}$$

implies the existence of a rectangular PBIB design with parameters

$$v = mn, b = b', r = r'r'', k = k'k''; \lambda_1 = r\lambda'',$$
$$\lambda_2 = r''^2\lambda_2', \lambda_3 = r''^2\lambda_3'; m, n.$$

Proof. Let the mn treatments be arranged in the rectangular array (3.1). We now construct m BIB designs with parameters (5.5) based on each of rows $\{i, m + i, ..., (n-1)m + i\}$ for $i = 1, 2, ..., m$, and then each block in the BIB designs constructed above can be represented by new treatments i_j, $j = 1, 2, ..., n'$. In this case, the following rectangle of m rows and n' columns as

$$\begin{matrix} 1_1 & 1_2 & \cdots & 1_{n'} \\ 2_1 & 2_2 & \cdots & 2_{n'} \\ \vdots & \vdots & & \vdots \\ m_1 & m_2 & \cdots & m_{n'} \end{matrix}$$

is arranged. For a rectangular PBIB design with parameters (5.4) based on the above array, each treatment i_j is replaced by any of the corresponding blocks of size k'' in the present BIB designs. It can be shown that this procedure yields the required design. □

Acknowledgements

The research was supported in part by Grand-in-Aid for Scientific Research, the Ministry of Education, Science and Culture under Contract Numbers (A)06302015 and (C)06640311, and by Research Fellowships of the Japan Society for the Promotion of Science for Young Scientists.

References

[1] A. Assaf, Modified group divisible designs, Ars Combin. 29(1990), 13-20.

[2] A. Assaf, An application of modified group divisible designs, J. Combin. Theory (A) 68(1994), 152-168.

[3] T. Beth, D. Jungnickel and H. Lenz, "Design Theory", Bibliographisches Institut, Zurich, 1985.

[4] Bhagwandas, K. Sinha and S. Kageyama, Constructions of PBIB designs based on nested group divisible association scheme, Utilitas Math. 41(1992), 169-174.

[5] X. Duan and S. Kageyama, Constructions of group divisible designs, preprint, 1994.

[6] S. Furino, Y. Miao and J. Yin, "Frames and Resolvable Designs", preprint, 1994.

[7] D. K. Ghosh and M. N. Das, Construction of two way group divisible designs, Sankhyā B 51(1989), 331-334.

[8] R. C. Mullin and R. G. Stanton, Balanced weighing matrices and group divisible designs, Utilitas Math. 8(1975), 303-310. Corrigenda: ibid 9(1976), 347.

[9] Raghavarao, D. (1971). "Constructions and Combinatorial Problems in Design of Experiments". Wiley, New York.

[10] J. Seberry and M. Yamada, "Hadamard Matrices, Sequences and Block Designs", Contemporary Design Theory (eds. Dinitz and Stinson), Wiley, 1992, 431-560.

[11] B. V. Shah, An upper bound for the number of disjoint blocks in certain PBIB designs, Ann. Math. Statist. 35(1964), 398-407.

[12] M. N. Vartak, On an application of Kronecker product of matrices to statistical designs, Ann. Math. Statist. 26(1955), 420-438.

[13] M. N. Vartak, The non-existence of certain PBIB designs, Ann. Math. Statist. 30(1959), 1051-1062.

ON CONNECTED DOMINATION NUMBER OF A GRAPH.

Kang Liying
(Dept. of Basic Courses, Shijiazhuang
Railway Institute, Shijiazhuang, Hebei 050041, P. R. China)
Shan Erfang
(Dept. of Math. Shijiazhuang Normal
College, Shijiazhuang, Hebei 050041, P. R. China)

Abstract. Let G and $\overline{G}$ be connected graphs, Where $\overline{G}$ is the complement of G. A domination set D of G is called a connected dominating set if the induced subgraph $G[D]$ is connected. $\gamma_c(G)$, $ir(G)$ and $d_c(G)$ denote the connected domination number, irredundance number and connected domatic number of a graph G, respectively. The purpose of this paper is to investigate the relations between these domination parameters, the following two results are proved.

(1) $\gamma_c(G) \leq 3ir(G) - 2$

(2) If $\gamma_c(\overline{G}) \neq 2$, then $\gamma_c(G) \leq 3d_c(\overline{G})$

and above inequalities are the best possible.

Key Words: Graph, Dominating Set, Irredundance number, Domatic number.

1. Introduction

For graph theory terminology not presented here, We follow [1]. All graph will be undirected and finite without any loops or multiple edges.

We denote by $V(G)$ and $E(G)$ the vertex set and the edge set of G, respctively, $\overline{G}$ is the complement of G. The neighbourhood $N_G(v)$ of a vertex v is the set of vertices adjacent to v in G, While the closed neigbourhood $N_G[v]$ of v is defined as $N_G(v) \cup \{v\}$, more generally, $N_G(X) = \cup_{x \in X} N_G(x)$ for a subset X of $V(G)$, and $N_G[X] = \cup_{x \in X} N_G[x]$. If $X \subseteq V(G)$, then $G[X]$ (resp. $G - X$) denotes the subgraph of G induced by X. (resp. $V(G) - X$).

A dominating set of G is a subset D of $V(G)$ such that each vertex of $V(G) - D$ is adjacent to at least a vertex in D. A dominating set D is called a connected dominating set if the induced subgraph $G[D]$ of G is connected. The concept of connected domination was introduced by E. Sampathkumar and H. B. Walikar [2] and it has been studied in [3], [4]. A vertex x of a subset X of $V(G)$ is redundant in X if $N_G[x] \subseteq N_G[X - \{x\}]$, and X is irredundant if it contains no redundant vertices. The cardinality $\gamma(G) = |D|$ of a minimum dominating set D is called the domination number of G, and the connected domination number $\gamma_c(G)$ is the minimum cardinality among all connected dominating sets of G. A maximal irredundant set X of minimum cardinality is called an ir-set, and $ir(G)=|X|$ is called the irredundance number of G. Zelinka[5] introduced the connected domatic number of a connected graph, denoted by $d_c(G)$, which equals the maximum order of a partition $\{V_1, V_2, \cdots, V_n\}$ of $V(G)$ such that every set V_i is a connected dominating set of G.

In this paper, the relations will be discussed between these domination parameters. For connected graphs G and $\overline{G}$, we show that $\gamma_c(G) \leqslant 3ir(G)-2$, and the inequality can not be improved. This result improves a result due to L. Sun[6]. We also show that $\gamma_c(G) \leqslant 3d_c(\overline{G})$ if $\gamma_c(\overline{G})$ is not equal to 2.

2 Relations Between Connected Domination Number And Irredundance Number

Let X be an irredundant set. For any $x \in X$ the set $N_G[x]-N_G[X-\{x\}]$ is denoted by $PN_G(x)$, and let U be those vertices of G not dominated by X.

Lemma 1. Let X be a maximal irredundant set, and $U \neq \varnothing$, then for any $u \in U$, there exists at least a vertex $x \in X$ such that $d_{G[X]}(x) \geqslant 1$ and $PN_G(x) \subseteq N_G(u)$.

proof. for each $u \in U$, define

$$X_u = \{x \in X \mid PN_G(x) \subseteq N_G(u)\},$$

then $X_u \neq \varnothing$. Assume to the contrary, that for any $x \in X$, $PN_G(x) \nsubseteq N_G(u)$. Since u is not dominated by X, $X \cup \{u\}$ must be an irredundant set, contrary to hypothesis.

We claim that there is a $x \in X_u$ with $d_{G[X]}(x) \geqslant 1$. Suppose that X_u is an isolated vertex set in $G[X]$. For any $x \in X$, then $N_G[X \cup \{u\}-\{x\}] \not\supseteq N[x]$. Therefore $X \cup \{u\}$ is an irredundant set of G, a contradiction. This completes the proof of the lemma 1.

Theorem 1. If G is a connected graph, then $\gamma_c(G) \leqslant 3ir(G)-2$.

proof. Let $X=\{v_1, v_2, \cdots, v_m\}$ be a maximal irredundant set such that $|X|=ir(G)$.

If $U=\varnothing$. then X is a dominating set of G, since G is a connected graph, there must exist vertices $y_1 \in N_G[v_1]$ and $z_1 \in N_G[v_2] \cup N_G[v_3] \cup \cdots \cup N_G[v_m]$ such that $y_1 = z_1$ or $y_1 z_1 \in E(G)$. Without loss of generality, assume $z_1 \in N_G[v_2]$. Similarly, there must exist $y_2 \in N_G[v_1] \cup N_G[v_2]$ and $z_2 \in N_G[v_3] \cup \cdots \cup N_G[v_m]$ such that $y_2 = z_2$ or $y_2 z_2 \in E(G)$, assume $z_2 \in N_G[v_3]$. Continued the process, we can obtain a vertex set

$$Y=\{y_1, y_2, \cdots, y_{q-1}, z_1, z_2, \cdots, z_{q-1}\}$$

where $q \leqslant ir(G)$. Obviously, $G[X \cup Y]$ is a connected subgraph of G, thus $\gamma_c(G) \leqslant |X \cup Y| \leqslant 3ir(G)-2$.

If $U \neq \varnothing$. Now define

$$M=\{v \in X \mid \text{there exist } u \in U \text{ such that } d_{G[X]}(v) \geqslant 1 \text{ and } PN_G(v) \subseteq N_G(u)\}.$$

By lemma 1, $M\neq\emptyset$. Without loss of generality, assume

$$M=\{v_1,v_2,\cdots,v_s\},$$

where $1\leqslant s\leqslant m$. For each $v_i\in M$, it follows that $PN_G(v_i)\neq\emptyset$ and $v_i\notin PN_G(v_i)$. Take $u_i\in PN_G(v_i)$, put $D=X\cup\{u_1,u_2,\cdots,u_s\}$. We claim that D is dominating set of G such that $|D|\leqslant ir(G)+s$. Assume to the contrary, there is a vertex $x\in V(G)-D$ such that x is not dominated by D, obviously, $x=\in U$ and $N_G[x]\nsubseteq N_G[X]$. For some $v_i\in X$, if $N_G[v_i]\subset N_G[X-\{v_i\}]\cup N_G[x]$, then $PN_G(v_i)\subseteq N_G[x]$ and $d_{G[X]}(v_i)\geqslant 1$, that is, $v_i\in M$, hence $u_i\in N_G[x]$. This means that x is adjacent to $u_i\in D$, a contradiction. Thus, for any $v_i\in X$, we have.

$$N_G[v_i]\not\subset N_G[X-\{v_i\}]\cup N_G[x]$$

Therefore, $X\cup\{x\}$ is an irredundant set. This is a contradiction with maximality of X.

Let $X_1,X_2,\cdots,X_p$ denote the vertex sets of all connected components in $G[D]$, then $1\leqslant p\leqslant ir(G)-\frac{s}{2}$. Define

$$Y_i=\{x\mid x\in X_i \text{ or } x \text{ is adjacent to at least a vertex in } X_i\},$$

then $G_i=G[Y_i]$ is a connected subgraph of G, and $G=\bigcup_{i=1}^{p}V(G_i)$. If $p=1$, then $G[D]$ is connected, this theorem is immediatly proved. We suppose $p\geqslant 2$. Since G is a connected graph, there must exist vertices $y_1\in Y_1$ and $z_1\in Y_2\cup Y_3\cup\cdots\cup Y_p$ such that $y_1z_1\in E(G)$ or $y_1=z_1$. Without loss of generality, assume $z_1\in Y_2$, similarly, there must exist vertices $y_2\in Y_1\cup Y_2$ and $z_2\in Y_3\cup Y_4\cup\cdots\cup Y_p$ such that $y_2z_2\in E(G)$ or $y_2=z_2$, assume $z_2\in Y_3$. Continued the process as such, we can obtain a vertex set $Y=\{y_1,y_2,\cdots,y_{q-1},z_1,z_2,\cdots,z_{q-1}\}$, where $q\leqslant p$. Obviously, $G[D]$ is a connected subgraph of G. So we have

$$\gamma_c(G)\leqslant|D\cup Y|\leqslant ir(G)+s+2(q-1)\leqslant 3ir(G)-2.$$

This finishes the proof

We notice that the inequality established in Theorem 1 can not be improved. Suppose $H_1=H_2=C_4$, let $G=H_1\cup H_2\cup\{v_1v_2\}$, where $v_1\in V(H_1)$, $v_2\in V(H_2)$. It is easy to see that $\gamma_c(G)=4$ and $ir(G)=2$, hence $\gamma_c(G)=3ir(G)-2$.

3. Relations Between Connected Domination Number of G And Connected Domatic Number of $\overline{G}$.

We begine by stating two known results in [1] and [4].

Lemma 2[1]. If G is a connected graph, then G contains at least two vertices which are not cut-vertex of G.

Lemma 3[4]. If G is a connected graph of order n, $\triangle(G)$ denote the maximum degree of G, then $\gamma_c(G)\leqslant n-\Delta(G)$.

An immediate corollary now follows by Lemma 3.

Corollary 1. Let G and $\overline{G}$ be connected graphs such that $\delta(\overline{G})\leqslant 2$, then $\gamma_c(G)\leqslant 3d_c(\overline{G})$.

Furthermore, we have the following theorem.

Thorem 2. Let G and $\overline{G}$ be connected graphs such that $\gamma_c(\overline{G})\neq 2$, then $\gamma_c(G)\leqslant 3d_c(\overline{G})$.

Proof. Let $d_c(\overline{G})=k$, and $D_1, D_2, \cdots, D_k$ be pairwise disjoint minimal connected dominating sets of $\overline{G}$. Since G is a connected graph and $\gamma_c(\overline{G})\neq 2$, this implies $\gamma_c(\overline{G})\geqslant 3$. By lemma 2, $\overline{G}[D_i]$ has at least two vertices, say x_i and y_i, which are not cut-vertex in $\overline{G}[D_i]$. Therefore, each of the sets $D_i-\{x_i\}$ and $D_i-\{y_i\}$ is not a dominating set of $\overline{G}$, that is, there exist two vertices $x_i^*, y_i^*\in V(\overline{G})-D_i$ such that x_i^* (y_i^*) is adjacent to no vertex of $\overline{G}[D_i]$ but $x_i(y_i)$ in $\overline{G}$. Define

$$A=\{x_i, y_i \mid 1\leqslant i\leqslant k\}$$
$$B=\{x_i^*, y_i^* \mid 1\leqslant i\leqslant k\}$$

For any $v\in D_1\cup\cdots\cup D_k$, it is easy to see that v is adjacent to at least one vertex of B in G.

If $\overline{G}-(D_1\cup D_2\cdots\cup D_k)=\varnothing$. then B is a dominating set of G, and $G[B]$ contains no isolated vertex. In other words, $G[B]$ has at most k connected components. Let $H_1, H_2, \cdots, H_q$ ($q\leqslant k$) be all connected component of $G[B]$. Take a vertex $v_i\in H_i$, for each $i\in\{1,2,\cdots,q\}$. As $\overline{G}$ satisfies $\gamma_c(\overline{G})\geqslant 3$, there must exist a vertex $u_i\in V(\overline{G})$ such that $u_iv_i, u_iv_{i+1}\notin E(\overline{G})$, where $1\leqslant i\leqslant q-1$. So $u_iv_i, u_iv_{i+1}\in E(G)$. Put $D=B\cup\{u_1, u_2, \cdots, u_{q-1}\}$, then D is a connected dominating set of G. Consequently, $\gamma_c(G)\leqslant |D|=2k+q-1\leqslant 3k$.

If $\overline{G}-(D_1\cup\cdots\cup D_k)\neq\varnothing$, we consider two cases.

Case 1: If $\overline{G}-(D_1\cup\cdots\cup D_k)$ is a connected subgraph of $\overline{G}$, then the set $V(\overline{G})-(D_1\cup\cdots\cup D_k)$ is not dominating set of $\overline{G}$, that is, there exists a vertex $x\in D_1\cup\cdots\cup D_k$ not dominated by $V(\overline{G})-(D_1\cup\cdots\cup D_k)$.

(a) $k=1$. Then $x\neq x_1^*, x\neq y_1^*$. This shows that $D=\{x, x_1^*, y_1^*\}$ is a connected dominating set of G, hence $\gamma_c(G)\leqslant |D|=3$.

(b) $k\geqslant 2$. Note that the set $B_1=B\cup\{x\}$ is a dominating set of G, and the induced subgraph $G[B_1]$ in G has at most k connected components. Let q be the connected component number of $G[B_1]$. Since $\gamma_c(\overline{G})\geqslant 3$, it is easy to realize that we choose the suitable vertices $u_1, u_2, \cdots, u_{q-1}$ in $V(G)$ such that $D=B\cup\{u_1, u_2, \cdots, u_{q-1}\}$ is a connected dominating set of G. So we have

$$\gamma_c(G)\leqslant |D|=2k+q\leqslant 3k.$$

Case 2: If $\overline{G}-(D_1\cup D_2\cup\cdots\cup D_k)$ is not connected. Let $S_1, S_2, \cdots, S_t$ be all connected components of $\overline{G}-(D_1\cup\cdots\cup D_k)$. If $k=1$, the veracity of the theorem is easy to be verified. Assume $k\geqslant 2$ in the following proof.

If there exist $x_i^*\in V(\overline{G})-(D_1\cup\cdots\cup D_k)$ in B. Assume $x_i^*\in V(S_1)$, take any $u_1\in V(S_2)$, then $G[B\cup\{u_1\}]$ has at most k connected components. As case 1, we can choose $q-1$ vertices

$u_2, u_3, \cdots, u_q$, where q is the number of connected components of $G[B \cup \{u_1\}]$. Thus, the set $D = B \cup \{u_1, u_2, \cdots, u_q\}$ is a connected dominating set of G. So we have $\gamma_c(G) \leqslant |D| \leqslant 3k$. Otherwise, $B \subset D_1 \cup D_2 \cup \cdots \cup D_k$. Then $G[B]$ contains no isolated vertex, that is , $G[B]$ has at most k connected components. Since G is connected, there exist $z_1 \in D_1 \cup \cdots \cup D_k, z_2, z_3 \in V(G) - D_1 \cup \cdots \cup D_k$ such that $z_1 z_2 \in E(G), z_2 \in V(S_i)$ and $z_3 \in V(S_j)$, where $i \neq j$, We consider two subcases.

Case 2.1: If some D_i contains three elements x^*, y^*, z^* of B. Let $A = \{x_i^*, y_i^*, x^*, y^*, z^*\}$, then $G[A]$ is a connected subgraph of G, hence $G[B]$ has at most $k-2$ connected conponents. As above, we can easily choose at most $q-3$ vertices $u_1, u_2, \cdots, u_{q-3}$ in G, where $q \leqslant k$. Put $D = B \cup \{u_1, u_2, \cdots, u_{q-3}, z_1, z_2, z_3\}$, then D is a connected dominating set of G. Consequently, $\gamma_c(G) \leqslant |D| \leqslant 3k$.

Case 2.2: Otherwise, every set D_i contains exactly two elements of B. If $G[B]$ has at most $k-2$ connected components, then this is similar to case 2.1 and is therefore omitted. We assume $G[B]$ has either k or $k-1$ connected components.

We first consider that $G[B]$ has exactly k connected components. Right now, it is easy to observe that $A=B$ and each conponent of $G[B]$ is isomorphic to K_2. Take a vertex v_i in each component of $G[B]$, put $B_1 = \{v_1, v_2, \cdots, v_k\}$, then B_1 must be a dominating set of $G[D_1 \cup \cdots \cup D_k]$. Thus, we can choose $k-1$ vertices $u_1, u_2, \cdots, u_{k-1}$, such that $D = B_1 \cup \{u_1, u_2, \cdots, u_{k-1}, z_1, z_2, z_3\}$ is a connected dominating set of G, hence $\gamma_c(G) \leqslant |D| \leqslant 3k$.

Now, we consider that $G[B]$ has exactly $k-1$ connected components. Let $H_1, H_2, , H_{k-1}$ be all connected components of $G[B]$. When $A=B$, discuss it as above. We assume $A \neq B$, suppose $x_1 \notin B$, $x_i^*, y_j^* \in D_1$, it is seen at once that $G[x_1^*, y_1^*, x_i^*, y_j^*]$ is only one component of $G[B]$ with order more than 3, without loss of generality, assume $H_1 = G[x_1^*, y_1^*, x_i^*, y_j^*]$, then $H_i \cong K_2$, for each $2 \leqslant i \leqslant k-1$. Take a vertex v_i in every H_i but H_1, Put $B_1 = \{x_1^*, y_1^*, x_i^*, y_j^*, v_2, v_3, \cdots, v_{k-1}\}$, then B_1 must be a dominating set of $G[D_1 \cup \cdots \cup D_k]$. Thus , we are able to easily choose $k-2$ vertices $u_1, u_2, \cdots, u_{k-2}$ such that $D = B_1 \cup \{u_1, u_2, \cdots, u_{k-2}, z_1, z_2, z_3\}$ is a connected dominating set of G. Consequently, $\gamma_c(G) \leqslant |D| \leqslant 3k$. This completes the proof of theorem 2.

By the proof of theorem 2, it is not hard to established the following two corollaries.

Corollary 2. Let G and $\overline{G}$ be connected graphs such that $d_c(\overline{G}) = 1$, then $\gamma_c(\overline{G}) \leqslant 3$.

Corollary 3. If G is a connected graph, then $\gamma(\overline{G}) \leqslant 2(d_c(G)+1)$.

Finally, we notice that the inequality in Theorem 2 is best possible but $\gamma_c(\overline{G}) = 2$. Let $G = \overline{C}_5$, it is easy to see that $\gamma_c(G) = 3d_c(\overline{G}) = 3$. Moreover, when $\gamma_c(\overline{G}) = 2$, it is probable that the inequality established in Theorem 2 holds. Perhaps corollary 3 can be improved to $\gamma(\overline{G}) \leqslant 2d_c(G)+1$.

Acknowledgement

The authors thank the referees for helpful comments and valuable suggestion. We are grateful to Professor L. Sun, for providing the problem.

References

1. J. A. Bondy and U. S. R. Murty. Graph Theory Wtih Application, North—Holland, Amsterdam, 1976.

2. E. Sampathkumar and H. B. Walikar, The connected domimation number of a graph, Math. Phys. Sci. 13(1979), 607—613.

3. R. Laskar and K. Peters, Connected domination of complementary graphs, Technical Report 429, Clemson Univ. (1983).

4. S. T. Hedetniemi and R. Laskar, Connected domination in graphs, Graph Theory and Combinatorics, (1984), 209—217, Academic Press London.

5. B. Zelinka, Connected domatic number of a graph, Math. Slovaca 36(1986), 387—392.

6. L. Sun, Some results on connected domination in graphs, Math. Applicata 5(1992), 29—34.

An Algorithm for Constructing a Rectilinear Embedding of a Given Graph in the Plane*

Serge Lawrencenko[a], Xin Liu[b], and Yanpei Liu[c]
[a]Department of Mathematics & Computer Center, Northern Jiaotong University, Beijing, 100044
[b]Institute of Applied Mathematics, Academia Sinica, Beijing, 100080
[c]Department of Mathematics, Northern Jiaotong University, Beijing, 100044

Abstract We suggest an algorithm to construct a rectilinear embedding of a given planar graph, with every nede of valency 4, embedded in the plane $R^2(x, y)$; the resulting rectilinear embedding satisfies the following conditions: (1) the constructed embedding be topologically equivalent to the given one, (2) all nodes be represented by points having both coordinates x and y integer, (3) one of the apriory chosen nodes on the exterior face boundary be represented by a point having a minimum ordinate, and the other maximum, and (4) all edges of the constructed embedding be with at most 2 bends each, exept possibly either one edge with 4 bends, or at most two edges with 3 bends.

Key words rectilinear embedding; plane graph; VLSI circuit design.

1. Introduction

The problem that we solve here is that of constructing a rectilinear embedding a given 4-valent planar graph G in the Euclidean plane $R^2 = R^2(x, y)$ with some additional restrictions. This problem has emerged from the VLSI (Very Large-Scale Integrated) circuit design; the nodes of the graph correspond to electronic components, and its edges to the conducting lines connecting that components.

The Problem is formally stated in Section 3. The algorithm to sove it consists of nine procedures, correspondingly to Sections 5-13; in Section 4 we enunciate how we store the data. Our algorithm is a development of the approach of Liu, Morgana, and Simeone [1, 2].

2. Plane graphs

In this paper we deal with both undirected graphs and directed ones (digraphs). The term "edge" will mean an undirected edge of the graph, while "arc" will mean a directed edge of the digraph in question. By $[u, v]$ we will denote an edge joining u and v, while by (u, v) we will denote the arc directed (or "oriented") from node u to node v. When we use "edge" for the arcs of some digraph, it means that we refer to the edges of the underlying (undirected) graph. By $V(\cdot)$ we denote the node set of the graph or the digraph in question.

We will distinguish between the terms "planar graph" and "plane graph"; the former means an abstract graph which can be embedded in R^2, while the latter will mean an abstract graph G which is actually embedded in R^2. By G we will genote a given plane

* The research was partly supported by Lu Jiashi Fund of the Academia Sinica, and partly by the Northern Jiaotong University, Beijing 100044, P.R. China.

graph, and by N its order; $N = |V(G)|$. Thus, a plane graph corresponds to some (topological) plane embedding of a planar graph. Such an embedding (and thus the plane graph itself) can be given by explicitly listing the faces (regions); or alternatively by its *rotation scheme*, that is, a collection of N cyclic sequences corresponding to the nodes of G, that specify the cyclic order in which the neighbors of each node are placed clockwise around that node. Two plane embeddings are said to be *topologically equivalent* if they can be defined by the same rotation scheme. Note that given a plane embedding one and only one of the faces is unbounded, and is thus distinguished as the *exterior face*, denoted by F_{ext} and its boundary by ∂F_{ext}.

3. The Problem

This is to embed a given 4-valent graph G in the plane $R^2 = R^2(x, y)$ in such a way that all edges of G are represented by broken line segments, with each link (a straight line segment) parallel to either $x = 0$ or $y = 0$. Such an embedding is called a *rectilinear embedding*. The break points of a line representing an edges of G is called the *bends* of that edge, and an edge with n bends will be called an *n-bended edge*. Formally the Problem is stated as follows:

Given: a 2-connected 4-valent plane graph G of order N with two of the nodes on ∂F_{ext} chosen as the "bottom" and the "top".

Problem: construct a rectilinear embedding of G in the plane $R^2(x, y)$, satisfying the following conditions:

(i) the resulting embedding be topologically equivalent to the given plane embedding of G;

(ii) every node be represented by a point (x, y) with x and y both integer;

(iii) some bend of a ceratin edge e_{bot} [resp., e_{top}] incident to the bottom [resp., the top] is to have a minimum [resp., maximum] ordinate among the points of the rectilinear embedding;

(iv) the edges of the rectilinear embedding are to have at most 2 bends each, except possibly either one 4-bended edge (which will be e_{bot}, and will coincide with e_{top} in that case), or else one or two 3-bended edges (one of which will be e_{bot} or e_{top}, while the other, if such exists, will be e_{top} or e_{bot}, respectively).

4. Storage of the data

The following are the main variables:

- *src* and *snk* -the source and the sink, respectively;

- *src*$_{im}$ and *snk*$_{im}$ -the imaginary source and the imaginary sink, respectively. Initially these (integer) variables are assigned the value of −1, which means that the imaginary source and imaginary sink do not exist until new values are assigned to them explicitly.

The following is a list of the main arrays:

- *rot* = *rot*(*u*, *i*) -this array specifies the rotation at node $u \in V(G)$, where *i* takes the values in the set {0,1,2,3} of the ordering numbers of the neighbors of *u* (i.e., the nodes adjacent to *u*) under some numbering in clockwise order around *u*. As matter of notation, "*rot*(*u*, *i*)" will be often shortened to "u_i". Thus u_0, u_1, u_2, and u_3 are the four neighbors of *u*, in cyclic order they apper as one proceeds clockwise around *u*. Addition of the indices will be always assumed to be modulo 4; thus, the node $u_{i+1} = u_{i+1 (\mathrm{mod}\ 4)}$ is the one occurring next to u_i clockwise around *u*.

Convention. When inputting the array *rot*, the initial neighbors u_0 are to be chosen arbitrarily for the nodes *u* which are not on ∂F_{ext}; but if *u* is on ∂F_{ext}, as u_0 ought to be chosen that of the two neighbors of *u* on ∂F_{ext} which has both u_0 and u_3 on ∂F_{ext}.

The procedure "Extension" (Section 9) extends the digraph G_d to the so-called *extension* of G_d, which graph will be denoted by ext(G_d); that procedire is to update the array *rot* to accomodate two new nodes, the imaginary source and the imaginary sink.

- *pointer*(*u*, *i*) - this specifies the pointer at *u* towards u_i, for $u \in V(\mathrm{ext}(G_d))$.

Given an edge [*u*, *v*] of *G* (or G_d, or ext(G_d)), associated with that edge are two *semiedges* [*u*, *i*) and [*v*, *j*), with indices *i* and *j* such that *rot*(*u*, *i*) = *v* and *rot*(*v*, *j*) = *u*, corresponding to the two ordered pairs of the nodes incident to that edge. The procedure "Assignment of pointers" (Section 7) establishes a special mapping from the set {"N", "E", "S", "W"} of the so-called pointers to the set of semiedges of G_d. That mapping will be extended to ext(G_d), by the procedure "Extension". To assign a pointer to a given semiedge [*u*, *i*) we will set *pointer*(*u*, *i*) to a desired value. As matter of notation, it will be sometimes convenient to shorten "*pointer*(*u*, *i*) for *i* such that *rot*(*u*, *i*) = *v*" (as it is actually realized in our computer program) to "*pointer*($u \to v$)", and say that it is the pointer *at* node *u towards* node *v*, or that the pointer is assigned to the semiedge [*u*, *i*).

- *orn*(*u*, *v*) -this is the array specifying the bipolar orientation constructed by the procedure "Determining a bipolar orientation" (Section 6), which changes the given undirected graph *G* into a bipolarly oriented digraph G_d; if that procedure orients edge [*u*, *v*] from *u* to *v*, then it set *orn*(*u*, *v*) to 1, and to −1 if in the opposite direction.

- $adm_H(u)$ and $adm_V(u)$ -these arrays specify the ordering numbers of that H-class and the V-class, respectively, to which belongs node u; these arrays are assigned values as a result by the "H- and V-Partitions" procedure (Section 8); these classes are then extended to $V(\text{ext}(G_d))$ by the "Extension" procedure.

- $arc_H(i, j)$ and $arc_V(i, j)$ are in fact adjacency matrices of the H- and V-digraphs, respectively, constructed by the procedure "H- and V-Digraphs" (Section 10).

- $adm_{add}(u, i)$ -the array specifying the additional V-classes introduced by the procedure "H- and V-Digraphs"; $adm_{add}(u, i) \geq 0$ indicates that a pair of additional nodes is added to the edge $[u, u_i]$. This pair constitutes an additional V-class having number $adm_{add}(u, i)$; the numbering of additional V-classes continues that of original V-classes; $adm_{add}(u, i) = 0$ indicates the absence of any additional node on $[u, u_i]$.

- $x(u)$ and $y(u)$ -these are for the coordinates of node $u \in V(\text{ext}(G_d))$;

- $x_{bend}(u, i)$ and $y_{bend}(u, i)$ -these are for the coordinates of that node of the additional V-class on $[u, u_i]$ (if such introduced) which is adjacent to u.

5. The main procedure

Assign:
src = "bottom",
snk = "top".
Call *Determining a bipolar orientation* (Section 6).
Call *Assignment of pointers* (Section 7).
Call *H- and V-Partitions* (Section 8).
Call *Extension* (Section 9).
Call *H- and V-Digraphs* (Section 10).
1st call of *Computation of coordinates* (Section 11).
Call *Elimination of crossings* (Section 12).
2nd call of *Computation of coordinates.*
Stop.

The first (Section 3.3) changes G into a bipolar digraph G_d, that is, an acyclic digraph with precisely one sourse and precisely one sink. In sections 3.5--3.7, associated with G_d are consctucted the so-called horizontal and the vertical digraphs, denoted by D_H and D_V. These digraphs will be heavily used in Section 3.8 for computing the coordinates x and y of the nodes of G, and obtain a desired embedding.

6. Determining a bipolar orientation

Our first procedure converts G to a bipolar digraph G_d. More accurately, node v of D whose incident arcs are all directed away from [resp., toward] v is said to be a *source* [resp., *sink*]; a digraph D is said to be *acyclic* if it contains no directed cycles, and a *bipolar digraph* or alternatively a *digraph with a bipolar orientation* if it is an acyclic digraph with precisely one sourse and one sink. Now we can state formally what is to be done in this section: Given two nodes, *src* and *snk*, of G on ∂F_{ext} we suggest a procedure to assign orientations to the edges of G so that the resultant digraph, G_d, is bipolar and has *src* as the source, and *snk* as the sink.

We shall now describe the procedure, but first we explain the idea. By $P = P(u, v)$ we denote the path of G having u and v as its endpoints. The procedure orients G path by path; the edges oriented by a current moment make up, together with the incident nodes, what we shall call the COS, for "current oriented subgraph" (of G). The nodes of the COS are in the natural partial order: "$u \nearrow v$" (and we say "v is further than u") if there is a directed path of the COS from u to v. At Step 3 we find a path $P = P(u, v)$ with just its terminal nodes, u and v, in the COS. For this, starting from $[u, u_i]$, we trace a path $P = P(u, v)$ (yet unoriented) through the nodes $p_0(=u)$-$p_1(=u_i$-$)$-...-$p_l(=v)$, proceeding clockwise around the boundary of the face meeting both $[u, u_i]$ and $[u, u_{i+1}]$ until a first node, v, in the COS is reached.

Step1. Initially, for two nodes u and v, set $u \nearrow v$ if $u = src$ and $v = snk$; otherwise u and v are to be incomparable. Let COS initially be the path $P = P(u, v)$ proceeding clockwise around ∂F_{ext} from $u = src$ to $v = snk$. Go to Step 4.

Step 2. Pick an edge $[u, u_i]$ (if any) such that neither (u, u_i) nor (u_i, u) is not an arc of the COS but either (u, u_{i+1}) or (u_{i+1}, u) is. Stop if there is no such an edge left.

Step 3. Set $p_0 = u$, $p_1 = u_i$. Then, repeatedly for $l = 2, 3, \ldots$ until $p_l \in V(\text{COS})$, determine k from the equation $\text{rot}(p_{l-1}, k+1) = p_{l-2}$, and set $p_l = \text{rot}(p_{l-1}, k)$. Set v to p_l corresponding to the last value of l. Add to the COS the path $P(u, v)$ $= p_0$-p_1- ... -p_l (yet unoriented).

Step 4. Orienting the path $P(u, v)$: If $u \nearrow v$, we orient P from u to v, and in the opposite direction otherwise (including the case u and v are incomparable).

Step 5. Extending the partial order "$\nearrow$" from $V(\text{COS}) - V(P(u, v))$ to $V(P(u, v))$.

Case 1: $P(u, v)$ is directed from n to m. Then, for n and m both in $V(P(u, v))$, set $n \nearrow m$; for n in $V(P(u, v))$ but m not, set $n \nearrow m$ if $v \nearrow m$, and $m \nearrow n$ if $m \nearrow v$.

Case 2: $P(u, v)$ is directed from m to n. Then, for n and m both in $V(P(u, v))$, set $m \nearrow n$; for n in $V(P(u, v))$ but m not, set $m \nearrow n$ if $v \nearrow m$, and $n \nearrow m$ if $m \nearrow v$.

Return to Step 2.

7. Assignment of pointers

In this section we establish a special mapping from the set {"N", "E", "S", "W"} of four *pointers*, conventionally associated with "North", "East", "South", and "West", to the set of semiedges of G_d.

Step 1. For each node *u* other than the source and the sink, the assignment of pointers proceeds as follows. First, compute $\sigma = \Sigma_{i=0}{}^3\, orn(u, u_i)$. If $\sigma = 2$, we identify such that $orn(u, u_i) = -1$, after which we assign: *pointer*(*u*, *i*) = "S", *pointer*(*u*, *i*+1) = "W", *pointer*(*u*, *i*+2) = "N", and *pointer*(*u*, *i*+3) = "E". If $\sigma = -2$, we identify *i* such that $orn(u, u_i) = 1$, and assign: *pointer*(*u*, *i*) = "N", *pointer*(*u*, *i*+1) = "E", *pointer*(*u*, *i*+2) = "S", and *pointer*(*u*, *i*+3) = "W". If $\sigma = 0$, we identify *i* such that $orn(u, u_i) = -1$ but $orn(u, u_{i+1}) = 1$, and assign: *pointer*(*u*, *i*) = "S", *pointer*(*u*, *i*+1) = "W", *pointer*(*u*, *i*+2) = "N", and *pointer*(*u*, *i*+3) = "E".

Step 2. Assigning the pointers at *src*. If *pointer*(src_3, 0) ≠ "E", then we assign: *pointer*(*src*, 0) = "W", *pointer*(*src*, 1) = "N", *pointer*(*src*, 2) = "E", and *pointer*(*src*, 3) = "S"; otherwise we assign: *pointer*(*src*, 0) = "S", *pointer*(*src*, 1) = "W", *pointer*(*src*, 2) = "N", and *pointer*(*src*, 3) = "E".

Step 3. Assigning the pointers at *snk*. If *pointer*(snk_0, 3) = "E", and $snk_3 \neq src$, then we assign: *pointer*(*snk*, 0) = "E", *pointer*(*snk*, 1) = "S", *pointer*(*snk*, 2) = "W", and *pointer*(*snk*, 3) = "N"; otherwise we assign: *pointer*(*snk*, 0) = "N", *pointer*(*snk*, 1) = "E", *pointer*(*snk*, 2) = "S", and *pointer*(*snk*, 3) = "W";

Remark: The conditions at Steps 2 and 3, somewhat cryptic at first sight, become more clear if the reader recalls the Convention (Section 4) which imposes the direction of rotation around the nodes on ∂F_{ext}; thus, *rot*(src_3, 0) = *src*, and *rot*(snk_0, 3) = *snk*.

8. H- and V-Partitions of the node set

Here we expound two procedures; one is to partition $V(G_d)$ into so-called horizontal classes ("H-classes"), and the other partitions $V(G_d)$ into vertical classes ("V-classes"). We will refer to such partitions as the *H-partition* and *V-partition*, respectively. (The recursive construction of our procedures may serve as inductive definitions of these classes.) We shall now give the procedures, but first we introduce some new concepts and make several motivating observations.

Below is the procedure to obtain an H-partition of $V(G)$. We will say that a node is *H-admitted* if it is already in some H-class. One other specification for the current node, namely, being *H-checked*, appears from the following construction. For the current H-admitted node *u*, let $v = u_i$ be such a node that *pointer*(*u*, *i*) = "W". Let *j* be such that rot(*v*, *j*) = *u*. If pointer(*v*, *j*) = "E", we admit *v* to that H-class with number

$adm_H(u)$ (i.e., to which u belongs). Then automatically we will have $pointer(u, i+2) =$ "E". If $pointer(u_{i+2} \to v) =$ "W", we also admit the node u_{i+2} to the same H-class, $adm_H(u)$. After this operation, u is regarded as *H-checked*, and we will refer to this operation as the *H-enlargement* around a given node u.

Step1. Initially, we pick an arbitrary node and indent our first H-class consisting (at that moment) of that node only.

Step 2. While there remains at least one node of G which is not H-admitted, repeatedly apply the two following steps.

Step 3. While there remains a node which is H-admitted but not H-checked, repeatedly pick such nodes and H-check them (one by one).

Step 4. Pick any node which is not H-admitted yet, and make that node H-admitted by indenting a new H-class consisting only of that node (at the moment of indenting).

Our procedure to determine a V-partition of $V(G)$ is almost the same as for the H-partition (with "H" replaced by "V"), slightly differing at the stage of V-enlargement around node u; this process is essentially the same as the H-enlargement around u, with "W" replaced by "N", and "E" by "S", except do not take the edge [*snk*, *src*] into account for V-enlargement around $u \in \{src, snk\}$ in case *src* and *snk* are adajacent with $rot(src, i) = snk$, $rot(snk, j) = src$, $pointer(src, i) =$ "S", and $pointer(snk, j) =$ "N".

To scan the semiedges of the graph G, we always apply our standard scanning. Technically, this is realized by two nested "for" loops: one with index u running through the vertex numbers, and the other with index i running from 0 to 3. The found H-partition is stored in array $adm_H(u)$, which specifies the ordering number of the H-class containing u; similarly store the V-partition (use "H" in place of "V").

9. Constructing the extension of the graph G_d

In this section we shall subdivide certain edges of the graph G_d by two new nodes which we shall call the "imaginary source", denoted by src_{im}, and the "imaginary sink", denoted by snk_{im}. We will call the so-modified graph the *extension* of G_d, and denote it by ext(G_d). We won't regard the extension as a digraph; in fact, the orientation of G_d was needed only for the assignment of pointers, and what we really need now is to extend the assignment of pointers (rather than orientation), and also the H- and V-partitions, from G_d to ext(G_d). More explicitly, we shall modify the array *rot*, introducing the imaginary source, as the one with number N; and the imaginary sink, with number $N+1$. Furthermore, we shall also modify the array *pointer*, and extend accordingly the H- and V-partitions. We have several cases to consider. Identify i and j such that $rot(src, i) = snk$, and $rot(snk, j) = src$.

Case 1: src and *snk* are adjacent with $pointer(src, i) =$ "S" but $pointer(snk, j) =$ "N".

Then we introduce src_{im} and snk_{im} by replacing the edge [*src*, *snk*] with these three: [*src*, src_{im}], [src_{im},snk_{im}], and [snk_{im}, *snk*]. For this we set:

$$
\begin{aligned}
&src_{\text{im}} &&= N && \text{(introducing the imaginary source)},\\
&snk_{\text{im}} &&= N+1 && \text{(introdusing the imaginary sink)},\\
&rot(src, i) &&= src_{\text{im}}, &&\\
&rot(snk, j) &&= snk_{\text{im}}; &&
\end{aligned}
$$

$$
\begin{aligned}
&rot(src_{\text{im}}, 0) = src, &&\\
&rot(src_{\text{im}}, 1) = -1, && \text{(indicates a fictitious edge)},\\
&rot(src_{\text{im}}, 2) = -1, && \text{(indicates a fictitious edge)},\\
&rot(src_{\text{im}}, 3) = snk_{\text{im}}; &&
\end{aligned}
$$

$$
\begin{aligned}
&rot(snk_{\text{im}}, 0) = snk, &&\\
&rot(snk_{\text{im}}, 1) = -1, && \text{(indicates a fictitious edge)},\\
&rot(snk_{\text{im}}, 2) = -1, && \text{(indicates a fictitious edge)},\\
&rot(snk_{\text{im}}, 3) = src_{\text{im}}. &&
\end{aligned}
$$

Set i' to 0 or 3 according to whether $i = 3$ or 0, respectively; in case $i \notin \{0, 3\}$ send an error message ("impossible case"). Next set the variable X to "W" or "E" according to whether *pointer*(*src*, i') = "E" or "W", respectively; in case *pointer*(*src*, i') $\notin$ {"E", "W"} send an error message ("impossible case"). Then assign pointers at the new nodes:

$$
\begin{aligned}
&pointer(src_{\text{im}}, 0) = \text{"N"}, &&\\
&pointer(src_{\text{im}}, 1) = \text{"\textbackslash 0"} && \text{(indicates a fictitious node)},\\
&pointer(src_{\text{im}}, 2) = \text{"\textbackslash 0"} && \text{(indicates a fictitious node)},\\
&pointer(src_{\text{im}}, 3) = X; &&
\end{aligned}
$$

$$
\begin{aligned}
&pointer(snk_{\text{im}}, 0) = \text{"S"}, &&\\
&pointer(snk_{\text{im}}, 1) = \text{"\textbackslash 0"} && \text{(indicates a fictitious node)},\\
&pointer(snk_{\text{im}}, 2) = \text{"\textbackslash 0"} && \text{(indicates a fictitious node)},\\
&pointer(snk_{\text{im}}, 3) = X. &&
\end{aligned}
$$

To extend the H-partition, we indent two new H-classes, one consisting of src_{im} only, and the other only of snk_{im}. Technically, this is realized by these assignments: $adm_{\text{H}}(N) = \#\text{H} + 1$, and $adm_{\text{H}}(N+1) = \#\text{H} + 2$, where #H specifies the total number of H-classes in G_{d}. To extend the V-partition, we set: $adm_{\text{V}}(N) = adm_{\text{V}}(src)$, and $adm_{\text{V}}(N+1) = adm_{\text{V}}(snk)$.

Case 2: Case 1 does not take place.

In this case, for brevity, we will omit the details as to how the array *rot* is modified; this is to be done similarly as shown above in Case 1.

First, consider the source. Let [*src*, *u*] be the edge of G_d having pointer "S" at *src*. (Note the Convention of Section 4 implies that [*src*, *u*] lies on ∂F_{ext}, and so that $u \in \{src_0, src_3\}$.)

> *Case 2-1:* If *pointer*(*src*, 0) = "S", then we introduce the imaginary source. For this, we replace the edge [*src*, src_0] by two, [*src*, src_{im}] and [src_{im}, src_0], and then set: *pointer*($src_{im} \rightarrow src$) to "N", *pointer*($src_{im} \rightarrow src_0$) to "W", *pointer*($src \rightarrow src_{im}$) to "S", and *pointer*($src_0 \rightarrow src_{im}$) to *pointer*($src_0 \rightarrow src$).

> *Case 2-2:* If *pointer*(*src*, 3) = "S", then we introduce src_{im}. For this, we replace the edge [*src*, src_3] by two, [*src*, src_{im}] and [src_{im}, src_3], and set: *pointer*($src_{im} \rightarrow src$) to "N", *pointer*($src_{im} \rightarrow src_3$) to "E", *pointer*($src \rightarrow src_{im}$) to "S", and *pointer*($src_3 \rightarrow src_{im}$) to *pointer*($src_3 \rightarrow src$).

Next consider the sink. Let [*snk*, *v*] be the edge of G_d having pointer "N" at *snk*. (By the Convention of Section 4, we have $v \in \{snk_0, snk_3\}$.)

> *Case 2-3:* If *pointer*(*snk*, 0) = "N", then we introduce the imaginary sink. For this, we replace the edge [*snk*, snk_0] by two, [*snk*, snk_{im}] and [snk_{im}, snk_0], and set: *pointer*($snk_{im} \rightarrow snk$) to "S", *pointer*($snk_{im} \rightarrow snk_0$) to "E", *pointer*($snk \rightarrow snk_{im}$) to "N", and *pointer*($snk_0 \rightarrow snk_{im}$) to *pointer*($snk_0 \rightarrow snk$).

> *Case 2-4:* If *pointer*(*snk*, 3) = "N", then we introduce snk_{im}. For this, we replace the edge [*snk*, snk_3] by two, [*snk*, snk_{im}] and [snk_{im}, snk_3], and set: *pointer*($snk_{im} \rightarrow snk$) to "S", *pointer*($snk_{im} \rightarrow snk_3$) to "W", *pointer*($snk \rightarrow snk_{im}$) to "N", and *pointer*($snk_3 \rightarrow snk_{im}$) to *pointer*($snk_3 \rightarrow snk$).

We thus have obtained the *extension* ext(G_d). To extend the H- and V-partitions of *V*(*G*) to those of ext(G_d), taking into account the two new nodes, src_{im} and snk_{im}, we update the arrays adm_H and adm_V as follows. As long as src_{im}, or snk_{im}, are introduced, we open a new H-class consisting of only src_{im}, or snk_{im}; furthermore, we add src_{im} to the V-class containing *src*, and snk_{im} to the one containing *snk*.

10. H- and V-Digraphs

Here we construct the so-called the horizontal digraph, or briefly the H-digraph, which we will denote by D_H, and the vertical digraph, or briefly the V-digraph, D_V. These two digraphs will be constructed on the base of the graph ext(G_d), together with the H- and V-partitions of its node set.

We begin with constructing D_H. Its nodes correspond to the H-classes constructed in Sections 3.5, and updated in the preceding section. To obtain its arcs, we apply our standard scanning through the semiedges [*u*, *i*) of ext(G_d). Set $v = rot(u, i)$, and identify *j* with $rot(v, j) = u$. Under the scanning, we add an arc ($adm_H(u)$, $adm_H(v)$) whenever *pointer*(*u*, *i*) = "N" and *pointer*(*v*, *j*) ∈ {"S", "E", "W"}; and an arc ($adm_H(v)$, $adm_H(u)$) whenever *pointer*(*u*, *i*) = "S" and *pointer*(*v*, *j*) ∈{"N", "E", "W"}.

Now we construct D_V. The nodes of D_V correspond to the V-classes. But, unlike the case of D_H, before defining its arcs we need to introduce some additional V-classes on certain edges of ext(G_d), in addition to the original ones constructed in Sections 8 and 9. Initially, we set all entries of the array $adm_{add}(u, i)$ to −1, which is to indicate the absence of additional V-classes on [*u*, *v*]. Then we change the value of −1 for the edges having either

$$pointer(u, i) = pointer(v, j) = \text{“E”, or} \quad (1)$$
$$pointer(u, i) = pointer(v, j) = \text{“W”}. \quad (2)$$

Namely, we add two nodes, bend(*u*, *i*) and bend(*v*, *j*), to every edge [*u*, *v*] satisfying either (1) or (2), replacing it by three edges: [*u*, bend(*u*,*i*)], [bend(*u*,*i*), bend(*v*,*j*)], and [bend(*v*, *j*), *v*]. Every pair of nodes {bend(*u*,*i*), bend(*v*,*j*)} is declared to be an additional V-class, whose ordering number is what we assign as value to $adm_{add}(u, i)$, and the same value to $adm_{add}(v, j)$. The numbering of the additional V-classes continues that of the original ones; i.e., it begins with the largest number of an original V-class incremented by 1.

Now we proceed to construct the arcs of D_V. Firstly, we shall add the arcs incident with the nodes corresponding to the additional V-classes. Let {bend(*u*, *i*), bend(*v*, *j*)} be an additional V-class. If in ext(G_d) *pointer*(*u*, *i*) = *pointer*(*v*, *j*) = "W" (resp., "E"), then we add to D_V two arcs: one from [resp., to] $adm_{add}(u, i)$ to [resp., from] $adm_V(u)$ and $adm_V(v)$.

Secondly, like in the above construction of the H-digraph, we examine the semiedges [*u*, *i*) of ext(G_d), skipping the edges on which additional V-classes were founded, and also skipping those having both endpoints in a common V-class, we add an arc from $adm_V(u)$ to $adm_V(v)$ provided that either *pointer*(*u*, *i*) = "E" (then automatically *pointer*(*v*, *j*) ≠ "E" because $adm_{add}(u, i) = -1$), or *pointer*(*u*, *i*) ∈ {"S", "N"} but *pointer*(*v*, *j*) ≠ "E".

11. Computation of coordinates

In fact, the problem of computing the coordinates reduces to that of finding a longest path in a given digraph between two given nodes. Consider first the digraph D_{H}.

Step 1 (converting D_{H} into a bipolarly oriented digraph). Add to D_{H} two nodes, which we will call the *imaginary source* and the *imaginary sink of* D_{H}, and denote them by s_h and t_h, respectively. Join s_h and, resp., t_h to the sources, resp., sinks, of D_{H} by the arcs directed from s_h, resp., to t_h. Thus we obtain a bipolarly oriented digraph which we denote by $[D_{\mathrm{H}}]$.

Step 2 (finding a directed path of maximum possible length from s_h to a given node of $[D_{\mathrm{H}}]$).

Step 2-1 (initialization). First, set

activated(h) = 0, and
$dist_{\mathrm{H}}(h) = 0$, the distance from s_h to h,

for each node h of $[D_{\mathrm{H}}]$. Next make the following assignments:

meter = 0,
active(*meter*) = s_h, and
done = 'NO'.

Step 2-2. This procedure, whose body is enclosed in braces, should be repeatedly implemented while the boolean variable *done* = 'NO'. We will be done as soon as *done* = 'YES' (which will mean that all paths have finally terminated in t_h). The integer variable *meter* specifies the size of the array *active*(m) (m = 1, ..., *meter*), the array of so-called *active* (currently) nodes of $[D_{\mathrm{H}}]$.

{*Step 2-2-1.* For m = 1 through *meter*, for each node h of $[D_{\mathrm{H}}]$ incident to the arc directed from node *active*(m), assign:

$dist_{\mathrm{H}}(h) = dist_{\mathrm{H}}(active(m)) + 1$, and
activated(h) = 1.

Step 2-2-2 (re-initialization). For m = 1 through *meter*, set

active(m) = 0,

after which set

meter = −1.

Step 2-2-3. For every node h of $[D_H]$, whenever $activated(h) = 1$, increment *meter* by 1 ($meter = meter + 1$), and then assign

$$active(meter) = h, \text{ and}$$
$$activated(h) = 0.$$

Step 2-2-4. If $meter = 0$ and $active(0) = t_h$ (which means that only one node, t_h, is currently activated), set

$$done = \text{'YES'}.\}$$

Stop.

We next implement exactly the same procedure for the vertical digraph D_V (replacing "H" by "V").

Thus we come to the coordinates:

$$x(u) = dist_V(adm_V(u)) - 1, \text{ and}$$
$$y(u) = dist_H(adm_H(u)) - 1,$$

for nodes u of ext(G_d).

Finally, we compute the coordinates of additional V-classes. Scanning the semiedges $[u, i)$, whenever $adm_{add}(u,i) \geq 1$ (which means that an additional V-class was founded on the corresponding edge), we set:

$$x_{bend}(u, i) = dist_V(adm_{add}(u, i)) - 1, \text{ and}$$
$$y_{bend}(u, i) = dist_H(adm_H(u)) - 1.$$

Thus, the process of computing the coordinates is complete.

12. Elimination of crossings

The coordinates computed in the preceding section define some representation of the given graph G, which is not necessarily an embedding because some edges may meet not only at common endpoints, namely, they may cross or touch one another. The procedure of this section serves to add some arcs to the H- and V-digraphs D_H and D_V, after which those digraph will already generate the coordinates of an actual embedding.

The idea of the procedure is that we add some arcs to D_H and D_V in order to eliminate crossings between (1) two 2-bended edges, (2) a 2-bended edge and a 0- or 1-bended edge, and (3) "rectangle-shaped" crossing, that is, the two edges, $[u, u_1]$ and $[v, v_1]$, together bound a rectangle defined by two of their bends (one of $[u, u_1]$, and the other of $[v, v_1]$) and two points of their intersection in such a way that the bends corresponding to the vertices of the rectangle do not lie on the same side. (When equality holds in the inequalities related to coordinates, the corresponding crossing degenerates into a "touching".)

References

[1] Y.P. Liu, Some combinatorial optimization problems arising from VLSI circuit design, *Applied Math. (JCU)* **8** (1993), 218-235.

[2] Y.P. Liu, A. Morgana and B. Simeone, A linear algorithm for finding 3-embeddings of a graph, Research Report--Series A, Department of Statistics, University "La Sapienza" of Rome, 1992.

On the maximal elements in the poset of graphical sequences[1]

Li Jiong-Sheng and Zhang Zhong-Liang

Department of Mathematics
University of Science and Technology of China
Hefei, Anhui 230026
The People's Republic of china

ABSTRACT A nonincreasing sequence of n nonnegative integers is graphical if it is the degree sequence of some simple undirected graph of order n. All graphical sequences of length n with trace f and even sum $2m$ form a poset $G(n,m,f)$ under majorization. For given $\pi \in G(n,m,f)$, there exists a maximal element π' in $G(n,m,f)$ which majorizes π. In this paper, we give an algorithm for finding such an element π'.

Key words graphical sequences; partially ordered sets; maximal elements; algorithm

1. Introduction

Let $\pi=(d_1,d_2,\cdots,d_n)$ be a nonincreasing sequence of n nonnegative integers with even sum $d_1+d_2+\cdots+d_n=2m$. Denote $f=f(\pi)=\max\ \{i{:}d_i\geq i$ and $1\leq i\leq n\}$. then f is said to be the trace of π. The set of all nonincreasing sequences of n nonnegative integers with trace f and even sum $2m$ is denoted by $NS(n,m,f)$. For given $\pi=(d_1,d_2,\cdots,d_n)\in NS(n,m,f)$, the off-diagonal maximal matrix of π is denoted by $\overline{A}(\pi;n)$, where the i-th row r_i of $\overline{A}(\pi;n)$ is equal to the vector $\varepsilon_1+\varepsilon_2+\cdots+\varepsilon_{i-1}+\varepsilon_{i+1}+\cdots+\varepsilon_{d_i+1}$ for each $i=1,2,\cdots,f$ and $\varepsilon_1+\varepsilon_2+\cdots+\varepsilon_{d_i}$ for each $i=f+1,f+2,\cdots,n$, and ε_j is a n-dimensional unit row vector with j-th component 1. Clearly, the row sum vector and the column sum vector of $\overline{A}(\pi;n)$ are π and $\overline{\pi}$ respectively, where $\overline{\pi}=(\overline{d_1},\overline{d_2},\cdots,\overline{d_n})$ is the corrected conjugate sequence of π. In addition, for each $k=1,2,\cdots,f$, let $\Delta_{BG}(k;\pi)=\sum_{i=1}^{k}\overline{d}_i-\sum_{i=1}^{k}d_i$, and let

[1]Supported by National Natural Science Foundation of China under grant No. 19371074

$\Delta_{BG}(\pi)=(\Delta_{BG}(1;\pi), \Delta_{BG}(2;\pi), \cdots,\Delta_{BG}(f;\pi))$. A sequence $\pi=(d_1,d_2,\cdots,d_n)\in NS(n,m,f)$ is graphical if π is the degree sequence of some simple undirected graph G of order n, and such a graph G is a realization of π. The set of all graphical sequences in $NS(n,m,f)$ is denoted by $G(n,m,f)$.

Berge [1] gave a criterion for a nonincreasing sequence of nonnegative integers being graphical. His result is the following

Theorem 1.1 Let $\pi\in NS(n,m,f)$. Then $\pi\in G(n,m,f)$ if and only if $\Delta_{BG}(k;\pi)\geq 0$ for each $k=1,2,\cdots,n$.

J.S.Li [2] further proved the following result.

Theorem 1.2 $\pi\in NS(n,m,f)$ is graphical if and only if the sequence $\Delta_{BG}(\pi)$ is nonnegative.

Let $x=(x_1,x_2,\cdots,x_n)$ and $y=(y_1,y_2,\cdots,y_n)$ be two sequences of real numbers. We say that x majorizes y, denoted by $x\succ y$, if the sum of k largest components is at least those of y for each $k=1,2,\cdots,n-1$ and $\sum_{i=1}^{n} x_i=\sum_{i=1}^{n} y_i$. It is easy to see that $G(n,m,f)$ is a poset under majorization. U.N.Peled and M.K.Srinivasan [3] and J.S.Li [4] determined the maximal elements in $G(n,m,f)$. Their result is as follows:

Theorem 1.3 $\pi\in G(n,m,f)$ is maximal if and only if $\overline{\pi}=\pi$.

For given $\pi\in G(n,m,f)$, it is easy to see that there is a maximal element π' in $G(n,m,f)$ which majorizes π. In this paper, we give an algorithm for finding such a maximal element π'.

2. Preliminaries

We require the following lemma.

Lemma 2.1 Let $\pi=(d_1,d_2,\cdots,d_n)\in G(n,m,f)$ in which $d_{f+1}<f$. Then $d_{f+1}\geq d_{f+2}\geq 1$.

Proof If otherwise, then $d_{f+2}=0$. By the off-diagonal maximal matrix $\overline{A}(\pi;n)$ of π, we have $\overline{d}_1\leq f$. Since π is graphical, $\Delta_{BG}(1;\pi)=\overline{d}_1-d_1\geq 0$ by Theorem 1.1. Hence $f\geq\overline{d}_1\geq d_1\geq d_2\geq\cdots\geq d_f\geq f$. In other words, $\overline{d}_1=d_1=d_2=\cdots=d_f=f$. Consequently, by the off-diagonal maximal matrix $\overline{A}(\pi;n)$ of π, $d_{f+1}\geq 1$. Assume $d_{f+1}=k$, $1\leq k<f$. Then $\overline{d_1}=\overline{d_2}=\cdots=\overline{d_k}=f$ and $\overline{d_{k+1}}=\cdots=\overline{d_f}=f$. Therefore

$$\Delta_{BG}(f;\pi)=\sum_{i=1}^{f}(\overline{d_1}-d_1)=-(f-k)<0$$

which contradicts to Theorem 1.1.

Theorem 2.2 Let $\pi=(d_1,d_2,\cdots,d_n)\in G(n,m,f)$ and $d_1\geq d_2\geq\cdots\geq d_f>k=d_{f+1}\geq t=d_{f+2}=\cdots=d_l>d_{l+1}\geq\cdots\geq d_n$. If $\pi'=(d_1',d_2',\cdots,d_n')=\pi+\varepsilon_{f+1}-\varepsilon_l$, then $\pi'\in G(n,m,f)$ and $\pi'\succ\pi$.

Proof Clearly, $\pi'\in NS(n,m,f)$ and $\pi'\succ\pi$. Hence it suffices to prove that $\pi'\in G(n,m,f)$. Observe that the off-diagonal maximal matrix $\overline{A}(\pi';n)$ is obained by deleting the 1 on the position (l,t) from the off-diagonal maximal matrix $\overline{A}(\pi;n)$ and adding a 1 to the position $(f+1,k+1)$ of $\overline{A}(\pi;n)$. Consequently $\overline{\pi'}=\overline{\pi}-\varepsilon_t+\varepsilon_{k+1}$. It follows that $\Delta_{BG}(j;\pi')=\Delta_{BG}(j;\pi)\geq 0$ for each $j=1,2,\cdots,t-1$, $\Delta_{BG}(j;\pi')=\Delta_{BG}(j;\pi)-1$ for each $j=t,t+1,\cdots,k$ and $\Delta_{BG}(j;\pi')=\Delta_{BG}(j;\pi)\geq 0$ for each $j=k+1,\cdots,n$. Since $\Delta_{BG}(k+1;\pi)=\Delta_{BG}(j;\pi)-(d_{f+1}-f)-\cdots-(d_{k+1}-f)-1\geq 0$ for each $j=t,t+1,\cdots,k$, we have $\Delta_{BG}(j;\pi)\geq 1$. Thus $\Delta_{BG}(j;\pi')\geq 0$ for each $j=t,t+1,\cdots,k$. In other words, the sequence $\Delta_{BG}(\pi')$ is nonnegative. By Theorem 1.1, $\pi'\in G(n,m,f)$.

Now suppose that $\pi=(d_1,d_2,\cdots,d_n)\in G(n,m,f)$ and $d_{f+1}=f$. It is easy to see by the off-diagonal maximal matrix $\overline{A}(\pi;n)$ of π that $\overline{d_1}\geq\overline{d_2}\geq\cdots\geq\overline{d_f}\geq f=\overline{d_{f+1}}$. For each $j=1,2,\cdots,f$, denote

$$g(j;\pi)=\sum_{i=1}^{j}\overline{d_i}+(f+j)-\sum_{i=1}^{f}d_i\ ,$$

i.e.,

$$g(j;\pi)=\sum_{i=1}^{j}(\overline{d_i}-f)-\sum_{i=1}^{f}(d_i-f).$$

Then $g(j+1;\pi)-g(j;\pi)=\overline{d_{j+1}}-f$ for each $j=1,2,\cdots f-1$. Hence $g(1;\pi)\leq g(2;\pi)\leq\cdots\leq g(f;\pi)$. Since $\pi\in G(n,m,f)$, we have $g(f;\pi)=\Delta_{BG}(f;\pi)\geq 0$. Denote

$$\omega=\omega(\pi)=\min\{\, j{:}g(j;\pi)\geq 0 \ \text{ and } \ 1\leq j\leq f\,\}.$$

Then $g(1;\pi)\leq g(2;\pi)\leq\cdots\leq g(\omega-1;\pi)\leq g(\omega;\pi)\leq\cdots\leq g(f;\pi)$. Consequently

$$\sum_{i=1}^{\omega-1}\overline{d_i}+f(f-\omega+1)<\sum_{i=1}^{f}d_i\leq\sum_{i=1}^{\omega}\overline{d_i}+f(f-\omega)\ .$$

Define a sequence $\pi'=(d_1',d_2',\cdots,d_n')$ as follows:

$$d_j'=\begin{cases}\overline{d_j} & \text{if } 1\leq j\leq\omega-1\\ f-g(\omega-1;\pi) & \text{if } j=\omega\\ f & \text{if } \omega+1\leq j\leq f\\ d_j & \text{if } f+1\leq j\leq n.\end{cases}\qquad(2)$$

Clearly, $d_1'\geq d_2'\geq\cdots\geq d_{\omega-1}'$. Moreover $d_{\omega-1}'\geq d_\omega'$, since $d_{\omega-1}'-d_\omega'=\overline{d_{\omega-1}}-f+g(\omega-1;\pi)\geq\overline{d_\omega}-f+g(\omega-1;\pi)=g(\omega;\pi)\geq 0$. In addition, $d_\omega'-d_{\omega+1}'=-g(\omega-1;\pi)>0$. Hence $d_\omega'\geq d_{\omega+1}'=d_{\omega+2}'=\cdots=d_f'=f\geq d_{f+1}\geq\cdots\geq d_n$. Consequently π' is nonincreasing. Finally, $f(\pi')=f$, and

$$\sum_{i=1}^{n}d_i'=\sum_{i=1}^{\omega-1}\overline{d_i}+(f+\sum_{i=1}^{f}d_i-\sum_{i=1}^{\omega-1}\overline{d_i}-f(f-\omega+1))+f(f-\omega)-\sum_{i=f+1}^{n}d_i=\sum_{i=1}^{n}d_i=2m\ .$$

Thus $\pi'\in NS(n,m,f)$.

Theorem 2.3 Let $\pi=(d_1,d_2,\cdots,d_n)\in G(n,m,f)$ and $d_{f+1}=f$. If $\pi'=(d_1',d_2',\cdots,d_n')$ is defined by (2), then $\pi'\in G(n,m,f)$, $\Delta_{BG}(\pi')$ is non-decreasing and $\pi'\succ\pi$.

Proof It follows from the off-diagonal maximal matrix $\overline{A}(\pi';n)$ of π' that $\overline{d_j'}=\overline{d_j}$ for each $j=1,2,\cdots,f$. Hence $\Delta_{BG}(j;\pi')=0$ for each $j=1,2,\cdots,\omega-1$. Clearly, $\Delta_{BG}(\omega;\pi')=\overline{d_\omega}-d_\omega'=g(\omega;\pi)\geq 0$, In addition, $\Delta_{BG}(j;\pi')-\Delta_{BG}(j-1;\pi')=\overline{d_j}-f\geq 0$ for each $j=\omega+1,\omega+2,\cdots,f$. Consequ-ently $\Delta_{BG}(\pi')$ is nondecreasing and nonnegative. By Theorem 1.2, $\pi'\in G(n,m,f)$. Finally, it is clear that, for each $j=1,2,\cdots,\omega-1$,
for each $j=\omega,\omega+1,\cdots,f$,

$$\sum_{i=1}^{j} d_i' - \sum_{i=1}^{j} d_i = \Delta_{BG}(j;\pi) = 0,$$

$$\sum_{i=1}^{j} d_i' - \sum_{i=1}^{j} d_i = \sum_{i=\omega+1}^{j} (d_i - f) \geq 0$$

and for each $j = f+1, f+2, \cdots, n$,

$$\sum_{i=1}^{j} d_i' - \sum_{i=1}^{j} d_i = \sum_{i=1}^{f} (d_i' - d_i) \geq 0$$

with equality holding when $j=n$. Thus $\pi' \succ \pi$.

Theorem 2.4 Suppose that $\pi = (d_1, d_2, \cdots, d_n) \in G(n,m,f)$ is nonmaximal, $d_{f+1} = f$ and $\Delta_{BG}(\pi)$ is nondecreasing. If $\Delta_{BG}(f;\pi) > 0$, then there exists a sequence $\pi' = (d_1', d_2', \cdots, d_n') \in G(n,m,f)$ such that $\Delta_{BG}(\pi')$ is nondecreasing, $\Delta_{BG}(f;\pi') < \Delta_{BG}(f;\pi)$ and $\pi' \succ \pi$.

Proof Since $d_{f+1} = f$, we have that $\overline{d}_1 \geq \overline{d}_2 \geq \cdots \geq \overline{d}_f \geq f = \overline{d_{f+1}}$ by the off-diagonal maximal matrix $\overline{A}(\pi;n)$ of π. Since

$$\Delta_{BG}(f;\pi) = \sum_{i=1}^{f} \overline{d}_i - \sum_{i=1}^{f} d_i = (f(f-1) + \sum_{i=f+1}^{n} d_i) - (2m - \sum_{i=f+1}^{n} d_i)$$

$$= f(f-1) + 2(\sum_{i=f+1}^{n} d_i - m)$$

i.e, $\Delta_{BG}(f;\pi)$ is a non-zero even integer, we may assume that $\Delta_{BG}(f;\pi) = 2t$, $t > 0$ and

$$0 = \Delta_{BG}(1;\pi) = \Delta_{BG}(2;\pi) = \cdots = \Delta_{BG}(k-1;\pi) < \Delta_{BG}(k;\pi) \leq \cdots$$

$$\leq \Delta_{BG}(l-1;\pi) < \Delta_{BG}(l;\pi) = \cdots = \Delta_{BG}(f;\pi).$$

Then for each $j = 1, 2, \cdots, k-1$,

$$\sum_{i=1}^{j}\overline{d_i}=\sum_{i=1}^{j}d_i$$

and for each $j=l,l+1,\cdots,f$,

$$\sum_{i=1}^{j}\overline{d_i}=\sum_{i=1}^{j}d_i+2t.$$

Hence $\overline{d_i}=d_i$ for each $i=1,2,\cdots k-1,l+1,l+2,\cdots f$. Since $0<\Delta_{BG}(k;\pi)<2t$, we have that $d_k<\overline{d_k}<d_k+2t$. But $d_{k-1}=\overline{d_{k-1}}\geq\overline{d_k}$. So $d_{k-1}>d_k$. In addition, $0<\Delta_{BG}(l-1;\pi)<\Delta_{BG}(l;\pi)=2t$. Consequently $\overline{d_l}\geq d_l+1$. If $\overline{d_{l+1}}=\overline{d_l}$, then $d_l\geq d_{l+1}=\overline{d_{l+1}}\geq d_l+1$, a contradiction. Therefore $\overline{d_l}>\overline{d_{l+1}}$. Denote $\pi'=\pi+\varepsilon_k-\varepsilon_{\overline{d_l}+1}$. Clearly $\pi'\in NS(n,m,f)$, $d'_{f+1}=d_{f+1}=f$ and $\pi'>\pi$. By the off- diagonal maximal matrix $\overline{A}(\pi';n)$ of π', we have that $\overline{d'_{d_k+2}}=\overline{d_{d_k+2}}+1$, $\overline{d'_l}=\overline{d_l}-1$ and $\overline{d'_i}=\overline{d_i}$ for other i . It follows that $\Delta_{BG}(j;\pi')=0$ for each $j=1,2,\cdots,k-1$, $\Delta_{BG}(j;\pi')=\Delta_{BG}(j;\pi)-1$ for each $j=k,k+1,\cdots,l-1$ and $\Delta_{BG}(j;\pi')=\Delta_{BG}(j;\pi)-2$ for each $j=l,l+1,\cdots,f$. Since $\Delta_{BG}(l;\pi')-\Delta_{BG}(l-1;\ \pi')=\overline{d'_l}-d'_l=\overline{d_l}-1-d_l\geq 0$, we have $\Delta_{BG}(l;\pi')\geq\Delta_{BG}(l-1;\pi)$.Thus $\Delta_{BG}(\pi')$ is nondecreasing and nonnegative. By Theorem 1.2, $\pi'\in G(n,m,f)$.

3. Algorithm

Suppose that $\pi=(d_1,d_2,\cdots,d_n)\in G(n,m,f)$ is non -maximal. If $d_{f+1}<f$, then by Theorem 2.2. there exists a sequence $\pi'=(d'_1,d'_2,\cdots,d'_n)\in G(n,m,f)$, $d'_{f+1}=d_{f+1}+1>d_{f+1}$ such that $\pi'>\pi$. By using repeatly Theorem 2.2, we have that there exists a sequence $\pi''=(d''_1,d''_2,\cdots,d''_n)\in G(n,m,f)$ so that $d''_{f+1}=f$ and $\pi''>\pi'>\pi$. Using Theorem 2.3 for the sequence π'', there exists a sequence $\pi^{(3)}=(d^{(3)}_1,d^{(3)}_2,\cdots,d^{(3)}_n)\in G(n,m,f)$ with $d^{(3)}_{f+1}=f$ for which $\Delta_{BG}(\pi^{(3)})$ is nondecreasing and $\pi^{(3)}>\pi''$. Now applying repeatly Theorem 2.4, we obtain a sequence $\pi^{(4)}=(d^{(4)}_1,d^{(4)}_2,\cdots,d^{(4)}_n)\in G(n,m,f)$, $d^{(4)}_{f+1}=f$ for which $\Delta_{BG}(\pi^{(4)})$ is nondecreasing and $\Delta_{BG}(f;\pi^{(4)})=0$. Hence $\Delta_{BG}(\pi^{(4)})$ is a zero sequence. Consequently, $\overline{d^{(4)}_k}=d^{(4)}_k$ for each $k=1,2,\cdots,f$.It follows that the off-diagonal maximal matrix $\overline{A}(\pi^{(4)};\pi)$ of $\pi^{(4)}$ is symmetric. So $\overline{d^{(4)}_k}=d^{(4)}_k$ for each $k=f+1,\cdots,n$. Thus by Theorem 1.3, $\pi^{(4)}$ is a maximal element in $G(n,m,f)$. Since $\pi^{(4)} \succ \pi^{(3)} \succ \pi'' \succ \pi' \succ \pi$, we have $\pi^{(4)} \succ \pi$. Thus we have the following algorithm.

Algorithm 3.1: For given $\pi \in G(n,m,f)$, find a maximal element $\pi' \in G(n,m,f)$ such that $\pi' \succ \pi$.

Step 1: If $d_{f+1}=f$, then go to Step 3. If otherwise, then go to Step 2.

Step 2: If $d_1 \geq d_2 \geq \cdots \geq d_f \geq f > d_{f+1} \geq t = d_{f+2} = \cdots = d_l > d_{l+1} \geq \cdots \geq d_n$, then set $\pi^{(1)} = (d_1^{(1)}, d_2^{(1)}, \cdots, d_n^{(1)}) = \pi + \varepsilon_{f+1} - \varepsilon_l$ and go back to Step 1.

The sequence obtained by applying repeatly Step 1 and 2 is denoted by $\pi^{(s)} = (d_1^{(s)}, d_2^{(s)} \cdots, d_n^{(s)})$ for which $d_{f+1}^{(s)} = f$.

Step 3: Find the off-diagonal maximal matrix $\overline{A}(\pi^{(s)};\pi)$ of $\pi^{(s)}$ and the sequence $\Delta_{BG}(\pi^{(s)})$. If $\Delta_{BG}(\pi^{(s)})$ is a zero sequence, then go to Step 5 . If $\Delta_{BG}(\pi^{(s)})$ is non-zero and nondecreasing, then go to Step 4 . If $\Delta_{BG}(\pi^{(s)})$ is non-zero but not nondecreasing, then find $g(j;\pi^{(s)}) = \sum_{i=1}^{j} \overline{d_i^{(s)}} + f(f-j) - \sum_{i=1}^{f} d_i^{(s)}$ for each $j=1,2,\cdots f$ and $\omega(\pi^{(s)})$. Set $\pi_1 = (\tilde{d}_1, \tilde{d}_2, \cdots, \tilde{d}_n)$, where $\tilde{d}_j = \overline{d_j^{(s)}}$ for each $j=1,2,\cdots,\omega(\pi^{(s)})-1$, $\tilde{d}_{\omega(\pi^{(s)})} = f - g(\omega(\pi^{(s)})-1;\pi^{(s)})$, $\tilde{d}_j = f$ for each $j = \omega(\pi^{(s)})+1, \cdots f$ and $\tilde{d}_j = d_j$ for each $j=f+1,\cdots,n$ and go to Step 4.

Step 4: If $0 = \Delta_{BG}(1;\pi^{(s)}) = \cdots = \Delta_{BG}(k-1;\pi^{(s)}) < \Delta_{BG}(k;\pi^{(s)}) \leq \cdots \leq \Delta_{BG}(l-1;\pi^{(s)}) < \Delta_{BG}(l;\pi^{(s)}) = \cdots = \Delta_{BG}(f;\pi^{(s)})$, then set $\pi_2 = \pi^{(s)} + \varepsilon_k - \varepsilon_{\overline{d_l^{(s)}}+1}$ and go back to step 3.

Step 5: Stop.

The following are some examples.

Example 3.2 For given $\pi = (5,5,3,2,2,2,1) \in G(7,10,3)$, find a maximal element $\pi' = (d_1', d_2', \cdots, d_7') \in G(7,10,3)$ such that $\pi' \succ \pi$.

Using Algorithm 3.1 we have

Step 1: $d_{f+1} = d_4 = 2 < 3 = f$. Go to Step 2.

Step 2: $f = 3 > d_{f+1} = d_4 = d_5 = d_6 = 2 > 1 = d_7$. Set $\pi^{(1)} = \pi + \varepsilon_4 - \varepsilon_6 = (5,5,3,3,2,1,1)$. Go back to Step 1.

Step 1: $d_{f+1}^{(1)} = d_4^{(1)} = 3 = f$. Go to Step 3.

Step 3: The off-diagonal maximal matrix $\overline{A}(\pi^{(1)};7)$ of $\pi^{(1)}$ is as follows:

$$\overline{A}(\pi^{(1)};7)=\begin{bmatrix}0&1&1&1&1&1&0\\1&0&1&1&1&1&0\\1&1&0&1&0&0&0\\1&1&1&0&0&0&0\\1&1&0&0&0&0&0\\1&0&0&0&0&0&0\\1&0&0&0&0&0&0\end{bmatrix}.$$

Hence $\overline{\pi^{(1)}}=(6,4,3,3,2,2,0)$. Consequently $\Delta_{BG}(1;\pi^{(1)})=1$, $\Delta_{BG}(2;\pi^{(1)})=\Delta_{BG}(3;\pi^{(1)})=0$. In other words, $\Delta_{BG}(\pi^{(1)})$ is not nondecreasing. It is easy to find that $g(1;\pi^{(1)})=-1$, $g(2;\pi^{(1)})=g(3;\pi^{(1)})=0$. Thus $\omega(\pi^{(1)})=2$. Set $\pi_1=(d_1^{(1)}, f-g(1;\pi^{(1)}), f, d_4^{(1)},\cdots,d_7^{(1)})$ $=(6,4,3,3,2,1,1)$. Go to Step 4.

Step 4: The off-diagonal maximal matrix $\overline{A}(\pi_1;7)$ of π_1 is as follows:

$$\overline{A}(\pi_1;7)=\begin{bmatrix}0&1&1&1&1&1&1\\1&0&1&1&1&0&0\\1&1&0&1&0&0&0\\1&1&1&0&0&0&0\\1&1&0&0&0&0&0\\1&0&0&0&0&0&0\\1&0&0&0&0&0&0\end{bmatrix}.$$

Hence $\overline{\pi_1}=(6,4,3,3,2,1,1)=\pi_1$. Stop.

Thus $\pi_1=(6,4,3,3,2,1,1)$ is a maximal element in $G(7,10,3)$ which majorizes π.

Example 3.3 For given $\pi=(3,3,2,2,2,2)\in G(6,7,2)$, find a maximal element $\pi'=(d_1',d_2',\cdots,d_6')\in G(6,7,2)$ such that $\pi'\succ\pi$.

Applying Algorithm 3.1 we have

Step 1: $d_{f+1}=d_3=2=f$. Go to Step 3.

Step 3: The off-diagonal maximal matrix $\overline{A}(\pi;6)$ is as follows:

$$\overline{A}(\pi;6)=\begin{bmatrix}0&1&1&1&0&0\\1&0&1&1&0&0\\1&1&0&0&0&0\\1&1&0&0&0&0\\1&1&0&0&0&0\\1&1&0&0&0&0\end{bmatrix}.$$

Hence $\overline{\pi}=(5,5,2,2,0,0)$, and $\Delta_{BG}(\pi)=(2,4)$. In other words, $\Delta_{BG}(\pi)$ is non-zero and nondecreasing. Go to Step 4.

Step 4: Since $0<\Delta_{BG}(1;\pi)<\Delta_{BG}(2;\pi)$, we have $k=1$ and $l=2$. Set $\pi_1=\pi+\varepsilon_1-\varepsilon_6=(4,3,2,2,2,1)$. Go back to Step 3.

Step 3: The off-diagonal maximal matrix $\overline{A}(\pi_1;6)$ of π_1 is as follows:

$$\overline{A}(\pi_1;6)=\begin{bmatrix}0&1&1&1&1&0\\1&0&1&1&0&0\\1&1&0&0&0&0\\1&1&0&0&0&0\\1&1&0&0&0&0\\1&0&0&0&0&0\end{bmatrix}.$$

Hence $\overline{\pi_1}=(5,4,2,2,1,0)$, and $\Delta_{BG}(\pi_1)=(1,2)$. Since $\Delta_{BG}(\pi_1)$ is non-zero and nondecreasing, go to Step 4.

Step 4: Since $0<\Delta_{BG}(1;\pi_1)<\Delta_{BG}(2;\pi_1)$, we have $k=1$ and $l=2$. Set $\pi_2=\pi_1+\varepsilon_1-\varepsilon_5=(5,3,2,2,1,1)$. Go back to Step 3.

Step 3: The off-diagonal maximal matrix $\overline{A}(\pi_2;6)$ is as follows:

$$\overline{A}(\pi_2;6)=\begin{bmatrix}0&1&1&1&1&1\\1&0&1&1&0&0\\1&1&0&0&0&0\\1&1&0&0&0&0\\1&0&0&0&0&0\\1&0&0&0&0&0\end{bmatrix}.$$

Hence $\overline{\pi_2}=(5,3,2,2,1,1)=\pi_2$.

Thus $\pi_2=(5,3,2,2,1,1)$ is a maximal element in $G(6,7,2)$ which majorizes π.

References

1. C. Berge, Graphs, Elsevier Science Pubishers B.V., North-Holland , 1985.
2. Jiong-Sheng Li, On the criteria for graphic sequences, submitted.
3. U.N.Peled and M.K.Srinivasan, The polytope of degree sequences, Linear Algebra and its Applications, 114/115(1989), 349-377.
4. Jiong-Sheng Li, Maximal elements in the poset of grahic sequences, in Combinatorics and Graph Theory (H.P.Yap et, eds), World Scientific, Singapore, 1993.

Geodetic block with some structrual properties*

Li xuechao
Department of Mathematics,Huazhong Normal University
Wuhan 430070, China

keywords: geodetic graph, block

ABSTRACT. We consider simple geodetic block G,give some structual properties for one kind of geodetic block,and give edges bounds for one kind of geodetic block of diameter 2. In the second part, we give some properties on one kind of geodetic block of diameter $d(\geq 2)$.

Introduction

We consider simple geodetic block G in which there are no loops ,no multiple edges , and G is two-connected in which any pair points are joined by a unique distance path.The shortest distance between the points u, v is denoted by $d(u, v)$ and the ith neithbourhood of a point v by $N_i(v)$. For general definition and terminology see [5] . J.G.Stemple and Mao jinzhong have studied geodetic block of diameter 2 and got some good results[1][4]. Some results are listed as follow:

1 some properties on geodetic graph of $diam(G) = 2$.

proposition 1. *Suppose G is a geodetic block with $diam(G) = 2$ and $cl(G) = k \geq 3$. Then*

(1) If U_k,U_j are cliques of G then $| V(U_k) |=| V(U_j) |$

(2) $xy \in E(G)$ and x, y are both clique-vertices, then there is a clique containing xy.

(3) $x \in V(G), d(x) = \Delta(G)$ if and only if either x is a clique-vertex, or the distance between x and any clique-vertex is two.

(4) If x is a non-clique-vertex and x is adjacent to some clique-vertex, then $d(x) = \Delta(G) - k + 2 = \delta(G)$.

(5) If $\Delta(G) \geq 3$, then $\Delta(G) \geq 2(k-1)$.

(6) $| V(G) |= n = \Delta(G)\delta(G) + 1$.

*Supported by the National Natural Science Foundation of China

Proposition 2.. *(Mao jingzhong[4]) G is a geodetic block with diameter two, $\delta(G) \geq 3$, and G contains at least one clique.If for every vertex a of G, a belong to at most one clique, then $\Delta(G) = 2k-1$, G contains $k+1$ cliques .*

Proposition 3.. *(Mao jingzhong[4]) $a \in V(G)$ and a belongs to exactly λ cliques, $\lambda \geq 2$ (here λ is the maximum,that is: if another point b belongs to β cliques, then $\beta \leq \lambda$). Let $d(a) = \Delta(G) = \lambda(k-1) + \mu$. And*

$$\Gamma(a) = \{y_{1,1}, \cdot y_{1,k-1}, y_{2,1}, \cdots, y_{2,k-1}, \cdots, y_{\lambda,1}, \cdots, y_{\lambda,k-1}, z_1, \cdots, z_\mu\}$$

$H_i \cup \{a\}$ generates a clique of G, here $H_i = \{y_{i,1}, \cdots, y_{i,k-1}\}, Y_{i,j} = \Gamma(y_{i,j})$ $Z_m = \Gamma(z_m) \backslash \{a\}$,and z_j is a non-clique-vertex, $1 \leq j \leq \mu$. Then we have

(1) $x \in Y_{i,j_1}, y \in Y_{i,j_2}, xy \notin E(G). 1 \leq i \leq \lambda$, $1 \leq j_1 \leq j_2 \leq k-1$, and $j_1 \neq j_2$.

(2) $x \in Y_{i,j}, \mid \Gamma(x) \cap Y_{i_1,j_1} \mid = 1, i \neq i_1, j \neq j_1. \mid \Gamma(x) \cap Z_m \mid = 1, 1 \leq m \leq \mu.$ $\mid \Gamma(z_s) \cap Z_m \mid = 1, \quad 1 \leq s \leq \mu, s \neq m. z_s \in Z_s. \mid \Gamma(z) \cap Y_{i,j} \mid = 1.$

(3) $\Gamma(y_{i,j}) \cap H_j = \emptyset, 1 \leq i \leq \lambda, i \neq j.$

(4) $\Gamma(y_{i,j}) \cap Z_m = \emptyset, 1 \leq m \leq \mu.$

(5) $x \in \Gamma(y_{i,j}), \Gamma(x) \cap H_s = \emptyset, s \neq i.$

(6) $\Gamma(z_j) \cap Z_m = \emptyset, j \neq m.$

(7) $\mid Y_{1,1} \mid = \mid Y_{i,j} \mid = \mid Z_m \mid = r = (\lambda-1)(k-1) + \mu, 1 \leq i \leq \lambda.$

(8) suppose $x \in Y_{i_1,j_1}, y \in Y_{i_2,j_2}, i_1 \neq i_2$, and $xy \in E(G)$. Then either (i) x belong to a clique of $Y_{1,1}$, or (ii) y belong to a clique of $Y_{\lambda,1}$.

Proof. (1) If $xy \in E(G)$,then $x, y, y_{i,j_1}, y_{i,j_2}$ generates a 4-clique.

(2) If $\mid \Gamma(x) \cap Y_{i_1,j_1} \mid \geq 2$, suppose $u, v \in \Gamma(x) \cap Y_{i_1,j_1}$, then x, u, v, y_{i_1,j_1} generates a 4-clique. Others results is true by the same reason as above.

(3) If $y \in \Gamma(y_{i,j}) \cap H_j$,then $H_i \cup \{y\} \cup \{a\}$ generates a new $(k-1)$-clique.

(4) The result is true by same reason as in (3).

(5) The result is true by same reason as in (3).

(6) The result is true by same reason as in (3).

(7) The result is true by (1),(2),(3),(4),(5),(6).

(8) Without loss of generality, we can suppose that $x \in Y_{1,1}, y \in Y_{\lambda,1}$. Consider the constitution of the shortest path from x to each vertex of $Y_{\lambda,1}$. First, the length of the shortest path from x to y is 1. From x passing through the vertices of $\Gamma(u) \cap (\cup_{i=2}^{\lambda-1} \cup_{j=1}^{k-1} Y_{i,j} \cup \bigcup_{m=1}^{m=\mu} Z_m)$ to $(\lambda-2)(k-1)+\mu$ different vertices of $Y_{\lambda,1}$, there are shortest paths of length 2 connecting them. Therefore we have got shortest paths with length 1 or 2 from x to $(\lambda-2)(k-1)+\mu+1$ different vertices of $Y_{\lambda,1}$. But $\mid Y_{\lambda,1} \mid - \{(\lambda-2)(k-1)+\mu+1\} = k-2 \geq 1$. So either y belong to a clique of $Y_{\lambda,1}$ and we can pass through y to connect x and other $(k-2)$ vertices of $\Gamma(x) \cap Y_{\lambda,1}$ by the shortest paths of length 2; or x belong to a clique of $Y_{1,1}$ and we can pass through $(k-2)$ vertices of $\Gamma(u) \cap Y_{1,1}$ to connect x and other $(k-2)$ verteices of $Y_{\lambda,1}$ by shortest path of length 2. The result follows.

Lemma 1. *Every two* Y_{i_1,j_1} *and* Y_{i_2,j_2} *contains at least two cliques, and*

$$| E(N_2(a)) |\geq (\lambda+\mu)(k-1)+ r(k-1)\{(k-1)\lambda(\lambda-1)/2+\lambda\mu\}$$
$$\lambda\mu(\mu-1)/2+2(k-1)^2[\lambda/2]+2(k-1)[\mu/2]$$

Proof. We know that $Y_{i,j}$ connect to Y_{i_1,j_1} with exactly $(\lambda-1)(k-1)+\mu$ edges by **proposition 3(7)**. So $E(N_2(a))$ contains at least

$$\{| Y_{1,1} | + | Y_{1,2} | + \cdots + | Y_{1,k-1} |\} \cdot \{| H_2 | + | H_3 | + \cdots + | H_\lambda | + \mu\}$$

$$+\{| Y_{2,1} | + | Y_{2,2} | + \cdots + | Y_{2,k-1} |\} \cdot \{| H_3 | \cdots + | H_\lambda | + \mu\} + \cdots +$$

$$\{| Y_{\lambda,1} | + | Y_{\lambda,2} | + | Y_{\lambda,k-1} |\}\mu + | Z_1 | (\mu-1) + \cdots + | Z_2 | (\mu-2)$$

$$+ \cdots + | Z_{k-1} |$$

edges.

Suppose $u \in Y_{i,j}, y \in Y_{i_1,j_1}$ and $uv \in E(G)$, Consider the constitution of the shortest path from u to each vertex of $Y_{2,1}$. First the length of the shortest path from u to v is 1. From u passing through the vertices of $\Gamma(u) \cap (\cup_{i=1}^{\lambda-1} \cup_{j=1}^{k-1} Y_{i.j} \cup \bigcup_{m=1}^{m=\mu} Z_m)$ to $(\lambda-2)(k-1)+\mu$ different vertices of $Y_{2,1}$, there are shortest paths of length 2 connecting them. Therefore we have got shortest paths with length 1 or 2 from u to $(\lambda-2)(k-1)+\mu+1$ different vertices of $Y_{2,1}$. But $| Y_{2,1} | - \{(\lambda-2)(k-1)+\mu+1\} = k-2 \geq 0$, therefore either v is a clique-vertex and we can pass through v to connect u and other $(k-2)$ vertices of $\Gamma(v) \cap Y_{2,1}$ by the shortest paths of length 2; or u is a clique-vertex of $Y_{1,1}$, and we can pass through $(k-2)$ vertices of $\Gamma(v) \cap Y_{1,1}$ to connect u and other $(k-2)$ vertices of $Y_{2,1}$ by shortest paths of length 2. We suppose u is a clique-vertex of $Y_{1,1}$ and $u \in \bar{H}_1 \subset Y_{1,1}$,here H_1 is definited as in Proposition 3, $\bar{H}_1 = G \backslash H_1$. $| V(\bar{H}_1) | = k-1$, if $x \in Y_{1,1} - V(\bar{H}_1)$, then $\Gamma(x) \cap V(\bar{H}_1) = \phi$. Let $y \in Y_{2,1}$ and $xy \in E(G)$, by the same reason as above, we have got shortest paths with length 1 or 2 from x to $(\lambda-2)(k-1)+\mu+1$ different vertices of $Y_{2,1}$.And $| Y_{2,1} | - \{(\lambda-2)(k-1)+\mu+1\} = k-2 \geq 0$. Therefore either x is a clique-vertex, or y is a clique-vertex. If y is a clique-vertex, then both $Y_{1,1}$ and $Y_{2,1}$ contain one clique respectivly. If x is a clique-vertex, continuing doing this as before, then $Y_{1,1}$ and $Y_{2,1}$ contain at least three cliques. Hence we have at least two cliques for every two Y_{i_1,j_1} and Y_{i_2,j_2} $i_1 \neq j_1$. That is to say, every $Y_{i,j}$ should contain a clique.Thus

$$\begin{aligned}
| E(N_2(a)) | \geq & \{\lambda(k-1)+\mu\}(k-1)+ \{| Y_{1,1} | + | Y_{1,2} | + \cdots + | Y_{1,k-1} |\} \\
& \{| H_2 | + | H_3 | + \cdots + | H_\lambda | + \mu\} \\
& + \{| Y_{2,1} | + | Y_{2,2} | + \cdots + | Y_{2,k-1} |\} \{| H_3 | \cdots + | H_\lambda | \mu\} \\
& + \cdots + \{| Y_{\lambda,1} | + | Y_{\lambda,2} | + | Y_{\lambda,k-1} |\}\mu \\
& + | Z_1 | (\mu-1) + \cdots + | Z_2 | (\mu-2) + \cdot + | Z_{k-1} | \\
& = \{\lambda(k-1)+\mu\}(k-1) + (k-1)r\{(k-1)(\lambda-1)+\mu\} + (k-1)r \\
& \{(\lambda-2)(k-1)+\mu\} \\
& + \cdots + (k-1)r \ \{(\lambda-\lambda)(k-1)+\mu\} + r(1+2+\cdots+\mu-1)
\end{aligned}$$

$$= \{\lambda(k-1)+\mu\}(k-1)+(k-1)r\{(k-1)\lambda(\lambda-1)/2+\lambda\mu\}+r\mu(\mu-1)/2$$
$$= (k-1)\{\lambda(k-1)+\mu\}+(k-1)\{(\lambda-1)(k-1)+\mu\}$$
$$\{(k-1)+\lambda(\lambda-1)/2+\lambda\mu\}$$
$$+\{(\lambda-1)(k-1)+\mu\}\mu(\mu-1)/2$$
$$= (k-1)\{r+k-1\}+(k-1)r\{(k-1)+\lambda(\lambda-1)/2+\lambda\mu\}+r\mu(\mu-1)/2$$

where $r = (\lambda-1)(k-1)+\mu$

Theorem 1. *$a \in V(G)$ and a belong to λ cliques (λ is maximum). Let $d(a) = \Delta(G) = \lambda(k-1)+\mu$, $| Y_{i,j} |= r$, Then*

$$| E(G) | \geq \quad r+k-1+\lambda(k-1)+r(r+k-1)+(k-1)(r+k-1)+ r(k-1)\{(k-1)+\lambda(\lambda-1)/2+\lambda\mu\}+r\mu(\mu-1)/2$$

where $r = (\lambda-1)(k-1)+\mu$

Proof. Since

$$| d(a) |= \lambda(k-1)+\mu = r+k-1$$
$$| E(N_1(a)) |= \lambda(k-1)$$
$$| E(N_1(a), N_2(a)) |= r(r+k-1)$$
$$| E(N_2(a)) |= (k-1)(r+k-1)+r(k-1)\{(k-1)+\lambda(\lambda-1)/2+\lambda\mu\} +r\mu(\mu-1)/2$$

Sum up the equations above , complete the claim.

This lower bound is best.We can give a example as follows:

Example. *$d(a) = \lambda(k-1)+\mu$. Here $\lambda = 3, \mu = 0, k = 4$.*

$$\Gamma(a) = \{y_{1,1}, y_{1,2}, y_{1,3}, y_{2,1}, y_{2,2}, y_{2,3}, y_{3,1}, y_{3,2}, y_{3,3}\}$$

We denote $H_i = \{y_{i,1}, y_{i,2}, y_{i,3}\}.1 \leq i \leq 3$. (as in **Proposition 3** *),then $H_i \cup \{a\}$ generates a clique of G. $Y_{i,j} = \Gamma(y_{i,j}) - (H_i \cup \{a\})$, $| Y_{i,j} |= (\lambda-1)(k-1) = 6$. Every $< Y_{i,j} >$ contains k_3 and three independent-vertices. The set of edges with one endvertex in A and one endvertex in B is denoted by $E(A, B)$.*

Let

$$Y_{i,j} = \{1_{y_{ij}}, 2_{y_{ij}}, \cdots, 6_{y_{ij}} \mid 1 \leq i, j \leq 3\}$$
$\{1_{y_{1j}}, 2_{y_{1j}}, 3_{y_{1j}}\}$ *are clique-vertex set of $Y_{1,j}$*
$\{4_{y_{ij}}, 5_{y_{ij}}, 6_{y_{ij}}\}$ *are clique-vertex set of $Y_{i,j}$ $i = 2, 3$*
$$E(Y_{1,j}, Y_{2,j}) = \{k_{y_{1j}} k_{y_{2j}} \mid 1 \leq j \leq 3. \quad 1 \leq k \leq 6\}$$
$$E(Y_{1,1}, Y_{3,j}) = \{k_{y_{11}} k_{y_{3j}} \mid 1 \leq j \leq 3. 1 \leq k \leq 6\}$$

$$E(Y_{1,2}, Y_{3,j})$$
$$= \{2_{y_{12}} 1_{y_{3j}}, 3_{y_{12}} 2_{y_{3j}}, 1_{y_{12}} 3_{y_{3j}}, 5_{y_{12}} 4_{y_{3j}}, 6_{y_{12}} 5_{y_{3j}}, 4_{y_{12}} 6_{y_{3j}} \mid j = 1, 2, 3\}$$

$$E(Y_{1,3}, Y_{3,j})$$
$$= \{3_{y_{13}} 1_{y_{3j}}, 1_{y_{13}} 2_{y_{3j}}, 2_{y_{13}} 3_{y_{3j}}, 6_{y_{13}} 4_{y_{3j}}, 4_{y_{13}} 5_{y_{3j}}, 5_{y_{13}} 6_{y_{3j}} \mid j = 1, 2, 3\}$$

$$E(Y_{2,1},Y_{3,1}) = \{4_{y_{21}}1_{y_{31}}, 5_{y_{21}}2_{y_{31}}, 6_{y_{21}}3_{y_{31}}, 1_{y_{21}}4_{y_{31}}, 2_{y_{21}}5_{y_{31}}, 3_{y_{21}}6_{y_{31}}\}$$

$$E(Y_{2,2},Y_{3,1}) = \{5_{y_{22}}1_{y_{31}}, 6_{y_{22}}2_{y_{31}}, 4_{y_{22}}3_{y_{31}}, 2_{y_{22}}4_{y_{31}}, 3_{y_{12}}5_{y_{31}}, 1_{y_{22}}6_{y_{31}}\}$$

$$E(Y_{2,3},Y_{3,1}) = \{6_{y_{23}}1_{y_{31}}, 4_{y_{23}}2_{y_{31}}, 5_{y_{23}}3_{y_{31}}, 3_{y_{23}}4_{y_{31}}, 1_{y_{23}}5_{y_{31}}, 2_{y_{23}}6_{y_{31}}\}$$

$$E(Y_{2,1},Y_{3,2}) = \{6_{y_{21}}1_{y_{32}}, 4_{y_{21}}2_{y_{32}}, 5_{y_{21}}3_{y_{32}}, 2_{y_{21}}4_{y_{32}}, 3_{y_{21}}5_{y_{32}}, 1_{y_{21}}6_{y_{32}}\}$$

$$E(Y_{2,2},Y_{3,2}) = \{4_{y_{22}}1_{y_{32}}, 5_{y_{22}}2_{y_{32}}, 6_{y_{22}}3_{y_{32}}, 3_{y_{22}}4_{y_{32}}, 1_{y_{22}}5_{y_{32}}, 2_{y_{22}}6_{y_{32}}\}$$

$$E(Y_{2,3},Y_{3,2}) = \{5_{y_{23}}1_{y_{32}}, 6_{y_{23}}2_{y_{32}}, 4_{y_{23}}3_{y_{32}}, 1_{y_{23}}4_{y_{32}}, 2_{y_{23}}5_{y_{32}}, 3_{y_{23}}6_{y_{32}}\}$$

$$E(Y_{2,1},Y_{3,3}) = \{5_{y_{21}}1_{y_{33}}, 6_{y_{21}}2_{y_{33}}, 4_{y_{21}}3_{y_{33}}, 3_{y_{21}}4_{y_{33}}, 1_{y_{21}}5_{y_{33}}, 2_{y_{21}}6_{y_{33}}\}$$

$$E(Y_{2,2},Y_{3,3}) = \{6_{y_{22}}1_{y_{33}}, 4_{y_{22}}2_{y_{33}}, 5_{y_{22}}3_{y_{33}}, 1_{y_{22}}4_{y_{33}}, 2_{y_{12}}5_{y_{33}}, 3_{y_{22}}6_{y_{31}}\}$$

$$E(Y_{2,3},Y_{3,3}) = \{4_{y_{23}}1_{y_{33}}, 5_{y_{23}}2_{y_{33}}, 6_{y_{23}}3_{y_{33}}, 2_{y_{23}}4_{y_{33}}, 3_{y_{23}}5_{y_{33}}, 1_{y_{23}}6_{y_{33}}\}$$

Corollary. *If* $\lambda = 2, \mu = 0. then$

$$\mid E(N_2(a)) \mid= 2(k-2)^2[\lambda/2] + 2(k-1)[\lambda/2]$$

Proof. We know that $\mid Y_{i,j} \mid= k-1.\quad 1 \le i \le 2$ by **Lemma 1**.

and there are exactly $k-1$ edges to connect $Y_{i,j}$ and $Y_{2,j}$ by **lemma 1** . Just only one of $Y_{1,j}$ and $Y_{2,j}$ has a $(k-1)-$ clique. Let $Y_{1,1}$ contains $(k-1)-$clique, then every $Y_{1,j}$ contains $(k-1)-$clique and every $Y_{2,j}$ is independent set. So

$$\mid E(N_2(a)) \mid=\mid Y_{i,j} \mid (k-1)^2$$

2. Some properties on a kind of geodetic graph of $diam(G) = d$.

We consider only ordinary graphs without loops and multiple line. For graph G, the set of points is $V(G) = V$ and the set of lines is $E(G) = E$.The induced graph on the point set $U \subseteq V$ is denoted by $< U >$. For $v \in V$ we denote the set $\{u \in V \mid d(v,u) = i\}$ by $N_i(N)$ and call it the ith neighbourhood of v.

If $C \subseteq N_i(v)$, then

$P = \{u \in N_{i-1}(v) \mid w \in C \text{ with } wu \in E(G)\}$ is called the predecessor set of C and

$S = \{u \in N_{i+1}(v) \mid w \in C \text{ with } wu \in E(G)\}$ is called the successor set C. The $(i-j)$th ancestral set and progeny set are defined analogously.

Some general results on geodetic blocks.

proposition 2.1[3]. *For every $v \in V(G), N_1(v)$ is a union of cliques and for any $v \in V$,every point of $N_1(v)$ is adjacent to at least one pointd of N_2, that means $\mid N_1(v) \mid \leq \mid N_2(v) \mid$*

proposition 2.2[3]. *For each vertex v of G with diameter 2, $< N_2(v) >$ is connected.*

proposition 2.3[3]. *A graph G is geodetic iff for each $v \in V(G)$,every $u \in N_i(v)$ has a unique predecessor, for $2 \leq i \leq e(v)$.*

proposition 2.4[3]. *In a geodetic block G for every $v \in V(G)$ with eccentricity e, $d(v) \leq \frac{1}{4}(p - 2e + 3)$.*

proposition 2.6[3]. *If G is a central geodetic block on p points with diameter $d \geq 2$, then $q \leq \frac{1}{4}p(p - 2e + 3)$.*

Some main results.

Lemma 2.1. *If geodetic block G has a central point v with eccentricity $e(v)$ and $N_2(v)$ contains at least two components , then every point in $N_2(v)$ is adjacent to at least one point of $N_3(v)$.*

Proof. Let $u \in N_2(v)$ be a point without a successor. Let $u_0 \in N_1(v)$ is a predecessor of u. Let C_1^2 be the component of $< N_2(v) >$ containing u. C_1^1 is the predecessor set of C_1^2 which is complete subgraph by **proposition 2.1**. There is at least one more component C_2^1 of $< N_1(v) >$ which is also complete subgraph. The progeny sets of C_1^1 and C_2^1 are disjoint (otherwise there creat 4-cycles). Since G is a block, there exist C_1^i and C_2^i which is the progeny set of C_1^1 and C_2^1 for $i \geq 2$, and $x_i \in C_1^i$ and $y_i \in C_2^i$ are connected in $< N_i(v) >$ by a path x_i,t_1, t_2, $\cdots$, t_k,y_i $(t_1 \neq x_i)$. Then $d(u, t_1) = i$, $d(u_0, t_1) = i + 1$ and we have two $(i+1)$-path between u_0 and t_1 on through u and x_i and another through v,contradicting geodeticity of G, This establish the claim.

Corollary 2.1. *In any geodetic block G, v is a point with $e(v)$, and every component of N_i contains at least two components, then $\mid N_2(v) \mid \leq \mid N_3(v) \mid$. And $d(v) =\leq (1/3)(p-2e+5)$.*

Proof. Since eccentricity of $v = e$, by proposition 2.1, $\mid N_1(v) \mid \leq \mid N_2(v) \mid$. And $\mid N_2(v) \mid \leq \mid N_3(v) \mid$ by Lemma 2.1. But we have $\mid N_1(v) \mid + \mid N_2(v) \mid + \cdots + \Sigma_{i=3}^{d} \mid N_i(v) \mid = p-1$. Hence we get $3 \mid N_1(v) \mid \leq p-1-2(e-3)$. Thus the claim.

Corollary 2.2. *In any geodetic block G, v is a point with $e(v)$, $N_i(v)$ contains at least two components, then $q \leq (1/6)p(p-2e+5)$.*

Theorem 2. *G is a geodetic block, $v \in V(G), e(v) = d(G)$, $\mid N_1(v) \mid = \mid N_2(v) \mid$, $\mid N_i(v) \mid = 2$, $(i \geq 3)$, then $N_1(v)$ must consist of at most 3 independent points or consist of one clique and two isolate point.*

Proof. Let $N_1(v) = \{u_1, u_2, \cdots, u_k\}, (k \geq 3)$. Since $\mid N_1(v) \mid = \mid N_2(v) \mid$, we can say that each $v_i \in N_2(v)$ has predecessor u_i respectively. Without loss of generity, we let v_1, v_2 have successor in $N_3(v)$. And $v, u_1, v_1, \cdots, s$ is a d-path from v to s. $v, u_2, v_2, \cdots, l$ is a d-path from v to l. $\{s, l\} \subset N_d(v)$. We know that each $N_i(v)$ are independent set. Otherwise, we get even-cycle and some point must have two predecessor contradicting proposition.

case 1. $N_1(v)$ is an independent set of k $(k \geq 4)$. Since $d(s, u_3) \leq d$, then $v_1v_3 \in E(G)$. We get a d-path $s, \cdots, v_1, v_3, u_3$ to connect s and u_3. and the same reason for points s and u_4, we has $v_1v_4 \in E(G)$. Repeating the argument for distance from l to v_3, and v_4. we see $\{v_2v_3, v_2v_4\} \subset E(G)$. Thus $v_1v_3v_2v_4v_1$ is a 4-cycle,contraditing geodeticity of G.So $k \leq 3$.

case 2. $N_1(v)$ is a clique. then $v_1v_2 \in E(G)$ and $\{sl\} \in E(G)$,$u_1, v_1, \cdots, s, l, \cdots, v_2, u_2, v_1$ is an even-cycle contraditing geodeticity of G.

case 3.1. $N_1(v)$ is consist of one clique $H = \{u_2, u_3, \cdots, u_k\}$ $(k \geq 3)$ and one isolate point u_1. Since diameter of G is d, $d(l, v_3) \leq d$, then $v_2v_3 \in E(G)$ creating a 4-cycle $v_2v_3u_3u_2v_2$.

case 3.2. $N_1(v)$ is consist of one clique $H = \{u_3, u_4, \cdots, u_k\}$ $(k \geq 4)$ and two isolate points u_1, u_2. Let v_1, v_2 have successor. Since $d(s, v_i) \leq d, d(l, v_i) \leq d$. $(3 \leq i \leq 4)$, the same reason as in case 1, we get $\{v_1v_3, v_3v_2, v_2v_4, v_1v_4\} \subset E(G)$. Thus $v_1v_3v_2v_4v_1$ is a 4-cycle.

case 3.3.$N_1(v)$ is consist of one clique $H = \{u_1, u_2, \cdots, u_k\}$ $(k \geq 5)$ and three isolate points u_2, u_3, u_4. Since $d(s, v_i) \leq d, d(l, v_i) \leq d(3 \leq i \leq 4)$, the same reason as in case 1, we can get 4-cycle $v_1v_3v_2v_4v_1$.

From the case 2 to case 3.3, we know that if $N_1(v)$ contains one clique, then $N_1(v)$ also contains exactly two isolate points.

case 4. Let $N_1(v)$ contains two cliques $H_1 = \{u_1, u_2, \cdots, u_k\}$ $(k \geq 2)$ and $H_2 = \{u_{k+1}, u_{k+2}, \cdots, u_p\}$ $(p \geq 2)$. Let I is an independent subset of $N_1(v)$, let C is a clique set of $N_1(v)$.Here $\mid C \mid = 2$. Each u_i has a successor v_i in $N_2(v)$.We let v_1 and v_{k+1} have successor in $N_3(v)$.

case 4.1. $|I|=0$. Since v_1 and v_{k+1} have successor in $N_3(v)$. s and l are two points in $N_d(v)$. Since $d(s,u_{k+2}) \leq d$ and $d(l,u_2) \leq d, v_1$ adjacent to v_{k+2} and v_{k+1} adjacent to v_2, Thus $v_1v_{k+2}v_{k+1}v_2v_1$ is a 4-cycle , contraditing geodeticity of G.

case 4.2. $|I|=1$. Let $I=\{u_0\}$. And u_0 has a successor v_0 in $N_2(v)$. Repeating the argument as in case 4(1), we must have another path connecting v_1 and v_{k+1} which length less than 4. Since $d(l,u_0) \leq d, d(s,u_0) \leq d, d(l,v_2) \leq d, d(s,v_2) \leq d$, we get $v_0v_1, v_0v_{k+1} \subset E(G)$. There are two 3-paths between v_1 and v_2 on through u_1 and another through v_{k+1} and v_0,contraditing geodeticity of G. So we have $v_0v_2 \in E(G)$. The same reason as in case 1, we get $v_0v_3 \in E(G)$. Hence there is a 4-path $v_2v_0v_3v_{k+1}v_2$, contraditing geodeticity of G.

From case 4 we can see that $N_1(v)$ can not have more than one cliques. Thus $|C|=1$ and $|I|=2$, establishing the claim in the theorem.

References

[1] J.G.Stemple, Geodetic Graph of Diameter Two. J.Comb.(B)17(1974) 266-280.

[2] K.R.Pathasarayhy and N.Srinivasan, Geodetic Blocks of Diameter Two: Some Structural Properties, Ars Comb. 20(1985) 49-60.

[3] K.R.Pathasarayhy and N.Srinivasan, An Extremal Problem in Geodetic Graphs, Dis. Math. 49(1984) 151-159.

[4] J.Z.Mao, The Structures of Geodetic Blocks with Diameter Two, J.Graph theory, 16(1992) 247-255.

[5] F.Harary, Graph Theory, Addison-Wesley,Reading,Mass,(1969).

ON THE DETERMINATION PROBLEM FOR P_3–TRANSFORMATION OF GRAPHS *

XUELIANG LI
Institute of Mathematics and Physics
Xinjiang University
Urumchi, Xinjiang 830046
P. R. China

ABSTRACT.
Broersma and Hoede studied the P_3-transformation of graphs and claimed that it is an open problem to characterize all pairs of nonisomorphic connected graphs with isomorphic connected P_3-graphs. In this paper, we solve the problem to a great extent by proving that the P_3-transformation is one-to-one on all graphs with minimum degree greater than two. The only cases that remain open are cases where the minimum degree is at most two, and examples indicate that the problem seems to be difficult in these cases. This in some sense can be viewed as a counterpart with respect to P_3-graphs for Whitney's result on line graphs.

Key words path graph; P_3-isomorphism; star-preserving.

I. Introduction.

Broersma and Hoede [1] generalized the concept of line graphs and introduced the concept of path graphs. We follow their terminology and give the following definition. Denote by $\Pi_k(G)$ the set of all paths of G on k vertices ($k \geq 1$). Note that a path does not have repeated vertices. The path graph

* Supported by NSFC, the Third World Academy of Sciences and the Institute of Mathematics, National Autonomous University of Mexico .

$P_k(G)$ of a graph G has vertex set $\Pi_k(G)$ and edges joining pairs of vertices that represent two P_k-paths if and only if the union of which forms either a path P_{k+1} or a cycle C_k in G. The way of describing a line graph stresses the adjacency concept, whereas the way of describing a path graph stresses the concept of path generation by consecutive paths.

For a graph transformation, Grünbaum [2] refers to two general problems. We state them here for the P_3-transformation.

Characterization Problem: Characterize the graphs that are P_3-graphs.

Determination Problem: Determine which graphs have a given graph as their P_3-graphs.

Broersma and Hoede [1] studied the P_3-transformation and, among other results, presented a solution to the characterization problem. H. Li and Y. Lin [5] have recently found, and corrected, a flaw in that paper. For the Determination Problem, Broersma and Hoede [1] found two pairs of and two classes of nonisomorphic connected graphs with connected P_3-graphs. It is not difficult to find more pairs of that kind of graphs. These examples show that Whitney's result [7] on line graphs (i.e. if G and G' are connected and have isomorphic line graphs, then G and G are isomorphic unless one is $K_{1,3}$ and the other is K_3) has no similar counterpart with respect to P_3-graphs. They claimed that it is an open problem to characterize all pairs of nonisomorphic connected graphs with isomorphic connected P_3-graphs. Recently, we proved [6] that the P_3-transformation is one-to-one on all graphs with minimum degree greater than three as well as on many having minimum degree three. In this paper, we obtain a stronger result that the P_3-transformation is one-to-one on all graphs with minimum degree greater than two, i.e. they are completely determined by their P_3-graphs. This can be regarded as best possible in the sense $P_3(C_6) \cong C_6 \cong P_3(S(K_{1,3}))$ (see [1]).

II.Preliminaries.

In what follows, all graphs are undirected, connected and simple with at least four vertices. As usual, $d(u)$ denotes the degree of a vertex u and $N(u)$ denotes the neighborhood of u. For a non-negative integer d, we denote by $\mathcal{G}_d$ the class of all connected graphs with minimum degree at least d.

We will follow Beineke-Hemminger [3] treatment of Whitney's Theorem, which in turn reflects Jung's ideas in [4] to introduce the following notations and obtain the corresponding results.

A graph isomorphism from G to G' is a bijection $f : V(G) \to V(G')$ such that two vertices are adjacent in G if and only if their images are adjacent in G'. We let $\Gamma(G, G')$ denote the set of all isomorphisms of G to G'. Thus, $\Gamma(G, G)$ is equal to the automorphism group $\Gamma(G)$ of G.

We shorten $\Gamma(P_3(G), P_3(G'))$ to $\Gamma_3(G, G')$ and call the members P_3-isomorphisms from G to G'. One easily sees that under a P_3-isomorphism, two P_3-paths in G form a P_4-path if and only if their images to the same (that is, two P_3-paths forming a P_4-path cannot map to P_3-paths forming a C_3-cycle in G').

For $f \in \Gamma(G, G')$, define $f^* : \Pi_3(G) \to \Pi_3(G')$ by $f^*(uvw) = f(u)f(v)f(w)$, and call f^* the mapping induced by f. We let $\Gamma^*(G, G') = \{f^* | f \in \Gamma(G, G')\}$.

Note that f^* is not defined for a connected graph in general unless it has at least three vertices. Also note that the two isomorphisms of the graph P_3 induce the same $*$-function; however, under our assumptions, G is connected with at least four vertices and so the following results are immediate.

Theorem 1([6]). *Let $G, G' \in \mathcal{G}_2$. Then we have*

(1) $\Gamma^*(G, G') \subseteq \Gamma_3(G, G')$.

(2) the mapping $T : \Gamma(G, G') \to \Gamma^*(G, G')$ given by $T(f) = f^*$ is one-to-one.

The following definitions are needed.

If $P_3 = uvw$, then v is called the middle vertex of the P_3 and $uvw = wvu$.

The set of all the P_3-paths with a common middle vertex v is denoted by $S(v)$ and any subset of $S(v)$ is called a star at v. A mapping $f : \Pi_3(G) \to \Pi_3(G')$ is called star-preserving if the set $f(S(v))$ is a star in G' for every vertex v of G.

From [6], we have the following results.

Theorem 2. *Let $G, G' \in \mathcal{G}_2$ and let $f : \Pi_3(G) \to \Pi_3(G')$ be a bijective mapping. Then f is induced by an isomorphism from G to G' if and only if both f and f^{-1} are star-preserving P_3-isomorphisms.*

Lemma 3. *Let $G, G' \in \mathcal{G}_2$ and let f be a P_3-isomorphism from G to G'. Then f is star-preserving if and only if for every edge uv of G $f(x_1uv), \cdots, f(x_ruv)$ have a common middle vertex and $f(uvy_1), \cdots, f(uvy_s)$ have a common middle vertex, where $x_1, \cdots, x_r$ and $y_1, \cdots, y_s$ are neighbors of u and v, respectively.*

III. Main Results.

Lemma 4. *Let G and G' belong to $\mathcal{G}_3$ and let f be a P_3-isomorphism from G to G'. If f has the property that for an edge uv of G and $N(u) - v = \{x_1, x_2, \cdots, x_r\}$ we have that $f(x_1uv), f(x_2uv), \cdots, f(x_ruv)$ have a common middle vertex in G', then $f(S(u))$ is a star of G'.*

Proof. Since $f(x_1uv), f(x_2uv), \cdots, f(x_ruv)$ have a common middle vertex, we need only to show that for any $x_iux_j \in S(u), f(x_iux_j)$ and $f(x_iuv)$ have a common middle vertex.

Since $G \in \mathcal{G}_3$, x_i has at least two neighbors p and q other than u. Since x_iuv is adjacent to both px_iu and qx_iu in $P_3(G)$ and f is a P_3-isomorphism from G to G', we know that $f(x_iuv)$ is adjacent to both $f(px_iu)$ and $f(qx_iu)$ in $P_3(G')$. There are only two cases that may happen.

Case 1. $f(px_iu)$ and $f(x_iuv)$ have a common edge other than the common edge of $f(qx_iu)$ and $f(x_iuv)$.

Case 2. One of the two edges of $f(x_iuv)$ is the common edge of $f(px_iu)$ and $f(qx_iu)$.

Now we consider x_iux_j, which is adjacent to both px_iu and qx_iu in $P_3(G)$. Since f is a P_3-isomorphism from G to G', $f(x_iux_j)$ must be adjacent to both $f(px_iu)$ and $f(qx_iu)$ in $P_3(G')$. Thus $f(x_iux_j)$ must have a common edge with each of the $f(px_iu)$ and $f(qx_iu)$.

If Case 1 happens, then $f(x_iuv)$, $f(x_iux_j)$,$f(px_iu)$ and $f(qx_iu)$ must form a C_4 in G'. If we let $f(x_iuv) = a'b'$ (adjacent) and $f(x_iux_j) = c'd'$, then $f(px_iu) = a'c'$ and $f(qx_iu) = b'd'$. Consider x_iuv. Since $f(x_juv)$ has a common middle vertex with $f(x_iuv)$, we know that $f(x_juv) = e'a'$ or $e'b'$ for some edge e' of G'. This implies that $f(x_juv)$ is adjacent to $f(px_iu)$ or $f(qx_iu)$ in $P_3(G')$. Since f is a P_3-isomorphism from G to G', x_juv is adjacent to px_iu or qx_iu in $P_3(G)$, which will bring about contradictions. Therefore, only Case 2 can happen. Then since $f(x_iux_j)$ has a common edge with each of the $f(px_iu)$ and $f(qx_iu)$, we know that in this case $f(x_iux_j)$ must has a common edge with $f(x_iuv)$, i.e., $f(x_iux_j)$ has a common middle vertex with $f(x_iuv)$, which completes the proof.

The next lemma is crucial in this paper.

Lemma 5. *Let $G, G' \in \mathcal{G}_3$ and let uv be an edge of G. If f is a P_3-isomorphism from G to G', then one of $f(S(u))$ and $f(S(v)$ is a star of G' if and only if the other is a star of G'.*

Proof. Let $f(S(v))$ be a star of G' and let $y_1, \cdots, y_s$ be the neighbors of v. Then the set of $P_3's$ $\{f(uvy_i)| \quad i = 1, 2, \cdots, s\}$ have a common middle vertex.

Let $x_1, \cdots, x_r$ be the neighbors of u. Then for any $i = 1, 2, \cdots, r$, $f(x_iuv)$ is adjacent to $f(uvy_j)$ for any $j = 1, 2, \cdots, s$. Since $G \in \mathcal{G}_3$ and so both $s \geq 2$ and $r \geq 2$, we get that $f(x_1uv), \cdots, f(x_ruv)$ have a common edge $s't'$ and $s't'$ is also the common edge of $f(uvy_1), \cdots, f(uvy_s)$. By Lemma 4, we know that

$f(S(u))$ is a star of G'. The proof is complete.

Theorem 6. *Let $G, G' \in \mathcal{G}_3$. If f is a P_3-isomorphism from G to G' then f is star-preserving if and only if there is a vertex v of G such that $f(S(v))$ is a star of G'.*

Proof. The necessity is obvious. For the sufficiency, let u be any vertex of G. If $u = v$, then we already know that $f(S(u)) = f(S(v))$ is a star of G'. Otherwise, $u \neq v$. Since G is connected, we have a path in G connecting v and u, say $v = v_1 v_2 \cdots\cdots v_k = u$. Since $f(S(v)) = f(S(v_1))$ is a star in G', from Lemma 5 we know that $f(S(v_2)), \cdots, f(S(v_k)) = f(S(u))$ are stars of G', which completes the proof.

The following results are much stronger than those in [6].

Lemma 7. *Let $G, G' \in \mathcal{G}_3$ and the maximum degree $\Delta(G)$ of $G \geq 4$. If f is a P_3-isomorphism from G to G', then f is star-preserving.*

Proof. From Theorem 4 of [6] and Theorem 6 it follows immediately.

Lemma 8. *Let $G, G' \in \mathcal{G}_3$ and G have a triangle. If f is a P_3-isomorphism from G to G', then f is star preserving.*

Proof. If $\Delta(G) \geq 4$, then from Lemma 7, the conclusion follows.

Now let G be 3-regular. Let the vertices x_1, x_2 and u induce a triangle of G. Since G is 3-regular, we know that u has another neighbor v different from x_1 and x_2. Let $N(v) - u = \{y_1, y_2\}$. The remainder of the proof can be obtained from the proof of Theorem 8 [6] and Theorem 6.

Theorem 9. *Let $G, G' \in \mathcal{G}_3 - \{K_{3,3}\}$. Then f is a P_3-isomorphism from G to G' if and only if f is induced by an isomorphism of G onto G'. Furthermore, for $G, G' \in \mathcal{G}_3$, $P_3(G)$ is isomorphic to $P_3(G')$ if and only if G is isomorphic to G'.*

Proof. From Theorem 2, we need to prove that both f and f^{-1} are star-preserving. Since G' has the same property as G, we only need to show that f is star preserving.

The "if" part is obvious. For the "only if" part, we distinguish the following cases.

Case 1. $\Delta(G) \geq 4$.

By Lemma 7, f is star-preserving.

Case 2. G is 3-regular.

Subcase 2.1. G has a triangle.

By Lemma 8, f is star-preserving.

Subcase 2.2. G is triangle-free.

Let uv be any edge of G and let $N(u)-v = \{x_1, x_2\}$ and $N(v)-u = \{y_1, y_2\}$. Then x_1, x_2, u, v, y_1 and y_2 must be six different vertices and no two of them are adjacent. By contrary, if f is not star-preserving, then from Theorem 6, f does not preserve any stars of G. Thus, $f(x_1uv), f(x_2uv), f(uvy_1)$ and $f(uvy_2)$ must form a C_4 of G'. If we let $f(x_1uv) = a'b'$ and $f(x_2uv) = c'd'$, then $f(y_1uv) = b'c'$ and $f(y_2uv) = a'd'$. Consider $N(x_1) - u = \{z_1, z_2)$ and $N(x_2) - u = \{w_1, w_2\}$. Since both $f(z_1x_1u)$ and $f(z_2x_1u)$ are adjacent to $f(x_1uv) = a'b'$, whereas both $f(w_1x_2u)$ and $f(w_2x_2u)$ are adjacent to $f(x_2uv) = d'c'$, we know that, no matter how they are adjacent, there are only two cases may occur: (1) The C_4 of G' formed by the four edges a', b', c' and d' must have a vertex with degree≥ 4 and so $\Delta(G') \geq 4$. Since f^{-1} is a P_3-isomorphism from G' to G and $\Delta(G') \geq 4$, by Lemma 7, we know that f^{-1} is star-preserving and therefore $\Delta(G) \geq 4$, a contradiction. (2) Similarly considering $N(y_1) - v = \{s_1, s_2\}$ and $N(y_2) - v = \{t_1, t_2\}$, we obtain that G' must be isomorphic to $K_{3,3}$, again a contradiction to that $G' \in \mathcal{G}_3 - \{K_{3,3}\}$. The two contradictions show that f is star-preserving.

For the second part, we need only to consider f^{-1}, the P_3-isomorphism from

G' to G, and we arrive at the point that $G \cong K_{3,3}$. Hence, $G \cong G'$.

Corollary 10. *Let $G \in \mathcal{G}_3 - \{K_{3,3}\}$. Then the automorphism group of $P_3(G)$ is isomorphic to that of G.*

Proof. From Theorems 1 and 9, it follows immediately.

Acknowledgement: The author is indebted to Prof. R. L. Hemminger [8] for correcting the original of Theorem 9.

References

[1]. H. J. Broersma and C. Hoede, J. Graph Theory **13(4)** (1989), 427–444.

[2]. B. Grünbaum, *Incidence patterns of graphs and complexes*, Lecture Notes in Mathematics (ed. G. Chartrand and S.F.Kapoor) **Vol.110** (1969), 115–128, Springer-Verlag, Berlin. MR #4152.

[3]. R. L. Hemminger and L. W. Beineke, *Line graphs and line digraphs*, in: Selected Topics in Graph Theory (ed. L.W.Beineke and R.J.Wilson) (1978), 271–305, Academic press, London.

[4]. H. A. Jung, *Zu einem isomorphiesatz von H. Whitney für graphen*, Math. Ann. **164** (1966), 270–271.

[5]. H. Li and Y. Lin, *On the characterization of path graph*, J. Graph Theory **17** (1993), 463–466.

[6]. X. Li, *Isomorphisms of P_3-graphs, accepted for publication in J. Graph Theory.*

[7]. H. Whitney, *Congruent graphs and connectivity of graphs*, Amer. J. Math. **54** (1932), 150-168.

[8]. R. L. Hemminger, *Private communication.*

Modules of Linear Recurring Arrays over Finite Fields

Mulan Liu

(Institute of Systems Science

Academia Sinica, Biejing 100080 China)

Lei Hu

(Beijing Institute of Technology

Beijing 100081 China)

Abstract: In this paper the structure of linear recurring arrays over finite field F as $F[x,y]$-*module* is studied. The necessary and sufficient condition for that linear recurring arrays is cyclic module is given.

Key words: Linear recurring array, Gröbner basis, Cyclic module.

1. Introduction

Recently many people have paid attention to two dimensional arrays over finite fields, especially two dimensional recurring arrays. Because they can be used in two dimensional range-finding, scrambling and other application in communication and coding. In the paper we study the stucture of linear recurring arrays as module.

Let F be a finite field with characteristic p, Z the set of integers, Z_+ the set of non-negative integers, and $Z^2 = Z \times Z$, ${Z_+}^2 = Z_+ \times Z_+$. A infinite matrix $A = (a_k)_{k \in Z^2}$ is called a two-dimensional array over F. Let $R_1 = F[x,y]$ be the polynomial ring in two indeterminates x and y over F, $R = S^{-1}R_1$ the ring of fractions of R_1 with respect to S, where

$$S = \{ x^i y^i \mid i \geq 0,\ j \geq 0 \}.$$

Define the action of $f(x,y) = \sum f_{ij} x^i y^j \in R$ on an array $A = (a_{ij})$ by $f(x,y)A = (b_{uv})_{(u,v)\in Z^2}$, and

$$b_{uv} = \sum f_{ij} a_{u+i,v+j}.$$

Let W be the set of all arrays over F, and for an ideal I of R, let

$$G(I) = \{ A \in W \mid \forall f \in I, fA = 0 \}.$$

By ordinary addition and scalar multiplication the set $G(I)$ is a linear space over F. Obviously, $G(I)$ is an *R-module* and an *R/I-module*. For any ideal I of R, let $\sqrt{I}$ be the radical ideal of I.

Convention: From now on, any ideal of R shall mean zero-dimensional ideal. Define the total order $\preceq$ of Z^2 by the lexicographical order , *i.e.* $(a,b) \preceq (c,d)$ *iff* $b \leq d$, or $b = d$ and $a \leq c$. Define the partial order $\leq$ of Z_+^2 as the following:

$$(a,b) \leq (c,d) \text{ iff } a \leq c \text{ and } b \leq d.$$

Since the ring R is Noetherian, any ideal I of R has the reduced Gröbner Basis, denoted by $RGB(I)$. If I contains $x^r - 1$ and $y^s - 1$ for some possitive integers r and s, then

$$RGB(I) = \{\ f_0(x), f_1(x,y), \cdots, f_l(x,y)\ \}$$

with respect to the order $\preceq$, where $f(x,y) \in F[x,y]$. Furthermore if the leading term of $f_i(x,y)$ is $x^{m_i}y^{n_i}$, then

$$m_0 > m_1 > \cdots > m_l = 0$$

$$0 = n_0 < n_1 < \cdots < n_l.$$

Let

$$\Gamma = \{\ k \in Z_+^2 \mid k \not\geq (m_i, n_i),\ for\ i = 0,\ 1,\ \cdots,\ l\ \}$$

and $t = |\Gamma|$, the number of points of Γ. Γ is called the *Gröbner* window of I. Let $\left(\begin{array}{c} n \\ r \end{array}\right)$ denote the number of combination of n things taken r at a time modulo p, and let the sequence $S^{(k)} = \left(\left(\begin{array}{c} i \\ k \end{array}\right)\right)_{i \in Z}$ and the array

$$A^{(k,l)} = (\ S_i^{(k)} S_j^{(l)}\ )_{(i,j) \in Z^2}$$

Lemma 1:

$$(x-1)^i (y-1)^j A^{(k,l)} = \begin{cases} A^{(k-i,l-j)}, & \text{if } k \geq i,\ l \geq j \\ 0, & \text{otherwise.} \end{cases}$$

Theorem 1 [1] Let I be a primary ideal of R with the radical ideal $\sqrt{I} = \langle x-1, y-1 \rangle$, *i.e.* $\sqrt{I}$ is generated by $x-1$ and $y-1$. Then the vector space $G(I)$ over F has a basis $\{\ B^{(k)} \mid k \in \Gamma\ \}$, where

$$B^{(k)} = A^{(k)} + \sum_{l_x < k_x} a_l A^{(l)},\ k = (k_x, k_y),\ l = (l_x, l_y).$$

2. Cyclic *R-module* $G(I)$

[1] See [3]

In this section we will show the necessary and sufficient condition for that $G(I)$ is a cyclic *R-module* or cyclic *R/I-module*.

We call an element m of an *R-module* M torsion free if $m \neq 0$ and $rm = 0, r \in R$ implies $r = 0$.

Proposition 1: The *R/I-module* $G(I)$ is cyclic if and only if $G(I)$ contains a *torsion-free* element.

Proof: Consider the mapping $\pi : R/I \longrightarrow G(I)$ defined by $\pi(f) = fA$, where $A \neq 0$ and $A \in G(I)$.

Corollary 1: Suppose that I is a prime ideal of R with $dim\ I = 0$. Then $G(I)$ is cyclic *R/I-module*.

Proof: Notes that every non-zero array A of $G(I)$ is *torsion-free*.

Lemma 2: Let I be primary ideal of the ring R with the radical ideal $\sqrt{I}$ and N the set of all nilpotent elements of the ring R/I. Then $R/\sqrt{I} \cong (R/I)/N$.

Proof: Consider

$$R/\sqrt{I} \cong (R/I)/\ (\sqrt{I}/I)$$

and

$$x^l \equiv\ 0\ (mod\ I)\ \mathit{iff}\ x \equiv\ 0\ (\ mod\sqrt{I}\ ).$$

In fact, $\sqrt{I}$ is a maximal ideal and $R/\sqrt{I}$ is a field, since $\sqrt{I}$ is prime and zero-dimensional.

Lemma 3: Let U be the set of units of R/I.Then

(1) $R/I = \bigcup_{x \in U}(x + N)$.

(2) Suppose $\left|R/\sqrt{I}\right| = Q$. Then

$$|N| = \frac{1}{Q},\ |U| = \frac{Q-1}{Q}.$$

Proof: Since I is primary, any zero-divisor of R/I is nilpotent. On the other hand, since R/I is finite, a non-unit of R/I is a zero-divisor.Hence (1) holds. Since $(R/I)/N \cong R/\sqrt{I}$, (2) is true.

Proposition 2: Let $\Gamma = \Gamma(I)$ be the *Gröbner* window of the ideal I and $A \in G(I)$. Then A is a generator of $G(I)$ as *R/I-module*, that is $G(I) = (R/I)A$, *iff* the set $\{Z^k A \mid k \in \Gamma(I)\}$ is *F-linear independent*, *i.e.* $S = \{\ Z^k A \mid k \in \Gamma(I)\ \}$ is a basis of the vector space $G(I)$ over the field F.

Proof: The array $A \in G(I)$ is *torsion−free* *iff* $G(I) = (R/I)A$ and *iff* π is an isomorphism. It is easy to know that $\{Z^k \mid k \in \Gamma\}$ is a basis of R/I as F-vector space.

Corollary 2: For any polynomial $f(x) \in F[x]$, in fact, for the ideal of $F[x]$ the vector space $G(I)$ of linear recurring sequences determined by $f(x)$ is a cyclic $F[X]/\langle f(x)\rangle$*-module*.

Notice that it is not true for two-dimensional case. The following theorem shows us an example.

Theorem 2: Let $X = x-1, Y = y-1$. Suppose $m = m_0 > m_1 > \cdots > m_{r-1} > m_r = 0, 0 = n_0 < n_1 < \cdots < n_{r-1} < n_r = n$, where $m_i, n_j \in Z_+$. If the ideal $I = \langle X^m, X^{m_1}Y^{n_1}, \cdots, x^{m_{r-1}}Y^{n_{r-1}}, Y^n \rangle$, $r \geq 2$, then $G(I)$ is not cyclic *R/I-module*.

Proof: Noties that $\{A^{(k)} \mid k \in \Gamma(I)\}$ is a basis of $G(I)$. (see [2]).

Example: Let $F_q = F_2$ and $I = \langle x^2+1, xy+x+y+1, y^2+1\rangle = \langle X^2, XY, Y^2\rangle$. Then $G(I) = \{0, A^{(0,0)}, A^{(1,0)}, A^{(0,1)}, A^{(0,0)+(1,0)}, A^{(1,0)}+A^{(0,1)}, A^{(0,0)}+A^{(0,1)}, A^{(0,0)}+A^{(1,0)}+A^{(0,1)}\}$ has no *tortion-free* element.

Let $I(A) = \{ f \in R[x,y] \mid fA = 0 \}$. $I(A)$ is called an annihilating ideal of the array A, denoted by $Ann(A)$.

Proposition 3: $G(I)$ is a cyclic *R-module iff* there is an array $A \in G(I)$ such that $I = Ann(A)$.

Proof: $G(I)$ is a cyclic *R-module iff* there is an any $A \in G(I)$ such that π, $\pi(f) = fA$, is an isomorphsm, it implies that $I = Ann(A)$.

Theorem 3: Let I be a primary ideal of the ring R with its radical ideal $\sqrt{I} = \langle x-1, y-1\rangle$, and $RGB(I) = \langle X^{m_0}, f_1(X,Y), \cdots, f_r(X,Y)\rangle$, $deg_X f_i = m_i deg_Y f_i = n_i$. Let

$$\{ B^{(k)} = A^{(k)} + \sum_{l_x < k_x} C_l A^{(l)} \mid k \in \Gamma \}$$

is a basis of $G(I)$ and the point $P_1 = (m_0 - 1, n_1 - 1)$. Then if A is cyclic, $A = \sum a_k B^{(k)}$ is a generator of $G(I)$ as *R/I-module iff* $a_{P_1} \neq 0$.

Proof: (Omit.)

In order to study $G(I)$ for general ideal I, we extend the field F to a suitable extension field E of F such that the set $V(I)$ of zero points of I are contained into E^2 and let $\bar{R} = R \otimes_F E$.

Corollary 3: Let J be a primary ideal of the ring $\bar{R}$ with $\sqrt{J} = \langle x-\alpha, y-\beta\rangle$, where $\alpha, \beta \in E$. Let $I = \{f(\alpha x, \beta y) \mid f(x,y) \in J\}$. Then $G(J)$ is a cyclic *$\bar{R}/J$-module iff* $G(I)$ is a cyclic *$\bar{R}/I$-module*.

Proof: Consider the mapping $\xi : \bar{R} \longrightarrow \bar{R}$ defined by $\xi(f(x,y)) = f(\alpha x, \beta y)$. Then $\xi(J) = I$, and $\xi(\sqrt{J}) = \langle x-1, y-1\rangle$. It is easy to know that ξ induces an isomorphism from $G(J)$ to $G(I)$ by $\xi((A_k)_{k\in Z^2}) = (A_k \alpha^{-k_1}\beta^{-k_2})$, where $k = (k_1, k_2)$. Hence $G(J)$ is a cyclic *$\bar{R}/J$-module iff* $G(I)$ is a cyclic *$\bar{R}/I$-module*. In fact, it is known that $\{B^{(k)} \mid k \in \Gamma(I)\}$ is a basis of $G(\eta(J)) = G(I)$. Then

$$\{ \xi^{-1}B^{(k)} = (B^{(k)}_{(k_1,k_2)} \alpha^{k_1}\beta^{k_2})_{(k_1,k_2)\in Z^2} \mid k \in \Gamma \}$$

is a basis of $G(J)$. If $G(I)$ is a cyclic *R-module*, then there is an array A such that $A = \sum_k a_k B^{(k)}$ with $a_{p_1} \neq 0$ by Theorem 3. Hence $\eta(A) = \sum_k a_k \eta(B^{(k)})$ is a generator of the *$\bar{R}$-module* $G(J)$.

Corollary 4: Suppose that J is a primary ideal of the ring R and $J^e = \bigcap_{i=0}^{l} J_i$, where J^e is the extension ideal of J in $\bar{R}$, is the primary decomposition of J^e in the ring $\bar{R}$ and the set of points of $\sqrt{J_i}$ has only one point. Then $G(J)$ is cyclic *iff* $G(J_i)$ is cyclic for each i.

Colloraly 5: Let I be an ideal of R and

$$I = \bigcap_{i=1}^{r} I_i$$

is the primary decomposition of I in the ring R. Then $G(I)$ is cyclic *iff* for each i, $G(I_i)$ is cyclic.

Corollary 6: Let I be an ideal of R with $RGB(I) = \{f(x), g(x, y)\}$. Then $G(I)$ is cyclic *R-module* or *R/I-module*.

Proof: Supose $deg_x f(x) = m,\ deg_y g(x, y) = n$. Take $A \in G(I)$ with

$$A^{(k)} = \begin{cases} 0 & \text{if } k \in \Gamma(I) \setminus (m-1, n-1) \\ 1 & \text{if } k = (m-1, n-1) \end{cases}$$

Then $\{\ Z^k A \mid k \in \Gamma(I)\ \}$ is a basis of $G(I)$. Hence $G(I)$ is cyclic.

References

[1] D. Lin and M. Liu, Strucure and Properties of Linear Recurring m-Arrays, IEEE on IT, 39(1993),1758-1762.

[2] M. Liu and L. Hu, Gröbner Basis and Linear Recurring Arrays, Chinacrypt'94, Academic Press, Beijing, 1994.

[3] M. Liu and L. Hu, The Trace Expression of Doubly Priodic Array, Submitted to Chinese Science Bulletin, 1994.

[4] S. Sakata, General Theory of Doubly Priodic Arrays over an Arbitrary Finite Field and Its Applications, IEEE on IT, 24 (1978), 719-730.

[5] N. Zierler and W. Mills, Products of Linear Recurring Sequences, J. of Algebra, 27 (1973), 147-157.

Some Irreducible Graphs on the Surface of Genus n

Liu Ying and Liu Yanpei
(Institute of Applied Mathematics, Chinese Academy of Sciences
Beijing, 100080 China
&
Department of Mathematics, Northern Jaotong University
Beijing 100044, P.R.China)

Abstract

Irreducible graphs (sometimes called *critical* graphs) characterize those graphs which can not be embedded on a surface in the sense that a graph will fail to be embedded on a surface if and only if it contains an irreducible graph for the surface. In these terms, Kuratowski's theorem characterizes planar graphs by proving that each irreducible graph for the surface of genus 0, or in other work for the sphere, is homeomorphic to either the complete graph of 5 vertices, K_5, or the complete bipartite graph, $K_{3,3}$.

In the simplest non-orientable case, Glover, Huneke and Wang[1] have shown that there are 103 topologically distinct irreducible graphs for the real projective plane. It seems to be the case that the number of topologically distinct irreducible graphs for a surface increases very rapidly as a function of the genus of the surface.

Decker [2] has shown that there are a large number of topologically distinct irreducible graphs for the torus by generating a list of the "simplest" of these graphs. Each of these graphs will contain a subgraph L of a particular type which is called a K- *graph* and whose vertices are disjoint with that of Kuratowski graph (that is , a graph homeomorphic to K_5 or $K_{3,3}$). It is shown that there are exactly 259 such irreducible graphs for the torus. Furthermore, the purpose of this paper is to show that the number of irreducible graphs for a surface is an exponential function of the surface.

Terminology not mentioned here can be seen in [3].

I. Preliminary

A subgraph $H \subset G$ will be called a K – graph (of G) if there exists a graph $J, H \subset J \subset G$ such that the pair (J, H) is topologically equivalent to one of the following three pairs:

(i) $(K_{3,3}, K_{2,3})$ with one cubic vertex of $K_{3,3}$ not contained in $K_{2,3}$,

(ii) (K_5, K_4) with one valency 4 vertex of K_5 not contained in K_4,

(iii) (K, K_4) where K is obtained from K_5 by splitting a valency 4 vertex of K_5 into two adjacent cubic vertices neither of which is contained in K_4.

Theorem 1[2]. Let G be a graph and suppose that G contains disjoint subgraphs L and K where L is a K-graph of G and K is homeomorphic to a Kuratowski graph. Then G is not embedded on T.

Let $L = K_4$ or $K_{2,3}$ and let K be homeomorphic to a Kuratowski graph. Suppose that L and K are subgraphs of G, then L and K are said to be in standard position in G if

(i) $L \cap K = \emptyset$;

(ii) L is a K-graph of G ;

(iii) $H = G - K - L$ is contained in every graph $J \subset G$ such that L is a K-graph of J.

Generating a list of graphs which contain a K-graph and a Kuratowski graph in standard position we can obtain 343 graphs for the torus. There are only 259 graphs among them that are topologically distinct. We define that these 259 graphs form a set $\mathcal{A}_1$.

II. Main Result

Now,we define the set of $\mathcal{A}_n$ inductively.

Let $L = K_4$ or $K_{2,3}$ and let $K \in \mathcal{A}_{n-1}$. Suppose that L and K are subgraphs of G, then L and K are said to be in critical position in G if

(i) $L \cap K = \emptyset$;

(ii) L is a K-graph of G ;

(iii) $H = G - K - L$ is contained in every graph $J \subset G$ such that L is a K-graph of J.

The set of $\mathcal{A}_n$ contains these graphs that have a K-graph and a graph belonging to $\mathcal{A}_{n-1}$ while they are in critical position.

Theorem 1. Let G be a graph. Suppose G contains disjoint subgraphs L and K_1, where L is a K-graph of G and K_1 is homeomorphic to a graph which is an irreducible graph of S_{n-1}. Then G is not embedded on S_n.

Theorem 2. If $G \in \mathcal{A}_n$, then G is an irreducible graph on S_n.

Theorem 3. Let G contain a K-graph L and a graph $G_1 \in \mathcal{A}_{n-1}$ and let L and G_1 be in critical position. Furthermore, there is a Kuratowski graph $K_1 \subset G_1$ such that $[L \cup K_1] \in \mathcal{A}_1$. Then $G \in \mathcal{A}_n$.

It is known that $|\mathcal{A}_1| = 259$[2]. By the procedure of constructing these 259 graphs, we can see that K_5 has 113 distinct ways to connect a K-graph such that the resultant graph is irreducible on the torus. $K_{3,3}$ has 146 distinct ways to connect a K-graph such that the resultant graph is irreducible on the torus. So based on the theorems above we are allowed to count the minimum of $\mathcal{A}_i$ inductively. Now, we find that

$|\mathcal{A}_2| \geq 146 \times 146 + 113 \times 113$,

$|\mathcal{A}_n| \geq 146^n$, for $n \geq 3$.

Considering the theorem obtained by Robertson and Seymour[4],because all the irreducible graphs can be treated as minors of a graph, we see that this result shows that if one extends to find forbidden minors of a graph for the embeddability on a surface of genus n, the cardinality of the complete set of forbidden minors is an exponential function of n in the way explained here.

References

[1] H. Glover, J. Huneke, and C.S. Wang, 103 graphs that are irreducible for the projective plane. *J. Combinatorial Theory*, B27(1979), 332- 370.

[2] R. W. Decker, Ph.D Dissertation, *The Genus of Certain Graphs*, The Ohio State University, 1978.

[3] Liu Yanpei, *Embeddability of Graphs*, Science Press and Kuwer Academic Publishers, 1994.

[4] N. Robertson and P.D.Seymour, Graph Minors. VIII. A Kuratowski Theorem for General Surfaces, *Journal of Combinatorial Theory*, B48(1990), 255-288.

PROOF OF A CONJECTURE ON GRACEFULNESS OF GRAPHS

YUJI LIU
(*Department of Mathematics, Yueyang Normal College, Yueyang, Hunan, 414000*)

QUAN PENG
(*Department of Mathematics, University of Alberta*)
(*Edmonton, Alberta, Canada T6G 2G1*)

Abstract Recently S. Lu considered the graph called *k-teeth star graph*, denoted by $\bar{W}_n^k$, which is formed by suspending k-edges from each tooth of the gear graph, and proved that the double teeth star graph $\bar{W}_n^2$ is graceful. He conjectured that all k-teeth star graphs are graceful. In this paper, we generalize the concept of k-teeth star graph to allow that the number of edges suspended from each tooth can be arbitrary, we prove that the resulting graph is graceful. As an immediate consequence, we answer confirmatively the conjecture of Lu.

Keywords *Simple graph, Graceful graph, Graceful numbering, Graph* $\bar{W}_n^*$.

1. INTRODUCTION

Let $G = (V, E)$ be a *simple graph*, where V is the set of vertices and E is the set of edges. The graph G is *numbered* if each vertex u is assigned a non-negative integer $f(u)$, and each edge, denoted by u_iu_j, is assigned the absolute value of the difference of the numbers at its endpoints, that is, $f^*(u_iu_j) = |f(u_i) - f(u_j)|$.

Let $|E|$ denote the number of edges of G, $N = \{1, 2, \ldots, |E| + 1\}$, $N^* = \{1, 2, \ldots, |E|\}$. The *numbering* f is said to be *graceful* if f is an injection from V to N and the induced map f^* is a bijection from E to N^*. A graph which admits a graceful numbering is called a *graceful graph.* Note that our definition of graceful graph is essentially the same as that defined in Golomb [2], and Delorme et al. [1].

There is a well-known conjecture by G. Ringel from the Symposium in Smolenice in 1963, it states that all *trees* are graceful. This conjecture remains unsolved today. Rosa [8] proved that a *cycle* of n vertices is graceful iff $n \equiv 0$ or 3 (mod 4). Delorme et al. [1] showed that the graph consisting a cycle plus a chord is graceful. Hoede and Kuiper [4] showed that all wheels are graceful. A *wheel*, denoted by W_n, is a graph with $2n$ edges which is a join of a cycle C_n and the trivial graph K_1 (see F. Harary [3]). In a wheel, if a vertex is added between every pair of adjacent vertices on the circle, the resulting graph is called a *gear graph*, denoted by $\bar{W}_n$, the added vertices are called teeth. Ma and Feng [7] proved that all gear graphs are graceful. Recently Lu [6] considered the graph called *k-teeth star graph*, denoted by $\bar{W}_n^k$, which is formed by suspending k-edges from each tooth of the gear graph, where each suspended edge would not intersect with $\bar{W}_n^k$ except it has one endpoint rooted at a tooth. Lu [6] proved that in case $k = 2$, the double teeth star graph $\bar{W}_n^2$ is graceful, and he conjectured that all k-teeth star graphs $\bar{W}_n^k$ are graceful.

In this paper, we generalize the concept of k-teeth star graph considered by Lu [6] to allow that the number of edges suspended from the teeth can be arbitrary, the resulting graph is called *teeth star graph*, denoted by $\bar{W}_n(m_1, m_2, \ldots, m_n)$ or $\bar{W}_n^*$, where $m_i (i = 1, 2, \ldots, n)$ represents the number of edges suspended for the i-th tooth, (see **Figure 1**). It can be seen that $\bar{W}_n^*$ has $\sum_{i=1}^n m_i + 3n$ edges and $\sum_{i=1}^n m_i + 2n + 1$ vertices.

The main result of the paper is stated as follows, it says that $\bar{W}_n^*$ is graceful. As an immediate consequence, we answer confirmatively the conjecture of Lu [6] that all k-teeth star graphs are graceful.

Theorem 1. *If* $\min\{m_1, m_2, \ldots, m_n\} \geq 1$, *then the teeth star graph* $\bar{W}_n(m_1, m_2, \ldots, m_n)$ *is graceful.*

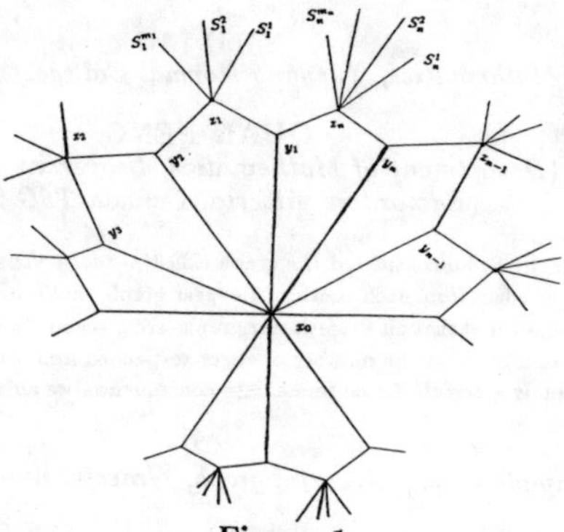

Figure 1

2. PROOF OF MAIN RESULT

In this section, we prove the main result of the paper. The proof of theorem 1 is devided into four parts corresponding to the cases when there is at least one even number among $m_1, m_2, \ldots, m_n$ (or all m_i, $i = 1, 2, \ldots, n$, are odd numbers), according to that n is odd (or even). We assign integers $f(\cdot)$ to the vertices of $W_n(m_1, m_2, \ldots, m_n)$ in each case. The proof that f is a one-one map and the induced map f^* is a bijection is straightforward. We omit the detailed verification.

Case i): There is at least one even number among $m_1, m_2, \ldots, m_n$, and n is odd. Without loss of generality, we may assume that m_n is even. We construct a numbering f of $\bar{W}_n^*$ as follows:

$$f(z_k) = k \qquad 1 \leq k \leq n$$

$$f(y_k) = \begin{cases} \sum_{i=1}^{n} m_i + 3n + 2 - k & 1 \leq k \leq \frac{n-1}{2} \\ \sum_{i=1}^{n} m_i + 3n - k & \frac{n+1}{2} \leq k \leq n. \end{cases}$$

$$f(x_0) = \sum_{i=1}^{n} m_i + 2n - 1 - \frac{1}{2} m_n$$

$$f(S^k_{\frac{n-1}{2}-1}) = \begin{cases} \sum_{i=1}^{n} m_i + \frac{5n+1}{2} & k = 1 \\ \frac{m_n}{2} + n + 2 + \sum_{i=1}^{\frac{n-1}{2}-2} (m_i + 1) + k & 2 \leq k \leq m_{\frac{n-1}{2}-1}. \end{cases}$$

$$f(S^k_{\frac{n-1}{2}}) = \begin{cases} \frac{m_n}{2} + n + 1 & k = 1 \\ \frac{m_n}{2} + n + \sum_{i=1}^{\frac{n-1}{2}-1} (m_i + 1) + 1 + k & 2 \leq k \leq m_{\frac{n-1}{2}}. \end{cases}$$

$$f(S^k_{\frac{n+1}{2}}) = \begin{cases} \frac{m_n}{2} + n + 3 & k = 1 \\ \frac{m_n}{2} + n + \sum_{i=1}^{\frac{n-1}{2}} (m_i + 1) + k & 2 \leq k \leq m_{\frac{n+1}{2}}. \end{cases}$$

$$f(S^k_i) = \begin{cases} \frac{m_n}{2} + n + 3 + \sum_{j=1}^{i-1} (m_j + 1) + k & 1 \leq i \leq \frac{n-1}{2} - 2,\ 1 \leq k \leq m_i \\ \frac{m_n}{2} + n + \sum_{j=1}^{i-1} (m_j + 1) + k & \frac{n+1}{2} + 1 \leq i \leq n - 1,\ 1 \leq k \leq m_i \end{cases}$$

$$f(S^k_n) = \begin{cases} n + k & 1 \leq k \leq \frac{m_n}{2} \\ n + \sum_{j=1}^{n-1} (m_j + 1) + k & \frac{m_n}{2} + 1 \leq k \leq m_n \end{cases}$$

where $\sum_{j=1}^{0} (m_i + 1) = 0$. To prove that $\bar{W}_n^*$ is graceful, we need to show that the numbering f is graceful. It remains to show that f is one-one map and f^* is a bijection. In order to show that f

is 1-1, we establish a series of inequalities as follows, where each term represents the maximum or minimum value of $f(\cdot)$ in a corresponding case or subcase of our definitions of $f(\cdot)$ above.

$$
\begin{aligned}
1 &< n < n+1 < n+\frac{m_n}{2} < n+\frac{m_n}{2}+1 < n+\frac{m_n}{2}+3 < \frac{m_n}{2}+n+4 \\
&< \frac{m_n}{2}+n+3+\sum_{j=1}^{\frac{n-1}{2}-3}(m_j+1)+m_{\frac{n-1}{2}-2} < \frac{m_n}{2}+n+4+\sum_{j=1}^{\frac{n-1}{2}-2}(m_j+1) \\
&< \frac{m_n}{2}+n+2+\sum_{j=1}^{\frac{n-1}{2}-2}(m_j+1)+m_{\frac{n-1}{2}-1} < \frac{m_n}{2}+n+3+\sum_{j=1}^{\frac{n-1}{2}-1}(m_j+1) \\
&< \frac{m_n}{2}+n+1+\sum_{j=1}^{\frac{n-1}{2}-1}(m_j+1)+m_{\frac{n-1}{2}} < \frac{m_n}{2}+n+2+\sum_{j=1}^{\frac{n-1}{2}}(m_j+1) \\
&< \frac{m_n}{2}+n+\sum_{j=1}^{\frac{n-1}{2}}(m_j+1)+m_{\frac{n+1}{2}} < \frac{m_n}{2}+n+1+\sum_{j=1}^{\frac{n+1}{2}}(m_j+1) \\
&< \frac{m_n}{2}+n+\sum_{j=1}^{n-2}(m_j+1)+m_{n-1} < \sum_{j=1}^{n} m_j+2n-1-\frac{1}{2}m_n \\
&< \frac{m_n}{2}+n+1+\sum_{j=1}^{n-1}(m_j+1) < n+\sum_{j=1}^{n-1}(m_j+1)+m_n < \sum_{j=1}^{n} m_j+2n \\
&< \sum_{j=1}^{n} m_j+\frac{5n-1}{2} < \sum_{j=1}^{n} m_j+\frac{5n+1}{2} < \sum_{j=1}^{n} m_j+\frac{5n+5}{2} < \sum_{j=1}^{n} m_j+3n+1.
\end{aligned}
$$

We compute the induced map f^* as follows,

$$
f^*(x_0y_k) = \begin{cases} \frac{m_n}{2}+n+3-k & 1 \le k \le \frac{n-1}{2} \\ \frac{m_n}{2}+n+1-k & \frac{n+1}{2} \le k \le n \end{cases}
$$

$$
f^*(y_kz_k) = \begin{cases} \sum_{j=1}^{n} m_j+3n+2-2k & 1 \le k \le \frac{n-1}{2} \\ \sum_{j=1}^{n} m_j+3n-2k & \frac{n+1}{2} \le k \le n \end{cases}
$$

$$
f^*(y_{k+1}z_k) = \begin{cases} \sum_{j=1}^{n} m_j+3n+1-2k & 1 \le k \le \frac{n-1}{2}-1 \\ \sum_{j=1}^{n} m_j+3n-1-2k & \frac{n-1}{2} \le k \le n-1 \end{cases}
$$

$$
f^*(y_1z_n) = \sum_{j=1}^{n} m_j+2n+1
$$

$$
f^*(S_i^kz_i) = \begin{cases} \frac{m_n}{2}+n+3+\sum_{j=1}^{i-1}(m_j+1)+k-i & 1 \le i \le \frac{n-1}{2}-2 \quad 1 \le k \le m_i \\ \frac{m_n}{2}+n+\sum_{j=1}^{i-1}(m_j+1)+k-i & \frac{n+1}{2}+1 \le i \le n-1 \quad 1 \le k \le m_i \end{cases}
$$

$$f^*(S^k_{\frac{n-1}{2}-1}z_{\frac{n-1}{2}-1}) = \begin{cases} \sum_{j=1}^n m_j + 2n + 2 & k=1 \\ \frac{m_n}{2} + \frac{n+5}{2} + 1 + \sum_{j=1}^{\frac{n-1}{2}-2}(m_j+1) + k & 2 \le k \le m_{\frac{n-1}{2}-1} \end{cases}$$

$$f^*(S^k_{\frac{n-1}{2}}z_{\frac{n-1}{2}}) = \begin{cases} \frac{m_n}{2} + \frac{n+3}{2} & k=1 \\ \frac{m_n}{2} + \frac{n+3}{2} + \sum_{j=1}^{\frac{n-1}{2}-1}(m_j+1) + k & 2 \le k \le m_{\frac{n-1}{2}} \end{cases}$$

$$f^*(S^k_{\frac{n+1}{2}}z_{\frac{n+1}{2}}) = \begin{cases} \frac{m_n}{2} + \frac{n+5}{2} & k=1 \\ \frac{m_n}{2} + \frac{n-1}{2} + \sum_{j=1}^{\frac{n-1}{2}}(m_j+1) + k & 2 \le k \le m_{\frac{n+1}{2}} \end{cases}$$

$$f^*(S^k_n z_n) = \begin{cases} k & 1 \le k \le \frac{m_n}{2} \\ \sum_{j=1}^{n-1}(m_j+1) + k & \frac{m_n}{2} + 1 \le k \le m_n \end{cases}$$

The above computations show that all edges of $\bar{W}_n^*$ are labelled from 1 to $\sum_{j=1}^n m_j + 3n$. To see taht f^* is a bijection, it remains to verify that f^* assigns distinct integer to each edge. The verification is straightforward and similar to the above arguement, we omit the details.

Case 2): m_n is even, and n is even. The numbering of $\bar{W}_n^*$ is given as follows:

$$f(z_k) = k \qquad 1 \le k \le n$$

$$f(y_k) = \begin{cases} \sum_{i=1}^n m_i + 3n + 2 - k & 1 \le k \le \frac{n}{2} \\ \sum_{i=1}^n m_i + 3n + 1 - k & \frac{n}{2} + 1 \le k \le n \end{cases}$$

$$f(x_0) = \sum_{i=1}^n m_i + 2n - \frac{1}{2}m_n$$

$$f(S^k_{\frac{n}{2}}) = \begin{cases} \frac{m_n}{2} + n + 1 & k=1 \\ \frac{m_n}{2} + n + \sum_{i=1}^{\frac{n}{2}-1}(m_i+1) + 1 + k & 2 \le k \le m_{\frac{n}{2}} \end{cases}$$

$$f(S^k_i) = \begin{cases} \frac{m_n}{2} + n + 2 + \sum_{j=1}^{i-1}(m_j+1) + k & 1 \le i \le \frac{n}{2} - 1,\ 1 \le k \le m_i \\ \frac{m_n}{2} + n + 1 + \sum_{j=1}^{i-1}(m_j+1) + k & \frac{n}{2} + 1 \le i \le n-1,\ 1 \le k \le m_i \end{cases}$$

$$f(S^k_n) = \begin{cases} n + k & 1 \le k \le \frac{m_n}{2} \\ n + 1 + \sum_{j=1}^{n-1}(m_j+1) + k & \frac{m_n}{2} + 1 \le k \le m_n \end{cases}$$

Case 3) $m_1, m_2, \ldots, m_n$ are odd numbers, and n is even. Without loss of generality, we may assume that $m_{\frac{n}{2}} = \max\{m_1, m_2, \ldots, m_n\}$, the numbering of $\bar{W}_n^*$ is given in the following five subcases:

i). If $m_{\frac{n}{2}} = 1$, $\bar{W}_n^*$ is reduced to $\bar{W}_n(1, 1, \ldots, 1)$, the numbering has been given in Liu [5].

ii) If $m_{\frac{n}{2}} \geq 3$, and $\frac{n}{2}-1 < m_n < n-1$, the numbering is given as follows:

$$f(z_k) = k \qquad 1 \leq k \leq n$$

$$f(y_k) = \begin{cases} \sum_{i=1}^{n} m_i + 3n + 2 - k & 1 \leq k \leq \frac{n}{2} \\ \sum_{i=1}^{n} m_i + 3n + 1 - k & \frac{n}{2}+1 \leq k \leq n \end{cases}$$

$$f(x_0) = \sum_{i=1}^{n} m_i + 2n$$

$$f(S_{\frac{n}{2}}^k) = \begin{cases} n+1 & k=1 \\ n + \sum_{i=1}^{\frac{n}{2}-1}(m_i+1) + k & 2 \leq k \leq m_{\frac{n}{2}} \end{cases}$$

$$f(S_i^k) = \begin{cases} n+2+\sum_{j=1}^{i-1}(m_j+1)+k & 1 \leq i \leq n-m_n-2,\ 1 \leq k \leq m_i \\ n+1+\sum_{j=1}^{i-1}(m_j+1)+k & n-m_n \leq i \leq \frac{n}{2}-1,\ 1 \leq k \leq m_i \\ n+\sum_{j=1}^{i-1}(m_j+1)+k & \frac{n}{2}+1 \leq i \leq n,\ 1 \leq k \leq m_i \end{cases}$$

$$f(S_{n-m_n-1}^k) = \begin{cases} \sum_{j=1}^{n} m_j - m_n + 2n - 1 & k=1 \\ n+1+\sum_{j=1}^{n-m_n-2}(m_j+1)+k & 2 \leq k \leq m_{n-m_n-1} \end{cases}$$

iii). $m_{\frac{n}{2}} \geq 3$ and $m_n = \frac{n}{2}-1$. $f(z_k),\ f(y_k),\ (1 \leq k \leq n)$ and $f(x_0)$ are same as ii).

$$f(S_{\frac{n}{2}}^k) = \begin{cases} n+1 & k=1 \\ \sum_{j=1}^{n} m_j + \frac{3}{2}n & k=2 \\ n + \sum_{i=1}^{\frac{n}{2}-1}(m_i+1) + k & 3 \leq k \leq m_{\frac{n}{2}} \end{cases}$$

$$f(S_i^k) = \begin{cases} n+2+\sum_{j=1}^{i-1}(m_j+1)+k & 1 \leq i \leq \frac{n}{2}-1,\ 1 \leq k \leq m_i \\ n+\sum_{j=1}^{i-1}(m_j+1)+k & \frac{n}{2}+1 \leq i \leq n,\ 1 \leq k \leq m_i \end{cases}$$

iv). $m_{\frac{n}{2}} \geq 3$, and $m_n < \frac{n}{2}-1$. $f(z_k),\ f(y_k),\ (1 \leq k \leq n)$ and $f(x_0)$ are same as ii).

$$f(S_{\frac{n}{2}}^k) = \begin{cases} n+1 & k=1 \\ n+1+\sum_{i=1}^{\frac{n}{2}-1}(m_i+1) + k & 2 \leq k \leq m_{\frac{n}{2}} \end{cases}$$

$$f(S_i^k) = \begin{cases} n+2+\sum_{j=1}^{i-1}(m_j+1)+k & 1 \leq i \leq \frac{n}{2}-1,\ 1 \leq k \leq m_i \\ n+1+\sum_{j=1}^{i-1}(m_j+1)+k & \frac{n}{2}+1 \leq i \leq n-m_n-2,\ 1 \leq k \leq m_i \\ n+\sum_{j=1}^{i-1}(m_j+1)+k & n-m_n \leq i \leq n,\ 1 \leq k \leq m_i \end{cases}$$

$$f(S_{n-m_n-1}^k) = \begin{cases} \sum_{j=1}^{n} m_j - m_n + 2n + 1 & k=1 \\ n+\sum_{j=1}^{n-m_n-2}(m_j+1)+k & 2 \leq k \leq m_{n-m_n-1} \end{cases}$$

v). $m_n \geq n-1$, and $m_{\frac{n}{2}} \geq m_n \geq n-1 > \frac{n}{2}$.

$$f(z_k) = k \qquad 1 \leq k \leq n$$

$$f(y_k) = \begin{cases} \sum_{i=1}^{n} m_i + 3n + 2 - k & 1 \leq k \leq \frac{n}{2} \\ \sum_{i=1}^{n} m_i + \frac{5n}{2} + 1 - k & \frac{n}{2}+1 \leq k \leq n \end{cases}$$

$$f(x_0) = \sum_{i=1}^{n} m_i + \frac{5n}{2} + 1$$

$$f(S_{\frac{n}{2}}^k) = \begin{cases} \sum_{j=1}^{n} m_j + \frac{5n}{2} + 1 - k & 1 \leq k \leq \frac{n}{2} \\ \frac{n}{2}+1+\sum_{i=1}^{\frac{n}{2}-1}(m_i+1) + k & \frac{n}{2}+1 \leq k \leq m_{\frac{n}{2}} \end{cases}$$

$$f(S_i^k) = \begin{cases} n+1+\sum_{j=1}^{i-1}(m_j+1)+k & 1 \leq i \leq \frac{n}{2}-1,\ 1 \leq k \leq m_i \\ \frac{n}{2}+1+\sum_{j=1}^{i-1}(m_j+1)+k & \frac{n}{2}+1 \leq i \leq n,\ 1 \leq k \leq m_i \end{cases}$$

Case 4) $m_1, m_2, \ldots, m_n$ are odd numbers, and n is odd. Without loss of generality, we may assume that $m_{\frac{n-1}{2}} = \max\{m_1, m_2, \ldots, m_n\}$, the numbering of $\bar{W}_n^*$ is given in the following five subcases:

i). If $m_{\frac{n-1}{2}} = 1$, $\bar{W}_n^*$ is reduced to $\bar{W}_n(1, 1, \ldots, 1)$, the numbering has been given in Liu [5].

ii) If $m_{\frac{n-1}{2}} \geq 3$, and $\frac{n-1}{2} < m_n < n-1$, the numbering is given as follows:

$$f(z_k) = k \qquad 1 \leq k \leq n$$

$$f(y_k) = \begin{cases} \sum_{i=1}^n m_i + 3n + 2 - k & 1 \leq k \leq \frac{n-1}{2} \\ \sum_{i=1}^n m_i + 3n - k & \frac{n+1}{2} \leq k \leq n \end{cases}$$

$$f(x_0) = \sum_{i=1}^n m_i + 2n - 1$$

$$f(S^k_{\frac{n-1}{2}}) = \begin{cases} n + k & 1 \leq k \leq 2 \\ \sum_{j=1}^n m_j + \frac{5n+3}{2} & k = 3 \\ n - 1 + \sum_{i=1}^{\frac{n-1}{2}-1} (m_i + 1) + k & 4 \leq k \leq m_{\frac{n-1}{2}} \end{cases}$$

$$f(S_i^k) = \begin{cases} n + 3 + \sum_{j=1}^{i-1} (m_j + 1) + k & 1 \leq i \leq n - m_n - 2,\ 1 \leq k \leq m_i \\ n + 2 + \sum_{j=1}^{i-1} (m_j + 1) + k & n - m_n \leq i \leq \frac{n-1}{2} - 1,\ 1 \leq k \leq m_i \\ n - 1 + \sum_{j=1}^{i-1} (m_j + 1) + k & \frac{n+1}{2} \leq i \leq n,\ 1 \leq k \leq m_i \end{cases}$$

$$f(S^k_{n-m_n-1}) = \begin{cases} \sum_{j=1}^n m_j - m_n + 2n - 2 & k = 1 \\ n + 2 + \sum_{j=1}^{n-m_n-2} (m_j + 1) + k & 2 \leq k \leq m_{n-m_n-1} \end{cases}$$

iii). $m_{\frac{n-1}{2}} \geq 3$ and $m_n = \frac{n-1}{2}$. $f(z_k)$, $f(y_k)$, $(1 \leq k \leq n)$ and $f(x_0)$ are same as ii).

$$f(S^k_{\frac{n-1}{2}}) = \begin{cases} n + k & 1 \leq k \leq 2 \\ \sum_{j=1}^n m_j + \frac{3n-3}{2} & k = 3 \\ n - 1 + \sum_{i=1}^{\frac{n-1}{2}-1} (m_i + 1) + k & 4 \leq k \leq m_{\frac{n-1}{2}} \end{cases}$$

$$f(S^k_{\frac{n-1}{2}-1}) = \begin{cases} \sum_{j=1}^n m_j + \frac{5n+1}{2} & k = 1 \\ n + 2 + \sum_{j=1}^{\frac{n-1}{2}-2} (m_j + 1) + k & 2 \leq k \leq m_{\frac{n-1}{2}-1} \end{cases}$$

$$f(S_i^k) = \begin{cases} n + 3 + \sum_{j=1}^{i-1} (m_j + 1) + k & 1 \leq i \leq \frac{n-1}{2} - 2,\ 1 \leq k \leq m_i \\ n - 1 + \sum_{j=1}^{i-1} (m_j + 1) + k & \frac{n+1}{2} \leq i \leq n,\ 1 \leq k \leq m_i \end{cases}$$

iv). $m_{\frac{n-1}{2}} \geq 3$, and $m_n < \frac{n-1}{2}$. $f(z_k)$, $f(y_k)$, $(1 \leq k \leq n)$ and $f(x_0)$ are same as ii).

$$f(S^k_{\frac{n-1}{2}}) = \begin{cases} n + k & 1 \leq k \leq 2 \\ \sum_{j=1}^n m_j + \frac{5n+3}{2} & k = 3 \\ n + \sum_{j=1}^{\frac{n-1}{2}-1} (m_j + 1) + k & 4 \leq k \leq m_{\frac{n-1}{2}} \end{cases}$$

$$f(S_i^k) = \begin{cases} n + 3 + \sum_{j=1}^{i-1} (m_j + 1) + k & 1 \leq i \leq \frac{n-1}{2} - 1,\ 1 \leq k \leq m_i \\ n + \sum_{j=1}^{i-1} (m_j + 1) + k & \frac{n+1}{2} \leq i \leq n - m_n - 2,\ 1 \leq k \leq m_i \\ n - 1 + \sum_{j=1}^{i-1} (m_j + 1) + k & n - m_n \leq i \leq n,\ 1 \leq k \leq m_i \end{cases}$$

$$f(S^k_{n-m_n-1}) = \begin{cases} \sum_{j=1}^n m_j - m_n + 2n - 2 & k = 1 \\ n - 1 + \sum_{j=1}^{n-m_n-2} (m_j + 1) + k & 2 \leq k \leq m_{n-m_n-1} \end{cases}$$

v). $m_n \geq n-1$, and $m_{\frac{n-1}{2}} \geq m_n \geq n-1 > \frac{n-1}{2}$.

$$f(z_k) = k \qquad 1 \leq k \leq n$$

$$f(y_k) = \begin{cases} \sum_{i=1}^{n} m_i + 3n + 2 - k & 1 \leq k \leq \frac{n-1}{2} \\ \sum_{i=1}^{n} m_i + \frac{5n+3}{2} - k & \frac{n+1}{2} \leq k \leq n \end{cases}$$

$$f(x_0) = \sum_{i=1}^{n} m_i + \frac{5n+3}{2}$$

$$f(S^k_{\frac{n-1}{2}}) = \begin{cases} \sum_{j=1}^{n} m_j + 2n + 1 + k & 1 \leq k \leq \frac{n-1}{2} - 1 \\ \frac{n+3}{2} + \sum_{i=1}^{\frac{n-1}{2}-1} (m_i + 1) + k & \frac{n-1}{2} \leq k \leq m_{\frac{n-1}{2}} \end{cases}$$

$$f(S^k_{\frac{n-1}{2}-1}) = \begin{cases} \sum_{j=1}^{n} m_j + \frac{5n+1}{2} & k = 1 \\ n + \sum_{i=1}^{\frac{n-1}{2}-2} (m_i + 1) + k & 2 \leq k \leq m_{\frac{n-1}{2}-1} \end{cases}$$

$$f(S^k_i) = \begin{cases} n + 1 + \sum_{j=1}^{i-1} (m_j + 1) + k & 1 \leq i \leq \frac{n-1}{2} - 2,\ 1 \leq k \leq m_i \\ \frac{n+3}{2} + \sum_{j=1}^{i-1} (m_j + 1) + k & \frac{n+1}{2} \leq i \leq n,\ 1 \leq k \leq m_i \end{cases}$$

Acknowledgment. The first author wishes to express his sincere gratitude to Professor J. Zhong for his valuable comments and encouragement. aluable comments and encouragement.

REFERENCES

[1] C. Delorme, M. Maheo, H. Thuillier, K. M. Koh, and H. K. Teo, Cycles with a chord are graceful, *Journal of Graph Theory,* 4:409-415, (1980).

[2] S. W. Golomb, How to number a graph, in *Graph theory and computing,* edited by R. C. Read, Academic Press, New York, 23-37, (1972).

[3] F. Harary, *Graph Theory,* Addison-Wesley, Reading, Mass., (1969).

[4] C. Hoede, and H. Kuiper, All wheels are graceful, *Utilitas Math.* 14:311, (1988).

[5] Y. Liu, On the gracefulness of single tooth star graphs (Chinese), *Changsha Shui Dian Shi Yuan Xue Bao, (Zi),* Zengkan, (1994).

[6] S. Lu, On the gracefulness of double teeth star graphs (Chinese), *J. Math. Res. Exposition,* 3:623-626, (1993).

[7] K. Ma, and C. Feng, On gracefulness of gear graphs (Chinese), *Math. Practice. Theory,* 4:72-73, (1984).

[8] A. Rosa, On certain valuations of the vertices of a grace, in *Theory of Graphs, Proc. International Symposium,* Rome, (1966).

The Reaction Graph of the Fano Plane

E. Keith LLOYD

(Faculty of Mathematical Studies, University of Southampton, SOUTHAMPTON SO17 1BJ, United Kingdom)

Abstract A construction in the chemistry literature produces graphs known as reaction graphs; it can also be applied to many non-chemical structures. One of the reaction graphs of the Fano plane proves to be isomorphic to the point-plane incidence graph of the three-dimensional projective geometry PG(3, 2) and an explicit correspondence emerges between the 3-subsets of a 7-set and the lines of PG(3, 2).

Keywords reaction graph; Fano plane; PG(3, 2); Kirkman schoolgirl problem; Steiner system; Room square.

1. Introduction

There is a construction in the chemistry literature which produces graphs termed *reaction graphs*. Mathematically the construction is usually equivalent or similar to the following procedure. Starting from a suitable molecule containing n atoms, all possible inequivalent labellings of those n atoms with n distinct symbols (usually 1, 2, ..., n or 0, 1, ..., n - 1) are considered. There are $m = n\,!/N$ such labellings where N is the order of the symmetry group of the molecule. A *reaction graph* is then defined to be a graph with m vertices where each vertex $V(b)$ corresponds to one of the labellings b. There is a directed edge from vertex $V(a)$ to vertex $V(b)$ if and only if labelling a can be converted into labelling b by a permutation of a fixed type. If the permutations have order two then undirected edges are used. Examples of reaction graphs can be found in Jones and Lloyd [4] and Lloyd and Jones [6] and the various references therein.

The above type of construction can be applied to a great many non-chemical structures in order to form "reaction graphs". In this paper a graph with thirty vertices is constructed in this way starting from the Fano plane. It proves to be isomorphic to the vertex-plane incidence graph of PG(3, 2) and an explicit correspondence emerges between the 3-subsets of a 7-set and the lines of PG(3, 2). The existence of such a correspondence has already been proved in the literature but in a more abstract setting.

N.B. Throughout the paper the terms *point* and *line* refer to projective planes or spaces and the terms *vertex* and *edge* refer to graphs. The terms 3-subset, 7-set, etc mean a subset of size 3, a set of size 7, etc.

2. The Fano Plane and its Reaction Graph

The Fano plane or PG(2, 2) is the smallest of the finite projective planes; it has seven points and seven lines with three points on each line. There is a standard picture of it in the euclidean plane (see Fig. 1) in which six of the triples forming lines are collinear in the euclidean sense but the seventh is represented by three points on a circle.

There are two standard ways in which to label the seven points of the Fano plane. In the first labelling, the residues mod 7 are used as the labels. One of the lines is labelled 1 2 4 and the other lines are obtained from it by repeatedly adding 1 (mod 7) to these three numbers (see Fig. 1), or, equivalently, by repeatedly applying the permutation (1234560) to 1 2 4, to give 2 3 5, 3 4 6, 4 5 0, 5 6 1, 6 0 2, 0 1 3. Sometimes it is more convenient to use the number 7 instead of 0 (see Fig. 2).

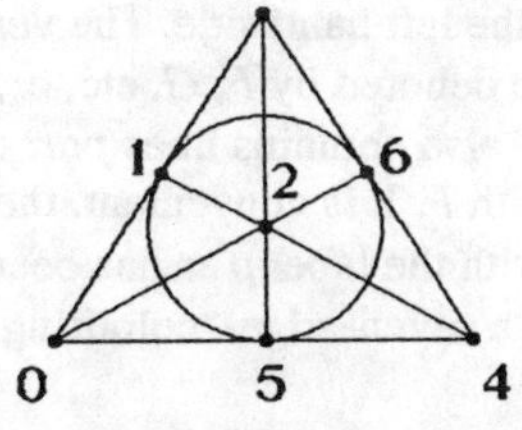

Fig. 1

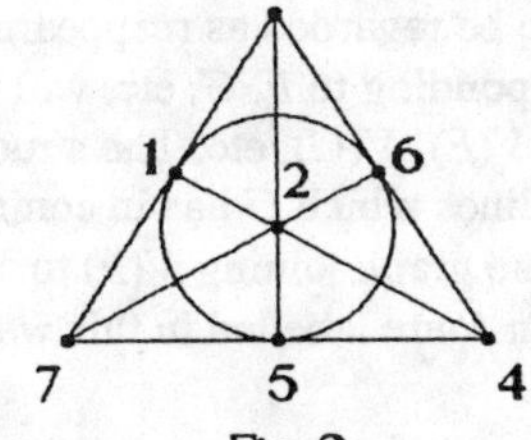

Fig. 2

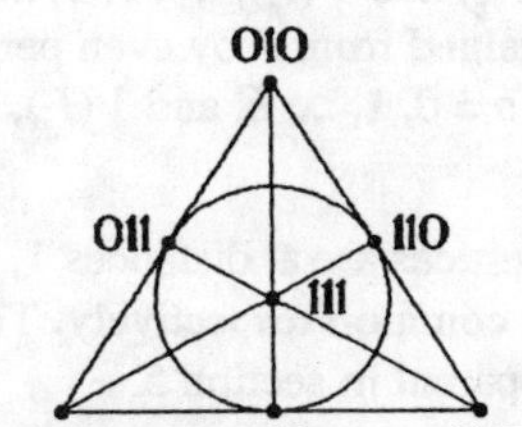

Fig. 3

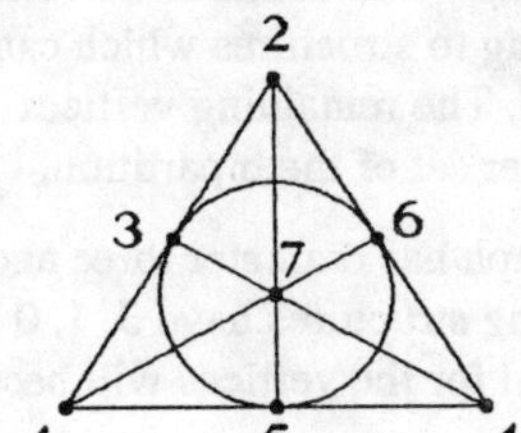

Fig. 4

In the second labelling the seven points are assigned homogeneous coordinates (*a*, *b*, *c*) over the Galois field GF(2). The coordinates can be abbreviated to *abc* (see Fig. 3) and then, if these numbers are regarded as binary numbers and converted to decimal numbers, the labelling shown in Fig. 4 is obtained.

In this paper all possible labellings of the vertices of the Fano plane by the numbers 1, 2, ..., 7 are considered. Since the automorphism group of the plane has order 168 there are only $7!/168 = 30$ inequivalent labellings. Each such labelled plane will be called a *Fano structure*.

Definition The *reaction graph R* of the Fano plane has 30 vertices, corresponding to the 30 inequivalent Fano structures; two vertices are adjacent if and only if the corresponding structures differ by a single transposition of two labels.

[Strictly the graph should be called *a* reaction graph since it is just one of several graphs which could be constructed in a similar way using different permutations in the definition of adjacency, but the other reaction graphs are much less interesting.]

Starting with the structure *F* in Fig. 4 there are 21 transpositions which can be applied to it but they lead to seven new labellings only. If *pqr*, *pst*, *puv* are the three lines through point *p* in *F* then each of the transpositions (*qr*), (*st*), (*uv*) leads to the same Fano structure *G* say. The reaction graph, therefore, must be at most seven-valent and, by drawing it, it is straightforward to show that it is precisely seven-valent. The graph is drawn in Fig. 5 where edges disappearing off the right hand side of the picture are to be regarded as reappearing on the left hand side. The vertices in the graph corresponding to *F*, *G*, etc, will also be denoted by *F*, *G*, etc, or, when precision requires, by $V(F)$, $V(G)$, etc. The structure *G* also contains lines *pqr*, *pst*, *puv* and these are the only lines which *G* has in common with *F*. It is convenient, therefore, to label the edge in the graph joining $V(F)$ to $V(G)$ with the label *p* as has been done in Fig. 5. If all edges in *R* are labelled in this way then a seven-edge-"colouring" of *R* would be obtained.

The graph is bipartite with 15 vertices in each set of the bipartition. The vertex set containing $V(F)$ also contains all vertices $V(H_n)$ and $V(I_n)$, $n = 0, 1, ..., 6$, corresponding to structures which can be obtained from *F* by even permutations of the points 1 to 7. The remaining vertices $V(G_n)$, $n = 0, 1, ..., 6$, and $V(J_n)$, $n = \infty, 0, 1, ..., 6$, form the other set of the bipartition.

The graph has diameter three and two vertices are at distances 1, 2, 3 apart if the corresponding structures have 3, 1, 0 lines in common respectively. The reason for the notation used for the vertices will become apparent in section 5.

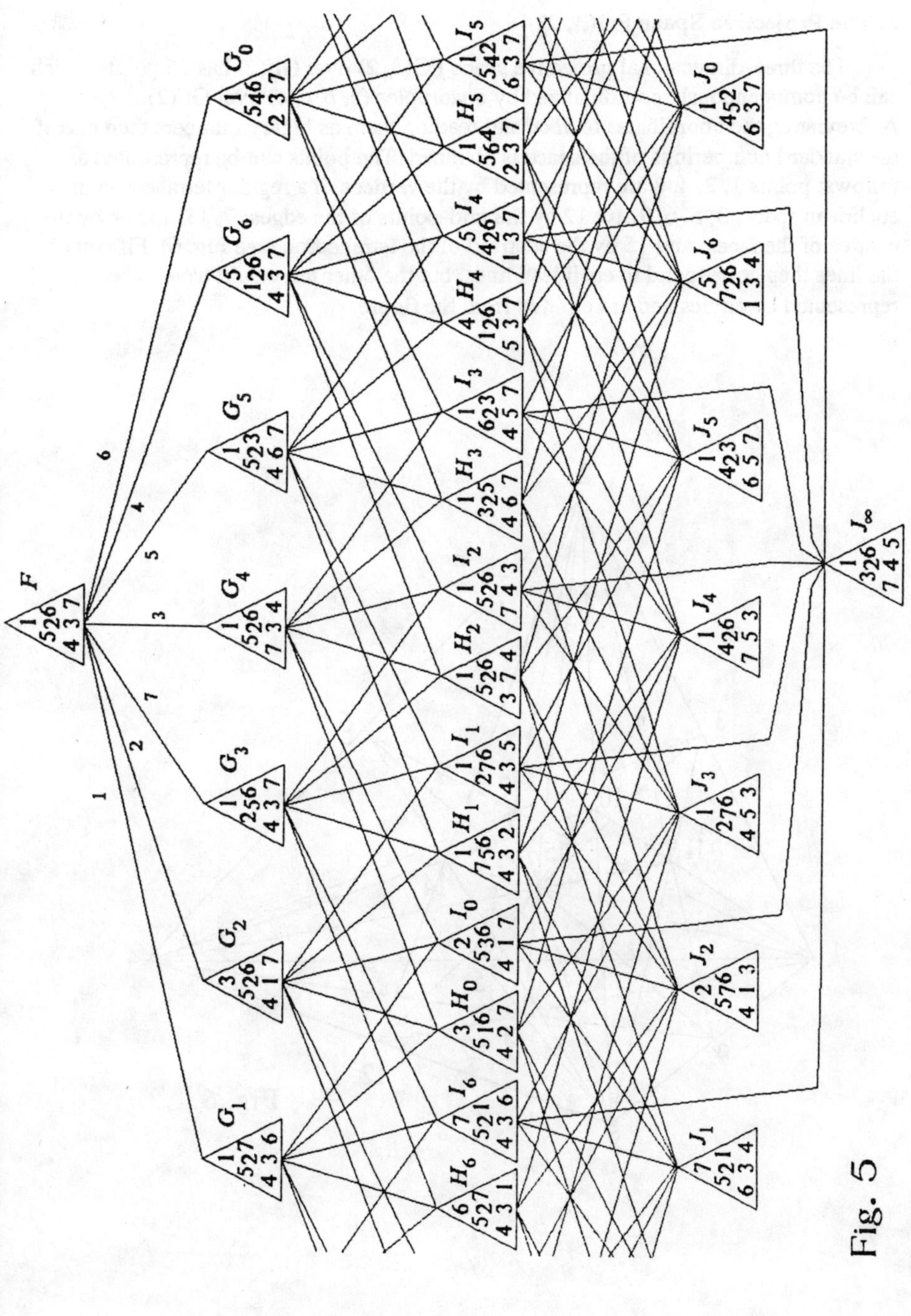

F 1 526 4 3 7
1
2
7
3
5
4
6
G_1 1 527 4 3 6
G_2 3 526 4 1 7
G_3 1 256 4 3 7
G_4 1 526 7 3 4
G_5 1 523 4 6 7
G_6 5 126 4 3 7
G_0 1 546 2 3 7
H_6 6 527 4 3 1
I_6 7 521 4 3 6
H_0 3 516 4 2 7
I_0 2 536 4 1 7
H_1 1 756 4 3 2
I_1 1 276 4 3 5
H_2 1 526 3 7 4
I_2 1 526 7 4 3
H_3 1 325 4 6 7
I_3 1 623 4 5 7
H_4 4 126 5 3 7
I_4 5 426 1 3 7
H_5 1 564 2 3 7
I_5 1 542 6 3 7
J_1 7 521 6 3 4
J_2 2 576 4 1 3
J_3 1 276 4 5 3
J_4 1 426 7 5 3
J_∞ 1 326 7 4 5
J_5 1 423 6 5 7
J_6 5 726 1 3 4
J_0 1 452 6 3 7

Fig. 5

3. The Projective Space PG(3, 2)

The three-dimensional projective space PG(3, 2) over GF(2) has 15 points which can be homogeneously coordinatized by quadruples (a, b, c, d) over GF(2). Abbreviating the coordinates to *abcd* and reading them as binary numbers then one of the standard numberings of the space is obtained. The points can be represented as follows: points 1, 2, 4, 8 are represented by the vertices of a regular tetrahedron in euclidean space; 3, 5, 6, 9, 10, 12 by the mid-points of the edges; 7, 11, 13, 14 by the centres of the faces; and 15 by the centroid of the tetrahedron (see Fig. 6). Fifteen of the lines then correspond to euclidean lines, but the other ten would have to be represented by circles and are omitted from the figure.

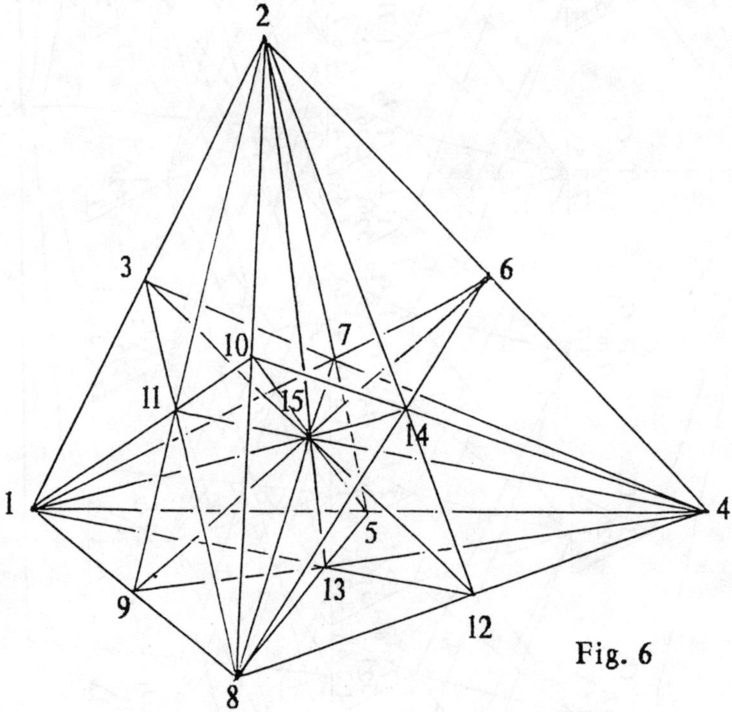

Fig. 6

Each plane has a linear equation over GF(2) of the form $aw + bx + cy + dz = 0$. The planes can be numbered <1> to <15> by reading the corresponding string *abcd* as a binary number - the brackets are used simply to distinguish point *p* from plane <*p*>. The space PG(3, 2) contains 35 lines and 15 planes. There are seven points in each plane and three points on each line and (by duality) seven planes through each point and three planes through each line. Any three non-collinear points determine a plane and each plane corresponds to a euclidean plane in Fig. 6.

Definition The *point-plane incidence graph* of PG(3, 2) is the bipartite graph with 15 vertices corresponding to points and 15 vertices corresponding to planes. A point-vertex is joined to a plane-vertex if and only if the corresponding point lies in the corresponding plane (see Fig. 7).

In this paper it proves to be more appropriate to regard a point-vertex in the incidence graph as corresponding not to the point but to the star of seven lines through that point and so it will be termed a *star-vertex.*

Theorem The reaction graph of the Fano plane and the point-plane incidence graph of PG(3, 2) are isomorphic.

Proof The proof consists of comparing Fig. 5 (the reaction graph) with Fig. 7 (the incidence graph). Since PG(3, 2) has a non-trivial automorphism group the isomorphism is, of course, not unique.

4. Correspondence between Triples and Lines

Each line in any of the 30 Fano structures is labelled with one of the 35 unordered triples (3-subsets) taken from the set {1, 2, 3, 4, 5, 6, 7} and each such triple occurs in precisely six Fano structures. The space PG(3, 2) contains 35 lines, each lies in three planes and each in three stars, so each line also occurs a total of six times.

In the graph, each vertex corresponds (a) to one of the Fano structures (Fig. 5) and (b) either to a plane or to a star in PG(3, 2) (Fig. 7). Comparison of the various lines in the corresponding structures and planes/stars establishes a consistent one-one correspondence between the 35 3-subsets of the 7-set and the 35 lines in PG(3, 2). The correspondence is listed in Table 1, from which it will be noted that given the first row of the table, the remainder can be generated by repeatedly applying the permutation

$$(1\ 2\ 7\ 3\ 5\ 4\ 6)(8)(9\ 10\ 15\ 11\ 13\ 12\ 14).$$

When applied to PG(3, 2) this permutation permutes the star of lines through point 8.

The five lines of PG(3, 2) in any row of Table 1 are disjoint and contain all points between them. Such a set is variously described as a set of *mutually skew lines* or a *parallel class.* The seven rows together partition the 35 lines into seven disjoint parallel

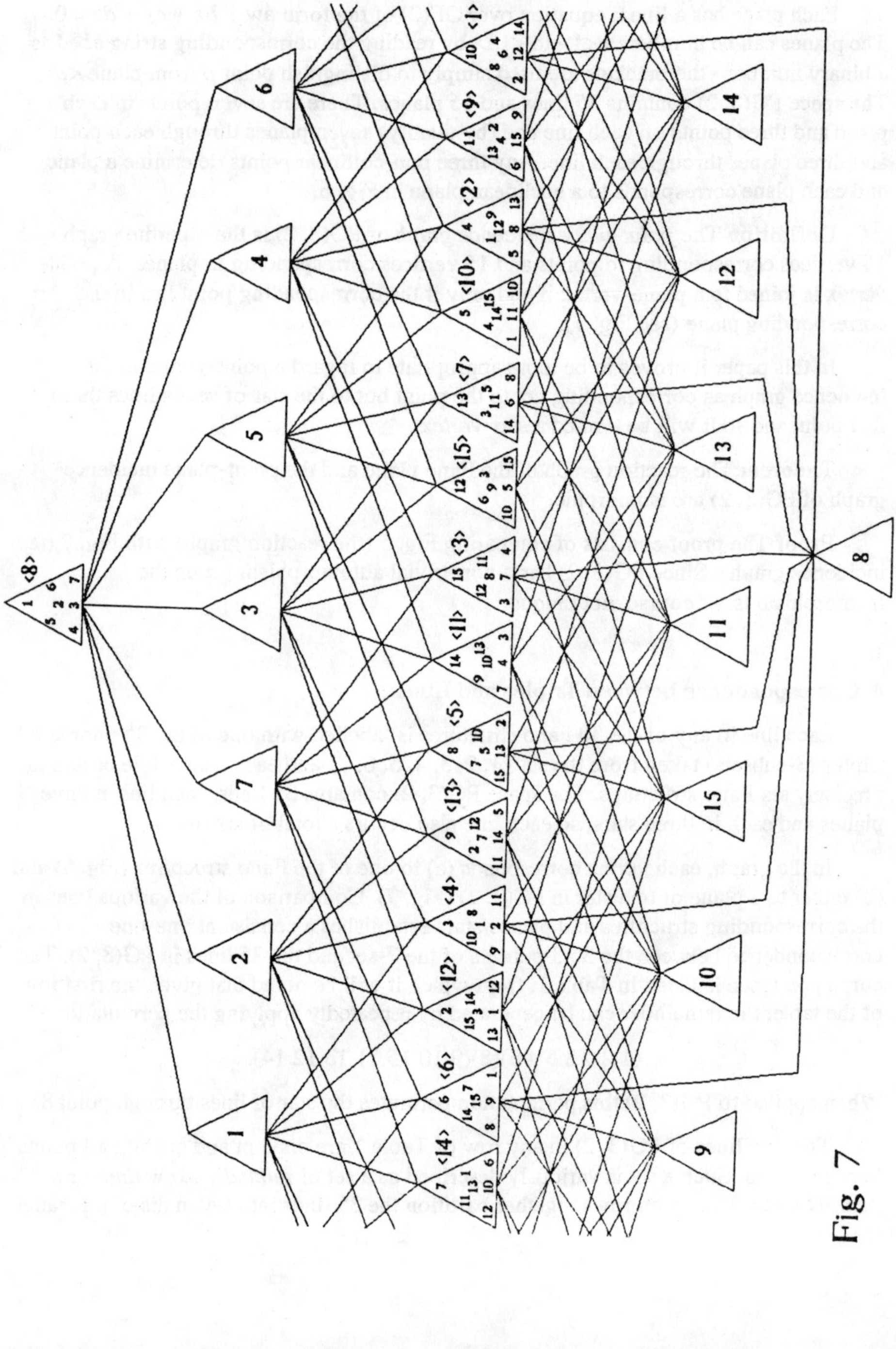

Fig. 7

classes. The partition is called a *Kirkman parallelism* since it forms a solution to the famous Kirkman schoolgirl problem posed in [5]: "Fifteen young ladies in a school walk out three abreast for seven days in succession: it is required to arrange them daily, so that no two shall walk twice abreast."

There are other ways to partition the lines of PG(3, 2). For example, the 15 lines marked § in Table 1 have the property that each point lies on precisely three of them; the 15 unmarked lines in rows two to seven have the same property. Further details of such partitions and other facts about PG(3, 2) are given by Hirschfeld [2, 3].

Table 1

1 2 3	1 2 3	5 6 7 4 9 13	4 7 6 5 10 15	4 5 7 6 8 14	4 6 5 7 11 12
2 7 5	2 7 5 §	4 1 3 6 10 12	6 3 1 4 15 11	6 4 3 1 8 9	6 1 4 3 13 14
7 3 4	7 3 4	6 2 5 1 15 14	1 5 2 6 11 13 §	1 6 5 2 8 10	1 2 6 5 12 9
3 5 6	3 5 6	1 7 4 2 11 9	2 4 7 1 13 12	2 1 4 7 8 15	2 7 1 4 14 10 §
5 4 1	5 4 1 §	2 3 6 7 13 10	7 6 3 2 12 14 §	7 2 6 3 8 11 §	7 3 2 6 9 15 §
4 6 2	4 6 2	7 5 1 3 12 15 §	3 1 5 7 14 9 §	3 7 1 5 8 13 §	3 5 7 1 10 11 §
6 1 7	6 1 7 §	3 4 2 5 14 11	5 2 4 3 9 10 §	5 3 2 4 8 12 §	5 4 3 2 15 13 §

The fact that the 35 triples can be identified with the 35 lines of PG(3, 2) is not a new result. Wagner [7] describes a construction which he attributes to A.M. Gleason. In the terminology of the present paper he considers the action of the alternating group A_7 on a fixed Fano structure (that in Fig. 2). This produces 15 structures which he calls "points". The 35 triples from {1, 2, 3, 4, 5, 6, 7,} he calls "lines". If the three symbols in a "line" are collinear in a "point" (Fano structure) then the "line" and "point" are called "incident". He then shows that the incidence structure forms a projective geometry, but up to isomorphism there is only one 15-point projective geometry, hence the incidence structure must be isomorphic to PG(3, 2). Hall [1] gives a similar construction but he considers all 30 Fano structures and the action on them of the stabilizer GL(3, 2) of one of them. This splits the 30 structures into two 15-orbits (corresponding to the fact that the reaction graph is bipartite). Hall then introduces essentially the same incidence structure with the one orbit forming the points of PG(3,2) and the other forming the planes; again the 3-subsets form the lines.

Unlike the present paper, neither Wagner nor Hall gives an explicit one-one correspondence between the 3-subsets and the lines of PG(3, 2). Hall does introduce two graphs Γ and Γ^* but they are not the same as the graphs in this paper since they have 35 vertices. The vertices of Γ are labelled by the 3-subsets, and two vertices are joined if and only if their labelling sets intersect in a 1-set. The vertices of Γ^* are labelled by the lines of PG(3, 2) and two vertices are joined if and only if the corresponding lines intersect. The graphs Γ and Γ^* are, of course, isomorphic and Hall uses the identification to give elementary proofs of various results related to the complete 3-design on seven points and PG(3, 2).

The correspondence in Table 1 can be used in the same way. For example, if P denotes the set of triples from {1, 2, 3, 4, 5, 6, 7} containing point p and Q the set containing point q then the following sets (where C denotes complementation) correspond to well-known configurations in PG(3,2):

$P \cap Q$	5 lines, 1 through each point in PG(3,2);
$C(P \cup Q)$	10 lines, 2 through each point;
P	15 lines, 3 through each point;
$C(P)$	20 lines, 4 through each point;
$P \cup Q$	25 lines, 5 through each point;
$C(P \cap Q)$	30 lines, 6 through each point.

Note, however, that some of the configurations above arise in other ways. The disjoint lines 1 2 3, 4 9 13, 5 10 15, 6 8 14, 7 11 12 in the first row of Table 1 correspond to the triples 1 2 3, 5 6 7, 4 6 7, 4 5 7, 4 5 6, i.e. each triple is identical to or disjoint from a fixed triple (1 2 3 in this example). Hall [1, §4.1] shows that any set of five mutually skew lines arises in one of the two ways described above.

Although any statement or observation about either the complete 3-design on seven points or about PG(3, 2) can be converted into a statement about the other, sometimes the one interpretation seems rather obscure. As already mentioned, the 15 lines in PG(3, 2) marked § in Table 1 have the property that each point lies on precisely three of them. The unmarked lines in rows two to seven form a second disjoint set with the same property. The remaining five lines (those in the first row) form a parallel class. In terms of the 3-subsets the partition can be arranged as follows:

(a) 1 2 3, 5 6 7, 4 6 7, 4 5 7, 4 5 6;

(b) 1 4 5, 1 6 7, 1 5 7, 1 2 5, 1 3 5, 2 3 5,
2 4 5, 2 6 7, 2 5 7, 1 2 7, 1 3 7, 2 3 7,
3 4 5, 3 6 7, 3 5 7;

(c) 1 4 7, 1 6 5, 1 4 6, 1 2 4, 1 3 4, 2 3 4,
2 4 7, 2 6 5, 2 4 6, 1 2 6, 1 3 6, 2 3 6,
3 4 7, 3 6 5, 3 4 6.

Set (b) comprises triples made up as follows: take any one of 1, 2, 3 with any one of the pairs 4 5, 6 7, 5 7 and then add triples formed by choosing any two of 1, 2, 3 with either of 5, 7. Set (c) is formed similarly but using pairs 4 7, 6 5, 4 6 and either of 4, 6.

5. Steiner Systems and Room Squares

Ian Anderson has kindly pointed out to me that the solution to the Kirkman schoolgirl problem in Table 1 is a well-known "2-rotational" solution which can be generated from a Steiner triple system and a Room square. To see this it is desirable to change the notation somewhat, so relabelling 1 → 1, 2 → 2, 7 → 3, 3 → 4, 5 → 5, 4 → 6, 6 → 0; 8 → ∞; 9 → 1', 10 → 2', 15 → 3', 11 → 4', 13 → 5', 12 → 6', 14 → 0' gives Table 2.

Table 2

124 124	503 61' 5'	630 52' 3'	653 0∞0'	605 34' 6'
235 235	614 02' 6'	041 63' 4'	064 1∞1'	016 45' 0'
346 346	025 13' 0'	152 04' 5'	105 2∞2'	120 56' 1'
450 450	136 24' 1'	263 15' 6'	216 3∞3'	231 60' 2'
561 561	240 35' 2'	304 26' 0'	320 4∞4'	342 01' 3'
602 602	351 46' 3'	415 30' 1'	431 5∞5'	453 12' 4'
013 013	462 50' 4'	526 41' 2'	542 6∞6'	564 23' 5'

The solution to the Kirkman problem now corresponds to the Steiner triple system and the Room square in Table 3. Each entry in the Room square contains either two or none of the symbols ∞, 0', 1', 2', 3', 4', 5', 6' and each of these eight symbols appears once in each row and once in each column. Adjoining the column heading to each of the pairs beneath it, 0 ∞ 0', 3 4' 6', 5 2' 3', 6 1' 5'..., gives 28 of the lines in PG(3, 2); the other seven lines come from the Steiner triple system, which is, of course, just the Fano plane in Fig. 1. Each triple lists the blank columns in the corresponding row of the Room square. The Room square itself is generated from its first row by adding one to each element (with the understanding that ∞ + 1 = ∞) and shifting cyclically to get the next row. Further information on Steiner systems and Room squares may be found in Wallis [8].

Table 3

Steiner system	Room square 0	1	2	3	4	5	6
1 2 4	∞0'	---	---	4'6'	---	2'3'	1'5'
2 3 5	2'6'	∞1'	---	---	5'0'	---	3'4'
3 4 6	4'5'	3'0'	∞2'	---	---	6'1'	---
4 5 0	---	5'6'	4'1'	∞3'	---	---	0'2'
5 6 1	1'3'	---	6'0'	5'2'	∞4'	---	---
6 0 2	---	2'4'	---	0'1'	6'3'	∞5'	---
0 1 3	---	---	3'5'	---	1'2'	0'4'	∞6'

The Room square structure can be discerned in the reaction graph and the vertex names in Fig. 5 have been chosen to reveal this. The 56 lines joining the fourteen vertices H_n and I_n ($n = 0, 1, ..., 6$) to the eight vertices J_n ($n = \infty, 0, 1, ..., 6$) are in one-one correspondence with the 56 symbols in the 28 non-empty entries of the Room square. Vertex H_2, for example, is joined to J_1, J_2, J_5, J_0 and I_2 is joined to J_3, J_4, J_6, J_∞. The column headed 2 in the Room square contains the symbols 1', 2', 5', 0' on the right and the symbols 3', 4', 6', ∞ on the left. The remaining edges can be identified with the remaining entries by repeatedly applying the permutation

(∞)(0 1 2 3 4 5 6)

to the subscripts on the vertices H_2, I_2 and J_n ($n = \infty, 0, 1, ..., 6$) and the permutation

(∞)(0' 1' 2' 3' 4' 5' 6')

to the entries in column 2 of the Room square.

6. Further Comments on the Graphs

Any drawing of a graph with a high degree of symmetry will bring out certain symmetries and hide others. The drawing of the reaction graph in Fig. 5 emphasises the fact that the stabilizer of vertex F contains the permutation (1 2 7 3 5 4 6). The vertex J_∞ is also stabilized by this permutation and that is why it appears to be different from the other vertices J_n ($n = 0, 1, ..., 6$). Vertex F is stabilized by 24 different 7-cycles, each of which stabilizes precisely one of the J_n. Any J_n can be moved to the bottom of the graph by a suitable reordering of the vertices G_n ($n = 0, 1, ..., 6$).

If seven identical objects are placed round a circle then there are five different ways in which three of them can be chosen; the five columns of 3-subsets of the 7-set in Table 1 correspond to these five ways. In Fig. 8 the shading marks the 3-subsets in the first row and the other rows are obtained by cyclically permuting the numbers. When the numbers are disregarded, three of the patterns in Fig. 8 are symmetrical but (a) and (d) are mirror images. Pattern (a) produces the lines in the Fano structure F and pattern (d) produces the lines in the Fano structure J_∞.

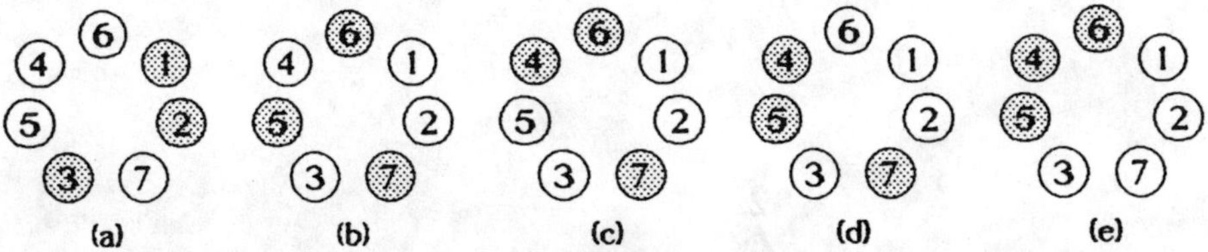

Fig. 8

Each edge in the reaction graph corresponds to a simple operation: interchanging two labels. The corresponding operation in the incidence graph is less easy to describe and is best understood by looking at a particular example such as the operation corresponding to the edge in the incidence graph joining star 12 and plane <2>. How are the seven lines passing through point 12 to be mapped on to the seven lines in plane <2>? The star and the plane have three lines in common: 1 12 13, 4 12 8 and 5 12 9 each of which occurs in a different row in Table 1. Choosing line 1 12 13 then the row in Table 1 containing it yields the five mutually skew lines

$$3\ \underline{5}\ 6,\ 2\ 11\ \underline{9},\ \underline{1\ 13\ 12},\ 7\ \underline{8}\ 15,\ \underline{4}\ 14\ 10 \qquad (1)$$

where the seven points underlined are those in plane <2>. In Fig.9 (which has been obtained from Fig. 6 by applying a suitable automorphism to the points) plane <2> forms the base of the tetrahedron, all lines of star 12 are shown and four of the lines in (1) are straight but 7 8 15 has been drawn as a bent line.

Algebraically each line in star 12 can be mapped to plane <2> by applying to it one of the permutations ξ, ξ^2, e where

$$\xi = (3\ 6\ \underline{5})(2\ 11\ \underline{9})(\underline{1\ 13\ 12})(7\ \underline{8}\ 15)(\underline{4}\ 14\ 10). \qquad (2)$$

Note that the first cycle in (2) is reversed when compared with (1). This gives:

$$e \quad : 1\ 12\ 13 \rightarrow 1\ 12\ 13,$$
$$\xi^2 \quad : 2\ 12\ 14 \rightarrow 9\ 13\ \ 4,$$
$$\xi^2 \quad : 3\ 12\ 15 \rightarrow 5\ 13\ \ 8,$$
$$e \quad : 4\ 12\ \ 8 \rightarrow 4\ 12\ \ 8,$$
$$e \quad : 5\ 12\ \ 9 \rightarrow 5\ 12\ \ 9,$$
$$\xi \quad : 6\ 12\ 10 \rightarrow 5\ \ 1\ \ 4,$$
$$\xi \quad : 7\ 12\ 11 \rightarrow 8\ \ 1\ \ 9.$$

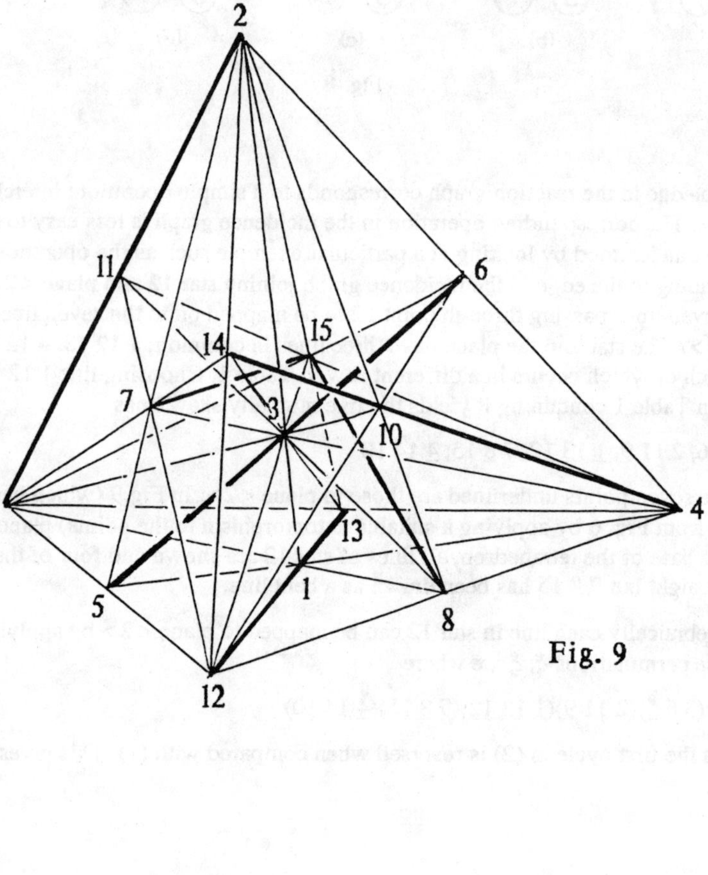

Fig. 9

A geometrical interpretation of the above procedure is possible. A line in star 12 which does not lie in plane <2> is of the form *p* 12 *q* where *p* and *q* lie on two different lines *j*, *k* in (1). Point *p* is projected along line *j* to plane <2> and point *q* is projected along line *k* to plane <2>. The point 12 is projected along line 1 12 13 but since this line lies in plane <2> the image of point 12 is not immediately clear; its image is obtained from the requirement that the three images of *p*, *q* and 12 are collinear.

If line 4 12 8 or 5 12 9 had been chosen instead of 1 12 13, then a different row of Table 1 would have been used leading to a different projection. The three lines common to the star and the plane would again be fixed but the images of the other four lines would have been permuted. The fact that there is a choice of three projections is to be expected, since there is a choice of three transpositions for each edge in the reaction graph.

Through point 12 there are seven planes: <1>, <2>, <3>, <5>, <12>, <13> and <14>. These planes together with the seven lines through point 12 form an incidence structure which is isomorphic to a Fano plane (see Fig. 10) where a plane is incident with a line if and only if the plane contains the line. Since two lines in star 12 lie in a unique plane and two lines in plane <2> intersect in a unique point, the mapping $\xi/\xi^2/e$ which maps the lines in star 12 to the lines in plane <2> induces a mapping from the seven planes through point 12 to the seven points in plane <2>. The induced mapping (compare corresponding points in Fig. 10 and Fig. 11) is:

$$<1> \to 4, <2> \to 12, <3> \to 8, <5> \to 5, <12> \to 13, <13> \to 9, <14> \to 1.$$

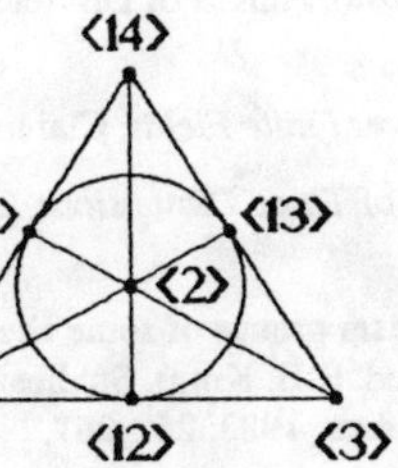

Fig. 10

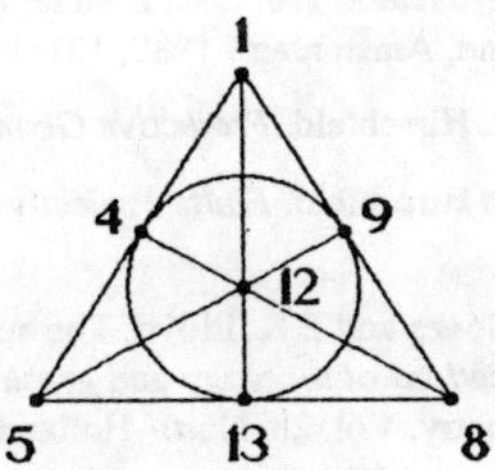

Fig. 11

The images of four of the planes can be obtained geometrically from Fig. 9 by using four of the skew lines in (1). For example, line 3 5 6 and point 12 lie in plane <5> and the image of plane <5> is the point 5 where line 3 5 6 meets plane <2>. Similarly:

line 2 11 9 lies in plane <13> which maps to point 9;

line 7 8 15 lies in plane <3> which maps to point 8;

line 4 14 10 lies in plane <1> which maps to point 4.

Each of the remaining three planes, namely <2>, <12> and <14>, contains the fifth skew line 1 13 12 so although these three planes map to these three points the order in which they do so cannot be obtained quite so simply.

7. Acknowledgements

I am grateful to the following people for helping in various ways. Gareth A. Jones provided the idea to look at the reaction graph of the Fano Plane. Ian Anderson read an early version of the paper and pointed out the relevance of Room squares. Terry Griggs also read an early version of the paper and made helpful comments. James W.P. Hirschfeld drew my attention to Hall's paper. Simon J. Cox found where to place the points in Fig. 6 so that all of the 25 lines included can be clearly seen.

References

[1] J.I. Hall, On identifying PG(3, 2) and the complete 3-design on seven points, in: *Topics on Steiner systems* (ed. C.C. Lindner and A. Rosa), Annals of Discrete Mathematics Vol. 7, North-Holland, Amsterdam, 1980, 131-140.

[2] J.W.P. Hirschfeld, *Projective Geometries over Finite Fields*, Clarendon Press, Oxford, 1979.

[3] J.W.P. Hirschfeld, *Finite Projective Spaces of Three Dimensions*, Clarendon Press, Oxford, 1985.

[4] G.A. Jones and E.K. Lloyd, The automorphism groups of some chemical graphs, in: *Chemical applications of topology and graph theory* (ed. R.B. King), Studies in Physical and Theoretical Chemistry, Vol. 28, North-Holland, Amsterdam, 1983, 252-267.

[5] T.P. Kirkman, VI. Query, *The Lady's and Gentleman's Diary*, 1850, 48.

[6] E.K. Lloyd and G.A. Jones, Reaction graphs, *submitted for publication.*

[7] A. Wagner, On collineation groups of projective spaces I, *Math. Zeitschr.* 76 (1961) 411-426.

[8] W.D. Wallis, *Combinatorial Designs*, Marcel Dekker, New York and Basel, 1988.

Kirkman schoolgirl problem with colouring of graphs

Lu Xinzhong Zhang zhongfu Han Jincang

(*Lanzhou Railway Institute. Lanzhon*, 730070)

Abstract In this paper, we difine the reciprocal chromatic number of graphs and study the relation of the reciprocal chromatic number of graphs with Kirkman schoolgirl problem.

Key words Kirkman schoolgirl problem; coluring of graphs.

1. Inrtoduction

In 1850, Rev. Thomas P. Kirkman put forward the following problem[1]:

Everyday, a woman teacher leds her fifteen schoolgirls have a walking, and she divides these fifteen schoolgirls into five lines (In each line, there are three schoolgirls), thus every schoolgirl has two follows in a common line. The question is whether there is a walking project for seven days in a row, which makes times that every schoolgirl with any two schoolgirls appear in a common line be not superior to one. (And it is just one). The solution of this problem is composed of thirty-five inordinate three-men-teams forming with fifteen schoolgirls, and any two schoolgirls appear and only appear in a common three-men-team. We can divide these thirty-five three-men-teams into seven partitions which is composed of five three-men-teams, and every partition correspond to a day. Then for erery schoolgirl is only in a three-men-team of every partitions.

After abstracted, above problem can be generalized as follows:

Let X be a set with a different elements. We divide a elements of X into b subsets (block) and construct a subset partition as follows:

(i) Every subset is just composed of k different elements,

(ii) every elements appears just in r different subsets,

(iii) any two different elements $x, y \in X$ appare in λ different subsets

This generalization is called a balanced incompleted block design, it is denoted by BIBD or by (a, b, k, r, λ)-design. Kirkman fifteen schoolgirls problem is just a $(15, 35, 3, 7, 1)$-design. It not only has theoretical meanings but also has applied value to dicide the solution numbers and the specific form of this kind of designs.

2. Reciprocal chromatic number of graphs

Definition 1[2-3] Let $G(V,E)$ be a simple graph, if

$\bigcup_{i=1}^{n} = V_i$, satisfies

(i) $V_i \cap V_i = \Phi \quad (i \neq j, i, j = 1,2,\cdots,n)$,

(ii) $E(G[V_i]) = \Phi \quad (i = 1,2,\cdots,n)$,

(iii) $E(G[V_i \cup V_j]) \neq \Phi \quad (i \neq j, i, j = 1,2,\cdots,n)$

Then, $\bigcup_{i=1}^{n} V_i$ is called a n -colour vertex partition of G. Here, Φ is an empty set, $G[S]$ is the induced subgraph by S.

Definition 2 $\chi(G) = \min\{n \mid \bigcup_{i=1}^{n} V_i$ is a n -colour vertex partition of $G\}$,

$\overline{\chi}(G) = \max\{n \mid \bigcup_{i=1}^{n} V_i$ is a n -colour vertex partition of $G\}$

are called chromatic number and reciprocal chromatic number of G, respectively.

By difinifion of $\overline{\chi}(G)$, we can easily prove that the following theorem 1 is true.

Theorem 1 If K_p is a complete graph of order p, then $\overline{\chi}(K_p) = p$

Theorem 2 If C_p is a cycle graph, of order $p(p = 2k^2 + k)$, then

$$\overline{\chi}(C_p) = 2k + 1$$

In order to prove theorem 2, We need the following lemma:

Lemma Let $K_{2n+1}(n \geqslant 1)$ is a complete graph of order $2n+1$, then, K_{2n+1} can be divided into n edge-disjionted Hamilton cycles.

Proof We use $0,1,\cdots, 2n$ to denote $2n+1$ vertices of K_{2n+1}. let 0 be at the centre of a circle and let the other $2n$ vertices be on the circumference of the circle uniformly. Let $C_1 = 012(2n)3(2n-1)4(2n-2)\cdots n(n+2)(n+1)0$, then, C_1 is a 2-factor of K_{2n+1}. By revolving C_1 through a vertex around 0 in the clockwise, we can get a new 2-factor of K_{2n+1}, so through n revolvings, we can get n 2-factors of $K_{2n+1} C_1, C_2, \cdots, C_n$. Obviously, $E(C_i) \cap E(C_j) = \Phi \ (i \neq j, i, j = 1,2,\cdots,n)$, and $K_{2n+1} = C_1 \cup C_2 \cup \cdots \cup C_n$.

Proof of theorem 2: First we prove $\overline{\chi}(C_{2k^2+k}) \leqslant 2k+1$. Suppose $\overline{\chi}(C_{2k^2+k}) > 2k+1$, let $C = \{1,2,\cdots,\overline{\chi}(C_{2k^2+k})\}$ be a colour set with $\overline{\chi}(C_{2k^2+k})$ different colours. As each vertex of C_{2k^2+k} is adjacent to two vertices, so the number of vertices with colour $i(1 \leqslant i \leqslant \overline{\chi}(C_{2k^2+k}))$ is at least $\frac{\overline{\chi}(C_{2k^2+k}) - 1}{2} > k$ in any $\overline{\chi}(C_{2k^2+k})$- colour vertex partition of C_{2k^2+k}. Hence, $V(C_{2k^2+k}) > 2k^2 + k$, it is a contradiction.

In order to prove $\overline{\chi}(C_{2k^2+k}) \geqslant 2k+1$, we construct a $(2k+1)$- colour vertex partition of C_{2k^2+k} as follows:

Let $1,2,\cdots,2k+1$ denote $2k+1$ different colours. We work out a complete graph K_{2k+1}, whose $2k+1$ vertices are also denoted by $1,2,\cdots,2k+1$. By lemma, K_{2k+1} has k edge-disjointed Hamilton cycles:

$$1 i_2 i_3 \cdots i_{2k+1} 1 \quad (i = 1,2,\cdots,k)$$

(Here $i_j i_{j+1}$ denote the edges between vertex i_j and vertenx i_{j+1})

Then, $11_2 1_3 \cdots 1_{2k+1} 1; 12_2 2_3 \cdots 2_{2k+1} 1; \cdots; 1k_2 k_3 \cdots k_{2k+1} 1$, is a $2k+1$ -colour vertex partition of C_{2k^2+k}. So we have $\overline{\chi}(C_{2k^2+k}) \geqslant 2k + 1$.

To sum up, we have $\chi(C_{2k^2+k}) = 2k + 1$.

Obviously, different Hamilton division of K_{2k+1} correspond to different $2k + 1$ -colour vertex partition of C_{2k^2+k} .

3. **The solution of Kirkman schoolgirl problem**

By 1 and 2, we have known that the solution of Kirkman schoolgirl problem, i. e. $(a, b, 3, r, 1)$ -design is one of colouring of $C_p (p = 3b = ar = 2k^2 + k, 2k + 1 \equiv 0 \pmod 3)$ with $\overline{\chi}(C_p)$ colours.

According to above discussion, from the colouring of C_p , we can get the solution of Kirkman schoolgirl problem. For this, we change difficult Kirkman schoolgirl problem into direct geometric problem. For instance, from a vertex-colouring of C_{36} in the clockwise (or counter clockwise) with 9 colours:

1,2,3,4,5,6,7,8,9,1,4,7,3,6,9,2,5,8,1,5,9,3,8,4,6,2,7,1,6,8,2,4,9,7,5,3

We can get a walking project of nine schoolgirls for four days in a row as follows:

(1,2,3) (4,5,6) (7,8,9); (1,4,7) (3,6,9) (2,5,8);

(1,5,9) (3,8,4) (6,2,7); (1,6,8) (2,4,9) (7,5,3).

(Where numbers 1,2,…,9 denote 9 different schoolgirls)

Similarly, from a vertex-colouring of C_{105} with 15 colours, we can also get a walking project of fifteen schoolgirls for seven days in a row as follows:

(1,10,4) (5,13,8) (7,14,6) (9,11,2) (3,12,15);
(1,5,11) (3,10,9) (13,14,12) (4,6,2) (7,15,8);
(1,14,15) (3,8,11) (10,6,12) (13,4,9) (2,5,7);
(1,6,8) (10,14,11) (15,13,2) (4,7,3) (9,5,12);
(1,2,12) (9,14,8) (5,3,6) (7,13,10) (15,4,11);
(1,3,13) (11,7,12) (10,2,8) (9,15,6) (5,14,4);
(1,9,7) (10,5,15) (2,14,3) (4,8,12) (11,6,13)

Using this method, some scholars are designing computer program to get the solutions of Kirkman schoolgirl problem.

Ackonwledgments:

This project is supported by the NNSF of China, GPNSF and RMNSF

References

[1] R. A Brualdi, Introductory Combinatorics, North—Holland, Inc, 1977

[2] F. Harary, Graph Theory, Addison—wesley, Reading(1969)

[3] F. Harary S. Hedetniemi and G. Prins, An interpolatism theorem for graphical homorphisms. Por Math. 26(1967), 435—462

Several Counting Results in the 18th–19th Centuries in Chinese Mathematics

Luo Jianjin

(Department of Mathematics, Northwest University, Xi'an 710069)

Abstract In the 18th–19th centuries Chinese mathematicians Ming, Dong and Xiang studied some counting structures, which are not well–known but are interesting. This paper will express the old form of their algorithms and results in modern form.

Key words the history of combinatorics, Chinese mathematics, enumeration, recurrence, reverting power series, Fibonacci numbers.

Chinese mathematicians in the Qing Dynasty independently put forward several counting functions and obtained some important achievements, such as Stirling numbers, Eulerian numbers, Euler numbers, Catalan numbers, Daixu's series numbers, and so on. There were other counting structures, which were not famous and perhaps few people know what their use is. This paper will express the traditional form of some algorithms and results in modern notation, in order to evoke some interests and discussion on it.

In the history of mathematics in China Ming Antu (明安图, 1692?–1763?), a mongolian scientist, in his posthumous work *Efficient Methods for the Precise Value of Circular Functions* (《割圜密率捷法》, 1839) [1], laid a foundation for the operation of infinite series. He was also one of the earliest mathematicians who studied the inverse functions of power series.

At the beginning of the 18th century, Petrus Jartoux (1668–1720), a missionary from France, passed on to Chinese mathematicians three formulae of infinite series, which had aroused Ming Antu's interest. One of the three formulae is the sine series ($a / r = \alpha$):

$$r\sin\alpha = a - \frac{a^3}{3!\ r^2} + \frac{a^5}{5!\ r^4} - \frac{a^7}{7!\ r^6} + \cdots\cdots \tag{1}$$

(J.Gregory, 1667)

It was adapted by Ming according to a concept in traditional Chinese mathematics–"Tong Xian" (通弦):

$$x = 2a - \frac{(2a)^3}{4 \cdot 3!} + \frac{(2a)^5}{4^2 \cdot 5!} - \frac{(2a)^7}{4^3 \cdot 7!} + \cdots\cdots \quad (2)$$

here "Tong Xian" $x = 2\sin\alpha,\ r = 1,\ a = \alpha r = \alpha$.

Using a very complicated method, Ming got an inverse function of the formula (2):

$$2a = x + \frac{1^2 x^3}{4 \cdot 3!} + \frac{1^2 \cdot 3^2 x^5}{4^2 \cdot 5!} + \frac{1^2 \cdot 3^2 \cdot 5^2 x^7}{4^3 \cdot 7!} + \cdots\cdots$$

$$= \sum_{n=0}^{\infty} [(2n-1)!!]^2 x^{2n+1} / 4^n (2n+1)! \quad (3)$$

here when n = 0, (2n−1)!! = 1.

In the process of operation Ming established and many times applied the following counting structure E_n^m (in modern notation):

$$E_n^m = \frac{(2m+2n+1)!}{(2m+1)!\ (2n+1)!} = \frac{(p+q-1)!}{p!\ q!} \qquad \binom{p = 2m+1,\ m \geq 0}{q = 2n+1,\ n \geq 0} \quad (4)$$

Clearly, this function has symmetry:

$$E_n^m = E_m^n, \quad \text{or} \quad \frac{1}{q}\binom{p+q-1}{p} = \frac{1}{p}\binom{q+p-1}{q} \quad (5)$$

Table 1

E_n^m (m \ n)	0	1	2	3	4	5	6
0	1	1	1	1	1	1	1
1	1	$1\frac{1}{3}$	7	12	$18\frac{1}{3}$	26	
2	1	7	$25\frac{1}{5}$	66	143		
3	1	12	66	$245\frac{1}{7}$			
4	1	$18\frac{1}{3}$	143				
5	1	26					
6	1						

(See [1] Vol.3, P.60a)

Ming gave 4 examples $E_1^1, E_2^2, E_2^3, E_3^3$ and got $3\frac{1}{3}, 25\frac{1}{5}, 66, 245\frac{1}{7}$.

When calculating the numerators of the coefficients in the expansion, Ming used a new recurrence formula (in modern notation):

$$\text{(i)} \quad S_n^0 \equiv 1, \quad S_0^m \equiv 1 \quad (m,\ n \geqslant 0)$$

$$\text{(ii)} \quad S_n^m = \sum_{k=0}^{n} E_k^{n-k} S_k^{m-1} \quad (m \geqslant 1) \tag{6}$$

Especially, when n = 1,

$$S_1^m = 1 + S_1^{m-1} = 1 + 1 + S_1^{m-2} = \cdots = m + 1 \tag{7}$$

Table 2

S_n^m (n \ m)	0	1	2	4	6	8	10
0	1	1	1	1	1	1	1
1	1	2	3	5	7	9	11
2	1	$5\frac{1}{3}$	13	$38\frac{1}{3}$	77	129	$194\frac{1}{3}$
3	1	16	$68\frac{1}{3}$	$378\frac{1}{3}$	$1117\frac{2}{3}$	2473	
4	1	$51\frac{1}{5}$	$402\frac{3}{5}$	4417	$19660\frac{1}{5}$		
5	1	$170\frac{2}{3}$	2555	58085			
6	1	$585\frac{1}{7}$	$17082\frac{1}{35}$				
Vol. 3		P. 60a	P. 61b	P. 62b	P.63b	P.64 a~b	

(see[1] Vol.3,pp.60~64)

Ming obtained an intermediate result:

$$x^{m+1} = \sum_{n=0}^{\infty} (-1)^n S_n^m (2a)^{2n+m+1} / 4^n (2n+1)! \tag{8}$$

Moreover, for the sake of convenience to express his algorithm, we set up another recurrence formula T_n^m:

$$\text{(i)} \quad T_n^0 \equiv 1,\ T_0^m \equiv 0 \quad (m \geqslant 1,\ n \geqslant 0)$$

$$\text{(ii)} \quad T_n^m = E_n^m S_n^{2m} T_1^{m-1} - T_{n+1}^{m-1} \tag{9}$$

Table 3

T_n^m \ m, n	0	1	2	3	4	5
0	1	0	0	0	0	0
1	1	9	225	11025	893025	108056025
2	1	90	7875	926100	147349125	30904023150
3	1	819	217350	56029050	16473632175	
4	1	7380	5618250	3020717700		
5	1	66429	142116975			
Vol. 3		P. 65b	P. 66b	P. 67b	P. 68b~69a	

(See[1] Vol.3, pp. 65~69)

The formula (9) is complicated, but all numbers T_n^m are integers. We know

$$T_2^m = \binom{2m+3}{3} T_1^m \tag{10}$$

$$(2m+1)^2 = \binom{2m+3}{3} - \binom{2m+1}{3} \tag{11}$$

$$T_1^m = [(2m+1)!!]^2 \tag{12}$$

Here we only give a proof of the formula (12).

Proof According to (9), (4), (7), when n = 1

$$\begin{aligned} T_1^m &= E_1^m S_1^{2m} T_1^{m-1} - T_2^{m-1} = \frac{(2m+3)!}{(2m+1)!3!}(2m+1)T_1^{m-1} - T_2^{m-1} \\ &= \binom{2m+3}{3} T_1^{m-1} - \binom{2m+1}{3} T_1^{m-1} = (2m+1)^2 T_1^{m-1} \\ &= (2m+1)^2(2m-1)^2 T_1^{m-2} = (2m+1)^2(2m-1)^2 \cdots 5^2 \cdot 3^2 \cdot 1^2 T_1^0 \\ &= [(2m+1)!!]^2. \end{aligned}$$

We have applied formulae (10) and (11) in the above proof.

We can adapt the formula (12) as $T_n = [(2n-1)!!]^2$ $(n \geqslant 0)$, hence the formula (3) can be rewritten as follows:

$$2a = x + \sum_{n=0}^{\infty} T_n x^{2n+1} / 4^n (2n+1)! \tag{3'}$$

Here $x = 2\sin\alpha$. When $\alpha = \frac{\pi}{6}$, from (3)

$$\pi = 3\left(1 + \frac{1^2}{4 \cdot 3!} + \frac{1^2 \cdot 3^2}{4^2 \cdot 5!} + \frac{1^2 \cdot 3^2 \cdot 5^2}{4^3 \cdot 7!} + \cdots\cdots\right) \quad (\textit{I.Newton}, 1676) \tag{13}$$

The formula (13) was passed on to China by P. Jartoux .

The third formula of infinite series is the versine series:

$$r versα = \frac{a^2}{2!r} - \frac{a^4}{4!r^3} + \frac{a^6}{6!r^5} - \cdots \quad (\ J.Gregory, 1667\ ) \tag{14}$$

Let $r = 1,\ y = 2\sin\frac{\alpha}{2}$, so $y^2 = 2versα$, because $versα = 1 - \cos\alpha = 2\sin^2\frac{\alpha}{2} = y^2/2$. Using a similar method to the above, Ming got an inverse function of the formula(14):

$$a^2 = 2\left[\frac{y^2}{2!} + \frac{y^4}{4!} + \frac{(2!)^2 y^6}{6!} + \frac{(3!)^2 y^8}{8!} + \cdots\right]$$
$$= 2\sum_{n=0}^{\infty} (n!)^2 y^{2n+2} / (2n+2)! \tag{15}$$

Ming established the counting structure F_n^m (in modern notation):

$$F_n^m = \frac{2(2m+2n-2)!}{(2m)!(2n)!} = \frac{2(p+q)!}{(p+1)!(q+1)!} \quad \left({p=2m-1,\ m\geq 1 \atop q=2n-1,\ n\geq 1}\right) \tag{16}$$

Clearly, this function has symmetry:

$$F_n^m = F_m^n, \quad \text{or} \quad \binom{p+q}{p+1} / \binom{q+1}{2} = \binom{q+p}{q+1} / \binom{p+1}{2} \tag{17}$$

Table 4

F_n^m m / n	1	2	3	4	5	6	7
1	1	1	1	1	1	1	1
2	1	$2\frac{1}{2}$	$4\frac{2}{3}$	$7\frac{1}{2}$	11	$15\frac{1}{6}$	
3	1	$4\frac{2}{3}$	14	33	$66\frac{11}{15}$		
4	1	$7\frac{1}{2}$	33	$107\frac{1}{4}$			
5	1	11	$66\frac{11}{15}$				
6	1	$15\frac{1}{6}$					
7	1						

(See [1] Vol.4,P. 32a)

Similarly, he used another recurrence formula of R_n^m (in modern notation):

(i) $R_n^1 \equiv 1,\ R_1^m \equiv 1 \quad (m,\ n \geqslant 1)$

(ii) $$R_n^m = \sum_{k=1}^{n} F_k^{n-k+1} R_k^{m-1} \quad (m \geqslant 2) \tag{18}$$

Ming obtained the intermediate result

$$y^{2m} = 2\sum_{n=1}^{\infty}(-1)^{n+1}R_n^m a^{2(m+n-1)} / (2n)! \tag{19}$$

and finally got the formula (15). This paper won't go into details.

The counting structures E_n^m and F_n^m are meaningful. E_n^m prevails odd property, but F_n^m — even. The formula (4) (or(16)) looks like an "odd" (or "even") combination. It seems that further research is needed.

After 1774 some mathematicians could read about the hand-written copy of Ming Antu's work. They carried on Ming's study and got many counting structures. For example, Dong Youcheng(董祐诚, 1791–1823) in his book (《割圜连比例法图解》, 1819) [2] established an expansion, which corresponds to the following formulae:

$$D_{2n} = \sum_{k=1}^{n}(-1)^{k+1}\binom{n+k-1}{2k-1}X^{2k-1} \tag{20}$$

$$D_{2n+1} = \sum_{k=1}^{n}(-1)^{k+1}\binom{n+k}{2k}X^{2k} \tag{21}$$

Xiang Mingda (项名达, 1789–1850) in his book (《象数一原》, 1843) [3] improved Dong's, and his can be written as:

$$G_{2n} = \sum_{k=0}^{n}(-1)^{k}\binom{n+k}{2k}X^{2k} \tag{22}$$

$$G_{2n+1} = \sum_{k=0}^{n}(-1)^{k}\binom{n+k+1}{2k+1}X^{2k+1} \tag{23}$$

in which coefficients are combination numbers. It is interesting to compare the coefficients with the formula of Fibonacci numbers:

$$F_{2n} = \sum_{k=0}^{n}\binom{n+k}{2k} \tag{24}$$

$$F_{2n+1} = \sum_{k=0}^{n}\binom{n+k+1}{2k+1} \tag{25}$$

Jia Xian (贾宪, *circa*1050 AD), a Chinese mathematican, advanced the arithmetical triangle (Pascal's triangle) for the first time in China. Using the method of progressive increase or decrease his successors obtained many results, which could be expressed in terms of combinations. The recurrence algorithm produced a great effect on the history of mathematics in China. The counting structures in 18th–19th centuries mentioned above are results of this influence.

References

[1] Ming Antu 明安图,《割圜密率捷法》, 1839.

[2] Dong Youcheng 董祐诚,《割圜连比例法图解》, 1819.

[3] Xiang Mingda 项名达,《象数一原》, 1843.

Translated by Luo Feijin

Six Even Subgraphs (3,4)–Cover of Graphs

Luo Youxue Huang Wenhua
Department of Mathematics, Fuzhou University Fuzhou fujian 350002 China

Abstract In this paper, We proved that every bridgeless graph admits a (3,4)–cover by six subgraphs.

key words Even subgraph Cover

1. Introduction

D. R. Fulkerson proposed the following conjectures [2]

conjecture A Every bridgeless cubic graph has a 2–cover by 6 perfect matching. This is equivalent to

conjecture A′ Every bridgeless cubic graph has a 4–cover by six even subgraphs.

In this paper a conjecture weaker than A′ has been proved.The graphs under study are finite and undirected, the nodefined concepts and terminologies refer to [1].

The 1–edge–cut of a graph is called a bridge of this graph. An even graph is one in which every vertex is of even degree. The k–regular graph is one in which every vertex is of k degree. The cover of a graph is a collection of subgraphs of this graph, these subgraphs covered together each edge of G at least once. A k–cover of graph G is one cover such that each edge of G is covered exactly k times by the subgraphs in cover, and (l,m)–cover if each edge is covered either exactly l or exactly m times.

Theorem A [3] For every bridgeless cubic graph G without loops and for every edge $e \in E(G)$, there exists 2–factor F of G such that:

(i) $e \in \overline{F}$,

(ii) F intersects every 3–edge–cut of graph in 2 edges.

Theorem B [3] Every 4–edge–connected graph has a nowhere–zero 4–flow.

Theorem C [4] Every graph with a nowhere–zero 4–flow has a 2–cover by 3 even subgraphs.

Theorem D [5] Suppose that G is a bridgeless graph without any loops, G has a vertex v with degree larger than 3. Then there exist two edges e_1 and e_2, which incident with v, such that deleting e_1 and e_2 and joining two different vertices of e_1 and e_2, we can obtain a bridgeless graph G′.

2. Proof of theorems

We first considered a graph with special construction and we called that a graph G admits Q construction, if

(1) G contains a even cycle c.

(2) The vertices of c take such a pairing: Every pair of vertices is connected by a path which internally–disjoint to c, and these paths disjoint each other.

(3) G is union of these paths and c (sea Fig.1)

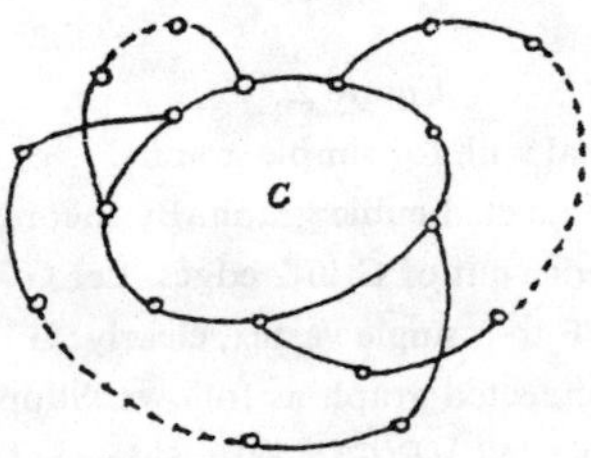

Figure 1

Lemma Suppose that G admits Q construction, we contract the cycle c to a single vertex, then obtained a new graph G^*. If G^* has a 1–cover then G has a (1,2)–cover, and the edges which are covered once exactly are just the edge of c.

Proof Suppose that $c = v_1v_2\text{---}v_{2k}$ i.e. c is directed as 2k section by vertices. The paths which connected pairing verties is indicated by p_1, p_2,---p_x. It is clear that G^* is an Euler graph, Scince G^* admits Euler tour i.e. G^* has closed trail set out from w and return to w which traverses each adge once exactly of p_1, p_2, ---, p_k. When G^* return to G, an Euler closed trail of G^* corresponds to a cycle or union of some disjoint cycles (i.e. even subgraph).

We can always choose an Euler tour s_1^* of G^* such that the cycles (or union of cycles) s_1 of G corresponding to s_1^* of G contain alternately some sections on c. We choose again an Euler closed trail s_2^* (may be s_1^*) of G^*, such that the cycle (or union of cycle) $s_2$of G corresponding to s_2^* contain exactly other sections on c (see Fig. 2). Hence we obtained a (1,2)–cover $\{s_1, s_2\}$ of G. and the edges on c is covered once exactly. ■

An example is shown in Fig.2. We take an Euler tour $s_1^* = wp_1wp_4wp_3wp_2w$ of G^*, it corresponds to a cycle. $s_1 = v_2v_1p_1v_3v_4p_4v_8v_7p_3v_5v_6p_2v_2$ of G. Then take an Euler tour $s_2^* = wp_1wp_4wp_4wp_3wp_2w$ of G^* again, it corresponds to a cycle $s_2 = v_1v_8p_4v_4v_5p_3v_7v_8p_2v_2v_3p_1v_1$ of G.

Theorem 1 Every 3–edge connected cubic graph G has a (3,4)–cover by 6–even subgraphs, and the edge which is covered 3 times constituted exactly on a 2–factor of G.

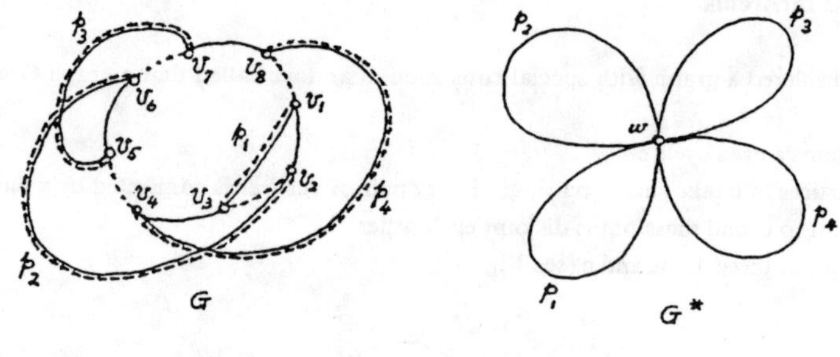

Figure 2

Proof It clearly suffices to deal with for simple graphs.

Suppose G is a simple 3–edge connected cubic graph. By theorem A, we choose a 2–factor F of G, such that F intersects every 3–edge cut of G in 2 edges. Let G^* be the graph obtained from G by contracting each component of F to a single vertex, clearly. G^* is a 3–edge connected graph. We will prove that G^* is a 4–edge connected graph as follows. Suppose S is a 3–edge cut of G^*, then S is a 3–edge cut of G too. Since $|S \cap (E(G) \setminus F)| = 3$, hence $|S \cap F| = 0$.However, this is a contradiction for a choice of F, thus G^* is a 4–edge connected graph.

By theorem B and C, G^* has a 2–cover by 3 even subgraphs, they are denoted by $Z^* = \left\{ Z_1^*, Z_2^*, Z_3^* \right\}$. Observe that each vertex of G^* corresponds to a cycle of G. Let $V(G^*) = \{f_1, f_2, ---, f_m\}$ and $F = \{F_1, F_2, ---, F_m\}$, f_i corresponds to $F_i(i=1,2,---,m)$. We may assume, without loss of generality, that $V(Z_1^*) = \{f_1, f_2, ---, f_r\} (r \leqq m)$. Since the degree of vertex $f_j (1 \leqq j \leqq r)$ in Z_1^* is even, hence each f_j may be regarded as w in lemma. We applicat lemma to all f_j, $j = 1,2,---,r$, then from Z_1^* may obtained two even subgraphs Z_1^1, Z_1^2 of G, and let $F' = F \setminus \bigcup_{j=1}^{r} F_j$. Thus from Z_1^* we obtained two even subgraphs $Z_1^1, Z_1^2 \cup F^1$. Similarly, four even subgraphs $Z_2^1, Z_2^2 \cup F^2$, $Z_3^1, Z_3^2 \cup F^3$ of G is obtained from Z_2^*, Z_3^*. Thus $Z = \left\{ Z_1^1, Z_1^2 \cup F^1, Z_2^1, Z_2^2 \cup F^2, Z_3^1, Z_3^2 \cup F^3 \right\}$ is a 6–even subgraph (3,4)–cover of G. obviously, the edges in F are covered three times exactly. The theorem follows. ■

Theovem 2 Every bridgeless graph admits a 4–cover by seven even subgraphs.

Proof. It is suffices to proof with for loopless and bridgeless graphs.

Suppose G is a loopless and bridgeless graph. By induction on $|E(G)|$,

When $|E(G)| = 2$, theorem clearly is true.

Suppose the theorem is true for all graphs which size is fewer than n.

Let $|E(G)| = n$, we will consider following three situations:

Case 1. If G has a vertex v with degrees larger than 3, then by theorem D, there exist two edges which incident with v: $e_1 = vx_1$, $e_2 = vx_2$, such that deleting e_1, e_2 and joining x_1, x_2 by a new edge we can obtain a new graph G′. By the induction hypothesis, G′ has a 7–even subgraphs 4–covers, an edge e in the even subgraphs which contain the edge e, is replaced by the path x_1vx_2. We obtained a 7–subgraphs 4–cover of G.

Case 2. Suppose G has 2–edge cut $\{e_1, e_2\}$. Let G′ is a graph obtained by contracting edge e_2 of G as a vertex v. By the induction hypothesis, G′ has 7–even subgraphs 4–cover, written as $Z' = \{Z'_i \quad i = 1 \sim 7\}$.

We may assume, without loss of generality, that four even subgraphs which cover e_1 is Z'_1, Z'_2, Z'_3, Z'_4. Since $\{e_1, e_2\}$ is 2–edges cut of G, thus the even subgraph which through edge e_1 must contain vertex v. Now, restore G′ to G, and let $Z_i = Z'_i \bigcup \{e_2\} (i = 1 \sim 4\}$, then Z_i, Z_2, Z_3, Z_4 are even subgraphs of G. therefore, $Z = \left\{ Z_1, Z_2, Z_3, Z_4, Z'_5, Z'_6, Z'_7, \right\}$ is a 7–even subgraphs 4–cover of G.

Case 3. Suppose G is a 3–edge connected simple cubic graph, by theorem 1, G has a 6–even subgraphs (3,4)–cover Z, and the edge which is covered 3 times constitute exactly on a 2–factor F of G. Therefore, the $Z \bigcup F$ is a 7–even subgraphs 4–cover of G.

By cases 1, 2 and and 3, every bridgeless graph admits 7–even subgraphs 4–cover. ■

Corollary. Every bridgeless graph admits 6–even subgraphs (3,4)–cover.

Proof. Suppose G is a bridgeless graph, by theorem 2, the G has a 7–even subgraphs 4–cover Z, we delete any one subgraph in Z, then obtained a 6–even subgraphs (3,4)–cover of G.

References

[1] J. A. Bondy, U.S.A. Murty. Graph Theorem with Applications. MacMillan Press LTP. 1976.

[2] D.R. Fulkerson. Blocking and Antiblocking pairs of Polyhedra. Math. Programming, 1971.

[3] F.Jaeger, Flows and Generalized Coloring Theorem in Graphs. J. Combinatorial Theory. Ser. B26. 1979.

[4] G.Fan. Integer Flows and Cycle Covers. J. Combinatorial Theory. Ser. B(54). 1992.

[5] H.Fleischner, Eine Gememsam Basic fur die Theorie der Eulerschen, Graphen und der standards 69 B (1965), 125–130.

The Properties and some Constructions of Geodetic Graphs with Diameter Three*

Mao Jingzhong

(Department of Mathematics, Central China Normal University)
(Wuhan , 430070, P.R.CHINA)

Abstract In this paper we have induced three types of the construction and investigate some properties of geodetic graphs with diameter three.

Key words Geodetic graph, Diameter, Affine plane

1. Introduction

Since Ore ([1]) have offered the concept of geodetic graph there were a large number investigate on the properties ([2],[3],[4]) and the constructions ([5],[6],[7],[8],[9]) of geodetic graphs. Especially for geodetic graphs with diameter two ([2]) which have received the more suficient investigate. But for geodetic graphs with diameter three the knowledge we have known is still very little. Especially, for the types of constrution of geodetic blocks with diameter three is almost unknown. We intend to consider this problem.

2. Basic Properties

*The work supported by the National Natural Science Fundation of China

We only consider 2-connected simple undirected finite graphs. For a graph $G = (V, E)$, we denote $N_D(x) = \{y|xy \in E, y \in D\}$ for $x \in V, D \subseteq V$. All other terminologies and notations without explanation are follow [11].

Suppose graph $G = (V, E)$ satisfies the conditions that, for each pair of distinct vertices $x, y \in V$, there is exactly one shortest path connecting them and G is 2−connected, then we say G is a geodetic block.

Suppose $G = (V, E)$ is a geodetic block with $diam(G) = 3$, and $cl(G) = k \geq 3$. Then we can take a maximal clique H_0 of G *s.t.*, $|V(H_0)| = |A| = k$. Now we can divide V as follows: Let

$$B = \{x| \quad \exists y \in A \quad s.t., \quad xy \in E \quad and \quad x \notin A\}$$

$$C = \{x| \quad \exists y \in B \quad s.t., \quad xy \in E \quad and \quad x \notin (A \cup B)\}$$

$$D = V \backslash (A \cup B \cup C)$$

By using these notations and from the fact that G is geodetic graph we can induce the following basic properties.

PROPERTY 1. *If* $\{y_1, y_2\} \subseteq B$ *s.t.*, $y_1 y_2 \in E$, *then there exists a unique vertex* $x \in A$ *s.t.*, $\{xy_1, xy_2\} \subseteq E$.

PROOF: Otherwise there would be a 4−cycle contained in G and contradicts to G is geodetic graph. ∎

PROPERTY 2. *If* $\{z_1, z_2\} \subseteq C$ *s.t.*, $z_1 z_2 \in E$, *then there exists unique vertex* $x \in A$ *s.t.*, $d(x, z_1) = d(x, z_2) = 2$.

PROOF: If $\exists y \in B$ s.t., $\{yz_1, yz_2\} \subseteq E$, because y can adjacent to only one vertex of A (noticed that , H_0 is a maximal clique of G and $A = V(H_0)$), then the conclusion is true. Otherwise , there will be two vertices y_1, y_2 of B s.t., $\{y_1 z_1, y_2 z_2\} \subseteq E$ and there is a vertex $x \in A$ s.t., $\{xy_1, xy_2\} \subseteq E$, then the conclusion is true also. Otherwise there would be a 6−cycle contained in G, contradicts to G is geodetic graph too. ∎

PROPERTY 3. *Take any vertex* $u \in D$ *for each vertex* $x \in A$ *there exists* $y \in B, z \in C$ *s.t.*, $\{xy, yz, zu\} \subseteq E$.

PROPERTY 4. *Take any two vertices* u, v *of* D, *if* $uv \notin E$, *then either (i) there exists exactly one vertex* $z \in C$ *s.t.*, $\{zu, zv\} \subseteq E$; *or there exists two vertices* z_1, z_2 *of* C *s.t.*, $\{z_1 z_2, z_1 u, z_2 v\} \subseteq E$.

PROOF: Otherwise it would induces $d(u, v) \geq 4$ contradicts to $diam(G) = 3$. ∎

PROPERTY 5. *If $D \neq \phi$, then for each $x \in A$ we always have $y \in B$ s.t., $xy \in E$.*

PROOF: Otherwise it would contradicts to $diam(G) = 3$. ∎

3. Three types of Basic Construction

In this paragraph we will discuss the most basic constructions of geodetic blocks with diameter 3 by using the partition of V described as above.

(3.1) G containes only one clique

THEOREM 1. *If $|B| = |C| = k$, then $|D| = 1$ and G is a pyramid with high equal to 3.*

PROOF: Without loss generality we can suppose $A = \{a_1, a_2, ..., a_k\}$. Since $D \neq \phi$, from property 5 we know for $\forall i, 1 \leq i \leq k$ we have $\exists b_i \in B s.t., a_i b_i \in E$. From $G[A]$ is a clique we know $b_i \neq b_j$ for $1 \leq i < j \leq k$. Now $|B| = k$, Therefore $B = \{b_1, b_2, ..., b_k\}$. Furthermore from property 1 we know $b_i b_j \notin E$ for $1 \leq i < j \leq k$.

Suppose $|D| \geq 2$. Take any two vertices u_1, u_2 of D. Let the shortest paths from a_1 to u_1 and u_2 are $a_1 b_1 y_1 u_1, a_1 b_1 y_2 u_2$ respectively. If $y_1 \neq y_2$ then for any $b \in B$ and $b \neq b_1$ we cann't have $\{by_1, by_2\} \subseteq E$. Otherwise it would be induce a 4−cycle $b_1 y_1 b y_2 b_1$ and this induces $b_1 b_2 \in E$ contradicts to property 1. Furthermore, for any two vertices b_i, b_j of B, we must have $|N_C(b_i) \cap N_C(b_j)| \leq 1$. From G is 2−connected we know $|N_C(u_1)| \geq 2, |N_C(u_2)| \geq 2$. Noticed property 2 we can induce $|C| \geq k + 1$. Contradicts to the suppose $|C| = k$. Therefore $|D| = 1$ and G is a pyramid. ∎

(3.2) G containes only two cliques

Construction (1) Let $A = \{a_1, a_2, ..., a_k\}$ and it generates a clique. $B = \bigcup_{i=1}^{m} B_i$ is a vertex-independent set. For $1 \leq j \leq m$ let $B_j = \{b_1^j, b_2^j, ..., b_k^j\}$ and for $1 \leq i \leq k$ we have $b_i^j a_i \in E$. $C = \{c_1, c_2, ..., c_m\}$ is a vertex-independent set and for $1 \leq j \leq m$, c_j is adjacent to all vertices of B_j. $D = \{u_1, u_2, ..., u_m\}$ and for $1 \leq j \leq m$, $c_j u_j \in E$. D generates a clique .

(2) Let $A = \{a_1, a_2, ..., a_k\}$ and it generates a clique. $B = \{b_1, b_2, ..., b_k\}$ is a vertex-independent set. For $1 \leq i \leq k$ we have $a_i b_i \in E$. For $1 \leq i \leq k$ let

$C_i = \{c_1^i, c_2^i, ..., c_m^i\}, C = \bigcup_{i=1}^k C_i$ **is vertex-independent set.** $D = \{u_1, u_2, ..., u_m\}$ **and for** $1 \leq i \leq k, 1 \leq j \leq m$ **we have** $b_i c_j^i \in E, u_j c_j^i \in E$. D **generates a clique.**

PROPERTY 6. *Suppose* $A = \{a_1, a_2, ..., a_k\}$ *and* $G[A]$ *is a clique.* $k \geq 3$. $B = \{b_1, b_2, ..., b_k\}$ *and* $a_i b_i \in E$, *for* $1 \leq i \leq k$. $D \neq \phi$. *Then for* $1 \leq i < j \leq k$ *we have* $N_C(b_i) \cap N_C(b_j) = \phi$ *and* B *is a vertex-independent set.*

PROOF: Since $G[A]$ is a complete graph, if for some $i, j, 1 \leq i < j \leq k$ we have $b_i b_j \in E$ then we would induce a $4-$ cycle $b_i a_i a_j b_j b_i$ and this induces $a_i b_j \in E$ and b_i, b_j both are adjacent to every vertices of A. this is a contradiction.

Obviously we cann't have $|N_C(b_i) \cap N_C(b_j)| \geq 2$. For otherwise it would be induces $b_i b_j \in E$ contradicts to the result described as above. If $N_C(b_1) \cap N_C(b_2) = \{y\}$ then for $\forall j, 3 \leq j \leq k$ we have $y b_1 a_1 a_j a_2 b_2 y$ is a $6-$cycle. this induces $d(a_j, y) \leq 2$. But obviously $d(a_j, y) \geq 2$. Therefore we must have $y b_j \in E$. Since $D \neq \phi$, if $|D| = 1$ and $u \in D$,then we should have paths from u to each vertex of B with length ≤ 3. Consider two cases.

Case 1. $uy \in E$.

From G is $2-$connected we know there is another vertex $c_1 \in C s.t., uc_1 \in E$. And since $c_1 \in C$, $\exists b_i \in B$ s.t., $c_1 b_i \in E$. This induces a $4-$cycle $b_i c_1 u y b_i$ and therefore it should have $b_i u \in E$, this is a contradiction.

Case 2. $uy \notin E$.

Therefore $\exists c_j \in C, c_j \neq y$ for $1 \leq j \leq k$ and $c_i \neq c_j$ for $1 \leq i < j \leq k$, s.t., $b_j c_j \in E$ and $c_j u \in E$. This induces a $6-$cycle $uc_1 b_1 y b_2 c_2 u$. From property 2 we have $c_1 y \notin E$ and $c_2 y \notin E$. Therefore we have two paths connecting u and y both are shortest and with length 3. this is a contradiction.

If $|D| \geq 2$, we take any $u \in D$ from $d(u, a_j) \leq 3$ for $1 \leq j \leq k$ we induce $N_C(u) \cap N_C(b_j) \neq \phi$. It is easy to induce the same contradiction as in the case $|D| = 1$. Therefore we must have $N_C(b_i) \cap N_C(b_j) = \phi$ for $1 \leq i < j \leq k$. ∎

PROPERTY 7. *Suppose* $A = \{a_1, a_2, ..., a_k\}$ *and* $G[A]$ *is a clique.* $B = \{b_1, b_2, ..., b_k\}$ *and* $a_i b_i \in E$ *for* $1 \leq i \leq k$. *If* $N_C(b_1) = C_1 = \{x | x \in C$ *and* $x b_1 \in E\}$ $= \{c_1^1, c_2^1, ..., c_m^1\}$, $m \geq 2$, *and we have* $N_C(b_i) \cap N_C(b_j) = \phi$ *for* $1 \leq i < j \leq k$. *Then we have* $|N_C(B_j)| = m$ *for* $1 \leq j \leq k$.

PROOF: Since $diam(G) = 3$, it should have $d(b_i, c_j^1) \leq 3$ for $1 \leq j \leq m, 2 \leq i \leq k$. But any two vertices of B cann't adjacent. The length of path $c_j^1 b_1 a_1 a_i b_i$ is 4. From property 1 we know the path connecting c_j^1 and b_i must pass through the vertex of $C \cup D$. Therefore we must have $u_j^1 \in D$ for $1 \leq j \leq m$ s.t.,$c_j^1 u_j^1 \in E$ and $N_C(u_j^1) \cap N_C(b_i) \neq \phi$. Since $m > 1$, obviously $u_\lambda^1 \neq u_\mu^1$ for $1 \leq \lambda < \mu \leq m$. Therefore $D \supseteq \{u_1^1, .u_2^1, ..., u_m^1\}$.

Furthermore for $1 \leq \lambda < \mu \leq m$ we cann't have $N_C(u_\lambda^1) \cap N_C(b_i) = N_C(u_\mu^1) \cap N_C(b_i) = \{c_i\}$. Otherwise $c_i u_\lambda^1 c_\lambda^1 b_1, c_i u_\mu^1 c_\mu^1 b_1$ are two paths connecting c_i and b_1 both are with length 3. Therefore $d(b_1, c_i) \leq 2$. But from property 2 we know $N(c_i) \cap C_1 = \phi, b_1 b_i \notin E$. It is impossible. Therefore we must have $m \leq |N_C(b_i)|$, for $2 \leq i \leq k$. Change the positions of b_1 and b_i with each other we should have $|N_C(b_i)| \leq m$. Therefore $|N_C(b_i)| = m$ for $\forall i, 1 \leq i \leq k$. ∎

PROPERTY 8. *On the assumption of property 7 we have, furthermore* , $|D| = m$ *and* $G[D]$ *is a clique,* C *is a vertex-independent set.*

PROOF: We using the same notation as in the proof of property 7 and from its proof we know $D \supseteq \{u_1^1, u_2^1, ..., u_m^1\}$ and we have $c_j^1 u_j^1 \in E$ for $1 \leq j \leq m$. We can denote $N_C(b_i) = C_i = \{c_1^i, c_2^i, ..., c_m^i\}$ and suppose $c_j^i u_j^1 \in E$ for $1 \leq j \leq m$, $1 \leq i \leq k$ without loss generality.

If $\exists u \in D \setminus \{u_1^1, u_2^1, ..., u_m^1\}$, then from there is a path from u to each vertex of A with length 3 we know $N_C(u) \cap N_C(b_i) \neq \phi$ for $1 \leq i \leq k$. If $\{c_1^1 u, c_1^2 u\} \subseteq E$ then we have a 4−cycle and it induces $c_1^1 c_1^2 \in E$ contradicts to property 2. If $\{c_1^1 u, c_2^2 u\} \subseteq E$ then we have a 6−cycle $c_1^1 u_1^1 c_1^2 b_2 c_2^2 u c_1^1$ and $c_1^1 c_1^2 \notin E$ (property 2) $c_1^1 b_2 \notin E$ (otherwise it would be a 4−cycle $c_1^1 b_2 c_1^2 u_1^1 c_1^1$, contradiction) $c_1^1 c_2^2 \notin E$ (property 2). Therefore $d(c_1^1, b_2) = 3$. But we have two paths connecting c_1^1 and b_2 both are with length 3. This is a contradiction. Therefore $|D| = m$, i.e., $D = \{u_1^1, u_2^1, ..., u_m^1\}$ and $c_j^i u_j^1 \in E$ for $1 \leq j \leq m$, $1 \leq i \leq k$.

Further, consider the shortest path connecting u_1^1, u_2^1. If $u_1^1 u_2^1 \notin E$, from property 4 we know $\exists x \in N_C(u_1^1)$ and a vertex $y \in N_C(u_2^1)$ s.t., $xy \in E$. Suppose $c_1^1 c_2^1 \in E$ without loss generality. It follows $c_1^i c_2^i \notin E$ *for* $2 \leq i \leq k$. Othere-wise it would be have a 6−cycle, for example, $c_1^2 c_2^2 \in E$ then $c_1^1 u_1^1 c_1^2 c_2^2 u_2^1 c_2^1 c_1^1$ is a 6−cycle and $c_1^1 c_2^2 \notin E$ (property 2),$c_1^1 u_2^1 \notin E$. This induces there are two paths connecting c_1^1 and c_2^2, a contradiction. But if $c_1^1 c_2^1 \in E$ and $c_1^i c_2^i \notin E$ for $2 \leq i \leq k$, then since $k \geq 3$, we would have the length of the shortest path connecting c_1^2 and c_2^k is 4 contradicts to $diam(G) = 3$.

Therefore we must have $u_1^1 u_2^1 \in E$. By the same reason we have $u_i^1 u_j^1 \in E$ for $1 \leq i < j \leq m$, i.e., $G[D]$ is a clique. ∎

Note. This type of construction is decided by the parameter m, k. When m and

k is given the order is $n = 2k + m(k+1)$, the degree sequence is $k, m+1, m, 2$.

PROPERTY 9. *Suppose $A = \{a_1, a_2, ..., a_k\}$ and $G[A]$ is a clique. $k \geq 3$. $B = \{b_1, b_2, ..., b_k\}$ and $a_i b_i \in E$ for $1 \leq i \leq k$. $D \neq \phi$. Then for $\forall i, 1 \leq i \leq k$ we have $C_i = \{c | cb_i \in E, c \in C\} \subseteq C$ cann't generate a clique, where $m = |C_i| \geq 2$.*

PROOF: From property 6 we know $N_C(b_i) \cap N_C(b_j) = \phi$ for $1 \leq i < j \leq k$ and B is a vertex-independent set.

Suppose the property is not true. $C_1 = \{c | cb_1 \in E, c \in C\} = \{c_1^1, c_2^1, ..., c_m^1\}$ generates a $m-$clique, $m \geq 2$. Then we have following results successively.

(1). $d(c_i^1, b_2) = d(c_i^1, b_3) = 3$ for $1 \leq i \leq m$ and the shortest paths connecting c_i^1 and b_2, b_3 must pass through the vertex of D and the neighbourhood of b_2, b_3 in C. This is because property 2 and the length of path $c_i^1 b_1 a_1 a_2 b_2$ is 4.

(2). Take any $u \in D$ we have $|N_C(u) \cap C_1| = 1$. Otherwise it would be have a $4-$cycle and induces $ub_1 \in E$. This is impossible.

(3). $|D| \geq m$. Since for $1 \leq i \leq m$ from each c_i^1 to b_2 we must have a shortest path with length 3 and pass through the vertex of D and from (2) we know that, for $\forall i, 1 \leq i \leq m$, $\exists u_i$ s.t., $c_i^1 u_i \in E$, and $u_i \neq u_j$ for $1 \leq i < j \leq m$. It follows $|D| \geq m$.

Now suppose $D = \{u_1, u_2, ..., u_l\}, l \geq m$, $c_i^1 u_i \in E$ for $1 \leq i \leq m$.

(4). For each vertex $c_1^2 \in N_C(b_2)$ there exists exactly one vertex $u_i \in \{u_1, u_2, ..., u_m\}$ s.t., $c_1^2 u_i \in E$. Otherwise if $c_1^2 u_1 \in E$ and $c_1^2 u_2 \in E$, then we get two paths connecting c_1^2 and b_1 both are with length 3, and from property 2 we know $c_1^2 c_1^1 \notin E$ and $c_1^2 c_2^1 \notin E$. this is a contradiction.

(5). By the fact that , from a_2 to each vertex of D there is have a shortest path with length 3 connecting them, we must have $N_C(b_2) = C_2 = \{c_1^2, c_2^2, ..., c_l^2\}$ and $c_j^2 u_j \in E$ for $1 \leq j \leq l$. Furthermore , by change the position of b_1 and b_2 in (3) we know $l = m$. By the same reason $|N_C(b_i)| = |C_i| = m$ for $3 \leq i \leq k$.

Now suppose $C_i = \{c_1^i, c_2^i, ..., c_m^i\}$ and $c_j^i u_j \in E$ for $1 \leq j \leq m$.

(6). C_i is vertex-independent set for $2 \leq i \leq k$.

Otherwise, suppose $c_1^2 c_2^2 \in E$, we would have a $6-$cycle $c_1^2 u_1 c_1^1 c_2^1 u_2 c_2^2 c_1^2$ and this induces two paths connecting c_1^1 and c_2^2 both are with length 3. Therefore we must have $d(c_1^1, c_2^2) \leq 2$. But from property 2 we know $d(c_1^1, c_2^2) \geq 2$. It follows that, $\exists w \in D$ s.t.,$\{c_1^1 w, c_2^2 w\} \subseteq E$. But this induces there are two paths connecting c_1^1 and b_2 both are shortest and with length 3, this is impossible.

(7). There are edges connecting the vertices of D.

Consider the length of the shortest paths from c_1^2 to $c_i^3, 2 \le i \le m$. By the same reason as in (1) we induce the shortest path from c_1^2 to $c_i^3, 2 \le i \le m$ must pass through the vertex of D. Now we have $c_1^2u_1 \in E$. By the same reason as in (2) we have $|N_C(u_1) \cap C_3| = 1$ and $c_j^3u_1 \in E$. By the same reason as in (6) we know there is no vertex w of D s.t., $c_1^2w \in E$ and $c_2^3w \in E$. Therefore we only can have $d(c_1^2, c_i^3) = 3$ and the shortest path from c_1^2 to c_i^3 are $c_1^2u_1u_ic_i^3, 2 \le i \le m$.

(8). $u_1u_2 \notin E$. Otherwise there is a 4−cycle $c_1^1u_1u_2c_2^1c_1^1$ contradicts to G is geodetic graph.

Because the results of (7) and (8) are contradicts to each other, therefore C_1 cann't generate a clique. ∎

(3.3) The distance between the cliques of G is 2.

Construction of third type. Suppose for positive integer k there is an affine plane. Take $(k+1)$ different pyramid with high 2 and its base is a clique of order k. Denote the top vertices of these pyramid by $a_1, a_2, ..., a_{k+1}$ and the vertex sets of their base cliques by $B_1, B_2, ..., B_{k+1}$. For $1 \le i \le k+1$ denote $B_i = \{b_1^i, b_2^i, ..., b_k^i\}$. Suppose the paths connecting a_i and b_j^i with length 2 is $a_iu_j^ib_j^i, 1 \le i \le k+1, 1 \le j \le k$. Denote $P_i = \{a_i\} \cup \{u_j^i | 1 \le j \le k\} \cup B_i, 1 \le i \le k+1$. Take another vertex set with k^2 vertices $L = \{c_{ij} | 1 \le i, j \le k\}$. $L \cap \bigcup_{i=1}^{k+1} P_i = \phi$. Using L as points set we construct an affine plane of order k. Denote the line family of this affine plane by $L_1, L_2, ..., L_{k+1}$. Suppose $L_i = \{l_1^i, l_2^i, ..., l_k^i\}, 1 \le i \le k+1$. Let b_j^i adjacent to every vertices of l_j^i for $1 \le i \le k+1, 1 \le j \le k$. Any two vertices of L are nonadjacent. Finally, we connect each pair of vertices in $A = \{a_1, a_2, ..., a_{k+1}\}$ to get a $(k+1)$−clique. It is easy to verify that , the graph G obtained in this way is a geodetic graph with $diam(G) = 3$. Any two different cliques have no common vertex and have no edge connecting the vertices belongs to these two cliques.

PROPERTY 10. *Suppose $A = \{a_1, a_2, ..., a_k\}$ and $G[A]$ is a clique. $k \ge 3$. $B_1 = \{b | ba_1 \in E, b \in B\} = \{b_1^1, .b_2^1, ..., b_m^1\}$. $m \ge 2$. Furthermore suppose $\exists c_1^1 \in C$ s.t., $b_1^1c_1^1 \in E$. $D \ne \phi$. Then we have (1) If we denote $B_i = \{b | ba_i \in E, b \in B\}$ for $2 \le i \le k$ then $|B_i| \ne \phi$. (2) If we suppose $N_C(B_i) \cap N_C(B_j) = \phi$ for $1 \le i < j \le k$, where $N_C(B_i) = \{c | c \in C$ and $\exists b \in B_i$ s.t., $cb \in E\}$, then we have $|B_i| > 1$ for $2 \le i \le k$ and $N_C(b_j^i) \ne \phi$ for $1 \le i \le k$, $1 \le j \le m$.*

PROOF: Since $D \ne \phi$, for any $u \in D$ and $1 \le i \le k$ there is a path from a_i

to u with length 3 and the inner vertices of this path must belongs to B and C respectively. This induces $B_i \neq \phi$ for $2 \le i \le k$.

Take any $b_j^i \in B_i (2 \le i \le k)$, since $d(b_j^i, c_1^1) \le 3$, from property 1 and 2 we cann't have $d(b_j^i, c_1^1) = 2$. Therefore it only can be $d(b_j^i, c_1^1) = 3$ and two inner vertices of the shortest path from b_j^i to c_1^1 must belongs to C, D respectively. That is $N_C(b_j^i) \neq \phi$.

If $|B_2| = ... = |B_k| = 1$, we denote $B_i = \{b_i\}$ for $2 \le i \le k$. Suppose $\{c_1^1, c_2^1, ..., c_m^1\} \subseteq C_1$ satisfies $c_j^1 b_j^1 \in E$ for $1 \le j \le m$. Take any $c_1^2 \in C$ s.t.,$c_1^2 b_2 \in E$. Consider the constitution of the shortest path from c_1^2 to b_j^1 for $1 \le j \le m$. From property 1 and assumption we have $d(c_1^2, b_j^1) = 3$ and this path must have the form $c_1^2 u_j^1 c_j^1 b_j^1$ where $u_j^1 \in D$. Furthermore since $d(a_1, u_j^1) = 3$ we have $u_j^1 \neq u_i^1$ for $1 \le i < j \le m$.

Now we consider the constitution of the paths from a_2 to $u_1^1, u_2^1, ..., u_m^1$ with length 3. Since $|B_2| = 1, a_2 b_2 \in E, b_2 c_1^2 \in E$ and there is a path from c_1^2 to b_j^1 for $1 \le j \le m$ with length 3 we induce $c_1^2 u_j^1 \in E \quad for \quad 1 \le j \le m$. Then we consider the constitution of paths from a_3 to $u_1^1, u_2^1, ..., u_m^1$ with length 3. Since $|B_3| = 1, a_3 b_3 \in E$, and $c_1^2 u_j^1 \in E$ for $1 \le j \le m (m \ge 2), d(a_3, u_j^1) = 3$. If the shortest path from a_3 to u_j^1 all pass through vertex b_3 then we would have at least $m(m \ge 2)$ shortest paths connecting c_1^2 and b_3. A contradiction. Therefore we must have $|B_j| > 1$ for $3 \le j \le k$.

Furthermore , there are at least two shortest paths from a_3 to $\{u_1^1, u_2^1, ..., u_m^1\}$ with length 3 contained no common inner vertices. Say,$a_3 b_1^3 c_1^3 u_1^1$, $a_3 b_2^3 c_2^3 u_2^1$ are two such paths in which $b_1^3, b_2^3 \in B_3$ and $b_1^3 \neq b_2^3$. Therefore $c_1^3 \neq c_2^3$ and $c_1^3, c_2^3 \in C$. Now consider the constitution of the shortest path from c_1^1 to b_2^3. By property 1 and 2 we have $d(c_1^1, b_2^3) = 3$. Therefore we must have $u \in D \setminus \{u_1^1, u_2^1, ..., u_m^1\}$ s.t.,$c_1^1 u \in E$ and $c_2^3 u \in E$. If the shortest path from a_2 to u with length 3 pass through vertex b_2, then it induces there are two paths with length 3 connecting c_1^1 and b_2. This is a contradiction. It follows $|B_2| > 1$. ∎

PROPERTY 11. *Suppose* $A = \{a_1, a_2, ..., a_k\}$ *and* $G[A]$ *is a clique.* $k \ge 3$. $B_1 = \{b | b \in B$ *and* $ba_1 \in E\} = \{b_1^1, .b_2^1, ..., b_m^1\}$, $m \ge 2$. $B_2 = \{b | b \in B$ *and* $ba_2 \in E\} = \{b_1^2, b_2^2, ..., b_l^2\}$, $l \ge 2$. $D \neq \phi$. $C_1 = \{c_1^1, c_2^1, ..., c_m^1\}$, $C_2 = \{c_1^2, c_2^2, ..., c_l^2\}$ *and* $b_i^1 c_i^1 \in E$ *for* $1 \le i \le m$, $b_j^2 c_j^2 \in E$ *for* $1 \le j \le l$. $|N_C(b_i^1)| = |N_C(b_j^2)| = 1$, *for* $1 \le i \le k$, $1 \le j \le m$. $C_1 \cap C_2 = \phi$. *Then* $|D| = ml$ *and* c_i^1 *have exact* l *neighbourhood vertices in* D *adjacent to* l *vertices of* C_2 *respectively for* $1 \le i \le m$. c_j^2 *have exact* m *neighbourhood vertices in* D *adjacent to* m *vertices of* C_1 *respectively for*

$1 \leq j \leq l$.

PROOF: Since for $1 \leq i \leq m$ from each c_i^1 and each vertex b_j^2 for $2 \leq j \leq l$ we have $d(c_i^1, b_j^2) \leq 3$. But according property 2 $c_i^1 c_j^2 \notin E$ for $1 \leq i \leq m$, $1 \leq j \leq l$. From property 1 $b_i^1 b_j^2 \notin E$ for $1 \leq i \leq m, 1 \leq j \leq l$. Therefore c_i^1 and c_j^2 must have common neighbourhood vertex in D. Because for $\forall u \in D$, we have $|N_{C_1}(u)| \leq 1, |N_{C_2}(u)| \leq 1$. For otherwise it would be two shortest paths from u to a_1 or a_2 with length 3 contradicts to G is geodetic. Therefore for $1 \leq i \leq m, c_i^1$ must have l destinct neighbourhood vertices in D adjacent to l defferent vertices of C_2 and $N_D(c_i^1) \cap N_d(c_j^1) = \phi$ for $1 \leq i < j \leq m$. It follows $|D| \geq ml$.

Now we suppose $D = \{u_{ij} | 1 \leq i \leq m, 1 \leq j \leq l\} \cup D_1$.

If $D_1 \neq \phi$, say, $w \in D_1$. Then we must have paths from w pass through the vertices of B_1, B_2 to a_1, a_2 with length 3 respectively. These paths have the form that $wc_i^1 b_i^1 a_1$ and $wc_j^2 b_j^2 a_2$. This induces there are two paths connecting c_i^1 with c_j^2 of length 2, one pass through the vertex of $D \backslash D_1$, the other pass through w. this is a contradiction. Therefore $D_1 = \phi$ and $|D| = ml$. ∎

Now suppose $D = \{u_{ij} | 1 \leq i \leq m, 1 \leq j \leq l\}$. And for $1 \leq i \leq m$ we can let c_i^1 is adjacent to every vertex of $D_i = \{u_{i1}, u_{i2}, ..., u_{il}\}$, for $1 \leq j \leq l$, c_j^2 is adjacent to every vertex of $D^{(j)} = \{u_{1j}, u_{2j}, ..., u_{mj}\}$. Using these notation we furthermore have

PROPERTY 12. *Under the assumption of property 11 and further assumption $b_i^1 b_j^1 \notin E$ for $1 \leq i, j \leq m$ then we have (1) $m = l$; (2) $k = m + 1$ and there is a affine plane of order k. (3) The construction of G is as described in the* **Construction of third type** .

PROOF: We complete this proof step by step.

1) Since $k \geq 3$ it follows $\exists b_3 \in B$ s.t., $a_3 b_3 \in E$ and $\exists c_3 \in C$ s.t., $b_3 c_3 \in E$

Now we consider the constitution of the shortest paths from c_3 to the vertices of $B_1 \cup B_2$ (length 3). It follows that, $|N(c_3) \cap D_i| = 1$ for $1 \leq i \leq m$, $|N(c_3) \cap D^{(j)}| = 1$ for $1 \leq j \leq l$. And if we denote $N(c_3) \cap D = \{u_{i_1 j_1}, u_{i_2 j_2}, ..., u_{i_p j_p}\}$ we must have $(i_1, i_2, ..., i_p)$ is a permutation of $(1, 2, ..., m)$, and $(j_1, j_2, ..., j_p)$ is a permutation of $(1, 2, ..., l)$. This can be occur only when $m = l$. From no on we denote $D = \{u_{ij} | 1 \leq i, j \leq m\}$.

2) Consider the constitution of the shortest path from c_i^1 to c_j^1. Notice $d(c_i^1, c_j^1) \leq 3$ for $1 \leq i < j \leq m$.

If there is an edge connecting the vertices of D_1 and D_2, say, $u_{11} u_{21} \in E$, then u_{22} can't adjacent to any vertex of $D_1 = \{u_{11}, u_{12}, ..., u_{1m}\}$. Otherwise, suppose u_{22} is adjacent to some $u_{1i}, 2 \le i \le m$, then since there is a 6−cycle $c_1^1 u_{11} u_{21} c_2^1 u_{22} u_{1i} c_1^1$, it would be induces $d(c_1^1, c_2^1) \le 2$. But $c_1^1 c_2^1 \notin E$(otherwise $c_1^1 c_2^1 u_{21} u_{11} c_1^1$ is a 4−cycle, contradiction). We only can have $d(c_1^1, c_2^1) = 2$. But this induces there are two paths connecting c_1^1 and b_2^1 both are with length 3 (one is pass through $b_1^1 a_1$, the other is pass through c_2^1). Therefore $d(b_2^1, c_1^1) \le 2$. This is impossible. Therefore $u_{22} u_{1i} \notin E$ for $2 \le i \le m$. But this induces there is no path with length ≤ 3 connecting u_{22} and b_1^1, contradicts to $d(b_1^1, u_{22}) \le 3$. this is still a contradiction.

Consequently, we have $c_1^1 c_2^1 \in E$. By the same reason we know $c_i^1 c_j^1 \in E$ for $1 \le i < j \le m$, i.e., $C_1 = \{c_1^1, c_2^1, ..., c_m^1\}$ generates a clique. Similarly we have $C_2 = \{c_1^2, c_2^2, ..., c_m^2\}$ generates a clique

3) Consider the constitution of the shortest path from u_{11} to $u_{2i}, 2 \le i \le m$. Because we already have two paths connecting u_{11} and u_{2i} both are with length 3, $u_{11} c_1^1 c_2^1 u_{2i}, u_{11} c_1^2 c_2^2 u_{2i}$, and obviously $u_{11} u_{2i} \notin E$ (otherwise it would be induces a 4−cycle $u_{11} c_1^1 c_2^1 u_{2i} u_{11}$, a contradiction). Therefore $d(u_{11}, u_{2i}) = 2$, and $\exists c^* \in C$ s.t., $u_{11} c^* \in E$ and $u_{2i} c^* \in E$, and $c^* \notin C_1 \cup C_2$. By the same reason we have $d(u_{i\lambda}, u_{j\mu}) = 2$ for $1 \le \lambda, \mu \le m$.

Now we consider the constitution of the shortest paths from c^* to the vertices of $B_1 \cup B_2$ (its length must be 3) and induce $|N(c^*) \cap D_i| = 1$ for $1 \le i \le m$, $|N(c^*) \cap D^{(j)}| = 1$ for $1 \le j \le m$. This is because there are paths from c^* to b_i^1 pass through a vertex of D_i and c_i^1, there are paths from c^* to b_j^2 pass through a vertex of $D^{(j)}$ and c_j^2. Therefore we have $|N_D(c^*)| = m$ and $|N_D(c^*) \cap D_i| = 1$, $1 \le i \le m$, and $|N_D(c^*) \cap D^{(j)}| = 1$, $1 \le j \le m$.

Then we consider the shortest paths from $c*$ to the vertices of A and induce $\exists b^* \in B$ s.t.,$b^* c^* \in E$. Furthermore, we have $a^* \in A$ s.t., $a^* b^* \in E$.

Therefore, for each vertex $c^* \in C \setminus (C_1 \cup C_2)$, there are m vertices of D have paths with length 2 pass through c^* connecting them with each other. Now $|D| = m^2$ and for every vertex $u_{ij} \in D$ except there are paths with length 2 connecting u_{ij} and the other $(m-1)$vertices of D_i pass thrpugh c_i^1 and there are paths with lemgth 2 connecting u_{ij} and the other $(m-1)$ vertices of $D^{(j)}$ pass through c_j^2, there must have another paths with length 2 connecting u_{ij} and the other $m^2 - 2(m-1) - 1 = (m-1)^2$ vertices of D. Therefore u_{ij} must adjacent to other $(m-1)$ vertices of $C \setminus (C_1 \cup C_2)$. It follows that, if we use $D \cup (C \setminus (C_1 \cup C_2))$ as vertex set and ignoreing the edges connecting the vertices of $C \setminus (C_1 \cup C_2)$

to get a bipartite graph , then the number of its edges is $m^2(m-1)$. Therefore $|C\setminus(C_1\cup C_2)| \geq \frac{m^2(m-1)}{m} = m(m-1)$. It follows that $|C| \geq m(m-1)+2m = m(m+1)$.

4) Now denote $C\setminus(C_1\cup C_2) = C^*$. Take $c_\lambda, c_\mu \in C^*$. Consider the constitution of the shortest paths conecting c_λ and c_μ. Since we have $d(u_{i\lambda}, u_{j\mu}) = 2$, therefore $d(c_\lambda, c_\mu) < 3$, there are two possibilities.

4.1)$c_\lambda c_\mu \notin E$ and $N_D(c_\lambda)\cap N_D(c_\mu) \neq \phi$. This induces $d(c_\lambda, c_\mu) = 2$ and obviously $|N_D(c_\lambda)\cap N_D(c_\mu)| = 1$. Otherwise there is a 4−cycle and induces $d(c_\lambda, c_\mu) = 1$, a contradiction.

4.2) $N_D(c_\lambda)\cap N_D(c_\mu) = \phi$.

Since we have $|N_D(c_\lambda)\cap D_i| = |N_D(c_\lambda)\cap D^{(j)}| = |N_D(c_\mu)\cap D_i| = |N_D(c_\mu)\cap D^{(j)}| = 1$ for $1 \leq i, j \leq m$. Let $\{u_{11}c_\lambda, u_{22}c_\lambda, u_{12}c_\mu\} \subseteq E$. Then $u_{12}c_1^1u_{11}c_\lambda$, $u_{12}c_2^2u_{22}c_\lambda$ are two paths connecting u_{12} and c_λ with length 3. therefore we have $d(u_{12}, c_\lambda) \leq 2$. But $u_{12}c_\lambda \notin E$ and c_λ cann't adjacent to c_1^1 or c_1^2(property 2). Therefore we must have $c_\lambda c_\mu \in E$, $\quad d(c_\lambda, c_\mu) = 1$.

Now we take any $c_\lambda \in C^*$ and consider the number of $c_\mu \in C^*$ satisfies $c_\lambda c_\mu \in E$.

Since $c_\lambda c_\mu \in E$ we cann't have $N_D(c_\lambda)\cap N_D(c_\mu) \neq \phi$. Otherwise if $w \in D$ s.t., $wc_\lambda, wc_\mu \in E$, then from property 2 we know there is exactly one vertex $a^* \in A$ s.t.,$d(a^*, c_\lambda) = d(a^*, c_\mu) = 2$. This induces there are two paths from a^* to w both with length 3 and this is impossible. Therefore we must have $N_D(c_\lambda)\cap N_D(c_\mu) = \phi$. From $|D| = m^2$ and each vertex $c_\lambda \in C^*$ is adjacent to m vertices of D and there exists only one vertex $a^* \in A$ s.t.,$d(a^*, c_\lambda) = 2$ and from each vertex of D to a^* there is a path with length 3 connecting them we induce that, besides c_λ we must have another $(m-1)$vertices $c_\mu \in C$ s.t., $d(a^*, c_\mu) = 2$ and these $(m-1)$ vertices c_μ are adjacent to the remained $m(m-1)$ vertices of D. That is to say, for a^* there are m vertices of C which are at distence 2 to a^*, and each vertex of them are adjacent to m distinct vertices of D . These amounts to m distinct vertices of C adjacent to m^2 distinct vertices of D . Therefore these m vertices generates a clique.

Now take any $c_\lambda \in C^*$ and denote the vertex set which generates a clique and containning c_λ by $C_\lambda = \{c_1^\lambda, c_2^\lambda, ..., c_m^\lambda\}$ then for any $b \in B$ we have $|N_C(b)\cap C_\lambda| \leq 1$. This is because $c_1^\lambda, c_2^\lambda, ..., c_m^\lambda$ are adjacent to the vertices of $D_1 = \{u_{11}, u_{12}, ..., u_{1m}\}$ respectively. If $|N_C(b)\cap C_\lambda| \geq 2$ then there are at least two paths wiht length 3 connecting b and c_1^1. This is contradicts to $d(b, c_1^1) = 3$.

Now take any $b \in B$ we have $|N_C(b)\cap C^*| = 1$. If not, suppose $|N_C(b)\cap C^*| \geq 2$,

c_λ, $c_\mu \in C^*$ s.t., $bc_\lambda, bc_\mu \in E$. If $c_\lambda c_\mu \in E$ then it contradicts to the results obtained as above; if $c_\lambda c_\mu \notin E$ then we must have $N_D(c_\lambda) \cap N_D(c_\mu) \neq \phi$ and this induces a 4−cycle, a contradiction.

Therefore from the results described as above we know that, for vertex c_λ there is a vertex set $B_\lambda \subseteq B$, $|B_\lambda| = m$ s.t., the vertices of B_λ are adjacent to the vertices of c_λ respectively and a^* is adjacent to every vertex of B_λ.

Now we prove that $|C^*| = m(m-1)$. Consider the constitution of the shortest paths from u_{11} to the vertices of $(D_2 \cup D_3 \cup ... \cup D_m)\backslash D^{(1)}$. Since two paths $u_{11}c_1^1c_i^1u_{ij}, u_{11}c_1^2c_j^2u_{ij}$ for $2 \leq i,j \leq m$ both are with length 3. Therefore $d(u_{11}, u_{ij}) = 2$. That is , there are $(m-1)$ vertices of C^* adjacent to u_{11} and $u_{ij}, 2 \leq i,j \leq m$. But these vertices of C^* cann't adjacent to each other because they have a common neighbourhood vertex u_{11}. This proves the number of vertex-cliques contained in C^* is at least $(m-1)$. If $|C^*| > m(m-1)$, i.e., there is a vertex $\hat{c}$ not belongs to the $(m-1)$ vertex-cliques each containes m vertices, then because there are paths with length 2 between any two vertices of D passing through the vertices of $C_1 \cup C_2$ and $m(m-1)$ vertices of C^* , if $\hat{c}$ adjacent to the vertices of D_i and $D^{(j)}$ it would induces a 4−cycle. This is a contradiction.

Now we have obtained the results that, $k = m+1, |C| = |B| = m(m+1), |D| = m^2$. C can be divided into family of m−vertex-clique: $\{C_1, C_2, ..., C_{m+1}\}$. The neighbourhood vertices sets of m vertices of each C_i in D form a partition of D into m blocks and each block contains exactly m verticves. And each pair of vertices of different subset C_i and C_j have exactly one common neighbourhood vertex in D. Therefore if we using the neighbourhood vertices set of each vertex of C as the line, then we get an affine plane, i.e., G is of the third type of construction. ∎

Up to now we have induce three types of construction of geodetic blocks with diameter 3 and the properties described the characterizations of these geodetic graphs. Obviously, there are a lot cases haven't considered .

REFERENCES

[1]O.Ore, *Theory of Graphs* , American Mathematics Society Providence, R.I., 1962

[2]J.G.Stemple, Geodetic Graphs of Diameter Two, *J. Combinatorial Theory*

(B).**17**, 266-280 (1974)

[3]J.Bosak, Geodetic Graphs,*Colloquia Mathematica Societatis Janos Bolyai* **18** *Combinatorics, Keszthely (Hungary), 151-172 (1976)*

[4]K.R.Parthasarathy and N.Srinivasan, Some general constructions of geodetic blocks, J. Combinatorial Theory (B), **33 (2)** *121-136 (1982)*

*[5]Ho-Jin Lee, A Note on Geodetic Graphs of diameter Two and Their Relation to Orthogonal Latin Square, J. Combinatorial Theory (B),***22***, 165-167 (1977)*

[6]J.Plesnik, A Construction of Geodetic Graphs Bassed on Pulling Subgraphs Homeomorphic to Complete Graphs, J. Combinatorial Theory (B) **36** *, 284-297 (1984)*

[7]K.R.Parthasarathy and N.Srinivasan, Geodetic Blocks of Diameter Two: Some Structural Properties, Ars Combinatorics **20** *,49-60 (1985)*

[8]R.J.Cook and D.G.Pryce, A Class of Geodetic Blocks,J. Graph Theory, **6** *, 157-168 (1982)*

[9]R.Scapellato, Geodetic Graphs of Diameter Two and Some Related Structures, J. Combinatorial Theory (B) **41**, *218-229 (1986)*

[10]K.R.Parthasarathy and N.Srinivasan, Geodetic Blocks of Diameter Three, Combinatorica **4(2-3)** *, 197-206 (1984)*

[11]B.Bollobas, Extremal Graph Theory, Academic Press, Inc. LTD, 1978

Weakly $5 - CI$ Abelian Groups

Jixiang Meng
Department of Mathematics of Xinjiang University
Urumuqi, Xinjiang, 830046, P.R.China.

Mingyao Xu
Inst. of Mathematics, Peking University
Peking 100871, P.R.China.

Abstract Let G be a group. An inverse-closed subset S of G is called a $CI-$subset of G if for any isomorphism $X(G,S) \cong X(G,T)$ of Cayley graphs, there is some $f \in AutG$ such that $T = f(S)$. Let m be a positive integer. G is called a weakly $m - CI$ group if any inverse-closed generating subset S of G with $| S |\leq m$ is a $CI-$subset. Here, the weakly $5 - CI$ abelian groups are characterized.

Key words Cayley graphs; $m - CI$ groups; ispmorphisms.

1. Introduction

Let G be a group and S an inverse-closed subset of G (that is, $S = S^{-1}$) not containing the identity element 1 of G. The Cayley graph of G with respect to S, denoted $X(G,S)$, is a graph with vertex set G and for x and y in G, x and y are adjacent if and only if $x^{-1}y \in S$.

A subset S not containing the identity 1 of G is called a Cayley subset. An inverse-closed Cayley subset S of G is called a $CI-$subset if for any $T \subseteq G$ with $X(G,S) \cong X(G,T)$, there is a group automorphism f of G such that $T = f(S)$. A group G is called a $CI-$group if any inverse-closed Cayley subset of G is a $CI-$subset.

The investigation of $CI-$groups was initialed by the well-known Adam's Conjecture[1] which states that finite cyclic groups are CI-groups. Although this conjecture was disproved by Elspas and Turner [2], this, however, was not the end of the investigation. Many researchers devote themselves to studying which groups are $CI-$groups. For results on this subject, see [3–5] for references.

Since $CI-$groups are very exceptional and very hard to determine. Mingyao Xu and Xingui Fang introduced the so called $m - CI$ groups. A group G is called a $m - CI$ group if any inverse-closed Cayley subset of G with $|S| \leq m$ is a $CI-$subset. In [6], the above refered authors characterized abelian $4 - CI$ groups. Before giving their theorem, let us first illustrate a notion they firstly used. A group is called homocyclic if it is the direct product of some copies of a cyclic group.

Theorem 1[6]. A finite abelian group G is $4-CI$ if and only if any Sylow subgroup of G is homocyclic and the Sylow 2-subgroup of G is cyclic or elementary abelian.

In [7], Boesch and Tindell conjectured that finite cyclic groups are $4 - CI$. In [8], Liang Sun confirmed this conjecture by showing that finite cyclic groups are $5 - CI$. Furthermore, Xingui Fang and Ming Wang characterized abelian $5 - CI$ groups in [9].

Theorem 2[9]. A finite abelian group is $5 - CI$ if and only if G is $4 - CI$.

The main purpose of investigating the $m-CI$ groups is to classify the Cayley graphs of small degrees. In carrying on this work, however, we are only interested in classifying the connected Cayley graphs. Based on this point, the second author of this paper proposed the following:

Definition 1. Let G be a group and m a positive integer. G is Called weakly $m - CI$ if any inverse-closed generating subset S of G with $|S| \leq m$ is a $CI-$ subset.

In our terminology, the main result of [10] can be rewritten as

Theorem 3[10]. Any finite abelian group is weakly $4 - CI$.

In this paper, we obtain the following:

Theorem 4. A finite abelian group is weakly $5 - CI$ if and only if it is not isomorphic to $Z_n \times Z_2$ for n dividable by 4.

Notations and definitions not defined here can be found in [11–12].

2. Proof of the main result

To prove of Theorem 4, we proceed to establish a sequence of lemmas.

In what follows, we always suppose that G is a finite abelian group.

Consider the Cayley graph $X(G,S)$. let $g \in G$. We use $N(g)$ to denote the neighbour set of g in $X(G,S)$. Let $x \in N(g)$, we call x a $M(g)$ element of $X(G,S)$ if for any $y \in N(g)$, we have $\mid N(x) \cap N(g) \mid \geq 2$. Clearly, if x is a $M(1)$ element , so is x^{-1}.

Lemma 1[6]. Suppose that $X(G,S) \overset{f}{\cong} X(G,T)$ with $f(0)=0$. Let S_1 be the set of $M(1)$ elements of $X(G,S)$. Then $f(S_1)=T_1$ is the set of $M(1)$ elements of $X(G,T)$ and $X(G,S_i) \overset{f}{\cong} X(G,T_i)(1 \leq i \leq 2)$, where $S_2 = S \setminus S_1$ and $T_2 = T \setminus T_1$.

Lemma 2. Let G be a finite abelian group and $K=\{a,b,-b\}$ or $\{b,-b,c,-c\}$ is a generating subset of G with $o(a)=2$ and $2b \neq \pm c$. If $X(G,K) \overset{f}{\cong} X(G,L)$ and $f(0)=0$, then $f \in AutG$ and $f(K)=L$.

The proof is easy. One can refer that of Lemma 2 in [6] for details.

Lemma 3. Let G be an abelian group and $S=\{a,\pm b,\pm c\}$ be a generating subset of G. If each element in S is a $M(1)$ element of $X(G,S)$ and $G \not\cong Z_n \times Z_2$ for any n dividable by 4, then S is a $CI-$ subset.

Proof. Suppose that $X(G,S) \overset{f}{\cong} X(G,T)$ with $f(0)=0$, where $T=\{a',\pm b',\pm c'\}$. Then T is the set of $M(1)$ elements of $X(G,T)$.

Since b is a $M(1)$ element of $X(G,S)$, we have that $\mid N(b) \cap N(-b) \mid \geq 2$. Then we can assume, without loss of generality that one of the following relations holds:

[1], $3b=a$. [2], $3b=c$. [3], $4b=0$.

Similarly, since c is also a $M(1)$ element of $X(G,S)$, we can assume, without loss of generality that one of the following relations is satisfied:

[1]′, $3c=a$. [2]′, $3c=b$. [3]′, $3c=-b$. [4]′, $4c=0$.

If $[i]$ and $[j]'$ are satisfied and $(i,j) \neq (3,4)$, then the order of G is at most 24 and G is either a cyclic group or the direct product of two cyclic groups Z_m and Z_p, where $p=2$ or 3. If G is cyclic, then it is $5-CI$ and therefore weakly $5-CI$. If $G=Z_m \times Z_p$, then $p^2 \mid m$, and so G is $5-CI$ by Theorem 2. Thus we need only to consider the case that $4b=4c=4b'=4c'=0$. Two cases are considered.

Case 1. The rank of G is 3.

In this case, $G=\langle a\rangle \times \langle b,c\rangle = \langle a'\rangle \times \langle b',c'\rangle$. If $2b=\pm 2c$, then $2b'=\pm 2c'$. Assume, without loss of generality that $2b = 2c$ and $2b' = 2c'$. Then $G = \langle a \rangle \times \langle b \rangle \times \langle b-c \rangle = \langle a' \rangle \times \langle b' \rangle \times \langle b'-c' \rangle$. If $2b \neq \pm 2c$, then $G = \langle a \rangle \times \langle b \rangle \times \langle c \rangle = \langle a' \rangle \times \langle b' \rangle \times \langle c' \rangle$, the result follows readily.

Case 2. The rank of G is 2.

In this case, $G = \langle b, c \rangle = \langle b', c' \rangle$. Since $G \not\cong Z_4 \times Z_2$, we have that $2b \neq \pm 2c$ and $2b' \neq \pm 2c'$. Thus $G = \langle b \rangle \times \langle c \rangle = \langle b' \rangle \times c' \rangle$. If $a = 2b$, then there exists a triangle in the induced subgraph $[S]$ in $X(G, S)$, and so there exists a triangle in the induced subgraph $[T]$ in $X(G,T)$. Thus $a' = 2b'$ or $2c'$. Similarly, if $a = 2b + 2c$, then $a' = 2b' + 2c'$. The result thus follows.

Lemma 4. Let $G = Z_n \times Z_2$, where $4 \mid n$. Then G is not a weakly $5 - CI$ group.

Proof. Set $S = \{a, \pm b, \pm c\}$ and $T = \{a', \pm b, \pm c\}$, where

$$\begin{cases} a = (n/2, 0) \\ b = (n/2 + 1, 0) \\ c = (1, 1) \\ a' = (0, 1) \end{cases}$$

Then, it is not difficult to prove that G has the following coset decompositions:

$$G = \bigcup_{j=0}^{n/2-1} (\langle a \rangle + jb) \bigcup \bigcup_{j=0}^{n/2-1} (\langle a \rangle + jb + c) = \bigcup_{j=0}^{n/2-1} (\langle a' \rangle + jb) \bigcup \bigcup_{j=0}^{n/2-1} (\langle a' \rangle + jb + c)$$

and the mapping f:

$$\begin{aligned} ia + jb &\mapsto ia' + jb \\ ia + jb + c &\mapsto ia' + jb + c \end{aligned}$$

is an isomorphism from $X(G, S)$ to $X(G, T)$. It is clear that no automorphism of G maps S into T. Thus S is not a $CI-$ subset and G is not weakly $5 - CI$.

Now we are prepared for the proof of Theorem 4.

The proof of Theorem 4:

The necessity follows readily from Lemma 4. Now we prove the sufficiency. In view of Theorem 3, it suffices to consider the inverse-closed generating subsets G with cardinality 5. Let S and T be two such subsets with $X(G, S) \overset{f}{\cong} X(G, T)$ and $f(0) = 0$. We will prove that there exists an automorphism of G mapping S into T.

Assume that there are m elements of order 2 in S. Then $m \in \{1, 3, 5\}$. Three cases are considered.

Case 1. $n = 5$.

In this case, G is elementary abelian, the result follows from Theorem 2.

Case 2. $n = 3$.

Let $S = \{a, b, c, d, -d\}$. Then a, b and c are all $M(1)$ elements of $X(G, S)$. Thus $a' = f(a)$, $b' = f(b)$ and $c' = f(c)$ are all $M(1)$ elements of $X(G, T)$. Assume that $T = \{a', b', c', d', -d'\}$. By lemma 1, we have the following:

$$X(G, \{a, b, c\}) \overset{f}{\cong} X(G, \{a', b', c'\}),$$
$$X(G, \{d, -d\}) \overset{f}{\cong} X(G, \{d', -d'\})$$

We consider two subcases.

Subcase 2.1. The order of $\langle a, b, c\rangle$ is 4.

Then the order of $\langle a', b', c'\rangle$ is also 4 and $G = \langle a, b\rangle\langle d\rangle = \langle a', b'\rangle\langle d'\rangle$. If $\langle a, b\rangle \cap \langle d\rangle = \{0\}$, then $\langle a', b'\rangle \cap \langle d'\rangle = \{0\}$, the result follows readily. If otherwise, assume that $\langle a, b\rangle \cap \langle d\rangle = \{e\}$ and $\langle a', b'\rangle \cap \langle d'\rangle = \{e'\}$, if in addition, $e = a$ or b, then $e' = a'$ or b', the result then follows. If $e = a + b = c$, then $e' = a' + b' = c'$, the result is clearly true in this case.

Subcase 2.2. The order of $\langle a, b, c\rangle$ is 8.

There is nothing to prove if $\langle a, b, c\rangle \cap \langle d\rangle = \{0\}$. Otherwise, suppose that $\langle a, b, c\rangle \cap \langle d\rangle = \{e\}$ and $\langle a', b', c'\rangle \cap \langle d'\rangle = \{e'\}$, where $o(e) = o(e') = 2$. Since $f(a + b + c) = a' + b' + c'$, $e' = a' + b' + c'$ if $e = a + b + c$, and so $G = \langle a, b\rangle \times \langle d\rangle = \langle a', b'\rangle \times \langle d'\rangle$ and the result follows. Similarly the result holds for other cases.

Now we deal with the last case.

Case 3. $n = 1$.

Assume that $S = \{a, b, -b, c, -c\}$ and $T = \{a', b', -b', c', -c'\}$. Then $G = \langle a\rangle\langle b, c\rangle = \langle a'\rangle\langle b', c'\rangle$. Suppose that there are t $M(1)$ elements of $X(G, S)$. Then there are t $M(1)$ elements of $X(G, T)$. Three subcases are considered.

Subcase 3.1. $t = 1$.

In this case, a and a' are the $M(1)$ elements of $X(G, S)$ and $X(G, T)$, respectively. Thus

$$X(G, S\setminus\{a\}) \overset{f}{\cong} X(G, T\setminus\{a'\}),$$
$$X(G,\{a\}) \overset{f}{\cong} X(G,\{a'\}).$$

Thus $|\langle b,c\rangle| = |\langle b',c'\rangle|$. If $a \in \langle b,c\rangle$, then $a' \in \langle b',c'\rangle$ and $G = \langle a\rangle \times \langle b,c\rangle = \langle a'\rangle \times \langle b',c'\rangle$, the result thus follows by Lemma 3. Otherwise, $a \in \langle b,c\rangle$ and $a' \in \langle b',c'\rangle$. If in addition, $2b = \pm 2c$, then $2b' = \pm 2c'$. Assume, without loss of generality that $2b = 2c$ and $2b' = 2c'$. Then $G = \langle b\rangle \times \langle b-c\rangle$ and the result follows from Theorem 2.

Subcase 3.2. $t = 3$

Assume, without loss of generality that $S_1 = \{a, b, -b\}$ and $T_1 = \{a', b', -b'\}$ are the sets of $M(1)$ elements of $X(G,S)$ and $X(G,T)$, respectively. Then

$$X(G, S_1) \overset{f}{\cong} X(G, T_1),$$
$$X(G, S\setminus S_1) \overset{f}{\cong} X(G, T\setminus T_1)$$

From this, we can deduce the following:

$$X(G,\{\pm b, \pm c\}) \overset{f}{\cong} X(G,\{\pm b', \pm c'\}),$$
$$X(G,\{a\}) \overset{f}{\cong} X(G,\{a'\})$$

The result follows by a similar proof as in Subcase 1.

Subcase 3.3. $t = 5$.

The result follows directly from Lemma 3.

This completes the proof.

References

[1] A.Adam, Research problem 2–10, J. Comb. Theory, 2(1967), 393.

[2] B.Elspas and J.Turner, Graphs with circulant adjacency matrices, J. Comb. Theory, 9(1970), 297–307.

[3] L.Babai and P.Frankl, Isomorphisms of Cayley graphs, Colloq. Math. Soc. J. Bolyai, 18, Combinatorics, Keszthely, 1976: North-Holland, Amsterdam, 1978, 35-52.

[4] P.P.Palfy, Isomorphism problem for relational structures with cyclic automorphism, European. J. Combin., 8(1987), 35–43.

[5] C.D.Godsil, On Cayley graph isomorphism, Ars Combinatoria, 15(1983), 231–246.

[6] X.G.Fang and M.Y.Xu, On isomorphisms of Cayley graphs of small valency, Algebra Colloq., 1:1(1994), 67–76.

[7] F.Boesch and R.Tindell, Circulants and their connectivities, J. Graph Theory, 8(1984), 487–499.

[8] L.Sun, Isomorphisms of circulant graphs, Chinese Annals of Mathematics, 9a:5(1988), 567–574.

[9] X.G.Fang and M.Wang, Isomorphisms of Cayley graphs with degree $m(\leq 5)$, Chinese Annals of Mathematics, 13A:(1992), 7-14.

[10] C.Delorme, O.Fvatron and M.Maheo, Isomorphism of Cayley multigraphs of degree 4 on finite abelian groups, European J. Combin, 13(1992), 5-7.

[11] N.Biggs, Algebraic Graph Theory, London, England, Cambridge University Press, 1974.

[12] J.A.Bondy and U.S.R.Murty, Graph Theory with Applications, North-Holland, New York, 1976.

On a Conjecture of Ringel

Keyi OuYang and Liyan Gao

Institute of Applied Mathematics, Academia Sinica,

Beijing 100080, China

Abstract This paper shows that there exists one spanning tree for each 2-connected maximal planar bipartite graph *(mpb)* such that its cotree is a tree. It was conjectured by Ringel[1]. Furthermore, for a 2-connected *mpb* graph G, the problem is turned to a simpler one of finding a decomposition of G into two forests without isolated vertices.

Keywords maximal planar bipartite graph, spanning tree, cotree.

1. Introduction

Throughout this paper, we deal with simple planar graphs embedded in the Euclidean plane. Readers can refer to [2] for terminologies and notations not explained here.

A graph is called *maximal planar bipartite graph*, denoted by *mpb*, if

(1) G is a planar graph;

(2) G is a bipartite graph;

(3) add an edge (without adding vertices) to G, and we get a graph which is not planar or bipartite.

A graph satisfying (1) and (2) is called *planar bipartite graph*.

For convienence, we treat planar graph mentioned in the paper as one of its planar embedding. It's easy to know that each 2-connected *mpb* graph divides the plane into quadrangle faces only.

Let H be an edge-induced subgraph of graph G. The complement of G, denoted by $\bar{H}$, is defined as the subgraph of G induced by $E(G)\backslash E(H)$. Especially, if T is a spanning tree of G, $\bar{T}$ s called cotree of T in G. The main results of reference [1] are:

Theorem A *For each 2-connected mpb graph of order v, there exists a tree T of order v such that $\bar{T}$ is also a tree.*

Theorem B *For each 2-connected mpb graph, there exists a tree T such that $\bar{T}$ is also a tree.*

In reference [1], Ringel gave the proof of Theorem A and regarded Theorem B as its corollary. Meanwhile Ringel proposed the following:

Conjecture *For each 2-connected mpb graph, there exists a spanning tree T such that $\bar{T}$ is also a tree.*

It's easy to prove (*q.v.* [1]), based on Theorem B, that the tree mentioned in theorem B is either a tree of order $(v-1)$ or a spanning tree in order that its cotree is a tree.

This paper shows that the conjecture is true. The main results are the following two theorems.

Theorem 1 *For each 2-connected mpb graph, there exists a spanning tree T such that $\bar{T}$ is a tree.*

Theorem 2 *For a 2-connected mpb graph G, the following two statements are equivalent.*

(1) There exists a forest F without isolated vertices in G, such that $\bar{F}$ is also a forest.

(2) There exists a spanning tree T in G, such that $\bar{T}$ is a tree.

This paper also provides an effective algorithm for finding a spanning tree satisfying statement (2) in Theorem 2 from a forest satisfying statement (1) in Theorem 1. And, it makes the problem of finding such a spanning tree simpler.

Before the proof, we introduce some definitions and notations.

We use $\lceil x \rceil$ to denote the minimal integral number which is not less than x. Use $v(G), \varepsilon(G)$ and $\omega(G)$ to denote the number of vertices, edges and connected components in graph G respectively.

Let $V_1 \subseteq V(G)$, we use $\langle V_1 \rangle_G$ to denote the subgraph of G induced by V_1, written by $\langle V_1 \rangle_G \prec G$. If V_1 and V_2 are two disjoint subsets of $V(G)$, then we write $[V_1, V_2]_G = \{(u,v) | (u,v) \in E(G), u \in V_1, v \in V_2\}$.

Let T be a spanning tree of $G, e \in E(\bar{T})$, then $T + e$ contains a unique cycle, denoted by C_e^T.

2. Proof of the main results

The proof of Lemma 1 bellow is easy and can be found in [1].

Lemma 1 *Let G be a planar bipartite graph, then $\varepsilon(G) \leq 2v(G) - 4$ and the equality is obtained iff $G \cong K_{1,2}$ or G is a 2-connected mpb graph.*

Lemma 2 (Nash-Williams [3]) *Let G be a nonempty graph, then $E(G)$ can be decomposed into*

$$\tau(G) = \max_{H \prec G} \lceil \frac{\varepsilon(H)}{v(H)-1} \rceil$$

disjoint subsets such that each subset induces a forest.

Corollary 1 *Let G be 2-connected mpb graph, then $\tau(G)$ =2.*

Proof: Notice that any induced subgraph H of a 2-connected *mpb* graph is planar bipartite graph. Then $\epsilon(H) \leq v(H) - 4$, by Lemma 1. Therefore, $\tau(G) \leq \max_{H \prec G} \frac{2v(H)-4}{v(H)-1} = 2$. However, $\tau(G) = 1 \iff G$ is a forest. The corollary is proved. □

Lemma 3 *Let G be a 2-connected mpb graph, then there exists a spanning tree T in G such that $\bar{T}$ is a forest.*

Proof: By corollary 1, there exists an edge-induced subgraph F in G such that F is a forest and $\bar{F}$ is also a forest. Let $T_1, T_2, \ldots, T_k$, which are all trees, denote all the connected components of F. Because G is connected, graph $H = (V(H), E(H))$ is connected, where $V(H) = \{T_i\}$, $E(H) = \{(T_i, T_j)|$ there exists a path with only two end-vertices in $V(T_i)$ and $V(T_j)$ respectively $\}$.

Let $T(H)$ be a spanning tree of H, we can always get only one path as above, corresponding to each edge of $T(H)$. Then add the path to F so that the resultant one, denoted by F_1, is a tree and that $\bar{F}_1$ is a forest. If F_1 is not a spanning tree of G, then there exists $v \in V(G) \setminus V(F_1)$ and $u \in V(F_1)$ such that $(u, v) \in E(\bar{F}_1)$. Add edge (u, v) to F_1, . . ., repeat the process, until that we get a spanning tree T of G at the end and that $\bar{T}$ is a forest. The Lemma is proved. □

Definition If a tree T is a spanning tree of G and $\bar{T}$ is a forest, then we say that T is statisfying *c.f.*-condition.

Lemma 4 *Let G be 2-connected mpb graph. For any tree T satisfying c.f.-condition, there is $|V(T) \setminus V(\bar{T})| \leq 2$ and $\bar{T}$ is a tree when the equality is obtained.*

Proof: By Lemma 1, $\varepsilon(G) = 2v - 4$. Furthermore, T is a spanning tree of G. Therefore, $\varepsilon(T) = v - 1$ and $\varepsilon(\bar{T}) = \varepsilon(G) - \varepsilon(T) = v - 3$. Since $\bar{T}$ is a forest, $v - 2 \leq v(\bar{T}) \leq v$

q.v. $|V(T) \setminus V(\bar{T})| \leq 2$. And when $v(\bar{T}) = v - 2$, *q.v.*$|V(T) \setminus V(\bar{T})| = 2$, $\bar{T}$ is a tree.

Proof of Theorem 1 By contradiction. If the Theoerm 1 is false, then we can choose such a tree T satisfying *c.f.*-condition, as

(1) $\omega(\bar{T})$ is as small as possible;

(2) under the condition (1), the edge number of the minimal connected component in $\bar{T}$ (here "minimal" means the number of edges) is as small as possible.

Furthermore, let $T_1, T_2, \ldots, T_k$ $(k \geq 2)$ be all the connected components of $\bar{T}$, among which T_1 is the minimal one. Denote $S = \{v | v \in V(T) \setminus V(\bar{T})$, and there exists $u \in V(T_1)$ such that $(u, v) \in E(G)\}$ and $U = V(T_1) \cup S$. Naturally, $S \subseteq V(T) \setminus V(\bar{T})$. Meanwhile, $|S| \leq 1$ by our assumption and Lemma 4. On the other hand, we know

$$\bigcup_{i=2}^{i=k} [V(T_1), V(T_i)]_{G\setminus S},$$

denoted by S_1, is nonempty and included in T.

If for any $e \in E(T_1), C_e^T \bigcap S_1 = \emptyset$, by the connectivity of T_1 we know that there is a path in $\langle U \rangle_T$, which joins x to y, for any $x, y \in V(T_1)$. Therefore, $\langle U \rangle_T$ is a spanning tree of planar bipartite graph $\langle U \rangle_G$. At the same time, T_1 is a cotree corresponding to $\langle U \rangle_T$, then we get that

$$\varepsilon(T_1) = \varepsilon(\langle U \rangle_G) - \varepsilon(\langle U \rangle_T). \tag{1}$$

When $|S| = 0$,

$$\varepsilon(\langle U \rangle_T) = v(T_1) - 1. \tag{2}$$

By Lemma 1,

$$\varepsilon(\langle U \rangle_G) \leq 2v(T_1) - 4. \tag{3}$$

Combining (2), (3) and (1), we get $\varepsilon(T_1) \leq v(T_1) - 3$. However, $\varepsilon(T_1) \leq v(T_1) - 3$, a contradition to the fact that T_1 is a tree.

When $|S| = 1$,

$$\varepsilon(\langle U \rangle_T) = v(t_1). \tag{4}$$

By Lemma 1,

$$\varepsilon(\langle U \rangle_G) \leq 2(v(T_1) + 1) - 4. \tag{5}$$

Combining (4), (5) and (1), we get $\varepsilon(T_1) \leq v(T_1) - 2$, also a contradiction to the fact that T_1 is a tree.

We know from discussion above: there exists some $e \in E(T_1)$ and

$$[V(T_1), V(T_i)]_{G\setminus S} (2 \leq i \leq k)$$

such that

$$C_e^T \bigcap [V(T_1), V(T_i)]_{G\setminus S} \neq \emptyset.$$

Let e' belong to the intersection of the two sets above and write $T^* = T - e' + e$. Clearly, T^* satisfies $c.f.$-condition. Therefore, we have either $\omega(\bar{T}^*) = \omega(\bar{T}) - 1$, which contradicts the choice (2) of T, or that the minimal connected component of $\bar{T}^*$ is strictly included in T_1, then we have $\omega(\bar{T}^*) = \omega(\bar{T})$, which contradicts the chioce (2) of T. Therefore, the theorem is true. □

Proof of Theorem 2: The deduction of (1) from (2) is trivial. And, we can get (2) from (1) by the procedure used in the proof of Lemma 3 and Theorem 1. □

3. Outline of the algorithm

For a given 2-connected *mpb* graph G, we give an outline of the algorithm for finding a spanning tree satisfying statement (2) in Theorem 2 from a forest satisfying statement (1) in Theorem 2.

Step 1. Suppose forest F statisfying statement (1), according to the proof process of Lemma 3, we can find a spanning tree T statisfying $c.f.$-condition.

Step 2.

(1) For the spanning tree T satisfying $c.f.$-condition, if $\omega(\bar{T}) = 1$ then goto Step 3; else

(2) Let $T_1, T_2, ..., T_k$ be all of the connected components of $\bar{T}$, among which T_1 is the minimal.

Let $S = \{v | v \in V(T) \setminus V(\bar{T})$, and there exists $u \in V(T_1)$ such that $(u, v) \in E(G)\}$.

Check the edges of T_1 one by one until we find such an edge e that

$$C_e^T \bigcup (\sum_{i=2}^{i=k} [V(T_1), V(T_i)]_{G \setminus S}) \neq \emptyset.$$

Let e' belong to the intersection of the two sets, and write $T^* = T - e' + e$. Replace T by T^*, then goto (1).

Step 3. stop.

By the proof of Theorem 1, we know that Step 2 is feasible.

It's easy to prove that the complexity of the algorithm is polynomial.

Acknowledgement: We thank Professor Yanpei Liu for carefully reading this paper and for giving helpful suggestions.

References

[1] Gerhard Ringel, Two trees in maximal planar bipartite graphs, *J. Graph Theory* **6**, 755-758(1993).

[2] Yanpei Liu, *Embeddability in Graphs* (Chinese Ed), Science Press, 1994.

[3] Nash-Williams, Decomposition of finite graphs into fortests, *J. London Math. Soc.* **39**, (1964).

Invariant graphs for a family of endomorphisms—a survey

Norbert POLAT

I.A.E., Université Jean Moulin (Lyon III), 15, quai Claude Bernard, 69239 Lyon Cedex 2, France.

Abstract

In this paper we give an overview of different fixed-point properties for graphs.

Introduction

Fixed-point theory was originally limited to topological spaces. More precisely, it was concerned with sufficient conditions for a continuous map f from a topological space into itself to admit a fixed point, that is, a point x such that f(x) = x. Later this study was extended to partially ordered sets, and more recently to graphs. Obviously a fixed-*point* property is too demanding in the discrete context of graphs. It is usually replaced by a property about invariant finite set of vertices, and more particularly about invariant regular subgraphs of special types, such as simplices, hypercubes, Hamming graphs, etc.

The first of such properties are due to Tits [35] who investigated the case of trees. Then Halin in [11] considered connected graphs with injective maps preserving the edges as morphisms, and proved that any endomorphism of a connected rayless graph (i.e., not containing a one-way infinite path) stabilizes a finite set of vertices, thus that any endomorphism of a rayless tree leaves a vertex or an edge invariant.

In the last few years, several similar, but more general results were obtained. They use contractions as morphisms, that is mappings that map each edge onto an edge or a single vertex. Also, other related properties were proved. These are concerned with subgraphs, usually finite, that are invariant under every element of some particular families of self-contractions. Until now two main kinds of such families have been considered: first the group of all automorphisms of a graph, with a first result due to Schmidt [31] who proved that any connected rayless graph has a finite subgraph that is invariant under any automorphism; then commuting families of endomorphisms, just like in two well-known classical results: the Markov-Kakutani Theorem, [17] and [18], for compact convex sets of locally convex linear topological spaces, and the Tarski Theorem [34] for complete lattices.

We survey, in this paper, most of those results through different sections dealing, each of them, with particular invariant subgraphs. Except for some improvement in Section 3 and new results in Section 4 which are proved, the other results are only stated.

1. Notation

1.1. The graphs we consider are undirected, without loops and multiple edges. A complete graph will be simply called a *simplex*. If $x \in V(G)$, the set $V(x; G) := \{y \in V(G) : \{x, y\} \in E(G)\}$ is the *neighborhood* of x. For $A \subseteq V(G)$ we denote by $G[A]$ the subgraph of G induced by A, and we set $G - A := G[V(G) - A]$. A *path* $P := \langle x_0, ..., x_n \rangle$ is a graph with $V(P) = \{x_0, ..., x_n\}$, $x_i \neq x_j$ if $i \neq j$, and $E(P) = \{\{x_i, x_{i+1}\} : 0 \leq i < n\}$. A *ray* or *one-way infinite path* $\langle x_0, x_1, ...\rangle$, and a *double ray* or *two-way infinite path* $\langle ..., x_{-1}, x_0, x_1, ...\rangle$ are defined similarly. A graph is *rayless* if it contains no ray. A path $\langle x_0, ..., x_n \rangle$ is called an x_0x_n-*path*, x_0 and x_n are its *endpoints*, while the other vertices are called its *internal* vertices.

1.2. The usual *distance in* G *between two vertices* x *and* y, that is the length of a shortest xy-path in G, will be denoted by $dist_G(x, y)$. The *diameter* of G is $diam(G) := \sup \{dist_G(x, y) : x, y \in V(G)\}$. A graph is *bounded* if its diameter is finite. A subgraph H of G is *isometric* if $dist_H(x, y) = dist_G(x, y)$ for all vertices x and y of H. If x is a vertex of G and r a non-negative integer, the set $B_G(x, r) := \{y \in V(G) : dist_G(x, y) \leq r\}$ is the *ball of center* x *and radius* r *in* G. The *interval* of two vertices x and y of a graph G is the set of vertices of all xy-geodesics (i.e., shortest xy-paths) in G. A graph G will be said *interval-finite* if all its intervals are finite, and it will be said *block-finite* if all its blocks (i.e. maximal 2-connected subgraphs) are finite. A block-finite graph is obviously interval-finite.

1.3. If G and H are two graphs, a map $f : V(G) \to V(H)$ is a *contraction* if f preserves or contracts the edges, i.e., if $f(x) = f(y)$ or $\{f(x), f(y)\} \in E(H)$ whenever $\{x, y\} \in E(G)$. A contraction f from G onto an induced subgraph H of G is a *retraction*, and H is a *retract* of G, if its restriction $f \mid H$ to H is the identity. A self-contraction f of G (i.e., from G into itself) *stabilizes* (resp. *strictly stabilizes*) a subgraph H of G, or H is *invariant* (resp. *strictly invariant*) under f, if $f(H) \subseteq H$ (resp. $f(H) = H$). Note that if H is finite and strictly invariant under f, then the restriction of f to V(H) is an automorphism of H. A subset A of V(G) is *invariant* (resp. *strictly invariant*) under a self-contraction f if so is the subgraph G[A].

2. Trees

First we consider the simplest connected graphs, that is trees. It is well known that, *in any finite tree, every self-contraction fixes a vertex or an edge, and that furthermore, there exists a vertex or an edge which is fixed by every automorphism of this tree.* The case may be quite different, however, if the tree is infinite. Take for example a double ray $\langle \ldots, x_{-1}, x_0, x_1, \ldots \rangle$. Then the automorphism $x_n \mapsto x_{n+1}$, for every n, is clearly fixed point-free. A 3-regular tree (i.e., a tree whose vertices are all of degree 3) is another example of an infinite tree with none of the preceding fixed point properties. Even if each of these two trees contains a double ray, one shall not think that the lack of double rays is a sufficient condition for the existence of a fixed point. There also exist fixed point-free trees that contains no double rays, as is shown by the following example. Choose a ray $R = \langle x_0, x_1, \ldots \rangle$ in the 3-regular tree T_3; and let T be the subgraph of T_3 induced by $\bigcup_{k<\omega} \{y : dist_{T_3}(x_k, y) \leq k\}$. Then T has neither double rays nor fixed elements.

Of course those three examples do not mean that every tree containing a ray cannot have a vertex or an edge fixed by any automorphism (cf Corollary 4.4), but they only show, with the following results, that the lack of rays in a tree, as well as in any other graph, is an essential condition to get such fixed point properties.

2.1. Theorem (Polat, Sabidussi [27]). *Any rayless tree* T *contains a vertex or an edge which is invariant under any automorphism of* T.

We will see a similar result for trees with rays in Section 4 (Corollary 4.4).

2.2. Theorem (Nowakowski, Rival [19]). *Any self-contraction of a graph* G *stabilizes a vertex* *or an edge if and only if* G *is a rayless tree*.

2.3. Theorem (Polat [26]). *Let* $\mathfrak{F}$ *be a commuting family of self-contractions of a rayless tree* T. *Then there exists a vertex or an edge of* T *which is fixed by every element of* $\mathfrak{F}$.

These results for trees will be differently extended to connected graphs according to various interpretations that can have a vertex or an edge, such as: a finite graph (Sections 3 and 4), a simplex (Section 5), an hypercube (Section 6).

3. Finite invariant sets of vertices in rayless graphs

We will recall some concepts and results in order to improve and complete some of the results of [22].

3.1. Let G be a graph. We will endow the vertex set of G with a topology. For $A \subseteq V(G)$ we denote by $\overline{A}$ the set of vertices of G which belong to A or which are infinitely linked to A in G, i.e., by Menger's theorem, $x \in \overline{A}$ if and only if A meets the component of x in the graph G - S for any finite subset S of V(G - x). By [22, 2.4], $A \mapsto \overline{A}$ is the closure operator of a topology on V(G). In the following we will suppose that V(G) is endowed with this topology. Note that V(G) is clearly a T_1-space, and that V(G) *is Hausdorff if and only if the vertices of* G *are pairwise finitely linked in* G. By [22, 3.2], V(G) *is compact if* G *is connected and rayless*.

3.2. We will now recall the concept of Cantor-Bendixon derivative. Let $\mathfrak{A}$ be a topological space. We denote by $\mathfrak{A}'$ the *derivative* of $\mathfrak{A}$, i.e., the set of cluster points of $\mathfrak{A}$. The *Cantor-Bendixon derivative of order* α of $\mathfrak{A}$, $\mathfrak{A}^{(\alpha)}$, is defined by induction as follows:

- $\mathfrak{A}^{(0)} := \mathfrak{A}$
- $\mathfrak{A}^{(\alpha+1)} := (\mathfrak{A}^{(\alpha)})'$
- $\mathfrak{A}^{(\alpha)} := \bigcap_{\beta<\alpha} \mathfrak{A}^{(\beta)}$ if α is a limit ordinal.

In account of cardinalities there exists an ordinal α such that $\mathfrak{A}^{(\alpha)} = \mathfrak{A}^{(\alpha+1)}$. The smallest of these ordinals, denoted by $r(\mathfrak{A})$, is the *Cantor-Bendixon rank* of A, and the set $\mathfrak{A}^{(r(A))}$ is the *perfect kernel* of $\mathcal{C}$. For $x \in \mathcal{C}$, if there exists an ordinal α such that $x \in \mathcal{C}^{(\alpha)} - \mathcal{C}^{(\alpha+1)}$, then α is called the *Cantor Bendixon rank* of x.

3.3. [22, 3.3 and 3.5] *If* G *is a rayless connected graph, then* r(V(G)) *is a successor ordinal with* $V(G)^{(r(V(G)))} = \emptyset$. Therefore, if $r(V(G)) = \rho + 1$, then $V(G)^{(\rho)}$ is a compact set whose derivative is empty, thus is finite.

3.4. Theorem. *Any rayless connected graph* G *contains a non-empty finite set of vertices which is strictly invariant under any injective self-contraction of* G.

Proof. Let f be an injective self-contraction of G. We claim that f stabilizes $V(G)^{(\alpha)}$ for any ordinal α. This is clear if $\alpha = 0$. Let $\alpha \geq 0$. Suppose that it holds for every $\beta < \alpha$. The result is obvious if α is a limit ordinal. Assume that $\alpha = \beta + 1$, and let x be a cluster point of $V(G)^{(\beta)}$. Then x is infinitely linked to a subset A of $V(G)^{(\beta)}$. Thus, since, by the induction hypothesis, f stabilizes $V(G)^{(\beta)}$, and since f is injective, f(x) is also infinitely linked to the subset f(A) of $V(G)^{(\beta)}$. Hence f(x) is a cluster point of $V(G)^{(\beta)}$. Therefore f stabilizes $V(G)^{(\alpha)}$, which proves the claim.

Now, since G is rayless and connected, $r(V(G)) = \rho + 1$ and the set $V(G)^{(\rho)}$, which is invariant under f by what precedes, is non-empty and finite by 3.3. And this completes the proof. ❑

3.5. Corollary (Schmidt [31]). *Any rayless connected graph* G *contains a non-empty finite set of vertices which is strictly invariant under any automorphism of* G.

3.6. Theorem (Polat [23]). *Any self-contraction of a rayless connected graph strictly stabilizes a non-empty finite set of vertices.*

This generalizes a consequence of a result of Halin [11, Theorem 4] that deals with injective self-contractions, instead of self-contractions. More generally we have:

3.7. Theorem (Polat [26]). *Let $\mathfrak{F}$ be a commuting family of self-contractions of a connected, interval-finite, rayless graph* G. *Then there exists a non-empty finite set of vertices which is strictly invariant under any element of* $\mathfrak{F}$.

Outline of the proof. For any $f \in \mathfrak{F}$ and $A \subseteq V := V(G)$ let $A_f := \{x \in A : f^i(x) \in A$ for every $i \geq 0$, and $f^n(x) = x$ for some $n > 0\}$. Note that, by Theorem 3.6, A_f is non-empty for any f. Due to the facts that the family $\mathfrak{F}$ is commuting, and the graph G is connected, interval-finite and rayless, one can prove that, for any non-empty finite $\mathfrak{H} := \{f_1, ..., f_n\} \subseteq \mathfrak{F}$, the set $V_{\mathfrak{H}} := \bigcap_{f \in \mathfrak{H}} V_f = (...(V_{f_1})...)_{f_n}$ is non-empty and such that $G[V_{\mathfrak{H}}]$ is isometric with G. Then, by topological arguments, and even if the V_f's are not necessarily closed with respect to the topology defined in 3.1, one can show that the set $V_{\mathfrak{F}} := \bigcap_{f \in \mathfrak{F}} V_f$ is non-empty. Then the subgraph $G[V_{\mathfrak{F}}]$ is rayless since so is G, and connected as each subgraph $G[V_{\mathfrak{H}}]$ is isometric and as G is interval-finite. Consequently, since the restriction of every $f \in \mathfrak{F}$ to $V_{\mathfrak{F}}$ is an automorphism of $G[V_{\mathfrak{F}}]$, Schmidt's theorem (3.5) implies that the set $V_{\mathfrak{F}}$ contains a non-empty finite subset which is strictly invariant under every element of $\mathfrak{F}$. ❑

4. Finite invariant sets of vertices in graphs with rays

4.1. The *ends* of a graph G (this concept was introduced by Freudenthal [9] and Hopf [13], and independently by Halin [10]) are the classes of the equivalence relation $\sim_G$ defined on the set of all rays of G by : $R \sim_G R'$ if and only if there is a ray R" whose intersections with R and R' are infinite. We will denote by $\mathfrak{T}(G)$ the set of all ends of G, and for an end τ and a finite set S of vertices, by $\mathfrak{C}_{G-S}(\tau)$ the component of $G - S$ that contains an element of τ. A vertex x is said to *dominate* an end τ, and we write $x \mathfrak{D} \tau$, if $x \in V(\mathfrak{C}_{G-S}(\tau))$ for any finite $S \subseteq V(G - x)$.

We will endow the end set $\mathfrak{T}(G)$ of a graph G with the topology for which the closure of a subset Ω of $\mathfrak{T}(G)$ is the set

$$\overline{\Omega} := \{\tau \in \mathfrak{T}(G) : \text{for any finite } S \subseteq V(G) \text{ there is } \tau' \in \Omega \text{ such that } \mathfrak{C}_{G-S}(\tau) = \mathfrak{C}_{G-S}(\tau')\},$$

i.e., $\overline{\Omega}$ is the set of all ends which cannot be finitely separated from Ω. This topology was introduced by Jung [16]. The Cantor-Bendixon rank of an end τ, when such a rank exists, will be called the *order* of τ. This corresponds exactly to the "order" of this end, as defined by Jung in

[15].

We recall that a subgraph H of G is said to be *end-respecting* if every end of G contains at most one end of H.

4.2. (Jung [15]) *If* G *contains no subdivision of the dyadic tree as an end-respecting subgraph, or equivalently* (see [21]) *if the space* $\mathfrak{T}(G)$ *is scattered, then every end of* G *has an order, and there is an end of maximal order.*

4.3. Theorem (Polat [22]). *Let* G *be a connected graph containing no subdivision of the dyadic tree as an end-respecting subgraph, and having one of the following properties*:

(i) G *has only one end of maximal order, and this end is dominated by finitely many vertices*;

(ii) G *has at least two ends of maximal order, and one of them is dominated*;

(iii) G *has at least three ends of maximal order.*

Then G *contains a non-empty finite set of vertices which is strictly invariant under any automorphism of* G.

4.4. Corollary (Polat [22]). *If a tree* T *contains no subdivision of the dyadic tree and has at least three ends of maximal order, then* T *contains a vertex or an edge which is invariant under any automorphism of* T.

4.5. In order to obtain an interesting consequence of this theorem, we will recall a few more concepts and results. Let G be an infinite graph.

4.5.1. For $\Omega \subseteq \mathfrak{T}(G)$ let $m(\Omega) := \sup\{|\mathfrak{R}| : \mathfrak{R}$ is a set of pairwise disjoint elements of $\bigcup\Omega\}$. If $\Omega = \mathfrak{T}(G)$ we write $m(G)$ for $m(\mathfrak{T}(G))$, and if $\Omega = \{\tau\}$ we write $m(\tau)$ for $m(\{\tau\})$.

4.5.2. (Polat [21, Proposition 2.2]) *If* $m(\tau)$ *is finite, then so is the set of vertices that dominate* τ.

4.5.3. Let τ, τ_0 and τ_1 be ends of G, and A a finite subset of V(G). We set:

$$\text{sep}(\tau_0, \tau_1) := \inf\{|S| : S \subseteq V(G) \text{ and } \mathcal{C}_{G-S}(\tau_0) \neq \mathcal{C}_{G-S}(\tau_1)\},$$

$$\text{sep}(A, \tau) := \inf\{|S| : S \subseteq V(G) \text{ and } A \cap \mathcal{C}_{G-S}(\tau) = \emptyset\}.$$

Furthermore we define a $\tau_0\tau_1$-*path* as a finite or infinite path (ray or double ray) W of G such that: (a) if W is a double ray, then $W = R_0 \cup R_1$ with $R_i \in \tau_i$ for $i = 0, 1$; (b) if W is a ray $\langle x_0, x_1, \ldots\rangle$, then $W \in \tau_i$ for some $i \in \{0, 1\}$ and $x_0 \ \mathcal{D}\ \tau_{1-i}$; (c) if W is a path $\langle x_0, \ldots, x_n\rangle$, then $x_0 \ \mathcal{D}\ \tau_i$ and $x_n \ \mathcal{D}\ \tau_{1-i}$ for some $i \in \{0, 1\}$. We denote by $\text{DISJ}(\tau_0, \tau_1)$ the set of all sets of pairwise disjoint $\tau_0\tau_1$-paths.

Similarly an $A\tau$-*path* is a ray $\langle x_0, x_1, \ldots\rangle \in \tau$ such that $V(\langle x_0, x_1, \ldots\rangle) \cap A = \{x_0\}$, or a path $\langle x_0, \ldots, x_n\rangle$ with $V(\langle x_0, \ldots, x_n\rangle) \cap A = \{x_0\}$ and $x_n \ \mathcal{D}\ \tau$. We denote by $\text{DISJ}(A, \tau)$ the set of all sets of pairwise disjoint $A\tau$-paths.

Finally we set:

$$\text{disj}(\tau_0, \tau_1) := \sup\{|\mathcal{W}| : \mathcal{W} \in \text{DISJ}(\tau_0, \tau_1)\},$$

$$\text{disj}(A, \tau) := \sup\{|\mathcal{W}| : \mathcal{W} \in \text{DISJ}(A, \tau)\}.$$

4.5.4. (By Polat [20, Théorème 4.7]) $\text{sep}(\tau_0, \tau_1) = \text{disj}(\tau_0, \tau_1)$ *and* $\text{sep}(A, \tau) = \text{disj}(A, \tau)$.

4.6. Theorem. *Let* G *be an* n*-connected graph, for some positive integer* n*, containing no subdivision of the dyadic tree as an end-respecting subgraph. If* $m(\tau) < n$ *for every end* τ*, then* G *contains a non-empty finite set of vertices which is strictly invariant under any automorphism of* G.

Proof. By 4.2, every end of G has an order and there is at least an end of maximal order. We distinguish three cases.

Case 1: G has only one end τ of maximal order.

Let $\mathcal{R}$ be a set of pairwise disjoint rays in τ such that $|\mathcal{R}| = m(\tau)$, and let A be the set of the endpoints of the elements of $\mathcal{R}$. Then, by 4.5.4,

$$\text{disj}(A, \tau) = \text{sep}(A, \tau) \geq n > m(\tau).$$

Thus, by the definition of an $A\tau$-path, this means that any element of $\text{DISJ}(A, \tau)$ of maximal cardinality has an element which is not a ray, hence that the end τ is dominated. Moreover, by 4.5.2, τ is dominated by finitely many vertices only.

Case 2: G has two ends τ_0 and τ_1 of maximal order.

Then, by 4.5.4,

$$\mathrm{disj}(\tau_0, \tau_1) = \mathrm{sep}(\tau_0, \tau_1) \geq n > \max\{m(\tau_0), m(\tau_1)\}.$$

Thus, by the definition of an $\tau_0\tau_1$-path, this means that any element of $\mathrm{DISJ}(\tau_0, \tau_1)$ of maximal cardinality has an element which is not a double ray, hence that at least one of the ends τ_0 and τ_1 is dominated.

Case 3: G has at least three ends of maximal order.

Then, for each of the three cases, the result is a consequence of Theorem 4.3. ❑

An immediate consequence of this theorem is the following result of Halin, which was proved in a quite different way.

4.7. Corollary (Halin [12]). *If* G *is an* n*-connected graph, for some positive integer* n, *such that* m(G) < n, *then* G *contains a non-empty finite set of vertices which is strictly invariant under any automorphism of* G.

This is clear, by 4.6, since G contains no subdivision of the dyadic tree as m(G) is finite, and furthermore $m(\tau) \leq m(G) < n$ for every end τ of G.

5. Invariant simplices

In this section we will consider three particular kinds of graphs, the first and main one being that of *strongly dismantlable graphs*. We recall : first that a vertex x is *dominated by a vertex* y *in a graph* G if y is adjacent with x and with all neighbors of x in G; then that a finite graph

G is *dismantlable* if its vertices can be linearly ordered $x_0, ..., x_n$ so that, for each $i < n$, the vertex x_i is dominated by a vertex $y \neq x_i$ in the subgraph of G induced by $\{x_i, ..., x_n\}$. The concept of dismantlability can be straightforwardly extended to infinite graphs as follows:

5.1. Definition. A graph G is said to be *dismantlable* if there is a well-ordering $\leq$ on V(G) such that any vertex x which is not the greatest element of $(V(G), \leq)$, if such a greatest element exists, is dominated by some vertex $y \neq x$ in the subgraph of G induced by the set $\{z \in V(G) : x \leq z\}$.

As this extension seems too much general to get interesting results, we introduced in [24] the following restricted concept :

5.2. Definition. A graph G is *strongly dismantlable* (*subretract-collapsible* in [24]) if there is a well-ordering $\leq$ on V(G) with a greatest element m such that, for every vertex $x \neq m$, there is a strictly increasing finite sequence $x = x_0 < \ldots < x_n = m$ where, for $0 \leq i < n$, the vertex x_i is dominated by x_{i+1} in the subgraph of G induced by the set $\{z \in V(G) : x_i \leq z\}$.

Note that *any strongly dismantlable graph is connected and dismantlable*. Furthermore, by [23, Theorem 4.4], *any rayless connected dismantlable graph is strongly dismantlable*. Thus in particular *the finite strongly dismantlable graphs are the dismantlable ones*.

5.3. Theorem (Polat [24]). *If* G *is a strongly dismantlable graph, then, for any self-contraction* f *of* G, *there exists a vertex* x *of* G *such that* x *and* f(x) *are equal or adjacent*.

5.4. Theorem (Poston [28] and Bandelt with van de Vel [4] for finite graphs, Polat [24] for infinite graphs). *If a strongly dismantlable graph* G *contains no infinite simplices, then any self-contraction of* G *strictly stabilizes a non-empty finite simplex*.

This result clearly holds if G is rayless, since any rayless graph contains no infinite simplices, but this stronger condition is essential for the following.

5.5. Theorem (Polat [24]). *Any rayless strongly dismantlable graph* G *contains a non-empty* *finite simplex which is strictly invariant under every automorphism of* G.

5.6. Theorem (Polat [26]). *Let* G *be a rayless, block-finite, strongly dismantlable graph. If* $\mathfrak{F}$ *is a commuting family of self-contractions of* G, *then there exists a non-empty finite simplex which is strictly invariant under any element of* $\mathfrak{F}$.

Note that in this result the term "block-finite" is only used as a short way to say that the set of paths joining any two vertices of the graph is finite. No use of the block-cutpoint tree is made to prove 5.6. As a matter of fact, the image of a block by a contraction is not necessarily included in a block, so a self-contraction of a graph does not generally induce a self-contraction of the block-cutpoint tree of this graph. Moreover we do not yet know if this condition of "being block-finite" is necessary.

For different ways of dealing with those invariant simplex properties for dismantlable graphs, and in particular for the finite ones, see Bélanger et al. [5].

5.7. The second class of graphs that we deal with is that of *ball-Helly graphs*. A *connected* graph G is called a *ball-Helly graph* (resp. *strongly ball-Helly graph*), if every finite family (resp. every family) of pairwise non-disjoint balls of G has a non-empty intersection. Note that we proved with Pouzet that *if a ball-Helly graph contains no infinite simplices, then it is a strongly ball-Helly graph*. There exists a link between strongly ball-Helly graphs and strongly dismantlable ones, which is given in the following result :

5.8. Proposition (Polat [24]). *A strongly ball-Helly graph is strongly dismantlable if and only if it has no isometric rays.*

The combination of 5.8 with Theorems 5.3, 5.4, 5.5 and 5.6 gives similar fixed point theorems for (strongly) ball-Helly graphs. The analogues of Theorems 5.3 and 5.5 for *finite* ball-Helly graphs were proved by Quillot [30]. The following result, which is a simple consequence of a more general result for strongly dismantlable graphs (see [24]), and which is substantially due to Quillot [29], deals with infinite simplices instead of finite ones.

5.9. Theorem (Jawhari et al. [14]). *Any self-contraction* f *of a bounded strongly ball-Helly graph* G *stabilizes a non-empty simplex* (i.e., there exists a non-empty (possibly infinite) simplex S of G such that $f(S) \subseteq S$).

5.10. The last class of graphs that we consider in this section is that of *chordal graphs*, that is of graphs that contain no induced cycles of length greater than three. The rayless chordal graphs can be characterized by means of strongly dismantlable graphs as follows.

5.11. Proposition (Anstee and Farber [1] for finite graphs, Polat [25] for infinite graphs). *Let* G *be a connected rayless graph. Then* G *is chordal if and only if every connected induced subgraph of* G *is strongly dismantlable.*

The association of 5.11 with Theorems 5.3, 5.4, 5.5 and 5.6 gives analogous results for chordal graphs. But we can avoid the restriction of "being block-finite" in 5.6, to get the following more general result :

5.12. Theorem (Polat [26]). *Let* $\mathfrak{F}$ *be a commuting family of self-contractions of a connected, rayless, chordal graph* G. *Then there exists a non-empty finite simplex which is strictly invariant under any element of* $\mathfrak{F}$.

Note that, as any tree is chordal, Theorems 2.1 and 2.3 are immediate consequences of the corresponding results for chordal graphs.

6. Invariant hypercubes and Hamming graphs

6.1. Definition. Let (u_1, u_2, u_3) be a triple of vertices of a graph G. A *quasi-median* of (u_1, u_2, u_3) is a triple of vertices (x_1, x_2, x_3) such that

- x_i, x_j lie on a u_iu_j-geodesic, $i, j \in \{1, 2, 3\}$;
- $dist_G(x_1, x_2) = dist_G(x_2, x_3) = dist_G(x_3, x_1) = k$;
- k is minimal with respect to those conditions.

If $k = 0$, then the quasi-median is reduced to a single vertex x, which is called a *median* of the triple (u_1, u_2, u_3).

6.2. Definition. A *median graph* is a graph in which every triple of vertices has a unique median.

6.3. Definition. A graph G is *quasi-median* if

(i) each triple of vertices of G has a unique quasi-median;

(ii) $K_{1,1,2}$ (i.e., K_4 minus an edge) is not an induced subgraph of G;

(iii) the convex hull of any isometric 6-cycle of G is a 3-cube.

Note that the median graphs are the bipartite quasi-median graphs.

6.4. Definition. A *Hamming graph* (resp. *hypercube*) is a cartesian product of simplices (resp. K_2).

The regular median (resp. quasi-median) graphs are precisely the hypercubes (resp. Hamming graphs). Note that while the (finite or infinite) median graphs were proved to be the retracts of hypercubes (Bandelt [2]), so far only the finite quasi-median graphs were proved to be the retracts of Hamming graphs (Wilkeit [36]).

6.5. Theorem (Tardif [32]). *Any median graph without isometric rays contains a finite hypercube which is strictly invariant under every automorphism.*

More generally one has:

6.6. Theorem (Chastand [6] for finite graphs, Chastand and Polat [7] for infinite graphs). *Any quasi-median graph without isometric rays contains a Hamming graph which is strictly invariant under every automorphism.*

Note that in Theorem 6.6, the strictly invariant Hamming graph may be infinite since it may contains an infinite simplex.

6.7. Theorem (Bandelt and van de Vel [4] for finite graphs, Tardif [32] for infinite graphs).

Every self-contraction of a median graph containing no isometric rays strictly stabilizes a finite hypercube.

6.8. Theorem (Chastand [6] for finite graphs, Chastand and Polat [7] for infinite graphs). *Every self-contraction of a rayless quasi-median graph strictly stabilizes a finite Hamming graph.*

Tardif's result (6.7) is stronger than the restriction of Theorem 6.8 to median graphs, since a median graph without isometric rays may contain a ray. Furthermore Tardif proves the following:

6.9. Theorem (Tardif [32]). *Let* G *be a median graph containing no isometric rays. If* $\mathfrak{F}$ *is a commuting family of self-contractions of* G, *then there exists a finite hypercube which is strictly invariant under any element of* $\mathfrak{F}$.

We complete this section with a result of Bandelt and Mulder, similar to Theorem 6.7, concerning another generalization of median graphs, the pseudo-median graphs.

6.10. Definition. Let u, v, w be three vertices of a connected graph G. Then the triple x, y, z is a *pseudo-median* of the triple u, v, w, if

- $\mathrm{dist}_G(u, x) + \mathrm{dist}_G(x, y) + \mathrm{dist}_G(y, v) = \mathrm{dist}_G(u, v)$
- $\mathrm{dist}_G(v, y) + \mathrm{dist}_G(y, z) + \mathrm{dist}_G(z, w) = \mathrm{dist}_G(v, w)$
- $\mathrm{dist}_G(w, z) + \mathrm{dist}_G(z, x) + \mathrm{dist}_G(x, u) = \mathrm{dist}_G(w, u)$
- $0 \leq \mathrm{dist}_G(x, y) = \mathrm{dist}_G(y, z) = \mathrm{dist}_G(z, x) \leq 1$.

So a pseudo-median of u, v, w is either a singleton or a triangle "lying between" the pairs of u, v, w.

6.11. Definition. A graph is *pseudo-median* if every triple of vertices has a unique pseudo-median.

6.12. Theorem (Bandelt and Mulder [3]). *Every self-contraction of a finite pseudo-median graph* G *strictly stabilizes some regular pseudo-median subgraph of* G.

Note that the regular pseudo-median graphs are the cartesian products of hypercubes with either simplices or hyperoctahedra (i.e., complete graphs minus a perfect matching). We still do not know if the last result holds for certain class of infinite pseudo-median graphs, in particular for that of rayless pseudo-median graphs.

7. Finite invariant box in the cartesian product of metric spaces

We will conclude with two nice results of Tardif which, even if they do not directly concern graphs, seem to be the expected generalizations of Theorems 6.5–6.8 for finite median or quasi-median graphs.

7.1. Definition. The *cartesian product* $\Pi_{i\in I}(X_i, \delta_i)$ of a finite family $((X_i, \delta_i))_{i\in I}$ of metric spaces is the space (X, δ) where $X = \Pi_{i\in I} X_i$, and $\delta = \Sigma_{i\in I} \delta_i \circ (pr_i \times pr_i)$ (pr_i is the projection on X_i). A map $\varphi : X \to X$ is *non-expanding* if $\delta(\varphi(x), \varphi(y)) \leq \delta(x, y)$ for all $x, y \in X$. For $R \subseteq X$, an idempotent non-expanding map $\varphi : X \to X$ is a *retraction* on R; if such a map exists, R is a *retract* of X. A subset S of $X = \Pi_{i\in I} X_i$ is a *box* of X if $S = \Pi_{i\in I} pr_i(S)$.

The finite median graphs and quasi-median graphs are the retracts of hypercubes and of Hamming graphs, respectively, which are particular instances of cartesian product of metric spaces. So it seems quite natural to search for fixed box properties for the cartesian product of metric spaces, similar to those given by Theorems 6.5–6.8.

7.2. Definition. Let (X, δ) be a metric space.

(i) For $x, y \in X$, the *interval* between x and y is the set

$$I(x, y) := \{z \in X\ ;\ \delta(x, y) = \delta(x, z) + \delta(z, y)\}.$$

(ii) A set $A \subseteq X$ is a *prefiber* of (X, δ) if for any $x \in X$, there exists a (necessarily unique) $p_A(x) \in A$ such that, for all $y \in A$, $p_A(x) \in I(x, y)$. The map $p_A : X \to A$ is the *projection* on A.

7.3. Theorem (Tardif [33]). *Let* (X, δ) *be a retract of a finite metric space* $\Pi_{i\in I}(X_i, \delta_i)$. *Then there exists a prefiber* $B \subseteq X$ *such that* $B = \Pi_{i\in I} pr_i(B)$ *and* $\varphi(B) = B$ *for each* *automorphism* φ *of* X.

7.4. Theorem (Tardif [33]). *Let* (X, δ) *be a finite metric space, and let* $\varphi : X \to X$ *be a non-expanding map. Then there exists a set* $S \subseteq X$ *such that* $S = \prod_{i \in I} pr_i(S)$ *and* $\varphi(S) = S$.

8. Open problems

There are obviously many problems that remain to solve. We will only mention a few of them which directly pertain to some of the preceding results.

8.1. Is it possible to delete the condition of "being interval-finite" in Theorem 3.7, and that of "being block-finite" in Theorem 5.6? These two conditions are inherent to the kind of proofs we made, but we have no example showing that they are necessary.

8.2. To improve Theorem 6.8 for quasi-median graphs, and in order to generalize Theorem 6.7 of Tardif, one should try to prove: first, that every self-contraction of a quasi-median graph without isometric rays stabilizes a Hamming graph (which might be infinite, even without isometric rays, for example an infinite simplex); then, if the answer was negative, that every self-contraction of a quasi-median graph without isometric rays and infinite simplices strictly stabilizes a finite Hamming graph.

8.3. Tardif showed in [32] that the lack of isometric rays in a median graph G corresponds to the condition of being compact for the space V(G) endowed with the topology generated by the convex sets as a subbase of closed sets. So, in order to get thorough generalizations of Theorems 6.5 and 6.7, it would be interesting to see if Theorems 7.3 and 7.4 still hold for compact metric spaces.

8.4. Is there a property analogous to Theorems 6.5 and 6.6 for the pseudo-median graphs? A much more interesting problem would be to search for invariant subgraph results that would be general enough to include median, quasi-median and pseudo-median graphs together.

8.5. Another generalization similar to the preceding one is also possible. Note that there is a strong similarity between strongly ball-Helly graphs and median or quasi-median graphs. As a matter of fact the strongly ball-Helly graphs are the retracts of the strong products of paths (we recall that the strong product is the product in the category of graphs with the retractions as morphisms), while the median (resp. quasi-median) graphs are the retracts of the cartesian powers of K_2 (resp. cartesian products of simplices). Furthermore, the simplices are the regular strongly ball-Helly graphs without isometric rays, while the hypercubes (resp. Hammimg graphs) are the regular median (resp. quasi-median) graphs. Therefore, it might be natural to search for some general results which would combine the invariant subgraph properties of these different classes of graphs.

8.6. In the statements, of Theorems 2.3, 3.7, 5.6, 5.12 and 6.9, a "commuting family" can be clearly replaced by an "abelian semigroup". Thus, since any abelian semigroup is left-amenable, it would be interesting to see if those results could be generalized by considering "left-amenable semigroups of self-contractions" instead of "commuting families of self-contractions", just as did Day [8] with the Markov-Kakutani Theorem.

References

[1] R.P. Anstee and M. Farber, On bridged graphs and cop-win graphs. *J. Combinatorial Theory Ser. B* **44** (1988), 22-28.

[2] H.J. Bandelt, Retract of hypercubes. *J. Graph Theory* **8** (1984), 501-510.

[3] H.J. Bandelt and H.M. Mulder, Regular pseudo-median graphs. *J. Graph Theory* **12** (1988), 533-549.

[4] H.J. Bandelt and M. van de Vel, A fixed cube theorem for median graphs. *Discrete Math.* **67** (1987), 129-137.

[5] M. F. Bélanger, J. Constantin et G. Fournier, Graphes et ordonnés démontables. *Discrete Math.* **130** (1994),

[6] M. Chastand, Graphes de Hamming invariants dans les graphes quasi-médians. Papier de recherche, 1992, n° 40 - IAE.

[7] M. Chastand and N. Polat, Invariant Hamming graphs in infinite quasi-median graphs. Papier de recherche, 1992, n° 42 - IAE.

[8] M.M. Day, Fixed-point theorems for compact convex sets. *Ill. J. Math.* **5** (1961), 585-590.

[9] H. Freudenthal, Über die Enden diskreter Räume und Gruppen, *Comment. Math. Helv.* **17** (1944), 1-38.

[10] R. Halin, Über unendliche Wege in Graphen. *Math. Ann.* **157** (1964), 125-137.

[11] R. Halin, Automorphisms and endomorphisms of infinite locally finite graphs. *Abh. Math. Sem. Univ. Hamburg* **39** (1973), 251-283.

[12] R. Halin, Fixed configurations in graphs with small number of disjoint rays. *Contemporary methods in graph theory*, 639-649, Bibliographisches Inst., Mannheim, 1990.

[13] H. Hopf, Enden offener Raüme und unendliche diskontinuierliche Gruppen. *Comm. Math. Helv.* **15** (1943), 27-32.

[14] E. Jawhari, D. Misane, and M. Pouzet, Retracts: Graphs and ordered sets from the metric point of view, *in "Combinatorics and ordered sets"*, Contemp. Math., Vol. **57**, p p. 175-226, Amer. Math., Providence, RI, 1986.

[15] H.A. Jung, Wurzelbäume und unendliche Wege in Graphen. *Math. Nachr.* **41** (1969), 1-22.

[16] H.A. Jung, Connectivity in Infinite Graphs, in: *Studies in Pure Mathematics*, pp. 137-143. Editor L. Mirsky. New York - London: Academic Press 1971.

[17] S. Kakutani, Two fixed-point theorems concerning bicompact convex sets. *Proc. Imp. Acad. Jap.* **14** (1938), 242-245.

[18] A.A. Markov, Quelques théorèmes sur les ensembles abéliens. *C. R. (Doklady) Acad. Sci. URSS (N. S.)* **1** (1936), 311-313.

[19] R. Nowakowski and I. Rival, Fixed edge theorem for graphs with loops. *J. Graph Theory* **3** (1979), 339-350.

[20] N. Polat, Aspects topologiques de la séparation dans les graphes infinis. I. *Math. Z.* **165** (1979), 73-100.

[21] N. Polat, Topological aspects of infinite graphs, in: *Cycles and Rays* (G. Hahn et al., eds), NATO ASI Ser. C, Kluwer Academic Publishers, Dordrecht 1990, pp. 197-220.

[22] N. Polat, Finite invariant sets in infinite graphs. Papier de recherche, 1992, n° 30 - IAE.

[23] N. Polat, Finite invariant simplices in infinite graphs. *Period. Math. Hungar.* **27** (1993), 125-136.

[24] N. Polat, Retract-collapsible graphs and invariant subgraph properties. To appear in *J. Graph Theory.*

[25] N. Polat, A Helly theorem for geodesic convexity in strongly dismantlable graphs. To appear in *Discrete Math.*

[26] N. Polat, Stable graphs for a family of endomorphisms. Papier de recherche, 1993, n° 35 - IAE.

[27] N. Polat and G. Sabidussi, Fixed elements of infinite trees. *Discrete Math.* **130** (1994), 97-102.

[28] T. Poston, Fuzzy geometry, Ph.D. thesis, University of Warwick (1971).

[29] A. Quilliot, Homomorphismes, points fixes, rétractions et jeux de poursuite dans les graphes, les ensembles ordonnés et les espaces métriques. Thèse de doctorat d'Etat, Univ. Paris VI (1983).

[30] A. Quilliot, On the helly property working as a compactness criterion on graphs. *J. Combin. Theory Ser. A* **40** (1985), 186-193.

[31] R. Schmidt, Ein Ordnungsbegriff für Graphen ohne unendliche Wege mit einer Anwendung auf n-fach zusammenhängende Graphen. *Arch. Math.* **40** (1983), 283-288.

[32] C. Tardif, On compact median graphs. Preprint, 1993.

[33] C. Tardif, A fixed box theorem for the cartesian product of graphs and metric spaces. Preprint, 1993.

[34] A. Tarski, A lattice theoretical fixed point theorem and its application. *Pacific J. Math.* **5** (1955), 285-309.

[35] J. Tits, Sur le groupe des automorphismes d'un arbre. *Essays on topology and related topics* (Mémoires dédiés à G. de Rham), Springer (1970), 188-211.

[36] E. Wilkeit, The retracts of Hamming graphs. *Discrete Math.* **102** (1992), 197-218.

Multiplier Conjecture and Multiplier Theorems for Non-Abelian Difference Sets*

Qiu Weisheng

(*Department of Mathematics, Peking University, Beijing 100871*)

Abstract In this paper we survey the results on the Multiplier Conjecture and multiplier theorems for non-abelian difference sets, including our method and results.

Key words difference sets, multipliers, multiplier theorems, multiplier conjecture, non-abelian difference sets, representation theory of finite groups, algebraic number theory.

§1. Introduction

Definition 1. Let G be a group of order v, written multiplicatively, and D a k-subset of G, with $1 < k < v-1$. Then D is called a (v,k,λ)-*difference set* if the list of "differences" xy^{-1} ($x, y \in D$) contains each nonidentity element of G exactly λ times. If G is abelian (resp. cyclic), then D is also called *abelian* (resp. *cyclic*). Set $n = k - \lambda$.

A symmetric (v,k,λ)-design with a regular automorphism group G is equivalent to a (v,k,λ)-difference set in G.

The most fundamental question about symmetric designs is:

Which triples (v,k,λ) are the parameters of a symmetric design?

Necessary conditions of existing a symmetric (v,k,λ)-design are as follows:

1. $\lambda(v-1) = k(k-1)$.
2. **(Schutzenherger theorem)** *If v is even, then n must be a square.*
3. **(Bruck-Ryser-Chowla theorem)** *If v is odd, then the equation*

$$nx^2 + (-1)^{\frac{v-1}{2}}\lambda y^2 = z^2$$

*The project supported by National Natural Science Foundation of China.

must have a solution in integers x, y, z, not all zero.

If a triple (v, k, λ) satisfies the above conditions, then we say "putative (v, k, λ)-difference set".

In 1947 Hall[4] introduced the concept of *multipliers* for cyclic (v, k, λ)-difference sets with $\lambda = 1$. This is quite useful in constructing or proving the nonexistence of a putative (v, k, λ)-difference set.

Definition 2. Let D be a difference set in a finite group G. An automorphism τ of G is called a *multiplier* of D if τ sends D onto Da for some $a \in G$. If G is abelian and if τ is given by $g \longmapsto g^t$, where t is an integer which is prime to the order of G, then τ is called a *numerical multiplier* of D. By abuse of language t itself is called a numerical multiplier.

It is easily seen that all multipliers of a difference set D in a finite group G form a group, it is called the full multiplier group of D. If G is abelian, then all numerical multipliers of D form a subgroup of the full multiplier group. If G is cyclic then every multiplier is numerical.

To apply multipliers to constructing or proving the nonexistence of a putative (v, k, λ)-difference set D, we need the following result:

Theorem 1.1 (McFarland and Rice[10], 1978). *Let D be an abelian (v, k, λ)-difference set in G. Then there exists a translate Dg of D that is fixed by every numerical multiplier of D.*

When considering a putative abelian (v, k, λ)-difference set D, by the theorem 1.1 we may assume that D is fixed by every numerical multiplier (otherwise, we replace D by a translate having this property). Obviously, D then must be the union of suitable orbits on G under any group M of numerical multipliers, and hence k is a sum of the lengths of these orbits.

Multiplier theorems may help us to find numerical multipliers.

First Multiplier Theorem . *Let D be an abelian (v, k, λ)-difference set, and let p be a prime dividing n but not v. If $p > \lambda$, then p is a numerical multiplier of D.*

The First Multiplier Theorem was first proved by Hall[4] for $\lambda = 1$ in the cyclic case, and later on, by Mann[8], Lander[6], Pott[15], and Qiu[16].

The following more general version is called Second Multiplier Theorem, it is due to Menon[12]. Somewhat simplified proofs appear in Mann[10].

Second Multiplier Theorem . *Let D be a (v, k, λ)-difference set in an abelian group G of order v, and let v_0 be the exponent of G (i.e. the least common* *multiple of the orders of the elements of G). Let n_1 be a divisor of n such that $(n_1, v) = 1$, and $n_1 > \lambda$. Suppose that t is an integer such that for every prime divisor p of n_1, there exists a nonnegative integer j with $t \equiv p^j \pmod{v_0}$. Then t is a numerical multiplier of D.*

We now give a example showing how to apply multiplier theorems to constructing or proving the nonexistence of a putative abelian (v, k, λ)-difference set D in G. If t is a numerical multiplier of D, then the multiplier group $M = < t >$ acts on G, and the length of the orbit Mx of an element x of order r in G equals the order of t modulo r. By the theorem 1.1 we may assume that D is fixed by M. Thus D must be the union of suitable orbits on G under M, and hence k is a sum of the lengths of these orbits.

Example 1. We seek a $(23, 11, 5)$-difference set D (its existence is guaranteed by Theorem 1.10 of Chapter VI in [2]). Since 23 is a prime, we necessarily have $G = Z_{23}$. Here $n = 6$. First Multiplier Theorem does not work since $2 < \lambda$ and $3 < \lambda$. Take $n_1 = n = 6$ and $t = 2$ in Second Multiplier Theorem. Since $3^7 \equiv 2 \pmod{23}$, thus all the hypotheses of Second Multiplier Theorem are satisfied. Hence 2 is a numerical multiplier of D. The multiplier group $M = < 2 >$ acts on Z_{23}. Since the order of 2 modulo 23 is 11, the length of the orbit Mx of x ($x \in Z_{23}$ and $x \neq 0$) equals 11. Since $k = 11$, the possible choice for D is $\{x, 2x, 4x, 8x, 16x, 9x, 18x, 13x, 3x, 6x, 12x\}$. Thus, up to equivalence, D is unique: $\{1, 2, 4, 8, 16, 9, 18, 13, 3, 6, 12\}$. It is easily checked that this is indeed a $(23, 11, 5)$-difference set.

All the known proofs of First Multiplier Theorem use the condition "$p > \lambda$" very crucially. However, for example, 2 is a numerical multiplier of $(23, 11, 5)$-difference set D in the example 1. In fact no example is known showing that this restriction is necessary. The following has been conjectured.

Multiplier Conjecture . *The First Multiplier Theorem holds without the assumption that $p > \lambda$.*

If the Multiplier Conjecture is true, then every divisor of n which is prime to v is a numerical multiplier of a (v, k, λ)-difference set D.

The example 1 shows that a stronger multiplier theorem can help us to find more multipliers, and help us to construct or prove the nonexistence of more putative (v, k, λ)-difference set. Hence studying the Multiplier Conjecture is very interesting.

Virtually all further multiplier theorem have arisen in an attempt to weaken the condition $p > \lambda$. The Second Multiplier Theorem may be viewed as a result where the condition $p > \lambda$ is replaced by $n_1 > \lambda$.

§2. The Results on the Multiplier Conjecture

In the present section, let D be an abelian (v, k, λ)-difference set of G; let n_1 be a divisor of $n = k - \lambda$ such that $(n_1, v) = 1$; Suppose that t is an integer such that for every prime divisor p of n_1, there exists a nonnegative integer j with $t \equiv p^j$ (mod v_0), where v_0 is the exponent of G.

§2.1 The Multiplier Conjecture for the case $n = 2p^r$ or $n = 2n_1$

Theorem 2.1 (Newman[13], 1963) . *If $n = 2p$, where p is an odd prime, and G is a cyclic group, then in the First Multiplier Theorem, the assumption $p > \lambda$ can be replaced by $(v, 7) = 1$.*

A little later, Turyn proved:

Theorem 2.2 (Turyn[25], 1964) . *If $n = 2p^r$, where p is an odd prime, then in the First Multiplier Theorem, the assumption $p > \lambda$ can be replaced by r is odd.*

It follows that

Corollary 2.1 (Turyn[25], 1964) . *If $n = 2p^r$, p is an odd prime and $7|v$, then in the First Multiplier Theorem, the assumption $p > \lambda$ can be replaced by "p is a quadratic residue modulo 7 ".*

Similarly, Turyn stated the following result:

Theorem 2.3 (Turyn[25], 1964) . *Let $n = 2n_1$, n_1 odd, and $7|v$. If t is a quadratic residue modulo 7, then the Second Multiplier Theorem holds without the assumption $n_1 > \lambda$.*

In 1990 Qiu Weisheng[17] gave a proof for the theorem 2.3.

The case where 7 divides v and t is not a quadratic residue modulo 7 has be considered but not completely settled by Mann and Zaremba.

Theorem 2.4 (Mann and Zaremba[9], 1969) . *Let $n = 2n_1$, n_1 odd. If t is a quadratic nonresidue modulo 7, and $7||v$, and $(\lambda, n_1) = 1$, and $n > \lambda$, then the Second Multiplier Theorem holds without the assumption $n_1 > \lambda$.*

Theorem 2.5 (McFarland[11], 1970). *Let $n = 2n_1$, n_1 odd. If $(v, 2 \cdot 7) = 1$, then the Second Multiplier Theorem holds without the assumption $n_1 > \lambda$.*

Theorem 2.6 (Qiu[20], 1993). *Let $n = 2n_1$, n_1 odd. then the Second Multiplier Theorem holds without the assumption $n_1 > \lambda$, provided that one of the following conditions holds:*

(1) $7^2|v$;

(2) *v has a prime divisor p which isn't congruent to 1 and -1 modulo 8;*

(3) *v has a prime divisor $p \equiv 1 \pmod 8$ or $p \equiv -1 \pmod 8$, such that $Ord_p(t)$ is even, where $Ord_p(t)$ denotes the order of t modulo p.*

§2.2 The Multiplier Conjecture for the case $n = 3n_1$

Theorem 2.5 (McFarland[11], 1970). *Let $n = 3n_1$. If $(v, 2\cdot3\cdot11\cdot13) = 1$, then the Second Multiplier Theorem holds without the assumption $n_1 > \lambda$.*

Theorem 2.7 (Qiu[21], 1993). *Let $n = 3n_1$. If $(v, 3) = 1$, then the Second Multiplier Theorem holds without the assumption $n_1 > \lambda$, provided that one of the following conditions holds:*

(1) $3|n_1$;

(2) *$(3, n_1) = 1$, and v has a prime divisor p which isn't congruent to 1 and -1 modulo 12;*

(3) *$(3, n_1) = 1$, and v has a prime divisor $p \equiv 1 \pmod{12}$ such that $Ord_p(3)$ does not divide $\frac{p-1}{4\cdot 2^c}$, where $4 \cdot 2^c || p - 1$;*

(4) *$(3, n_1) = 1$, and $(v, 11 \cdot 13) = 1$;*

(5) *$(3, n_1) = 1$, and v has a prime divisor $p \equiv 1 \pmod{12}$ or $p \equiv -1 \pmod{12}$ such that $Ord_p(t)$ is even, where $p \neq 11, 13$;*

(6) *$(3, n_1) = 1$, and $(v, 11) = 1$, and $13^2|v$;*

(7) *$(3, n_1) = 1$, and $(v, 11) = 1$, and $4|Ind_{13}(t)$, where $Ind_{13}(t)$ denotes the index of t modulo 13;*

(8) *$(3, n_1) = 1$, and $11^2|v$, and t is a quadratic nonresidue modulo 11;*

(9) *$(3, n_1) = 1$, and $(v, 13) = 1$ or $13^2|v$ or $4|Ind_{13}(t)$, and one of the following conditions holds:*

(9.1) *$t \equiv 3^j \pmod{11^e}$, where 11^e is the exponent of Sylow 11-subproup of G;*

(9.2) *The Sylow 11-subgroup of G is an elementary abelian group.*

Example 2. Using theorem 2.7 we can exclude the existence of a putative abelian $(299, 149, 74)$-difference set D. Here $n = 75 = 3n_1$, and $n_1 = 25$. Take $t = 5$. Since $Ord_{23}(5) = 22$, by theorem 2.7, 5 is a numerical multiplier of D. Consider the multiplier group $M =< 5 >$. We can assume that D is fixed by M. Since $v = 13\cdot23$, thus G is a cyclic group. Let $a, b, c \in G$, and $|a| = 13, |b| = 23, |c| = 299$. Since $Ord_{13}(5) = 4, Ord_{299}(5) = 44$, thus $|M(a)| = 4, |M(b)| = 22, |M(c)| = 44$.

It is easy to see that $1+4x+22y+44z=149$ and $4x+22y+44z=149$ are impossible. Hence no abelian $(299,149,74)$-difference set exists.

Remark.

In the example 2 the Second Multiplier Theorem does not apply since only possible case which satisfies the assumption $n_1>\lambda$ is $n_1=n=75$, in this case

$$\{t|t\equiv 3^j(\mod 299), j\geq 0\}\cap\{t|t\equiv 5^j(\mod 299), j\geq 0\}=<27>,$$

thus $<27>$ is a multiplier group of D. Using this multiplier group can no exclude the existence of D.

In the example 2 the McFarland's multiplier theorem does not work since $13|v$. Using the Mann Test or Lander's theorem 4.4 and theorem 4.5, it is impossible to rule out the existence of the above difference set since $5^2||n$ and no nonnegative integer j satisfies $5^j\equiv -1(\mod 299)$.

§2.3 The Multiplier Conjecture for the case $n=4n_1$

Theorem 2.5 (McFarland[11], 1970). *Let $n=4n_1$. If $(v,2\cdot 3\cdot 7\cdot 31)=1$, then the Second Multiplier Theorem holds without the assumption $n_1>\lambda$.*

Theorem 2.8 (Qiu[22], 1993). *Let $n=4n_1$. If $(v,2\cdot 7\cdot 31)=1$, then the Second Multiplier Theorem holds without the assumption $n_1>\lambda$, provided that one of the following conditions holds:*

(1) $2|n_1$;

(2) 15 *does not divide v;*

(3) $15|v$ *and n_1 is odd, and $t\equiv 2^m(\mod 3^a5^b)$ or $-1\equiv 2^m(\mod 3^a5^b)$ for some nonnegative integer m, where 3^a5^b is the exponent of P_1P_2, P_1 and P_2 are the Sylow 3-subgroup and Sylow 5-subgroup of G, respectively.*

§2.4 The Multiplier Conjecture for general cases

Theorem 2.5 (McFarland[11],1970). *The Second Multiplier Teorem remains true if the assumption "$n_1>\lambda$" is replaced by $(v,M(\frac{n}{n_1}))=1$, where $M(m)$ is an integral valued function on the set of positive integers: $M(1)=1$, $M(2)=2\cdot 7$, $M(3)=2\cdot 3\cdot 11\cdot 13$, $M(4)=2\cdot 3\cdot 7\cdot 31$; For $m\geq 5$, $M(m)$ is the product of the distinct prime factors of the numbers*

$$m, M(\frac{m^2}{p^{2e}}), p-1, p^2-1, \cdots, p^u-1,$$

where p is s prime dividing m with $p^e||m$, and $u=\frac{m^2-m}{2}$.

Theorem 2.9 (Lander[6], 1983). *The Second Multiplier Teorem remains true if the condition $t \equiv p^j (\mod v_0)$ is replaced by " $t \equiv p^j (\mod v_0)$ or $p^j \equiv -1 (\mod v_0)$ ".*

Theorem 2.10 (Qiu[19], 1992) . *If $n = dn_1$, d a prime not dividing n_1, and G is an elementary abelian group of order $v = p^s (s > 1)$, and the order w of d mod p is greater than 1, then the Second Multiplier Theorem holds without the assumption $n_1 > \lambda$, provided that one of the following conditions holds:*

(i) $w > d^2$;

(ii) $\frac{d^2}{3} < w \leq d^2$, *and* $\frac{p-1}{w} = 2$ *or* 4;

(iii) $\frac{d^2}{3} < w \leq d^2$, *and* $\frac{p-1}{2w} \geq 3$, *and t is a quadratic nonresidue mod p;*

(iv) $\frac{d^2}{3} < w \leq d^2$, *and* $\frac{p-1}{2w} \geq 3$, *and* $t \equiv x^{\frac{p-1}{2w}j}$ (*mod* p) $\quad (1 \leq j < 2w)$,

where x is a primitive root of p.

§3 The Method of Studying the Multiplier Conjecture

In 1992 Qiu[18] presented a method of studying the Multiplier Conjecture, using the representation theory of finite groups and the algebraic number theory. It consists of three parts.

§3.1 A necessary and sufficient condition for a multiplier

Let G be an abelian group of order v, and let d be a positive integer. In the group ring ZG the equations

$$\sum_{g \in G} c_g = d, \tag{1}$$

$$\xi \xi^{(-1)} = d^2 \cdot 1, \tag{2}$$

where $\xi = \sum_{g \in G} c_g g$, are called CH-equations of d for G. Obviously $\xi = dg (\forall g \in G)$ is a solution of the CH-equations of d for G. They are called trivial. The other solutions are called nontrivial. If a nontrivial solution ξ of the CH-equations of d for G has the form

$$\xi = g_j \sum_{i=1}^{m} c_i g_{l_i}, \quad c_i \neq 0 \quad (1 \leq i \leq m),$$

and $g_{l_i} (i = 1, \cdots, m)$ are in a cyclic group $< g_u >$ (g_j not necessarily in $< g_u >$), then ξ is called a cyclic solution; If the order of g_u is a prime p, then ξ is called a p-solution.

Remark. If $\xi = \sum_{g\in G} c_g g \in ZG$ satisfies (2), then $\sum_{g\in G} {c_g}^2 = d^2$.

Theorem 3.1. *Let D be a (v,k,λ)-difference set in an abelian group G, and let v_0 be the exponent of G. Let $n = dn_1 (n_1 > 1)$ and $(n_1, v) = 1$. Suppose that t is an integer co-prime with v such that for every prime divisor p of n_1, there exists a nonnegative integer j such that $p^j \equiv t(\mod v_0)$. Then t is a numerical multipler of D if and only if no nontrivial solution ξ of the CH-equations of d for G satisfies*

$$D^{(t)}D^{(-1)} = n_1\xi + \lambda G. \tag{3}$$

§3.2 A Method of Finding Solutions of the CH-Equations of q^e for G

Lemma 1. *Let G be an abelian group of order v. Let q be a prime, and $(q,v) = 1$. Suppose that $\xi \in ZG$ is a solution of the CH-equations of q^e for G, where $e \geq 1$. Then there exists $g_r \in G$ such that*

$$\xi^{(q)} = \xi g_r. \tag{4}$$

Remark. Let G be an abelian group of order v. Let q be a prime and $(q,v) = 1$. Given $g_j \in G$, set

$$\tau_{q,j}(g) := g^q {g_j}^{-(q-1)}, \quad \forall g \in G,$$

then $\tau_{q,j}$ is a permutation of the set G. A cycle of the decomposition into disjoint cycles of $\tau_{q,j}$ is called an orbit of $\tau_{q,j}$. We denote the orbit containing g by $Orb(g)$.

Lemma 2. *Let G be an abelian group of order v. Let q be a prime, and $(q,v) = 1$. Let $\xi = \sum_{l=0}^{v-1} c_l g_l \in ZG$. Suppose that there exists $g_r \in G$ such that $\xi^{(q)} = \xi g_r$. If $(q-1,v) = 1, (q^2-1,v) = 1, \cdots, (q^u-1,v) = 1$, let $g_j^{q-1} = g_r$, then for any $g_i (i \neq j)$ we have $Orb(g_i) = (g_i g_{i_1} g_{i_2} \cdots g_{i_u} \cdots)$. Thus $|Orb(g_i)| \geq u+1$, and*

$$c_i = c_{i_1} = c_{i_2} = \cdots = c_{i_u},$$

where $i, i_1, \cdots, i_u$ are pairwise distinct nonnegative integers.

Lemma 3. *Let G be an abelian group of order v. Let q be a prime and $(q,v) = 1$. Given $g_j \in G$. Consider $\tau_{q,j}$.*

If $|Orb(g_{l_1})| = s (s > 1)$, then there exists $g_u \in G$ such that $g_{l_i} = {g_u}^{q^{i-1}} g_j (1 \leq i \leq s)$ and the order m of g_u satisfies $Ord_m(q) = s$.

Conversely if p is a prime divisor of v and $Ord_p(q) = s$ $(s > 1)$. Then there exists an orbit of $\tau_{q,j}$ which length equals s.

§3.3 On Conditions for a Nontrivial Solution ξ Not Satisfying the Equation (3)

Using representation theory of finite groups and algebraic number theory we can find some conditions for a nontrivial solution ξ not satisfying the equation (3).

§4 Multiplier Theorems For Non-Abelian Differencs Sets

Very little is known about multipliers of non-abelian difference sets. Until 1993 the only known result is due to Ott([14], 1981) and Lander([6], 1983).

In 1981 Ott[14] introduced the concept of *weak multipliers*:

Definition 3. Let D be a difference set in a finite group G. We say that an integer t relatively prime to $|G|$ is a *weak multiplier* of D if the set $D^{(t)} := \{x^t | x \in D\}$ is a left translate gD of D.

We say that D is central (Ott called D invariant) if $gDg^{-1} = D$ for all $g \in G$.

Theorem 4.1 (Ott, 1981). *Let D be an invariant (v, k, λ)-difference set of a finite group G. Let e be the exponent of G. Let $n = p_1^{r_1} \cdots p_s^{r_s} > \lambda$, where p_i's are pairwise distinct prime divisors of n, and $(n, v) = 1$. Then every integer $t \equiv p_i^{e_i}($ mod $e)$ for suitable $e_i \geq 1$, $(i = 1, \cdots, s)$, is a weak multiplier.*

In 1983 Lander[6] also proved the above theorem, and he replaced the condition $n > \lambda$ by the condition $n_1 > \lambda$, where n_1 is a divisor of n and $(n_1, v) = 1$.

In 1994 we generalize our method of studying the Multiplier Conjecture to non-abelian case. We give a necessary and sufficient condition for a weak multiplier, i.e., theorem 4.2. We state and prove two multiplier theorems for central difference sets of non-abelian groups, i.e., theorem 4.3 and theorem 4.4. These improve Ott and Lander's results in the following ways:

(1) We replace the condition "$t \equiv p_i^{e_i}($ mod $e)$" by the condition "either $p_i^{e_i} \equiv t($ mod $e)$ or $p_i^{e_i} \equiv -1($ mod $e)$".

(2) We replace the condition "$n_1 > \lambda$" by the condition "$n_1 > \lambda$ or $(v, M(\frac{n}{n_1})) = 1$", where $M(\frac{n}{n_1})$ is defined as in theorem 2.5.

Theorem 4.2 (Qiu[24],1994). *Let D be a central (v, k, λ)-difference set in a finite group G, and let v_0 be the exponent of G. Let $n = dn_1 (n_1 > 1)$ and $(n_1, v) = 1$. Suppose that t is an integer co-prime with v such that for every prime divisor p of n_1, there exists a nonnegative integer j such that either $p^j \equiv t($ mod $v_0)$ or $p^j \equiv -1($ mod $v_0)$. Then t is a weak multipler of D if and only if no* *nontrivial solution ξ of the CH-equations of d for G satisfies*

$$D^{(t)}D^{(-1)} = n_1\xi + \lambda G. \tag{20}$$

Theorem 4.3 (Qiu[24], 1994) *Let D be a central (v, k, λ)-difference set in a finite group G, and let v_0 be the exponent of G. Let $n_1 > \lambda$ be a divisor of n and $(n_1, v) = 1$. Suppose that t is an integer co-prime with v such that for every prime divisor p of n_1, there exists a nonnegative integer j such that either $p^j \equiv t(\mod v_0)$ or $p^j \equiv -1(\mod v_0)$. Then t is a weak multipler of D.*

Theorem 4.4 (Qiu[24], 1994). *The theorem 4.3 remains true if the assumption $n_1 > \lambda$ is replaced by the condition $(v, M(\frac{n}{n_1})) = 1$, where $M(1) = 1$, $M(2) = 2\cdot 7, M(3) = 2\cdot 3\cdot 11\cdot 13$, $M(4) = 2\cdot 3\cdot 7\cdot 31$; for $m \geq 5$, $M(m)$ is defined as in theorem 2.5.*

Remark. Theorem 4.4 generalizes McFarland and Lander's multiplier theorems to non-abelian case.

References

[1] K.T.Arasu, Recent results on difference sets. In: D.Ray-Chaudhuri Eds., *Coding theory and Design theory Part II*, Springer, New York, 1990, 1-23.

[2] T.Beth, D. Jungnickel, and H. Lenz, *Design theory.* Cambridge University Press, Cambridge, 1986.

[3] C. W. Curtis and I. Reiner, *Representation theory of finite groups and associated algebras.* Wiley-Interscience, New York, 1962.

[4] M.Hall.Jr, Cyclic projective planes. *Duke Math. J.* 14(1947), 1079-1090.

[5] D.Jungnickel, Difference sets. In: J.H.Dinitz and D.R.Stinson Eds., *Contemporary design theory: A collection of surveys*, Wiley-Interscience, New Youk, 1992.

[6] E.S.Lander, *Symmetric designs: An algebraic approach.* Cambridge University Press , Cambridge, 1983.

[7] R.L.Long, *Algebraic number theory*, Dekker, New York, 1977.

[8] H.B.Mann, *Addition theorems.* Wiley, New York, 1965.

[9] H.B.Mann and S.K.Zaremba, On multipliers of difference sets. *Illinois J. Math.* 13(1969), 378-382.

[10] R.L.McFarland and D.F.Rice, Translates and multipliers of abelian difference sets. *Proc.AMS.* 68(1978), 375-379.

[11] R.L.McFarland, On multipliers of abelian difference sets. Ohio State University, Ph.D.dissertation, 1970.

[12] P.K. Menon, Difference sets in abelian groups. *Proc. Amer. Math. Soc.* 11(1960), 368-376.

[13] M. Newman, Multipliers of difference sets. *Canad. J. Math.* 15(1963), 121-124.

[14] U. Ott, Some remarks on representation theory in finite geometry, In: M. Aigner and D. Jungnickel, Eds., *Geometrices and Groups*, Lecture Notes in Mathematics, 893, (1981), 68-110.

[15] A.Pott, Applications of the DFT to abelian difference sets. *Arch. Math.* 51(1988), 283-288.

[16] Qiu Weisheng, Proving the multiplier theorem using the representation theory of groups. *Northeastern Math.J.* 9(1993), 169-172.

[17] Qiu Weisheng, On the multiplier conjecture. *Acta Mathematica Sinica, New Series.* 10(1994), 49-58.

[18] Qiu Weisheng, A Character approach to the multiplier conjecture and a new result on it. Submitted.

[19] Qiu Weisheng, The multiplier conjecture for elementary abelian groups. *J. Combinatorial Designs.* 3(1994), 117-129.

[20] Qiu Weisheng, Further results on the multiplier conjecture for the case $n = 2n_1$ (I). Submitted.

[21] Qiu Weisheng, Further results on the multiplier conjecture for the case $n = 3n_1$ (II). Submitted.

[22] Qiu Weisheng, The multiplier conjecture for the case $\frac{n}{n_1} = 4$. Submitted.

[23] Qiu Weisheng, New progress on the multiplier conjecture, J.Statistical Planning and Inference (Special issue), Proceedings of the Shanghai Conference On Designs, Codes and Finite Geometries(1993).

[24] Qiu Weisheng, Multiplier theorems for central difference sets of non-abelian groups, Submitted.

[25] R.J.Turyn, The Multiplier theorm for difference sets. *Canad. J. Math.* 16(1964), 386-388.

Counting Digraphs with Restrictions on the Strong Components

Robert W. Robinson*
Computer Science Department
415 Graduate Studies Building
The University of Georgia
Athens, GA 30602-7404, U.S.A.

Abstract

The principle of inclusion-exclusion is specialized in order to count labeled digraphs with separately specified out-components, in-components, and isolated components. Applications include counting digraphs with no in-nodes or out-nodes, digraphs with a source and a sink, and digraphs with a unique source and a unique sink.

*Research partially supported by the National Science Foundation under grant MCS-8302282.

1 Introduction

Standard definitions and terminology for graph theoretic concepts may be found in the book [1] by Harary. Two nodes are in the same *strong component* of a digraph D if each can be reached from the other by directed paths in D. It is a standard result that the strong components of any digraph D partition the nodes.

The groundwork for counting labeled digraphs with restricted strong components is laid in [3], [4], [5], and [6]. The notation of [1] and [6] is followed herein except that *node* is now preferred to point or vertex. Strong components are henceforth referred to simply as *components*. An *out-component* is a component to which no other component is adjacent. An *out-node* is a node to which no other node is adjacent. Clearly a node is an out-node if and only if it is an out-component. *In-component* and *in-node* are the dual notions. It is well known that every nonempty digraph contains at least one out-component and at least one in-component. A component which is both an out-component and an in-component is said to be *isolated*.

A node is a *source* if every other node is reachable from it by a directed path. The dual is a *sink*. Clearly the component of a source must be an out-component, and there can be no other out-component. Conversely, if a digraph, δ, contains just one out-component γ, then every node of γ is a source of δ. Of course the dual statements serve to characterize sinks as the nodes in a unique in-component.

2 A Specialization of Inclusion-Exclusion

Let U be a finite or countable set and ω a countably additive mapping

$$\omega : U \to M$$

where M is an additive group. In our applications, U will be some set of labeled digraphs, and M will be the ring Q of rational numbers or else some ring of power series over Q, such as $Q[[x]]$.

For any $A \subseteq U$, let

$$\omega(A) = \sum_{\chi \epsilon A} \omega(\chi).$$

We call $\omega(A)$ the *enumerator* for A, since it will always be some sort of numerical count or generating function in our applications.

In this setting, we can formulate the ordinary principle of inclusion-exclusion as follows. A set of properties (subsets) of U are given, say $P_1, \ldots, P_m$. The set $E_0 = U - P_1 - \ldots - P_m$ is the focus of interest. For $I \subseteq \{1, \cdots, m\}$, let $\omega(I) = \omega\left(\bigcap_{j\in I} P_j\right)$ with the understanding that if $I = \emptyset$ then the intersection is taken to be U. Then

$$\omega(E_0) = \sum (-1)^{|I|}\omega(I) \tag{1}$$

where the sum ranges over all subsets I of $\{1, \cdots, m\}$.

The proof is elementary. Because ω is additive on subsets of U, we can view each term in equation (1) as being the sum of contributions $\omega(\chi)$ for those χ in the relevant set. Suppose that χ is contained in exactly k of the sets $P_1, \cdots, P_m$ (so $0 \leq k \leq m$). The contribution of χ to $\omega(E_0)$ is just $\omega(\chi)$ if $k = 0$ and 0 if $k \geq 1$. Now consider the contribution of χ to the right side of (1). The contribution of χ to $(-1)^{|I|}\omega(I)$ will be 0 if $\chi \in P_i$ for some $i \in I$, and will be $(-1)^{|I|}\omega(\chi)$ if $\chi \in P_i$ for all $i \in I$. Since there are exactly $\binom{k}{j}$ sets of j properties all of which contain χ, the sum of all these contributions is $\omega(I)\sum_{j=0}^{k}(-1)^j\binom{k}{j}$. Now the sum is $(1-1)^k = 0^k$, and so is 1 if $k = 0$ and 0 if $k \geq 1$, showing that the contribution of χ to the right side of (1) is the same as its contribution to the left side. Summing this fact over all χ in U proves (1).

Now we consider a specialization of the general fact (1) to a situation in which the properties to be excluded come in dual pairs P_i and Q_i for $i = 1, \ldots, m$. Of course $\omega(E_0)$ could be expressed as a sum over the 2^{2m} subsets of the $2m$ properties using (1). One way of doing this might be as a double sum, over subsets of $\{P_1, \cdots, P_m\}$ and subsets of $\{Q_1, \cdots, Q_m\}$. However in many of our applications the evaluation of ω applied to a set of properties will depend on just what dual pairs are contained in that set. In that case, the situation can be most conveniently expressed as an ordered triple (I, J, K) of disjoint subsets of $\{1, \ldots, m\}$, the properties in the set consisting of P_i for $i \in I$, Q_j for $j \in J$, and both P_k and Q_k for $k \in K$. Let $\omega(I, J, K)$ be the enumerator for the intersection of these properties. Since there are $|I|+|J|+2|K|$ properties corresponding to (I,J,K), (1) becomes

$$\omega(E_0)=\sum(-1)^{|I|+|J|}\omega(I,J,K) \tag{2}$$

where the sum in (2) is over sequences of disjoint subsets $I,J,K\subseteq\{1,\ldots,m\}$.

3 Digraphs having no in-nodes or out-nodes

As an introductory problem, consider the number $Y[p]$ of labeled digraphs on p nodes which have no out-node. Ordinary inclusion-exclusion can be applied with U the set of all $2^{p(p-1)}$ labeled digraphs on p nodes, and property $P_i (1\le i\le p)$ that the node labeled i is an out-node. Letting $\omega(\chi)=1$ for each digraph, it is clear that for $I\subseteq\{1,\ldots,p\}$ and $|I|=m$ we have

$$\omega(I)=2^{(p-1)(p-m)}. \tag{3}$$

This is because for each of the $p-m$ nodes not represented in I there are 2^{p-1} possible sets of arcs into it, and these choices are independent. Given these $p-m$ choices of arc sets, the digraph being counted by $\omega(I)$ is completely determined, as no arcs can be directed to any of the m nodes for which the labels lie in I. Now ordinary inclusion-exclusion (equation (1)) gives that

$$\begin{aligned} Y[p]=\omega(E_0) &= \sum_{m=0}^{p}(-1)^m\binom{p}{m}2^{(p-1)(p-m)} \\ &= (2^{p-1}-1)^p \end{aligned} \tag{4}$$

Of course, we could have seen this more directly by observing that the set of arcs incident to any given node must not be empty, so there are $2^{p-1}-1$ choices for that set of arcs. A digraph on p nodes having no out-node is determined uniquely by p such choices, which are independent of each other.

Now consider the problem of counting the number of $W[p]$ of labeled digraphs having no out-node and no in-node. Our set U, enumerator ω, and properties $P_i(1\le i\le p)$ can all stay the same. However in addition we need to exclude the properties $Q_i(1\le i\le p)$, where Q_i holds if the node labeled i is an in-node. It is easy to see that for a set L of properties, $\omega(L)$ does not depend simply on $|L|$, or even just on the cardinalities of $L\cap\{P_1,\cdots,P_p\}$ and $L\cap\{Q_1,\cdots,Q_p\}$. This is because those i for which P_i and Q_i are both in L play a special role in determining $\omega(L)$. However equation (2) applies conveniently. Let I, J, and K be disjoint subsets of $\{1, \ldots, p\}$ with cardinalities k, m, and r, respectively. Then if $n = p - k - m - r$, we have

$$\omega(I, J, K) = 2^{n(n-1)+kn+mn+km} \tag{5}$$

This is because the arcs which may be chosen in constructing a digraph counted by $\omega(I, J, K)$ consist of the $n(n-1)$ arcs joining any node not in $I \cup J \cup K$ to any other, the kn arcs joining any node not in $I \cup J \cup K$ to any node in I, the mn arcs joining any node in J to any node not in $I \cup J \cup K$, and the km arcs joining any node in I to any in J.

In general, let $C(p; n_1, n_2, \ldots, n_q)$ denote the multinomial coefficient

$$\frac{p!}{n_1!n_2!\cdots n_q!} \tag{6}$$

when $p = n_1 + n_2 + \cdots + n_q$ and $n_1, \cdots, n_q$ are all non-negative integers, and 0 for any other set of arguments. Then $C(p; k, m, r, n)$ is precisely the number of disjoint ordered triples of subsets of $\{1, \ldots, p\}$ having cardinalities k, m, r (in order) when $n = p - k - m - r$. Since $\omega(E_0) = W[p]$ in this situation, equation (2) now gives

$$W[p] = \sum_{k,m,r} (-1)^{k+m} C(p; k, m, r, n) 2^{(k+n)(m+n)-n} \tag{7}$$

after rewriting the exponent in (5).

Note that keeping track of the number of edges in the digraphs being counted in equations (4) and (7) is straightforward. The appropriate enumerator function is then $\omega(\chi) = y^q$ where $q = q(\chi)$ is the number of edges in χ. Ordinary generating functions in y are obtained by replacing 2 with $(1 + y)$ in (4) and (7).

4 Exponential and special generating functions

The derivation of efficient recurrence relations for computing quantities such as $W[p]$ from the previous section is greatly facilitated by the use of exponential and special generating functions. Given a sequence $a = (a_0, a_1, a_2, \cdots)$ of rational numbers, the *exponential* generating function for a (in the variable x) is the infinite series

$$a(x) = \sum_{i=0}^{\infty} a_i x^i / i!$$

treated as a number of the ring $Q[[x]]$, i.e., without regard for questions of convergence. If $b(x)$ is the exponential generating function for $(b_0, b_1, b_2, \cdots)$, then the product $a(x)b(x)$ is the exponential generating function for $(c_0, c_1, c_2, \cdots)$ where

$$c_k = \sum_{i=0}^{k} \binom{k}{i} a_i b_{k-i}.$$

The reason is that $\binom{k}{i} = k!/i!(k-i)!$. This is very useful in counting labeled digraphs of various sorts, because $\binom{k}{i}$ is the number of ways to merge a linear ordering on i nodes with a linear ordering on $k-i$ disjoint nodes to produce a linear ordering on their union, all k node nodes together. This is exactly what must be done in order to construct a larger labeled digraph from the union of smaller ones. A labeling of the node set is equivalent to a linear ordering on the nodes, and the latter view is taken when digraphs are being constructed from other digraphs. An account of the uses of exponential generating function in graphical enumeration can be found in the text [2, Ch. 1].

Given a sequence $a = (a_0, a_1, a_2, \cdots)$, the special generating function for a is the infinite series

$$A(x) = \sum_{i=0}^{\infty} a_i x^i / i! 2^{i(i-1)/2}.$$

Denote by D the linear operator that transforms the exponential generating function for a to the special generating function. Then we may write $A(x) = \Delta(a(x))$ and $a(x) = \Delta^{-1}(A(x))$. If $B(x)$ is the special generating function for the sequence $b = (b_0, b_1, b_2, \ldots)$, then the product $A(x)B(x)$ is the special generating function for the sequence $d = (d_0, d_1, d_2, \ldots)$ where

$$d_k = \sum_{i=0}^{k} \binom{k}{i} 2^{i(k-i)} a_i b_{k-i}.$$

The reason is that in computing the power of 2 in the product we find

$$\binom{k}{2} - \binom{i}{2} - \binom{k-i}{2} = i(k-i).$$

Again, this is very useful in counting ways to construct digraphs, as $2^{i(k-i)}$ is the number of ways to select a set of edges directed from an i-set of nodes to a disjoint $(k-i)$-set of nodes, or *vice versa.*

If $D(x)$ is the exponential generating function for all labeled digraphs by number of nodes, then

$$D(x) = \sum_{i=0}^{\infty} 2^{i(i-1)} x^i / i! \tag{8}$$

and

$$\Delta D(x) = \sum_{i=0}^{\infty} 2^{i(i-1)/2} x^i / i! \tag{9}$$

Now, to adapt equation (2) to exponential generating functions, let $\omega(\chi) = x^p/p!$ whenever χ is a digraph on p nodes, so that the sum over all digraphs is $D(x)$. We adopt this particular enumerator for the remainder of the paper. To help keep track of the enumeration to be performed on the right side of (2) we form the mixed generating function

$$R(x,y,z,w) = \sum y^{|I|} z^{|J|} w^{|K|} \omega(I,J,K).$$

Here R is exponential in x but ordinary in y, z, and w. Then (2) can be written

$$\omega(E_0) = R(x,-1,-1,1). \tag{10}$$

In the case of digraphs without in-nodes or out-nodes, we have

$$R(x,y,z,w) = e^{wx} \Delta^{-1} \left(\Delta e^{yx} \Delta e^{zx} \Delta D(x) \right).$$

This is because

$$e^{yx} = \sum_{i=0}^{\infty} \frac{y^i x^i}{i!}$$

is the exponential generating function for sets of nodes which will be out-nodes, and which therefore have no edges between them. Likewise e^{zx} is the exponential generating function for sets of nodes which will be in-nodes, and e^{wx} for sets of nodes which will be both out-nodes and in-nodes (i.e., isolated). In the product

$$\Delta e^{yx}\Delta e^{zx}\Delta D(x)$$

we have the special generating function for labeled digraphs with specified in-nodes (counted by the power of y) and specified out-nodes (counted by the power of z), since $\Delta D(x)$ allows for all possible arcs among the unspecified nodes, while the products count in the number of ways to select edges joining unspecified nodes to in-nodes, and edges joining out-nodes to in-nodes and unspecified nodes. Applying Δ^{-1} gives the exponential generating function for these configurations, and the final product with e^{wx} counts the ways to interleave isolated nodes without adding any more edges. Thus by (10) the exponential generating function $W(x)$ for $(W[0], W[1], W[2], \cdots)$ can be expressed as

$$W(x) = e^{x}\Delta^{-1}((\Delta e^{-x})^2\Delta D(x)). \tag{11}$$

Coefficients can be calculated from (11) in three steps. Consider

$$T(x) = \Delta^{-1}((\Delta e^{-x})^2) \tag{12}$$

to be the exponential generating function for $T[i](i = 0, 1, 2, \cdots)$, and

$$Z(x) = \Delta^{-1}(\Delta T(x)\Delta D(x)) \tag{13}$$

to be the exponential generating function for $Z[i](i = 0, 1, 2, \cdots)$. Then

$$W(x) = e^{x}Z(x), \tag{14}$$

and in terms of coefficients the equations (12), (13) and (14) give

$$T[n] = (-1)^n \sum_{k=0}^{n} \binom{n}{k} 2^{k(n-k)}, \tag{15}$$

$$Z[n] = \sum_{k=0}^{n} \binom{n}{k} 2^{k(n-1)} T[n-k], \tag{16}$$

$$W[n] = \sum_{k=0}^{n} \binom{n}{k} Z[k]. \tag{17}$$

In deriving (16) we have made use of (8) and the simplification $k(n-k) + k(k-1) = k(n-1)$. A table of the first few values of these coefficients is given below.

n	$T[n]$	$Z[n]$	$W[n]$
0	1	1	1
1	-2	-1	0
2	6	2	1
3	-26	14	18
4	162	1,634	1,699
5	-1,442	583,934	592,260

In terms of efficiency, note that using equation (7) directly to calculate $W[p]$ requires $O(p^3)$ arithmetic operations, and to calculate the sequence $W[0], W[1], \cdots, W[p]$ requires $O(p^4)$. On the other hand, equations (15), (16), and (17) allow the same sequence to be calculated in $O(p^2)$ arithmetic operations.

5 Digraphs with Source and Sink

Let $C(x)$ be the exponential generating function for a set of labeled strongly connected digraphs, and let $G(x)$ be the exponential generating function for the set of all labeled digraphs for which every strong component lies in the set enumerated by $C(x)$. Then we can apply (10) to derive the relation

$$\Delta(\exp(-C(x))\Delta G(x) = 1, \tag{18}$$

by which $C(x)$ can be determined from $G(x)$ or *vice versa*. The properties being excluded are that the various possible strong digraphs counted by $C(x)$ are out-components. The enumerator in which the power of y gives the number of specified out-components is

$$\Delta(\exp(yC(x)))\Delta G(x).$$

Then term $y^i C(x)^i/i!$ counts the number of ways select a set of i disjoint copies of strong digraphs counted by $C(x)$, without adding any edges between them. The product

$$\Delta(y^i C(x)^i/i!)\Delta G(x)$$

then counts the ways in which such a disjoint set of strong components can be joined by arbitrary edges to a digraph counted by $G(x)$, resulting in a digraph which again is of the type counted by $G(x)$ but now has i out-components specified. There are no dual properties (one can think of them as always false), and so no terms in z or w. So, setting $y = -1$ gives us $\omega(E_0)$. Since only the empty graph has no out-component, $\omega(E_0) = 1$, which proves (18).

Since we had just one homogeneous set of properties to exclude, the out-components, equation (18) follows from ordinary inclusion-exclusion. Indeed, (18) has been derived previously in different ways; see [2], [5], and [6].

In this section, we need (18) in order to derive equations for $\exp(-S(x))$ and $S(x)$, where $S(x)$ is the exponential generating function for labeled strongly connected digraphs. As with (18), these equations have been given previously, but we need them here for the study of digraphs with a source and a sink.

When all labeled strong digraphs are allowed as components, all labeled digraphs can be constructed from them and so (18) takes the form

$$\Delta(\exp(-S(x))\Delta D(x) = 1. \tag{19}$$

From (19) and (8) we could derive explicit and efficient recurrence relations for the coefficients of $S(x)$. This is done in [6, equations (33) and (34)], obtaining recurrences which were derived earlier by Wright [7] based on a more complicated set of recurrences due to Liskovets [3].

Now, to count digraphs with a source and a sink our strategy is to enumerate configurations in which there is a designated out-component and a separate designated in-component, and then use our specialized form of inclusion-exclusion to eliminate all other out-components and in-components. This will be correct in counting digraphs with a source and a sink, i.e., with a unique out-component and a unique in-component, except for the case of a single strong component. That has been omitted in favor of two isolated components, one designated as an out-component and the other as an in-component. Once the correction of adding $S(x) - S(x)^2$ is made, the enumeration is valid because the designated in-component and out-component are actually the only ones present and therefore are distinguished intrinsically.

To enumerate with a designated out-component and a designated in-component, we simply include a factor of $S(x)$ for each, since we are willing for any strongly connected digraphs to be the out-component, and any to be the in-component. This gives

$$\exp(wS(x))\Delta^{-1}(\Delta(S(x)\exp(yS(x))\Delta(S(x)\exp(zS(x))\Delta D(x)),$$

combining the principles applied previously to deriving equations (11) and (19). Now, following (2) we set $w = 1$ and $y = z = -1$, which along with the correction mentioned earlier leads to the expression

$$S(x) - S(x)^2 + \exp(S(x))\Delta^{-1}((\Delta(S(x)\exp(-S(x))^2\Delta D(x)) \quad (20)$$

for the exponential generating function for all labeled digraphs having a source and a sink. It is then straightforward to derive efficient recurrence relations for those numbers based on equations (8) and (19), and expression (20).

To modify the above procedure to count labeled digraphs with a unique source and a unique sink, we simply replace $S(x)$ by x in the locations where $S(x)$ counts the possible designated out-component or in-component. For in this case we want the component to be a single node, for which x is the exponential generating function. Letting $L(x)$ denote the exponential generating function for $L[0], L[1], L[2], \cdots$ where $L(p)$ is the number of labeled digraphs with a unique source and a unique sink, we find

$$L(x) = x - x^2 + \exp(S(x))\Delta^{-1}(\Delta(x\exp(-S(x)))^2\Delta D(x)). \quad (21)$$

Coefficients can be efficiently extracted from (21) in the usual way. We find that $L[0] = 0, L[1] = 1, L[2] = 2$, and for $n \geq 3$ that

$$L[n] = 2n(n-1)S[n-1]. \quad (22)$$

The latter may be unexpected at first, but has an easy combinatorial proof. For $n \geq 2$, any digraph with a unique source v and a unique sink u has $v \neq u$. If u and v are identified, then the result is a strong digraph on $n-1$ nodes since every other node is reachable from the new node and *vice versa*. To see how many digraphs or order n result in a given strong digraph on $n-1$ nodes, consider the reverse process. We can choose any of $n-1$ nodes to split. The chosen node is split into an in-node and an out-node in a unique way with respect to the other nodes, but when $n \geq 3$ the edge from the new out-node to the new in-node is optional, resulting in a factor of 2 in counting the possibilities. If we consider that the new out-node inherits the label from the original node, we now have n choices for labeling the new in-node. The resulting digraph is now completely specified and labeled, and is obviously a digraph with a unique source (the new out-node) and a unique sink (the new in-node).

References

1. F. Harary, *Graph Theory*, Addison-Wesley, Reading, MA, 1969.

2. F. Harary and E. M. Palmer, *Graphical Enumeration*, Academic Press, New York, 1973.

3. V. A. Liskovets, On a recurrence method of counting graphs with labelled vertices, *Soviet Math. Dokl.* **10** (1969), pp. 242-246. (*Dokl. Akad. Nauk SSSR*, **184** (1969), No. 6)

4. V. A. Liskovets, Enumeration of rooted digraphs with a source, *Vesci Akad. Navuk BSSR Ser. Fiz.-Mat. Navuk* (1969) 23-32. (Russian)

5. V. A. Liskovets, A general enumeration scheme for labelled graphs, *Dokl. Akad. Nauk BSSR* **21** (1977) 496-499, 571. (Russian)

6. R. W. Robinson, Counting labeled acyclic digraphs, *New Directions in the Theory of Graphs* (F. Harary, ed.), Academic Press, New York, 1973, pp. 239-273.

7. E. M. Wright, The number of strong digraphs, *Bull. London Math. Soc.* **3** (1971), pp. 348-350.

ON FULLY OBSERVABLE GRAPHS

N. Zagaglia Salvi(*)
Dip. di Matematica
Politecnico di Milano
P.za L. da Vinci 32
Milano, Italy

ABSTRACT - A d - regular graph G is fully observable if its number of vertices is $\binom{obsG}{d}$, where obsG is the observability of G. An algorithm to obtain an obs-coloring for the fully observable cycles is given. Moreover it is proved that every 3-regular graph of order 10 is fully - observable.

(*) Work partially supported by the Italian Ministero della Universita' e della Ricerca Scientifica e Tecnologica.

1 - INTRODUCTION

Let G be a finite undirected graph without loops and multiple edges and let $E_v(G)$ be the set of all edges of G incident with $v \in V(G)$. Let f be a mapping of the set E(G) of the edges of G onto the k-set {1,2,...,k}. The color set of v induced by f is

$$\bigcup_{e \in E_v(G)} \{ f(e) \} .$$

The f coloring is proper if the edges of $E_v(G)$ are distinguished by their colors. If moreover the color sets are distinct, so that vertices of G are distinguished by their color sets, G is said observable through f or f-observable.

The **observability** of G, denoted obs G, is the minimum value of k by which G is f-observable.

Clearly an obs-coloring of G is possible only when G has not a component K_2 or more than one component K_1.

A d-regular graph G, having observability k, is **fully observable** if $| V(G) | = \binom{k}{d}$. So the color sets corresponding to the vertices are all the d-subsets of [1,k].

We give an algorithm for determining an obs - coloring of the fully observable cycles and get the generating function corresponding to the sequence of the orders of these cycles, together with some identities on particular binomial coefficients.

We prove that every 3 - regular graph of order 10 contains $3K_2$ or $P_4 \cup K_2$ as induced subgraphs of order 6, but only one case.

Using this result we prove that all the 3-regular graphs of order 10 are fully observable (Theorem 1). This result turns out to be a check in a particular case of the following conjecture [1]:

If $k \in [5,\infty)$, $d \in [3, k-2]$ and $\binom{k-1}{d-1} \equiv 0 \pmod 2$, then every d-regular graph on $\binom{k}{d}$ vertices is fully observable.

If u, v are adjacent vertices of G we denote $u \sim v$. For every edge uv of G we denote by G(uv) the subgraph of G induced by the ver-

tices not adjacent to u or v. That is the vertices of G(uv) have distance at least 2 from u and v.
Moreover we denote by N(uv) the subgraph of the vertices adjacent to u or v, distinct from u,v.

2 - FULLY OBSERVABLE CYCLES

First consider the problem of the existence and construction of fully observable cycles. If obs $C_n = k$ and C_n is fully observable, then $n = \binom{k}{2}$.
Because every color has to be adjacent to all the other ones and the number of the adjacencies is even, then the number of colors has to be odd.

In [1] it is proved that when k is odd and $n = \frac{k^2 - k}{2}$, then $\text{obs}C_n = k$.
Thus, when k is odd, C_n is fully observable.
Now we give an algorithm to determine a fully obs-coloring of C_n, where obs $C_n = k + 2$, by starting from a similar coloring of $C_{n'}$, where obs $C_{n'} = k$, $n' = \binom{k}{2}$ and k is odd.

Let $k = 2h + 1$ and $a, b_1, b_2, \dots, b_{2h}$ the colors, C a cycle of $\binom{k}{2}$ vertices and ϕ a fully obs-coloring of C.
As the color a has to be adjacent to all the other 2h colors, ϕ contains the h triples $b_1ab_2, b_3ab_4, \dots, b_{2h-1}ab_{2h}$.
Let x, y two new colors.
We replace $b_1\ a\ b_2$ by $b_1\ a\ x\ y\ a\ b_2\ x\ b_1\ y\ b_2$.
This new sequence, which adds 7 vertices to the preceding ones, starts and ends with the same colors b_1, b_2, contains all the adjacencies of x, y with respect to b_1, a, b_2 and also the color sets $\{b_1, a\}$, $\{b_2, a\}$ and $\{x, y\}$.
Then we replace $b_i a\ b_{i+1}$, $2 < i \le 2h-1$, by $b_i\ x\ b_{i+1}\ y\ b_i\ a\ b_{i+1}$.
This changement adds 4 vertices, contains all the adjacencies of x, y with respect to b_i, b_{i+1} and the color sets $\{b_i, a\}$, $\{b_{i+1}, a\}$.
We have obtained a fully obs - coloring of a cycle C_n, whose observability is k + 2.

Moreover we obtain $n = n' + 4(h - 1) + 7 = n' + 4h + 3$, that is the identity $\binom{2h+3}{2} = \binom{2h+1}{2} + 4h + 3$.

Denoted $c_h = \binom{2h+1}{2}$, we have $c_h = c_{h-1} + 4h - 1$.

This relation does not hold in a particular case; in fact when $h = 0$ we get 0 on the left and -1 on the right. So we add $1.[h = 0]$ on the right, where $[h = 0]$ means 1 for $h = 0$, 0 otherwise [5]:

$$c_h = c_{h-1} + 4h - 1 + 1.[h=0]. \qquad (1)$$

In this way we have a recurrence as a single equation for c_h. Now we want to transform the equation (1) into an equation for $G(z) = \sum_{h \geq 0} c_h z^h$, the generating function of the sequence $< c_h >$.

We multiply both sides of the equation (1) by z^h and sum over all h.

$$\Sigma\, c_h z^h = z\, \Sigma c_{h-1} z^{h-1} + 4z\Sigma h\, z^{h-1} - \Sigma\, z^h + 1.$$

Using some easy identities on the power series we have

$$(1 - z)\, G(z) = 4z.\frac{1}{(1 - z)^2} - \frac{1}{1 - z} + 1$$

that is

$$G(z)= \frac{5z}{(1 - z)^3} - \frac{1}{(1 - z)^3} + \frac{1}{1 - z} \qquad (2)$$

We expand in a power series every term on the right - hand side and obtain

$$G(z)= \Sigma\, [\, 5\binom{h+1}{2} - \binom{h+2}{2} + 1\,]\, z^h$$

Equating the coefficient of z^h in the above equation we get

$$\binom{2h+1}{2} = 5\binom{h+1}{2} - \binom{h+2}{2} + 1. \qquad (3)$$

From (2) we have

$$G(z) = \frac{z^2 + 3z}{(1 - z)^3}.$$

Then expanding again in a power series we obtain

$$\binom{2h+1}{2} = \binom{h}{2} + 3\binom{h+1}{2}. \qquad (4)$$

From (3) and (4) we have the identity

$$\binom{h+2}{2} = 2\binom{h+1}{2} - \binom{h}{2} + 1. \qquad (5)$$

3. PARTICULAR INDUCED SUBGRAPHS

Throughout this section G denotes a 3-regular graph of order 10. If G is not connected, then it is the union of K_4 and a 3-regular graph of order 6. It is well known that there are exactly two nonisomorphic 3 - regular graphs of order 6: $K_{3,3}$ and the "envelope" graph. Fig. 1 gives an example of a fully observable coloring of these graphs.

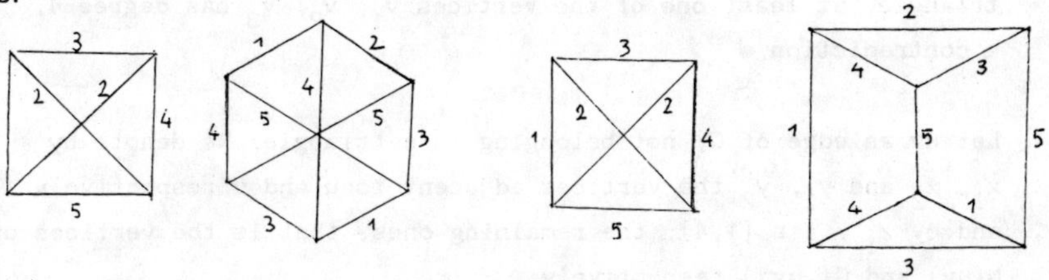

fig.1

Thus we assume G is connected.
In this section we consider the problem of the existence of certain induced subgraphs of order 6, which turn to be useful for proving the fully observability of G.
We prove that all the 3 - regular graphs of order 10, but only one case, have either $3K_2$ or $P_4 \cup K_2$ or both as induced subgraphs.
Let us consider the permutation graph H of fig. 2.

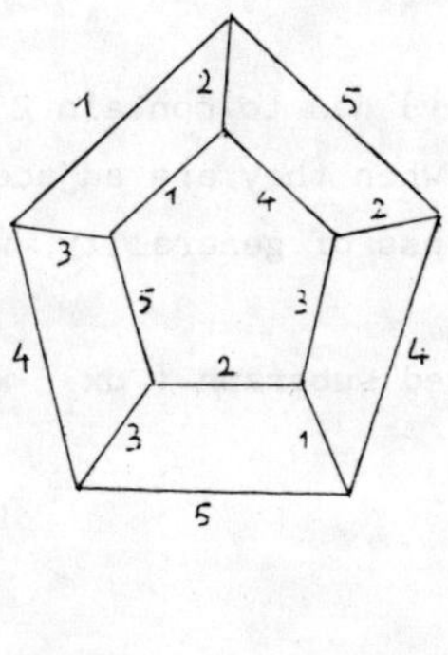

fig. 2

to which we have yet assigned a fully 5 - obs - coloring.
It is easy to see that H(uv) is isomorphic to C_4 or $K_{1,3}$ for every edge uv $\in$ E(H).
In the following Proposition 1, we shall prove that H is the unique 3 - regular graph of order 10 not having as induced subgraph $3K_2$ or $P_4 \cup K_2$.

LEMMA 1 - G has at least one edge not belonging to a triangle.

Proof. Let v_1, v_2, v_3 be a triangle and v_1', v_2', v_3' the vertices respectively adjacent to the preceding ones. These vertices can not coincide because otherwise we have a 3 - regular graph of order 4 and G is not connected.
Let $v_1' \neq v_2'$. As the edges v_1v_1', v_2v_2' have to belong to a triangle, at least one of the vertices v_1, v_2, v_3 has degree 4, a contradiction.■

Let uv an edge of G, not belonging to a triangle. We denote by x_1, x_2 and y_1, y_2 the vertices adjacent to u and v respectively, and by z_i , i $\in$ [1,4], the remaining ones, that is the vertices of N(uv) and G(uv) respectively.
Clearly G contains as induced subgraph $3K_2$ or $P_4 \cup K_2$ if and only if it contains an edge uv such that G(uv) coincides with $2K_2$ or P_4.
The subgraph $P_4 \cup K_2$ whose vertices are v_i, i $\in$ [1,6] will be represented by (v_1v_2, v_3v_4) $\cup$ v_5v_6.

LEMMA 2 - Suppose G $\not\cong$ H has an edge uv such that G(uv) $\cong$ C_4.
Then G contains $P_4 \cup K_2$ as induced subgraph of order 6.

Proof. By the assumption N(uv) has to contain 2 edges.
Then we have to distinguish when they are adjacent or not.
In the first case, whitout loss of generality, we can suppose $x_1 \sim x_2$, $x_1 \sim y_1$ and $x_2 \sim z_1$.
If $z_3 \sim y_2$, we have as induced subgraph (ux_1, x_2z_1) $\cup$ z_3y_2, as shown in fig.3.

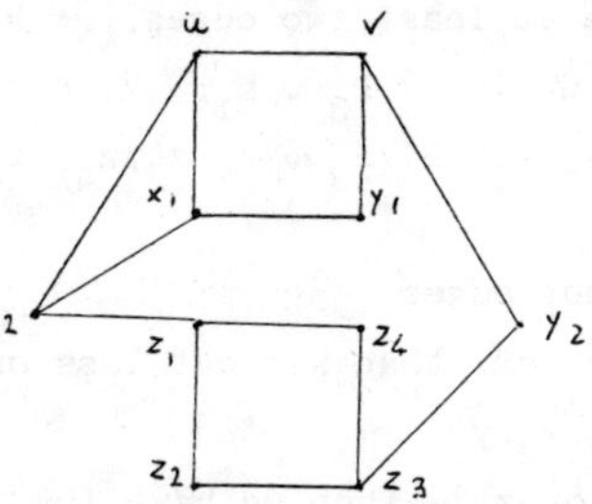

fig. 3

The missing edges can be whatever, provided that G is regular.
If $z_3 \sim y_1$, we have $(ux_2, vy_2) \cup z_2z_3$.
In the second case, there are two cases to consider.

1) Suppose $x_1 \sim x_2$, $y_1 \sim y_2$ and also $x_1 \sim z_1$.
Let $x_2 \sim w$, where $w \in \{z_2, z_3\}$; we can also assume $z_4 \sim y_1$.
Then G contains $(ux_2, vy_2) \cup z_1z_4$.

2) Suppose $x_1 \sim y_1$, $x_2 \sim y_2$ and again $x_1 \sim z_1$.
If $y_1 \sim z_4$ and $x_2 \sim z_2$, then $y_2 \sim z_3$ and we obtain a graph isomorphic to H (fig. 2).
If $y_1 \sim z_4$ and $x_2 \sim z_3$, then $y_2 \sim z_2$ and we have $(vy_2, ux_1) \cup z_3z_4$.

When $y_1 \sim z_3$ and $y_2 \sim z_2$, then there is $(ux_1, vy_2) \cup z_3z_4$.
If $y_1 \sim z_3$ and $y_2 \sim z_4$, we have $(ux_1, vy_2) \cup z_2z_3$, as shown in fig. 4.

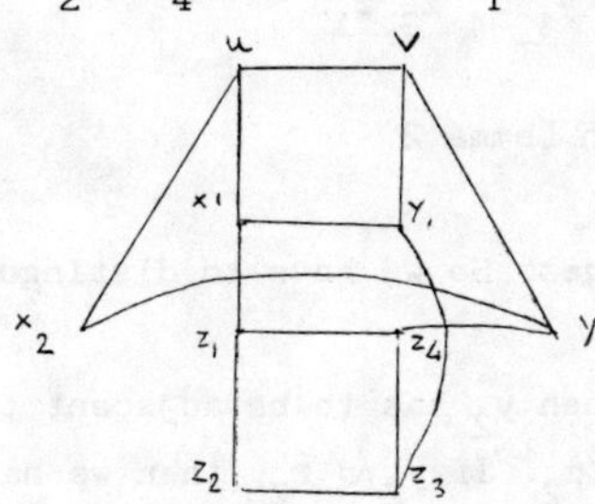

fig. 4

PROPOSITION 1 - Let $G \not\cong H$. Then G contains as induced subgraph $3K_2$ or $P_4 \cup K_2$.

Proof. It is sufficient to prove that there exists an edge uv such that G(uv) is isomorphic to $2K_2$ or P_4.
Suppose it is not the case.

Because G(uv) has to have at least two edges, we have to consider the following cases for G(uv): 1) $P_3 \cup K_1$; 2) $K_{1,3}$; 3) C_4; 4) $z_4 \sim z_1, z_2, z_4$ and $z_1 \sim z_2$; 5) $z_3 \sim z_1, z_2, z_4$, $z_4 \sim z_1, z_2$.

1) In this case N(uv) has not edges.

If z_1 is the isolated vertex, then without loss of generality we can suppose that $z_1 \sim x_1, x_2, y_1$.
If $x_1 \sim z_3$ and $x_2 \sim z_2$ [or z_4], then we have (uv, x_2z_1) ∪ z_3z_4 [or z_3z_2].
If $x_1 \sim z_3$ and $x_1 \sim z_4$ [or z_2], then we consider (uv, x_1z_1) ∪ z_2z_3 [z_4z_3].

2) In this case N(uv) contains an edge: so we have two cases to consider as well as $x_1 \sim x_2$ or $x_1 \sim y_1$. Let z_3 the vertex of degree 3 in G(uv).

a) Suppose $x_1 \sim x_2$ and $x_1 \sim z_1$. Let $y_1 \sim w$, where $w \in \{z_2, z_4\}$. Then, if $w = z_4$, we have (y_1z_4, z_3z_2) ∪ ux_1; if $w = z_2$, we have (y_1z_2, z_3z_4) ∪ ux_1.

b) Suppose $x_1 \sim y_1$ and $x_1 \sim z_1$. If $y_2 \sim z_4$[or z_2], but not to both of them, then we have (z_2z_3, z_4y_2) ∪ ux_1 [or (z_4z_3, z_2y_2)∪ ux_1.
If $y_2 \sim z_2, z_4$ and $x_2 \sim z_2, z_4$, then we obtain (x_2u, vy_2) ∪ z_1z_3.
If $y_2 \sim z_2, z_4$ and $x_2 \sim z_1, z_2$[or z_4], hence $y_1 \sim z_4$ [or z_2].
In this case G ≅ H (fig.2): the outer 5-cycle is u v y_2 z_2 x_2, the inner 5 - cycle is x_1 y_1 z_4 z_3 z_1.

3) This case is proved in Lemma 2.

4) N(uv) contains two edges. So we have to distinguish when they are adjacent or not.

a) Suppose $x_1 \sim x_2, y_1$. Then y_2 has to be adjacent to at least one of the vetices z_1, z_2. If $y_2 \sim z_2$, then we have (y_2z_2, z_3z_1) ∪ ux_1; if $y_2 \sim z_1$, then we obtain (y_2z_2, z_3z_1) ∪ ux_1.

b) Whitout loss of generality we can suppose $z_4 \sim y_1$.
Let $x_1 \sim x_2$ and $y_1 \sim y_2$. Then we have (x_2u, vy_2) ∪ z_3z_4.
If $x_1 \sim y_1$ and $x_2 \sim y_2$, then (x_1u, vy_2) ∪ z_3z_4.

5) In this case N(uv) has three edges, while the vetices have at most degree two. So N(uv) is isomorphic to P_4 or $K_3 \cup K_1$.

a) If $N(uv) \cong P_4$, we can suppose $x_1 \sim y_1$, $y_1 \sim y_2$ and $y_2 \sim x_2$. Then $x_1 \sim z_1$ and $x_2 \sim z_2$. So we obtain $(x_1u,\ vy_2) \cup z_3z_4$.

b) If $N(uv) \cong K_3 \cup K_1$, let x_1 the isolated vertex. So $x_1 \sim z_1, z_2$ and we have $(\ x_1u,\ vy_2\ ) \cup z_3z_4$.■

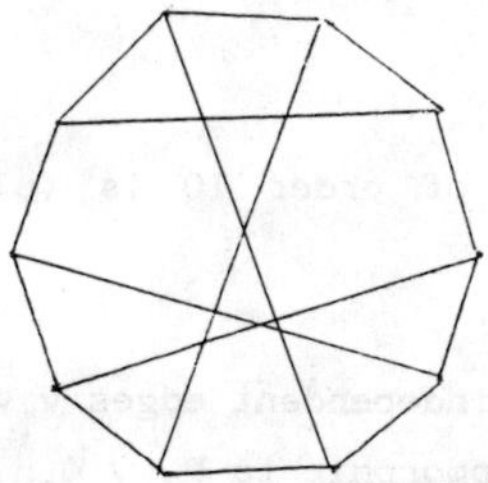

fig. 5

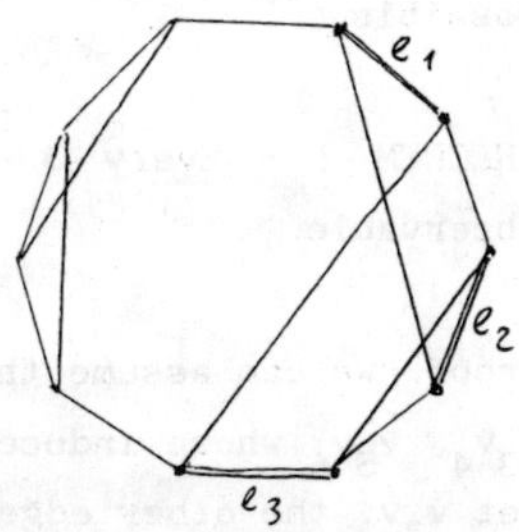

fig. 6

The graph of Fig. 5 does not contain $3K_2$ as induced subgraph, while contains $P_4 \cup K_2$. On the contrary the Petersen graph contains $3K_2$, but not $P_4 \cup K_2$.

4) FULLY OBSERVABLE 3 - REGULAR GRAPHS OF ORDER 10

Again in this section G denotes a 3 - regular graph of order 10. If G is fully observable, then every color is assigned to three independent edges.

However this condition is not arbitrary.

For example, if we assign the same color to the independent edges e_1, e_2, e_3 of the graph in fig.6 we can not obtain a fully observable coloring . In fact the subgraph induced by e_1, e_2, e_3 contains 5 edges, distinct from the preceding ones, to which we can assign only 4 distinct colors.

By Prop. 1, every graph G, but H, contains as induced subgraph $3K_2$ or $P_4 \cup K_2$. In [7] we proved that the Petersen graph is fully observable and is the unique graph $G \neq H$ not containing $P_4 \cup K_2$.

Thus asume that G contains $P_4 \cup K_2$; then we can determine 3 independent edges whose induced subgraph has only 1 edge besides the three assigned.

Denote by V the vertex set of this subgraph and by Z the set of the remaining vertices.

If an edge e connects vertices of V and Z, we call V - vertex [Z - vertex] the vertex of e belonging to V [Z].

In a partial coloring of G, an edge not yet colored is called **possible**.

THEOREM 2 - Every 3 - regular graph of order 10 is fully observable.

Proof. We can assume that G contains 3 independent edges v_1v_2, v_3v_4, v_5v_6 whose induced subgraph is isomorphic to $P_4 \cup K_2$; let v_2v_3 the other edge of P_4.

Then the subgraph induced by Z has to contain an edge, say z_1z_2. Every other edge of G connects vertices of V and Z.

We assign the color 1 to the independent edges v_1v_2, v_3v_4, v_5v_6 and the color 2 to the edges v_2v_3, z_1z_2 and an edge connecting vertices of V and Z, different from v_2, v_3, z_1, z_2, say z_3v_5 (fig.7).

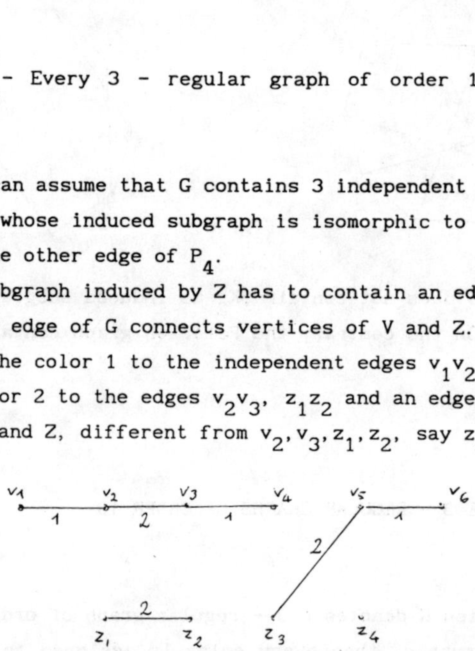

Fig 7

Step 1. We want determine three independent edges such that one V -vertex corresponds to the partial color set {12} and two Z - vertices correspond to {2}.

Consider the possible edges incident to v_2, v_3. We can suppose that at least one of them is incident to one of the vertices $\{z_1, z_2, z_3\}$. In fact, on the contrary, $v_2, v_3 \sim z_4$ and we remove the color 2 of z_3v_5 assigning it to the third edge incident z_4.

Let $e_1 = v_iz_h$, where $v_i \in \{v_2, v_3\}$ and $z_h \in \{z_1, z_2, z_3\}$.

Then consider an edge $e_2 \neq e_1$ incident to z_4 and to a vertex of V different from v_2, v_3, v_5. It exists because there are not possible edges incident to v_i, where $v_i \in \{v_2, v_3\}$ and z_4 has degree 3: say $e_2 = z_4 v_j$.

Then consider an edge e_3 incident to z_1 or z_2 and a vertex of V different from $\bar{v}_i$, v_j v_5, where $\bar{v}_i$ is the vertex of $\{v_2, v_3\}$ different from v_i.

Because the possible edges incident to these vertices are three, while that incident to z_1 or z_2 are 4, then e_3 exists.

We color by 3 the edges e_1, e_2, e_3. So we have obtained a partial coloring of G where the color sets of V are 1, 12, 12, 13, 13 123 and of Z are 2, 3, 23, 23.

Denote by v_i', $i \in [1,6]$ and z_j', $j \in [1,4]$, the vertices respectively corresponding to the above color sets.

Step 2. Now we want determine three independent edges whose V - vertices and Z - vertices correspond to 1, 12, 13 and to 2, 3, 23.

Denote by v_j' the vertex adjacent to z_3', where v_j' is a vertex of V different from v_6'. Let $f_1 = v_j' z_3'$.

a) Suppose $v_j' = v_1'$. Then among the 4 possible edges incident to v_2', v_3', v_4', v_5' one could be incident to z_4', while the other three are incident to z_1', z_2'. So it is possible determine 2 independent edges incident to z_1', z_2' and vertices corresponding to {12}, {13}.

b) Let $v_j' \neq v_1'$. We can suppose that also $z_4' \not\sim v_1'$, otherwise we change z_3' with z_4'. Then $v_1' \sim z_1'$, z_2'.

Let $f_1 = v_j' z_3'$.

Suppose v_j' correspond to {12}. Then among the possible edges incident to v_4', v_5' one could be incident to z_4', while the other one f_2 is incident to z_1', z_2', say z_1'. So we determine $f_3 = v_1' z_2'$.

In any case we obtain three independent edges satisfying the above conditions. We color by 4 these edges and obtain a partial coloring of G where the color sets corresponding to the vertices of V are 14, 12, 13, 124, 134, 123 and of Z are 24, 34, 23, 234.

From these sets it is clear that the remaining edges are independent. We color by 5 these edges and obtain a fully coloring of G.■

REFERENCES

[1] J. Černý, M. Horňák and R. Soták, Observability of a graph, to be appear.

[2] G. Chartrand and R. J. Wilson, The Petersen graph, Graphs and Applic. (F. Harary and J.S. Maybee eds), 1985, 69 - 100.

[3] F. Harary, Graph Theory, Addison - Wesley, Reading, 1969.

[4] F. Harary and M. Plantholt, The point - distinguishing chromatic index, Graphs and Applic. (F. Hararu and J.S. Maybee, eds), Wiley - Interscience, New - York, 1985, 147 - 162.

[5] R. Graham, D. Knuth and O. Patashnik, Concrete Mathematics, Addison Wesley, Reading, 1989.

[6] N. Zagaglia Salvi, On the critical observability of the cycles, submitted.

[7] N. Zagaglia Salvi, New characterizations of the Petersen graph, submitted.

The Recurrence Property Of Generating Function Of Graphical Sequences

Shao Xuecai

(Bijing Polytechnic University, Beijing, 100022)

Ye Xiuming

(Shanghai University, Shanghai, 201800)

Abstract All sequences corresponding in one-to-one to directed walkes in a directed graph G with element of points in G are called graphical sequences of G. This paper is on the basis of [1] to furtherly discuss some recurrence properties for generating function of graphical sequences, mainly including cut point theorem, cut edge theorem, delecte point formula. Using there properties can simplify the computation of generating function of graphical sequences.

Key Words Graphical Sequences; Generating Function

All sequences corresponding in one-to-one to directed walkes in a directed graph G with element of points in G are called graphical sequences of G. This paper is on the basis of [1] to furtherly discuss some recurrence properties of generating function of graphical sequences. Using these properties can simplify the computation of generating function of graphical sequences. The marks and terms in the paper are the same with [1],[2].

Theorem 1 Set $G_{-i} = G - \{i\} = H$. Assume the out-neighbourhood of point i is $N_G^+(i) = \{j_1, j_2, \cdots, j_r\}$, and the in-neighbourhood of point i is $N_G^-(i) = \{l_1, l_2, \cdots, l_p\}$, then

$$G^* = H^* + cx\left(1 + \sum_{s=1}^{r} H^{(j_s)} + \sum_{t=1}^{p} H_{(l_t)} + \sum_{s=1}^{r}\sum_{t=1}^{p} H_{(l_t)} H^{(j_s)}\right)$$

where,

$$c = \left(1 - a_{ii}x - \sum_{s=1}^{r}\sum_{t=1}^{p} H_{(l_t)}^{(j_s)} x\right)^{-1}$$

$A = (a_{ij})_{n\times n}$ is the adjacent matrix of G.

Proof Assume $i = 1$, and A_{-1} is the adjacent matrix of H,

$$I - Ax = \begin{bmatrix} 1 - a_{11} & -v'x \\ -ux & I - A_{-1}x \end{bmatrix}$$

$$u' = (a_{21}, a_{31}, \cdots, a_{n1}), \quad -ux = \sum_{t=1}^{p} e_{l_t} x$$

$$v' = (a_{12}, a_{13}, \cdots, a_{1n}), \quad -v'x = -\sum_{s=1}^{r} e'_{j_s} x.$$

Let

$$(I-Ax)^{-1}=\begin{bmatrix} c & \beta \\ \alpha & H \end{bmatrix}$$

It is easy to know:

$$\begin{aligned} c = \; & ((1-a_{11}x)-v'(I-A_{-1}x)^{-1}xux)^{-1} \\ = \; & \left((1-a_{11}x)-\sum_{s=1}^{r}\sum_{t=1}^{p}H_{(l_t)}^{(j_s)}x\right)^{-1} \\ H = \; & ((I-A_{-1}x)-u(1-a_{11}x)^{-1}xv'x)^{-1}=(I-A_{-1}x)^{-1} \\ & +\frac{(I-A_{-1}x)^{-1}u(1-a_{11}x)^{-1}xv'x(I-A_{-1}x)^{-1}}{1-v'(I-A_{-1}x)^{-1}xux(1-a_{11}x)^{-1}} \\ = \; & (I-A_{-1}x)^{-1}+c(I-A_{-1}x)^{-1}xuv'(I-A_{-1}x)^{-1}x \\ & \beta=(1-a_{11}x)^{-1}xv'H, \quad \alpha=(I-A_{-1}x)^{-1}xuc \end{aligned}$$

Thus

$$\begin{aligned} G^* = \; & 1+j'(I-Ax)^{-1}xj \\ = \; & 1+cx+j'(I-A_{-1}x)^{-1}xucx+j'(I-A_{-1}x)^{-1}xj \\ & +j'(I-A_{-1})^{-1}xuv'(I-A_{-1}x)^{-1}xjcx \\ & +(1-a_{11}x)^{-1}xv'(I-A_{-1}x)^{-1}xj \\ & +(1-a_{11}x)^{-1}xv'(I-A_{-1}x)^{-1}xuv'(I-A_{-1}x)^{-1}xjcx \\ = \; & H^*(x)+cx(1+\sum_{t=1}^{p}H_{(l_t)}+\sum_{s=1}^{r}\sum_{t=1}^{p}H_{(l_t)}H^{(j_s)} \\ & +(1-a_{11}x)^{-1}x\sum_{s=1}^{r}\sum_{t=1}^{p}H_{(l_t)}^{(j_s)}\sum_{s=1}^{r}H^{(j_s)})+(1-a_{11}x)^{-1}x\sum_{s=1}^{r}H^{(j_s)} \\ = \; & H^*(x)+cx\left(1+\sum_{t=1}^{p}H_{(l_t)}+\sum_{s=1}^{r}H^{(j_s)}+\sum_{s=1}^{r}\sum_{t=1}^{p}H_{(l_t)}H^{(j_s)}\right). \end{aligned}$$

Corollary 1 Set $G_{-i}=G-\{i\}\equiv H$. Assume $N_G^-(i)=\phi, N_G^+(i)=\{j_1,j_2,\cdots,j_r\}$. Then

$$G^*=H^*+\frac{x}{1-a_{ii}x}\left(1+\sum_{s=1}^{r}H^{(j_s)}\right)$$

Where $A=(a_{ij})_{n\times n}$ is the adjacent matrix of G.

Proof By theorem 1, let $H_{(l_t)}=H_{(l_t)}^{(j_s)}=0$, the result is obtained immediately.

Corollary 2 Let $G_{-i}=G-\{i\}\equiv H$. Assume that i is the undirected pendant point of G, $N_G^+(i)=N_G^-(i)=\{j\}$, then

$$G^*=H^*+cx\left(1+H_{(j)}+H^{(j)}+H_{(j)}\cdot H^{(j)}\right).$$

Where $c=\left(1-a_{11}x-H_{(j)}^{(j)}x\right)^{-1}$.

Theorem 2 Assume that i is the cut point of G. The graph G is obtained by joining S together T at cut point i. Let $V(S)\cap V(T)=\{i\}$, and point i isn't lie on any directed cycle in S. Then

$$\begin{aligned} G^* = \; & S^*+T^*-1-x+\tfrac{1}{x}(S_{(i)}-x)(T^{(i)}-x) \\ & +\tfrac{1}{x}(S^{(i)}-x)(T_{(i)}-x)+\tfrac{1}{x^2}(S_{(i)}-x)(T^{(i)}-x)(T_{(i)}^{(i)}-x) \end{aligned}$$

Proof The directed walk set of G can be divided into five unintersective types as follows:

(1) Its points (except i) are not included in T;

(2) Its points (except i) are not included in S;

(3) They starting at a point (except i) in S go through the cut point i to end one (except i) in T;

(4) They starting at a point (except i) in T go through i to end one (except i) in S;

(5) They starting at a point (except i) in S go through i and orbit around a directed cycle in T, then come back via the cut i to end one (except i) in S.

Notice that the directed walk of G corresponds to the graphical sequencesof G and that cut point i repeat two times in S^* and T^*. Thus the generating function for the graphical sequences corresponding to (1) and (2) shall be $S^* + T^* - 1 - x$. The graphical sequences correspoding to (3) is obtained by joining $S_{(i)}$ (not including single point sequence i) and $T^{(i)}$ (not including single point sequence i). As i is common when two parts are joined and the length of sequence reduces to 1, so its generating function shall be $\frac{1}{x}(S_{(i)} - x)(T^{(i)} - x)$. The same with it, the generating function for the graphical sequences coresponding to (4) and (5) shall be $(S^{(i)} - x)(T_{(i)} - x)$ and $\frac{1}{x}(S_{(i)} - x)(T^{(i)} - x)(T^{(i)}_{(i)} - x)$.

Corollary Let $G_{\dot{i}}$ be the graph obtained by deleting the loop at point i of graph G, and there be not any directed cycle through point i in $G_{\dot{i}}$, then

$$G^{\Gamma} = G^*_{\dot{i}} + \frac{1}{1-x} G_{\dot{i}(i)} G^{(i)}_{\dot{i}}$$

Proof Regarding i as cut point, according to theorem 2, $S = G_{\dot{i}}, T = \dot{K}_1$. Thus

$$\begin{aligned} G^* = \quad & G^*_{\dot{i}} + \tfrac{1}{1-x} - 1 - x + \tfrac{x}{1-x}(G_{\dot{i}(i)} - x) \\ & \tfrac{x}{1-x}(G^{(i)}_{\dot{i}} - x) + \tfrac{1}{1-x}(G_{\dot{i}(i)} - x)(G^{(i)}_{\dot{i}} - x) \end{aligned}$$

By arranging, the result is obtained immediately.

Theorem 3 Assume that $\{i, j\}$ is the point cut set of G, namely that the graph G is a union of S and T by only common points i and j, $V(S) \cap V(T) = \{i, j\}$, and that point i and point j do not lie on any directed cycle in S respectively or simultaneously. Then

$$\begin{aligned} G^* = \quad & S^* + T^* - 1 - 2x + \tfrac{1}{x}(S_{(i)} - x)(T^{(i)} - x) + \tfrac{1}{x}(S^{(i)} - x)(T_{(i)} - x) \\ & + \tfrac{1}{x^2}(S_{(i)} - x)(T^{(i)}_{(i)} - x)(S^{(i)} - x) + \tfrac{1}{x}(S_{(j)} - x)(T^{(j)} - x) \\ & + \tfrac{1}{x}(S^{(j)} - x)(T_{(j)} - x) + \tfrac{1}{x^2}(S_{(j)} - x)(T^{(j)}_{(j)} - x)(S^{(j)} - x) \\ & + \tfrac{1}{x^2}(S_{(i)} - x)(T^{(i)}_{(j)} - x)(S^{(j)} - x) + \tfrac{1}{x^2}(S_{(j)} - x)(T^{(j)}_{(i)} - x)(S^{(i)} - x) \end{aligned}$$

Proof The proof is similar with theorem 2.

Theorem 4 Assume that (i, j) and (k, l) are the cut set of G.

$$G - (i, j) - (k, l) \quad = G_1 \dot{+} G_2, i, l \in V(G_1); j, k \in V(G_2)$$

and that there is not any directed walk between i and l in G_1, then

$$G^* = G_1^* + G_2^* - 1 + G_{1(i)} G_2^{(j)} + G_{2(k)} G_1^{(l)} + G_{1(i)} G_{2(k)}^{(j)} G_2^{(l)}.$$

Proof The proof is similar with theorem 2.

Theorem 5 Assume that i is the cut point of G, namely that the graph G is a union of S and T by only common point i, $|V(S)| \geq 2, |V(T)| \geq 2, V(S) \cap V(T) = \{i\}$, and that there are directed cycles throughing i both in S and T, then

$$\begin{aligned} G^* = \quad & S^* + T^* - 1 - x \\ & + \tfrac{1}{1 - \frac{1}{x}(S^{(i)}_{(i)} - x)(T^{(i)}_{(i)} - x)} \left(\tfrac{1}{x}(S_{(i)} - x)(T^{(i)} - x) + \right. \\ & + \tfrac{1}{x}(S^{(i)} - x)(T_{(i)} - x) + \tfrac{1}{x^2}(S_{(i)} - x)(T^{(i)}_{(i)} - x)(S^{(i)} - x) + \\ & \left. + \tfrac{1}{x^2}(T_{(i)} - x)(S^{(i)}_{(i)} - x)(T^{(i)} - x) \right) \end{aligned}$$

Proof The directed walk set of G can be divided into following typies:

(1) to (5) are the same with theorem 2 and their proof are also similar with theorem 2;

(6) They starting at a point (except i) in S go through the i and orbit around some directed cycle in T, then come back via the i to S. They again orbit around some directed cycle in S, then through the i end at some point(except the i) in T;

(7) They starting at a point (except i) in T go through the cut point i and orbit around some directed cycle in S, then come back via i to T. They again orbit around some directed cycle in T, then through the i end at some point (except i) in S; $\cdots$, Repeating this process, they go and come back between S and T via the cut point i for 3 times, 4 times, $\cdots$. Thus

$$\begin{aligned}
G^* = \;& S^* + T^* - 1 - x + \tfrac{1}{x}(S_{(i)} - x)(T^{(i)} - x) \\
& + \tfrac{1}{x}(S^{(i)} - x)(T_{(i)} - x) + \tfrac{1}{x^2}(S_{(i)} - x)(T_{(i)}^{(i)} - x)(S^{(i)} - x) \\
& + \tfrac{1}{x^2}(T_{(i)} - x)(S_{(i)}^{(i)} - x)(T^{(i)} - x) \\
& + \tfrac{1}{x^3}(S_{(i)} - x)(T_{(i)}^{(i)} - x)(S_{(i)}^{(i)} - x)(T^{(i)} - x) \\
& + \tfrac{1}{x^3}(T_{(i)} - x)(S_{(i)}^{(i)} - x)(T_{(i)}^{(i)} - x)(S^{(i)} - x) \\
& + \tfrac{1}{x^4}(S_{(i)} - x)(T_{(i)}^{(i)} - x)(S_{(i)}^{(i)} - x)(T_{(i)}^{(i)} - x)(S^{(i)} - x) \\
& \tfrac{1}{x^4}(T_{(i)} - x)(S_{(i)}^{(i)} - x)(T_{(i)}^{(i)} - x)(S_{(i)}^{(i)} - x)(T^{(i)} - x) + \cdots \\
= \;& S^* + T^* - 1 - x + \tfrac{1}{x}(S_{(i)} - x)(T^{(i)} - x) \cdot \\
& \cdot\left(1 + \tfrac{1}{x^2}(S_{(i)}^{(i)} - x)(T_{(i)}^{(i)} - x) + \tfrac{1}{x^4}(S_{(i)}^{(i)} - x)^2(T_{(i)}^{(i)} - x)^2 + \cdots\right) \\
& + \tfrac{1}{x}(T_{(i)} - x)(S^{(i)} - x)\left(1 + \tfrac{1}{x^2}(S_{(i)}^{(i)} - x)(T_{(i)}^{(i)} - x)\right. \\
& \left. + \tfrac{1}{x^4}(S_{(i)}^{(i)} - x)^2(T_{(i)}^{(i)} - x)^2 + \cdots\right) + \tfrac{1}{x^2}(S_{(i)} - x)(T_{(i)}^{(i)} - x)(S_{(i)} - x) \cdot \\
& \cdot\left(1 + \tfrac{1}{x^2}(S_{(i)}^{(i)} - x)(T_{(i)}^{(i)} - x) + \tfrac{1}{x^4}(S_{(i)}^{(i)} - x)^2(T_{(i)}^{(i)} - x)^2 + \cdots\right) \\
= \;& S^* + T^* - 1 - x + \frac{1}{1 - \frac{1}{x}(S_{(i)}^{(i)} - x)(T_{(i)}^{(i)} - x)}\left(\tfrac{1}{x}(S_{(i)} - x)(T^{(i)} - x)\right. \\
& + \tfrac{1}{x}(T_{(i)} - x)(S^{(i)} - x) + \tfrac{1}{x^2}(S_{(i)} - x)(T_{(i)}^{(i)} - x)(S^{(i)} - x) \\
& \left. + \tfrac{1}{x^2}(T_{(i)} - x)(S_{(i)}^{(i)} - x)(T^{(i)} - x)\right)
\end{aligned}$$

Corollary Assume that $G_{\dot{i}}$ represents the graph obtained by deleting the loop at point i in G, and there is directed cycle through point i in $G_{\dot{i}}$, then

$$G^* = G_{\dot{i}}^* + \frac{1}{1 - x + x^2 - xG_{\dot{i}(i)}^{(i)}}\left(G_{\dot{i}(i)}G_{\dot{i}}^{(i)} + x^2 G_{\dot{i}(i)}^{(i)} - x^3\right)$$

Proof Regarding i as cut point, by theorem 5, $S = G_{\dot{i}}, T = \dot{K}_1$, then

$$\begin{aligned}
G^* = \;& G_{\dot{i}}^* + \tfrac{1}{1-x} - 1 - x \\
& + \frac{1}{1 - \frac{x}{1-x}(G_{\dot{i}}^{(i)} - x)}\left(\tfrac{x}{1-x}(G_{\dot{i}(i)} - x) + \tfrac{x}{1-x}(_{\dot{i}}^{(i)} - x)\right. \\
& \left. + \tfrac{1}{1-x}(G_{\dot{i}(i)} - x)(G_{\dot{i}}^{(i)} - x) + \tfrac{x^2}{(1-x)^2}(G_{\dot{i}(i)}^{(i)} - x)\right)
\end{aligned}$$

By arranging, the result is obtained immediately.

Similarly, we have

Theorem 6 Assume that (i, j) and (k, l) are the cut set of G, $G - (i, j) - (k, l) = S \dot{+} T, i, l \in V(S); j, k \in V(T)$, and there are directed cycles through i, j and k, l, respectively, both in S and T, then

$$\begin{aligned}
G^* \; & = S^* + T^* - 1 + \\
& + \frac{1}{1 - S_{(i)}^{(l)} T_{(k)}^{(j)}}\left(S_{(i)} T^{(j)} + T_{(K)} S^{(l)} + S_{(i)} T_{(k)}^{(j)} S^{(l)} + T_{(k)} S_{(i)}^{(l)} T^{(j)}\right)
\end{aligned}$$

References

[1] Ye Xiuming, The Generating Function of Graphical Sequences, Journal of Shanghai University of Science and Technology, No.1,1994.

[2] Ye Xiuming, Zhu Dingxiang, Gong Chongguang, Sequence Enumeration Methods For Count Of Cycles Of Rectangular Lattice Graphs Journal Of Nanjing University (Special Issue on Graph Theory), Vol.27,1991.

MAX 2SAT and Directed Multicut

Lijie Shi and Rongheng Li

(*Institute of Applied Mathematics, Academia Sinica, Beijing* 100080)

Abstract In this paper, first we give a new algorithm for the MAX 2SAT problem with ratio 3/4. Second we extend the undirected multicut problem to directed multicut in directed graghs. We prove that the directed multicut problem is NP-hard for any fixed $k \geq 3$ and MAX SNP-hard on graphs whose underlying simple graphs are trees for arbitrary k. We also show that for any given graph $G = (V, E)$, $|V(G)| = 2n$, finding the minimum n-cut for the set of node pairs $\{(u_i, v_i)\ 1 \leq i \leq n\}$ with $u_i \neq v_i$, $u_i \neq u_j$ and v_j for $i \neq j$, is MAX SNP-hard, even when G is a bipartite graph with the property: edge $[u_i, v_j] \in E(G)$ if and only if $[u_j, v_i] \in E(G)$. The corresponding directed n-cut problems is also SNP-hard. This means these problems do not have a polynomial time approximation scheme (PTAS) unless $P = NP$.

Keywords Approximation; Computational complexity; Directed multicut; MAX 2SAT ; MAX SNP-hard

1. Introduction

During the last few years, there have been some very significant developments in the study of approximation of NP-hard optimization problems. In 1988, Papadimitriou and Yannakakis[8] defined a special reduction that preserves certain approximability properties. Using this reduction and based on Fagin's syntactically definition of the class NP, they defined a complexity class MAX SNP. Several optimization problems have been shown to be MAX SNP-hard under L-reduction, such as, Max 2-satisfiability ,Maximum cut, Minimum bounded node cover, Maximum bounded 3-dimensional matching [7], Maximum bounded satisfiability for Horn function [10], etc. It is known that every problem in this class can be approximated with some constant factor.

Recently, Arora et al.[1] proved, using results in probabilistic checking of proof, that it is impossible for any MAX SNP-hard problem to have a polynomial time approximation scheme (PTAS) unless P=NP. This means that any MAX SNP-complete problem can be approximated with a constant but not with every constant.

Maximum Satisfiability (MAX SAT) is one of the basic MAX SNP-complete problem. Johnson [6] presented one appproximatin algorithm B_2 for MAX kSAT. For any boolean formula F, if every clause has at least k literals, then B_2 produces a truth assignment that satisfies at least $(1 - \frac{1}{2^k})|F|$. In the case of MAX 2SAT, Poljak and Turzik [9] gave a diferent algorithm with the same performance as B_2. Recently Yannakakis[12] presents a 3/4 approximation algorithm by the application of network flow techmiques. The crucial thing is to find a subset V_0 of the variables, such that the MAX 2SAT with exactly two literals in each clause on V_0 has the same approximation ratio for the MAX 2SAT (with at most two literals). By the above algorithm B_2 or Poljak and Turzik's, MAX SAT can be approximated with ratio 3/4.

The classical max-flow min-cut theorems play an important role in combinatorial optimization. Multicut, as an extension of two terminals, was proposed and studied in 1969[5]. Recently multicut problem draws more attention and gets a new development. From the stimulation of the greedy algorithm for MAX 2SAT, we extend the multicut to directed multicut in directed graphs. We prove that the directed multicut problem is NP-hard for any fixed $k \geq 3$ and MAX SNP-hard on graphs whose underlying simple graphs are trees for arbitrary k.

2. Definition

Let V=$\{x_1, \ldots, x_n\}$ be a set of n boolean variables, each of which takes values 1(true) or 0 (false). Let $\overline{V} = \{\bar{x}_1, \ldots, \bar{x}_n\}$ denote the set of boolean complements of these variables, i.e. $\bar{x}_j = 1 - x_j$ for $j = 1, \ldots, n$. The elements of $V \cup \overline{V}$ are called literals, denoted by ξ_i. A clause C is the disjunction $\xi_1 \cup \xi_2 \cup \ldots \cup \xi_r$ of a finite number of literals; we assume that $\xi_i \neq \xi_j, \bar{\xi}_j$, for all $i \neq j$. When r=1, C is called a unit clause.

A boolean formula ϕ (in conjunctive normal form(CNF)) is the conjunction of a finite number of distinct clauses

$$\phi = \phi(x) = C_1 \cap C_2 \cap \ldots \cap C_m.$$

The maximum satisfiability (MAX-SAT) problem is the following.

MAX-SAT

Instance: A set V= $\{x_1, \ldots, x_n\}$ of boolean variables and a formula ϕ in CNF.

Problem: Find the truth assignment for V that satisfies the greatest number of clauses which in ϕ.

The MAX kSAT is the restriction of the MAX SAT problem to instances that all clauses have lengh at most k. For any $k \geq 2$, MAX kSAT is MAX SNP-complete. This means MAX kSAT problem does not admit a PTAS.

In 1988, Papadimitriou and Yannakakis[8] defined the following complexity class MAX SNP and a reduction L-reduction

Definition 1. MAX SNP is the class of optimization problem P such that

$$OPT(I) = \max_s |\{\bar{x} : \Psi_p(\bar{x}, S, I)\}|$$

where Ψ_p is a quantifier free formula, I an instance of P encoded as a relation, $OPT(I)$ the optimal cost for I and S a relation describing the solution.

Minimization problems will be placed in the class through the following L-reduction to maximization problems. Such as the Minimum bounded node cover problem.

Definition 2. Let Π_1 and Π_2 be two optimization (i.e., maximization or minimization) problems, we say that Π_1 L-reduces to Π_2 if there are two polynomial time algorithms f and g and two positive constants α and β, such that for any instance I of Π_1,

(1) $OPT(f(I)) \leq \alpha \cdot OPT(I)$.

(2) Given any solution of f(I) with cost c_2, algorithm g produces in polynomial time a solution of I with cost c_1 such that

$|c_1 - OPT(I)| \leq \beta \cdot |OPT(f(I)) - c_2|$.

Clearly, if Π_1 L- reduces to Π_2 and problem Π_2 can be approximated in polynomial time within a factor of $1 + \epsilon$, then Π_1 can be approximated within the factor $1 + \alpha\beta\epsilon$.

A problem is MAX SNP-hard if every problem in MAX SNP can L-reduce to it. Since the composition of two L -reductions is also a L-reduction. A problem is MAX SNP-hard if it can be L-reduced to another MAX SNP-hard problem.

In 1969, T.C. Hu[5] raised the following question: Given a graph $G=(V,E)$, a set of node pairs $\{(u_i,v_i)|1\leq i\leq k\}$, how to find the minimum set of edges whose deletion leads each pair of nodes u_i and v_i not connected. This set of edges is called a multicut denoted by $C(u_i,v_i|1\leq i\leq k)$. This is a generalization of the former "min cut/ max-flow" problem. It has drawn much attention for many years. It was proved that the multicut problem is NP-hard for arbitrary graphs for any fixed k, when $k\geq 3$[3]. In this paper, we consider the following generalization of multicut in directed graphs. Given a directed graph G, how to find the set of edges E' such that each pair of nodes u_i and v_i is not in the same strongly cónnected component of the Graph $G-E'$. E' is called the directed multicut.

3. Greedy Algorithm for MAX 2SAT

Yannakakis [12] showed: if MAX 2SAT with exactly two literals per clause can be approximated with ratio r, then so can MAX 2SAT with at most two literals per clause. Therefore, we will only consider the MAX 2SAT with exactly two literals in the following discussion. Let F be such an instance of MAX 2SAT, $F=C_1\cap C_2\cap\cdots\cap C_m$. We construct a directed graph $G(F)$ as follows[2]:

(1) For each variable in F, we add two nodes x_i and $\overline{x}_i$ in $G(F)$.

(2) For each clause $C_i=(a\vee b)$, we add two directed edges $\overline{a}\rightarrow b$ and $\overline{b}\rightarrow a$, abbreviated as $(\overline{a},b)$ and $(\overline{b},a)$ respectively, to $G(F)$.

we say these two edges are correspondent each other and these two edges are the corresponding edges of C_i in $F(G)$. Sometimes we also say that F and $G(F)$ are corresponding, if there is no confusion. The transformation from F to $G(F)$ is denoted by h_1. In the following discussion, we denote, by $[a,b]$, an undirected edge e connected nodes a and b.

We will use the terms "literal" and "node " interchangeably. A graph is strongly connected if there is a path from any node to any other. The maximal strongly connected subgraphs of a graph are called its strong components. we can easily prove the following lemma.

Lemma 1 [2]. A boolean formula F of 2SAT is satisfiable if and only if no strong component of the graph $G(F)$ contains both a literal and its complement. (minimum directed multicut for node pairs $\{(x_i,\overline{x}_i)|1\leq i\leq n\}$ in $G(F)$ equals zero)

Let $[S_1,S_2]$ denote the set of edges between S_1 and S_2, (S_1,S_2) denote the set of edges from S_1 to S_2 where S_i are subsets of $V(G(F))$, $\overline{S}$ denote the literal complement set of literals in S. We give the following greedy algorithm to find the multicut for the node pairs $\{(x_i,\overline{x}_i)|1\leq i\leq n\}$ in $G(F)$ such that we get a set of satisfiable clauses.

Greedy algorithm:

(1) Set $S,\overline{S}=\phi$, $T=\{y|y$ is literal occuring in $F\}$

(2) If $T=\phi$, halt and return S.

(3) Let y be any literal in T; If $|(y,S)|\geq|(y,\overline{S})|$, set $S=S\cup\{y\}$, $\overline{S}=\overline{S}\cup\{\overline{y}\}$; otherwise, set $S=S\cup\{\overline{y}\}$, $\overline{S}=\overline{S}\cup\{y\}$.

(4) Set $T=T-\{y,\overline{y}\}$

(5) Goto 2.

Theorem 2. The MAX 2SAT problem can be approximated by greedy algorithm with ratio 3/4.

Proof. Let F be a given formula, $G(F)$ be the corresponding graph of F, $G_F(S)$ the subgraph induced by S in $G(F)$. From the greedy algorithm, obviously, $|[S,\overline{S}]|\leq|E(G_F(S))|+|E(G_F(\overline{S}))|$, so $|[S,\overline{S}]|\leq\frac{1}{2}|E(G(F))|$.

For any edge $e = (u, v) \in (S, \overline{S})$, $u \in S$, $v \in \overline{S}$, by the algorithm, we know $\overline{u} \in \overline{S}, \overline{v} \in S$, the corresponding edge of (u, v) is $(\overline{v}, \overline{u}) \in E(G(F))$. This means, for a clause (ξ_i, ξ_j), if one of its two corresponding edges in $G(F)$ belongs to $(S, \overline{S})$, then the other edge belongs to $(S, \overline{S})$ too. Let $E_0 = (S, \overline{S})$ or $(\overline{S}, S)$ with the less edge of the two edge sets. Apparently, there is no node pair $(x_i, \overline{x}_i)$ in the same strong component in $G - E_0$. Let $F_0 = \{C_i |$the corresponding edge of C_i in $G(F)$ belong to $E_0\}$. Then we have $|F_0| = \frac{1}{2}|E_0| \leq \frac{1}{4}|[S, \overline{S}]| \leq \frac{1}{8}|E(G(F))|$. Because $|E(G(F))| = 2|F|$, so $|F_0| \leq \frac{1}{4}|F|$. The residual clauses set $F_{sat} = F - F_0$ is satisfiable from Lemma 1 and $|F_{sat}| \geq \frac{3}{4}|F|$. ∎

The following example shows that the relation $|F_1| \geq \frac{3}{4}|F|$ is tight.

$$F = (\overline{x}_1 \vee x_2) \wedge (x_1 \vee x_2) \wedge (\overline{x}_1 \vee \overline{x}_2) \wedge (x_1 \vee \overline{x}_2).$$

4. Directed Multicut

Let $G(V, E)$ be a given undirected graph, we construct the corresponding directed graph $\overline{G} = (\overline{V}, \overline{E})$ by $\overline{V}(\overline{G}) = V(G)$, $\overline{E}(\overline{G}) = \{a \to b, b \to a | [a, b] \in E(G)\}$. That is, for each undirected edge of $E(G)$, there are two opposite edges to correspond in $\overline{G}$. These two opposite edges are called opposite pair. $\overline{G}$ is called the double directed graph of G.

Lemma 3. Let G' be a directed graph. Suppose $\overline{M}$ is a directed k-cut for the k node pairs $\{(s_i, \overline{s_i}) | 1 \leq i \leq k\}$ in graph G', If both (a, b) and (b, a) belong to $\overline{M}$, then either $\overline{M} - (a, b)$ or $\overline{M} - (b, a)$ is still a directed k-cut in G'.

Proof. We will prove it by contradiction. If neither $\overline{M} - (a, b)$ nor $\overline{M} - (b, a)$ is a directed k-cut, there exist some node pairs which are strongly connected in $G'_1 = (V', (E(G') \backslash \overline{M}) \cup (a, b))$ and some in $G'_2 = (V', (E(G') \backslash \overline{M}) \cup (b, a))$. If there are two different node pairs which are strongly connected in G'_1 and G'_2 respectively. Without loss of generality, suppose the two node pairs are $(s_1, \overline{s}_1)$ and $(s_2, \overline{s}_2)$. From the assumption, there exist two directed paths $P_1(s_1, \overline{s}_1)$, ${P_1}'(\overline{s}_1, s_1)$ in ${G_1}'$ that at least one of them passes edge (a, b) and $P_2(s_2, \overline{s}_2)$, ${P_2}'(\overline{s}_2, s_2)$ in ${G_2}'$ that at least one of them passes edge (b, a). Without loss of generality, let

$$\begin{aligned}
P_1 &= P_1(s_1, \overline{s}_1) = (s_1, u_{1,1}, \cdots, u_{1,r_1} = a, b = u_{1,r_1+1}, \cdots, u_{1,t_1}, \overline{s}_1) \\
P'_1 &= P'_1(\overline{s}_1, s_1) = (\overline{s}_1, v_{1,1}, \cdots, v_{1,q_1}, s_1) \\
P_2 &= P_2(s_2, \overline{s}_2) = (s_2, u_{2,1}, \cdots, u_{2,r_2} = b, a = u_{2,r_2+1}, \cdots, u_{2,t_2}, \overline{s}_2) \\
P'_2 &= P'_2(\overline{s}_2, s_2) = (\overline{s}_2, v_{2,1}, \cdots, v_{2,q_2}, s_2).
\end{aligned}$$

We will construct two directed pathes to connect node pair s_1 and $\overline{s}_1$ from P_1,P'_1,P_2, P'_2 without using edges (a, b) and (b, a).

If (b, a) is not on path P'_2, Let

$$\begin{aligned}
P_3 = \{&s_1, u_{1,1}, \cdots, a = u_{2,r_2+1}, \cdots, u_{2,t_2}, \overline{s}_2, v_{2,1}, \cdots, v_{2,q_2}, s_2, \\
&u_{2,1}, \cdots, b = u_{1,r_1+1}, \cdots, u_{1,t_1}, \overline{s}_1\}
\end{aligned}$$

From P_3, we get a directed path $P_3(s_1, \overline{s}_1)$ without passing directed edge (a, b).

If (b, a) is on path P'_2 , suppose $b = v_{2,i}$ for some $1 \leq i \leq q_2 - 1$, then $a = v_{2,i+1}$. Let

$$\begin{aligned}
P_3 = \{&s_1, u_{1,1}, \cdots, a = u_{2,r_2+1}, \cdots, u_{2,t_2}, \overline{s}_2, v_{2,1}, \cdots, v_{2,i} = \\
&b = u_{1,r_1+1}, \cdots, u_{1,t_1}, \overline{s}_1\}
\end{aligned}$$

Thus we also get a directed path $P_3(s_1, \overline{s}_1)$, without passing directed edge (a, b). So if (a, b) is not on path P'_1, then s_1 and $\overline{s}_1$ are strongly connected in $G' \backslash \overline{M}$ by P'_1 and

$P_3(s_1, \overline{s}_1)$. This is a contradiction. So (a, b) must appear in path P'_1. Suppose $a = v_{1,j}$, $b = v_{1,j+1}$, for some $1 \le j \le q_1 - 1$. Let

$$P'_3 = \{\overline{s}_1, v_{1,1}, \cdots, v_{1,j} = a = v_{2,i+1}, \cdots, v_{2,q_2}, s_2,$$
$$u_{2,1}, \cdots, b = v_{1,j+1}, \cdots, v_{1,q_1}, s_1\} \text{ for } (b, a) \in P'_2$$

or

$$P'_3 = \{\overline{s}_1, v_{1,1}, \cdots, v_{1,j} = a = u_{2,r_2+1}, \cdots, u_{2,t_2}, \overline{s}_2, v_{2,1}, \cdots, v_{2,q_2}, s_2, u_{2,1}, \cdots,$$
$$b = v_{1,j+1}, \cdots, v_{1,q_1}, s_1\} \text{ for } (b, a) \notin P'_2.$$

We can get a directed path $P'_3(\overline{s}_1, s_1)$ from P'_3 without use of directed edge (a, b). Obviously, $P_3(s_1, \overline{s}_1)$ and $P'_3(\overline{s}_1, s_1)$ made nodes s_1 and $\overline{s}_1$ strongly connected in $G' \backslash \overline{M}$. This is a contradiction.

If s_1 and $\overline{s}_1$ is the only strongly connected nodes pair in both graph G'_1 and G'_2, similar proof can lead the same contradiction. Therefore either $\overline{M} - (a, b)$ or $\overline{M} - (b, a)$ is still a directed k-cut in G'. ∎

Corollary 4. Let $\overline{M}$ be a directed k-cut for the k node pairs $\{(s_i, \overline{s_i}) | 1 \le i \le k\}$ in graph $\overline{G}$, there exists a directed k-cut $\overline{M}'$ such that $\overline{M}'$ contains no opposite edge pair and $|\overline{M}'| \le |\overline{M}|$.

The strong component of a directed graph can be found in polynomial time by an algorithm of Tarjan[11]. So, we can decide whether $\overline{M} - (a, b)$ is a directed k-cut in polynomial time. Similarly the directed k-cut $\overline{M}'$ in corollary 4 can be found in polynomial time.

Recently, Dahlhaus et al. [3] proved that the undirected multicut problem for arbitrary graph is NP-hard for any fixed $k \ge 3$ .We will prove the NP-hard result for the directed multicut problem by corollary 4.

Theorem 5. The directed multicut problem for arbitrary directed graph is NP-hard for any fixed $k \ge 3$.

Proof. Let G be any undirected graph, $\overline{G}$ is the double directed graph of G. Suppose M_0 and $\overline{M}_0$ are the minimum k-cut in G and the minimum directed k-cut in $\overline{G}$ for the node pairs $\{(u_i, v_i) | 1 \le i \le k\}$ respectively. Let $\overline{M} = \{(a, b), (b, a) | [a, b] \in M_0\}$. Obviously, $\overline{M}$ is a directed k-cut in $\overline{G}$. From corollary 4, we can find a directed k-cut $\overline{M}'$ in $\overline{M}$ such that $\overline{M}'$ does not contain any opposite edge pair. So $|\overline{M}'| = \frac{1}{2}|\overline{M}| = |M_0|$. Therefore $|\overline{M}_0| \le |M_0|$. Let $M = \{[a, b] \in E(G) | \text{either } (a, b) \text{ or } (b, a) \in \overline{M}_0\}$. Obviously , M is an undirected k-cut in G, so, $|M_0| \le |M|$. Because $\overline{M}_0$ is a minimum directed k-cut, $\overline{M}_0$ does not contain any opposite edge pair, therefore $|\overline{M}_0| = |M| \ge |M_0|$. We get $|\overline{M}'| = |\overline{M}_0| = |M_0|$. So M_0 is a minimun k-cut in G if and only if $\overline{M}'$ is the minimun directed k-cut in $\overline{G}$. ∎

The underlying simple graph of a given directed graph G is: for any directed edge (u,v) of $E(G)$, we replace all the edges connecting nodes u and v by only one undirected edge $[u, v]$ or uv. Garg et al. [4] proved that the undirected minimun multicut problem for arbitrary k is NP-hard and MAX SNP-hard for trees. By the above results, we prove the following theorem.

Theorem 6. The directed minimum multicut problem for arbitrary k is NP-hard and MAX SNP-hard for directed graphs whose underlying simple graphs are trees.

Proof. Let G be an undirected tree, $\overline{G}$ the double directed graph of G, k any integer. Suppose M_0 and $\overline{M}_0$ are the minimum k-cut in G and minimum directed k-cut in $\overline{G}$ for the node pairs $\{(u_i, v_i)|1 \le i \le k\}$ respectively. We know that $|M_0| = |\overline{M}_0|$ from the proof of Theorem 5.

Let $\overline{M}$ be any directed k-cut for $\{(u_i, v_i)|1 \le i \le k\}$ in $\overline{G}$. From corollary 4, we can find a directed k-cut $\overline{M}'$ such that $\overline{M}'$ does not contain any opposite edge pair in polynomial time. Let $M = \{[a, b] \in E(G)|\text{either } (a, b) \text{ or } (b, a) \in \overline{M}'\}$. Obviously M is also a k-cut for $\{(u_i, v_i)|1 \le i \le k\}$ in G , and $|M| = |\overline{M}'| \le |\overline{M}|$. So $|M| - |M_0| \le |\overline{M}| - |\overline{M}_0|$. Let $\alpha = 1$, $\beta = 1$, thus we have a L-reduction. ∎

From now, we will focus on proving the following theorem.

Theorem 7. For any undirected graph $G = (V, E)$, $|V(G)| = 2n$, the problem of finding the minimum n-cut for the set of node pairs $\{(u_i, v_i)|1 \le i \le n\}$ with $u_i \ne v_i$, $u_i \ne u_j$ and v_j for $i \ne j$ is MAX SNP-hard, even when G is a bipartite graph with the property: edge $[u_i, v_j] \in E(G)$if and only if $[u_j, v_i] \in E(G)$.

By the similar proof of Theorem 6, we get the following theorem from Theorem 7.

Theorem 8. For any directed graph $G = (V, E)$, $|V(G)| = 2n$, the problem of finding the minimum directed n-Cut for the set of node pairs $\{(u_i, v_i)|1 \le i \le n\}$ with $u_i \ne v_i$, $u_i \ne u_j$ and v_j for $i \ne j$ is MAX SNP-hard, even when G is a bipartite graph with the property: edge $(u_i, v_j)((v_i, u_j)) \in E(G)$if and only if $(u_j, v_i)((v_j, u_i)) \in E(G)$.

We will prove Theorem 7 by the following reduction, h, from the MAX cut to the multicut problem. The MAX cut problem is : given a graph , find a partition of the node set , that maximizes the number of edges in the cut. In this paper , sometimes we use its another expression – the minimum edge-deletion bipartite subgraph problem: given a graph G ,find the minimum edges E_0 such that $G - E_0$ is a bipartite graph.

Let G be a graph, we use x_i to label every node of $V(G)$. The graph H is constructed as follows:

(1) For each node x_i, we add two nodes x_i and $\overline{x}_i$ in H.

(2) For each edge $e = [x_i, x_j]$ of $E(G)$, add two undirected edges $[\overline{x}_i, x_j]$ and $[\overline{x}_j, x_i]$ to H.

This reduction is denoted by h_2.

For example: Let $G_0 = \{V_0, E_0\}$, $V_0 = \{x_1, x_2, x_3, x_4\}$, $E_0 = \{x_1x_2, x_2x_3, x_3x_4, x_4x_1, x_2x_4\}$, then $V(H) = \{x_1, \overline{x}_1, \cdots, x_4, \overline{x}_4)\}$. H is the following graph (Fig.1).

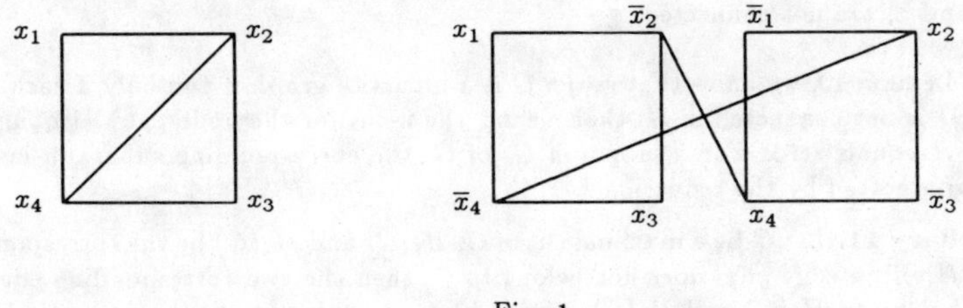

Fig. 1

Lemma 9. For any graph G, the graph H is a bipartite graph.

Proof. From the reduction, we know one edge $e = [u, v] \in E(H)$ if and only if $[x_i, x_j] \in E(G)$ with $u = x_i$ and $v = \overline{x}_j$ or $u = \overline{x}_i$ and $v = x_j$, not both positive or negative variables. That means any edge $e = uv \in E(H)$ have two expression $e = x_i\overline{x}_j$ or $\overline{x}_ix_j$. If H is not a bipartite graph, then there are odd cycles in G'. Suppose $R = \{u_1, u_2, \cdots, u_{2l}, u_{2l+1}\}$ is an odd cycle. Without loss of generality, we let $u_1 = x_1$, $u_i = x_i$

or $\overline{x}_i$ for $1 \leq i \leq 2l+1$, then $u_2 = \overline{x}_2$, $u_3 = x_3, \cdots, u_{2l} = \overline{x}_{2l}, u_{2l+1} = x_{2l+1}$. From above, we know there is no edge to connect x_1 and x_{2l+1} in G'. Thus H is a bipartite graph. ∎

Lemma 10. The connected graph G is a connected bipartite graph if and only if $H = H_1 \cup H_2$, where $H_i = (V_i, E_i)$ for $i = 1, 2$ are two isolated copies of graph G with $V_1 = \overline{V}_2 = \{\overline{x} | x \in V_2\}$.

Proof. Suppose G is a connected bipartite graph, first we will prove that for any i, nodes x_i and $\overline{x}_i$ are not connected in H. If not, without loss of generality, we can assume x_1 and $\overline{x}_1$ are connected by a path $P(x_1, \overline{x}_1)$ which is the shortest path for all connected pairs $(x_i, \overline{x}_i)$ in H. Let $P = (x_1, \overline{x}_2, x_3, \cdots, x_{2k-1}, \overline{x}_1)$. From the structure of H, we conclude that $(x_1, x_2, x_3, \cdots, x_{2k-1}, x_1)$ is an odd cycle of G. This is a contradiction.

Second, we will prove that any node pair x_r and $\overline{x}_r$ are connected differently with x_1 or $\overline{x}_1$ in H. Because G is connected, there is a path, $P = (x_1, x_2, \cdots, x_r)$, to connect x_1 and x_r in G. Then x_r is connected with x_1 by path $\overline{P} = (x_1, \overline{x}_2, \cdots, x_r)$ in H if r is odd number; x_r is connected with $\overline{x}_1$ by path $\overline{P} = (\overline{x}_1, x_2, \cdots, x_r)$ in H if r is even number. Obviously, if x_r is connected with x_1, then $\overline{x}_r$ is connected with $\overline{x}_1$ and if x_r is connected with $\overline{x}_1$, then $\overline{x}_r$ is connected with x_1. So H is two isolated connected graphs. Let $V_1 = \{x_i | \ x_i \text{ is connected with } x_1 \text{ in } H\}$, $V_2 = \{x_i | \ x_i \text{ is connected with } \overline{x}_1 \text{ in } H\}$, $H_1 = \{V_1, E_1\}$, $H_2 = \{V_2, E_2\}$, then $E(H) = E_1 \cup E_2$. We define the following two mappings:

$$f_1(x) = \begin{cases} x & for\ x \in V_1 \\ \overline{x} & for\ \overline{x} \in V_1 \end{cases} \qquad f_2(x) = \begin{cases} \overline{x} & for\ x \in V_1 \\ x & for\ \overline{x} \in V_1 \end{cases}$$

It is easy to see that f_1 is an isomorphic mapping from G to H_1 and f_2 an isomorphic mapping from H_1 to H_2. So the conclusion is proved.

Conversely, if a graph G satisfies $H = H_1 \cup H_2$, where $H_i = (V_i, E_i)$ for $i = 1, 2$ are two isolated copies of graph G with $V_1 = \overline{V}_2 = \{\overline{x} | x \in V_2\}$. If G is not a bipartite graph, then there exists an odd cycle $(x_1, x_2, \cdots, x_{2k-1}, x_{2k} = x_1)$ in G. From the structure of H, $P = (x_1, \overline{x}_2, \cdots, x_{2k-1}, \overline{x}_{2k} = \overline{x}_1)$ is a path in H. This is a contradiction, because for any i, x_i and $\overline{x}_i$ are not connected. ∎

From Lemma 10, we know that graph G is a bipartite graph if and only if each node pair $\{x_i, \overline{x}_i\}$ is not connected in H, that means, the n-cut for the node pairs $\{(u_i, v_i) | 1 \leq i \leq n\}$ in H equals zero. For a subgraph G_i of G, the corresponding subgraph in H is the subgraph gotten by the reduction h_2.

Corollary 11. Let C be a maximal cut of G, $H_1(C)$ and $H_2(C)$ be the corresponding graphs in H. If an edge [x,y] does not belong to C, then the two corresponding edges in H do not belong to $H_1(C)$ and $H_2(C)$, but nodes x and y belong to the same $H_i(C)$ for $i = 1$ *or* 2.(see Fig.2).

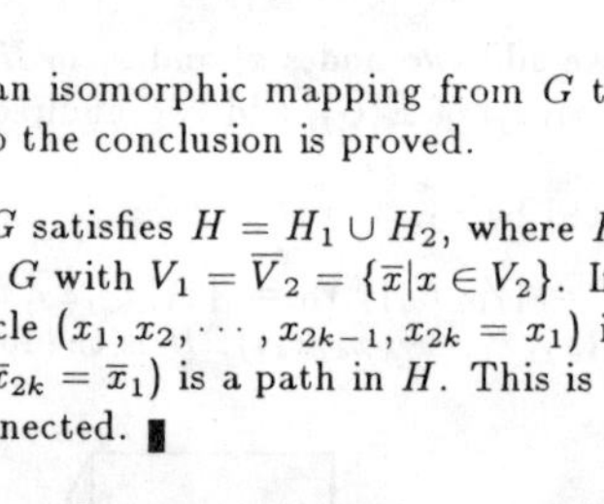

Fig. 2

Now we give the proof of Theorem 7:

Proof. Without loss of generality, we assume graph G is connected.

(1) Let C_0 be the maximum cut for the given graph $G = (V, E)$ with $|V(G)| = n$, M_0 be the minimum n-cut for the node pairs $\{(u_i, v_i)|1 \le i \le n\}$ in H. $C_0' = E(G) - C_0$ is called the residual edge for C_0. $H_1(C_0)$ and $H_2(C_0)$ are the corresponding graphs of $G(C_0)$ in H. $H(C_0')$ is the corresponding graph of $G(C_0')$ in H. From lemma 10 and corollary 11, $E(H(C_0'))$ is the set of edges between $H_1(C_0)$ and $H_2(C_0)$. Obviously, $[H_1(C_0), H_2(C_0)]$ is a n-cut and $|[H_1(C_0), H_2(C_0)]| = 2|C_0'|$. So, $|M_0| \le 2|C_0'| = 2(|E(G) - C_0|)$. Because $|E(G)| \le 2|C_0|$, then $|M_0| \le 2|C_0|$. Let $\alpha = 2$, the condition (1) of L-reduction is satisfied.

(2) Suppose M' be a given n-cut in H, let

$$E(M') = \{e \in E(G)|\text{at least one of } e\text{'s corresponding edges belongs to } M' \text{ in } H\},$$

and $C = E(G) - E(M')$, then $G(C) = (V, C)$ is a bipartite graph. We will prove that there exists a n-cut M_1 with $|M_1| = 2|E(M_1)| \le |M'|$. If $G(C)$ is a connected subgraph of G, then the corresponding edges of $E(M')$ are the edges between $H_1(C)$ and $H_2(C)$. Because each edge $e \in E(G)$ has two corresponding edges in H, so $|M'| = 2|E(M')|$. Let $M_1 = M'$. If $G(C)$ is not connected, let $G_1, \cdots, G_k$ be all the connected components of $G(C)$, then each G_i is a bipartite graph. From Lemma 10, let H_{i1}, H_{i2} for $1 \le i \le k$, be the corresponding graphs of G_i in H, $V(H_{i1}) = \overline{V}(H_{i2})$. We construct graph $\overline{H}$ as follow: $V(\overline{H}) = \{H_{i1}, H_{i2}|1 \le i \le k\}$, $E(\overline{H})$ contains all the corresponding edges of $E(M')$ in $\overline{H}$. By the greedy algorithm in section 3, we get two node sets S and $\overline{S}$, $|[S, \overline{S}]| \le \frac{1}{2}|E(\overline{H})| = \frac{1}{2}(2|E(M')|) = |E(M')|$. Let $M_1 = [S, \overline{S}]$, then $|M_1| = 2|E(M_1)|$. So we get a n-cut M_1 with $|M_1| = 2|E(M_1)|$. $C_1 = E(G) - E(M_1)$ is a new cut of G. We have

$$\begin{aligned}|C_0 - C_1| &\le (|E(G)| - \frac{1}{2}|M_0|) - (|E(G)| - |E(M_1)|)\\ &\le \frac{1}{2}(|M_1| - |M_0|)\\ &\le \frac{1}{2}(|M'| - |M_0|).\end{aligned}$$

Let $\beta = \frac{1}{2}$, the condition (2) of L-reduction holds. Thus we have a L-reduction. Because MAX cut problem is MAX SNP-complete, the conclusion is proved. ∎

5. Conclusions

From Lemma 3 and Corollary 4, we give a general method for generalizing the NP-hard and SNP-hard results of undirected multicut problem to the directed multicut problems. That is, if the multicut problems on some undirected graphs are NP-hard or MAX SNP-hard, then the corresponding directed graphs are NP-hard or MAX SNP-hard too. On the other hand, our results can easily be extended to the weighted cases.

For the multicut problem on trees, there is an approximation algorithm with ratio 2 [4]. For the multicut problem and directed multicut problem on bipartite graph with $[u_i, v_j] \in E(G) \nLeftrightarrow [u_j, v_i] \in E(G)$, the proof of Theorem 7 also means that the greed algorithm in section 3 is an approach way. We leave open the question whether the undirected or directed multicut problems on trees or these special bipartite graphs can be approximated with some constant factor.

6. Acknowledgement

We are grateful to our supervisor Professor Minyi Yue for his encouragement on this problem.

References

[1] S. Arora, C. Lund, R. Motwani, M. Sudan, and M. Szegedy, Proof verification and hardness of approximation problems, in: Proc. 33nd Ann. IEEE Symp. on Foundations of Computer Science (1992) 14-23.

[2] B. Aspvall, M.F. Plass and R.E. Tarjan, A linear algorithm for testing the truth of certain quantified boolean formulas, 8(3)(1979) 121-123.

[3] E. Dahlhaus, D.S. Johnson, C.H. Papadimitriou, P.D. Seymour and M. Yannakakis, The complexity of multiway cuts, Proceedings, 24th ACM Symposium on Theory of Computing(1992).

[4] N. Garg, V.V.Vazirani, and M. Yannakakis, Primal-dual approximation algorithms for integral flow and multicut in trees, with applications to matching and set cover, Proceedings, Intl. Coll. Automata, Languages and Programming(1993)

[5] T.C. Hu, Integer Programming and Network Flows, Addison-Wesley Publishing Co., Reading, MA, 1969.

[6] D.S. Johnson, Approximation algorithms for combinatorial problems, J. Comput. System Sci. 9(1974) 256-278.

[7] V. Kann, Maximum bounded 3-dimensional matching is MAX SNP-complete, Inform. Process. Lett. 37 (1991) 27-35.

[8] C.H. Papadimitriou and M. Yannakakis, Optimization, approximation, and complexity classes, J. Comput. System Sci. 43 (1991) 425-440.

[9] S. Poljak and D. Turzik, A polynomial algorithm for constructing a large bipartite subgraph, with an application to a satisfiability problem, Canadian J. of Mathematics 34(3)(1982) 519-524

[10] L.J. Shi, Maximum bounded satisfiability for Horn function is MAX SNP-hard, Technical Report-TR 94-038 (1994), Institute of Applied Mathematics, Academia Sinica.

[11] R.E. Tarjan, Depth first search and linear graph algorithms, SIAM J. Comput. 1(2)(1972) 146-160.

[12] M. Yannakakis, On the approximation of maximum satisfiability, Proceedings, 3rd Ann. ACM-SIAM Symp. on Disc. Algorithm.

ON CHROMATIC COEFFICIENTS OF UNLABELED GRAPHS

by

Hugo S. H. Sun, California State University, Fresno

Ixin Wen, Reedley College

Li Nian Zu, The Second Technological University of Shanghai

Abstract

The first four coefficients of the chromatic polynomial of an unlabeled graph are given here in terms of symmetries of the vertex subsets .

1. Introduction. Let G be an unlabeled graph with p vertices, g a labeled graph of G, A(g) the automorphism group of g, i.e., $A(g) = \{\pi \in S_p | \pi(g) = g\}$, where S_p is the symmetric group of degree p.

Let γ_p be the set of all labeled graphs with p vertices and Γ_p be the set of all unlabeled graphs with p vertices. For example, for n = 3,

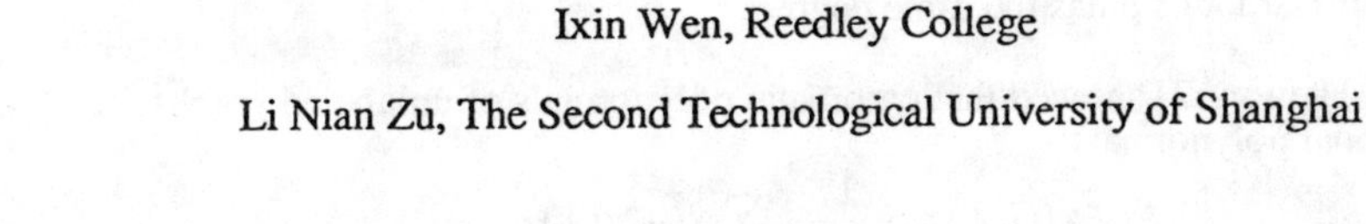

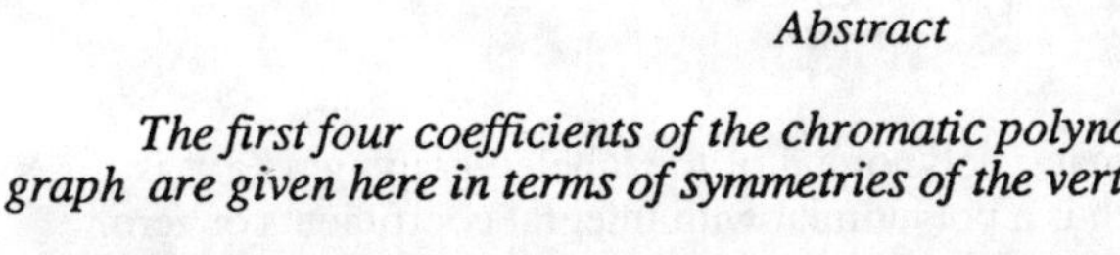

Therefore, $G_1 = \{g_1\}$, $G_2 = \{g_2, g_3, g_4\}$, $G_3 = \{g_5, g_6, g_7\}$, and $G_4 = \{g_8\}$. This implies that an unlabeled graph may be regarded as a subset of γ_p.

Let the vertex set of g be $V(g) = \{1, 2, 3, ..., p\}$, and $d_{g(i)}$, $1 \le i \le p$, the degree of the vertex i in V(g).

Lemma 1. If $\pi \in A(g)$, then $d_g(\pi(i))= d_g(i)$, for all $i \in V(g)$.

Proof. Let $j = \pi(i)$, and $\pi(g) = g_1$. Then $d_{g_1}(j) = d_g(i)$. If $d_g[\pi(i)] \neq d_g(i)$, i.e., $d_g(j) \neq d_g(i) = d_{g_1}(j)$, which means that $g_1 \neq g$. Therefore, $\pi \notin A(g)$, a contradiction.

Intituitively Lemma 1 implies that every vertex of a cycle of an automorphism π of g has the same degree.

P. Hanlon [2] showed that chromatic polynomials of unlabeled graphs are rational polynomials:

$$\chi(G,\lambda) = \frac{1}{|A(g)|} \sum_{\pi \in A(g)} \chi(g,\pi,\lambda),$$

where $\chi(g, \pi, \lambda)$ is the chromatic polynomial of the labeled graph $g{:}\pi$ ($g{:}\pi$ is the quotient graph of g) which is a polynomial with integral coefficients or zero.

In order to discuss the chromatic coefficients of unlabeled graphs, we assume $\chi(G,\lambda) = \sum_{i=0}^{p} a_i \lambda^i$ The polynomials have the following properties:

(1) The constant term of $\chi(G, \lambda)$ is zero.

(2) If $p = |V(G)|$, the coefficient of λ^p is $a_p = \dfrac{1}{|A(G)|}$.

(3) a_i/a_p, $i = 1, 2, 3, \ldots, p$, is an integer.

(4) $|A(g)| = p!/|G|$, where $|G|$ is the number of labeled graphs of G, and $g \in G$.

Since the constant term of every $\chi(g, \pi, \lambda)$ is zero, we have (1); there is an identical permutation in the automorphism group A(g), so we have (2); every $\chi(g, \pi, \lambda)$ is a polynomial with integral coefficients, so we have (3); G, as a subset of γ_p, is the orbit of the permutation group S_p on γ_p, which contains the labeled graph g, A(g) is the subgroup of S_p, fixing g in γ_p so we have (4).

2. The Results. Let N_2 be the number of vertex subsets $\{i, j\}$ of g satisfying the following conditions:

(1) $d_g(i) = d_g(j)$,

(2) The edge $\{i, j\} \notin E(g)$,

(3) The permutation $(ij)(k_1)(k_2)\ldots(k_{p-2}) \in A(g)$,

where $i, j, k_1, k_2, \ldots, k_{p-2}$ are vertices of g and E(g) is the edge set of g.

Clearly, condition (1) follows from condition (3) via Lemma 1.

Theorem 1. Let the chromatic polynomial of an unlabeled graph G be

$$\chi(G,\lambda) = \sum_{i=1}^{p} a_i\lambda^i = a_p[\lambda^p + \frac{a_{p-1}}{a_p}\lambda^{p-1} + \frac{a_{p-2}}{a_p}\lambda^{p-2} + \ldots + \frac{a_1}{a_p}\lambda].$$

Then $(a_{p-1}/a_p) = -q + N_2$, where $q = |E(g)|$.

Example 1. Assume G = 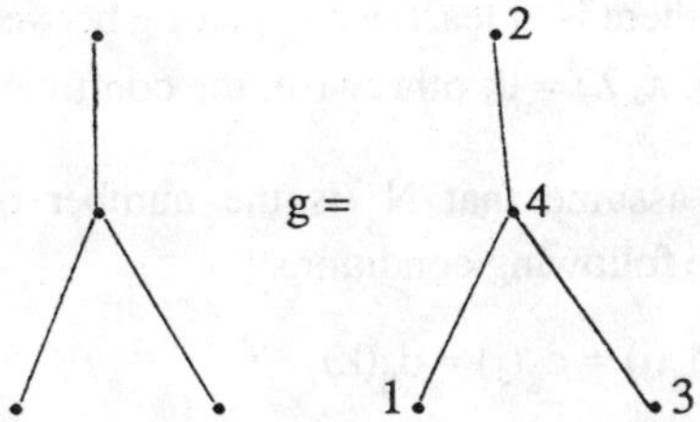

Since the vertex subsets of g satisfying the above conditions are {1, 2}, {1, 3}, and {2, 3}, we have $N_2 = 3$, therefore, $a_3/a_4 = -q + N_2 = -3 + 3 = 0$, which is the coefficient of λ^3.

Proof of Theorem 1. Consider two cases for a_{p-1}/a_p.

(a) Let I be the identity permutation in A(g), then $\chi(g, I, \lambda) = \chi(g, \lambda)$ is the chromatic polynomial of the labeled graph g of G in which the coefficient of λ^{p-1} is -q.

(b) Consider the cyclic form of a permutation π in A(g), $(ij)(k_1)(k_2)\ldots(k_{p-2})$, there are p-1 vertices in the quotient graph $g{:}\pi$ because there are p-1 distinct cycles in π. If edge $\{i, j\} \in E(g)$, then $\chi(g, \pi, \lambda) = 0$; otherwise, the coefficient of λ^{p-1} in $\chi(g, \pi, \lambda)$ is +1, that also means that the vertex subset {i, j} satisfies the above conditions. There are N_2 such permutations in A(g); therefore in this case the coefficient of λ^{p-1} is N_2, thus proving Theorem 1.

To calculate a_{p-2}/a_p, we consider the following four cases:

Case 1. In the chromatic polynomial $\chi(g, I, \lambda) = \chi(g, \lambda)$, the coefficient of λ^{p-2} is C(q, 2) - A, here C(q, 2) is the binomial coefficient, and A is the number of subgraphs of g which are triangles.

Case 2. Consider the cyclic form of the permutation π in A(g), $(ij)(k_1)(k_2)...(k_{p-2})$ and edge $\{i, j\} \notin E(g)$, the coefficient of λ^{p-2} in $\chi(g, \pi, \lambda)$ is $-|E(g:\pi)|$, where $|E(g;\pi)|$ is the number of edges of the quotient graph $g:\pi$ There are N_2 such permutations in A(g), so in this case, the coefficient of

λ^{p-2} is $\sum_{i=1}^{N_2} |E(g:\pi_i) = E.$

Case 3. If the cyclic form of the permutation π in A(g) is $(ijk)(l_1)(l_2)...(l_{p-3})$, and if there is at least one edge of g between any two of the vertices i, j, k we have $\chi(g, \pi, \lambda) = 0$; otherwise, the coefficient of λ^{p-2} in $\chi(g, \pi, \lambda)$ is +1.

Now assume that N_3 is the number of vertex subsets {i, j, k} satisfying the following conditions:

(4) $d_g(i) = d_g(j) = d_g(k)$,

(5) Edges $\{i, j\}, \{j, k\}, \{k, i\} \notin E(g)$,

(6) The permutation $(ijk)(l_1)(l_2)...(l_{p-3}) \in A(g)$, where $i, j, k, l_1, l_2,..., l_{p-3}$ are vertices of g.

Since $(ijk)(l_1)(l_2)...(l_{p-3}) \in A(g)$ which is a group, we also have $(ikj)(l_1)(l_2)...(l_{p-3}) \in A(g)$. Therefore, in this case, the coefficient of λ^{p-2} is $2N_3$.

Case 4. If the cyclic form of the permutation π in A(g) is $(ij)(kl)(s_1)(s_2)...(s_{p-4})$, and if edge $\{i, j\} \in E(g)$ or $\{k, l\} \in E(g)$ we have $\chi(g, \pi, \lambda) = 0$; otherwise, the coefficient of λ^{p-2} is + 1.

Let N_{2d} be the number of pairs of two-vertex subsets $\{\{i, j\}, \{k, l\}\}$ of g satisfying the following conditions:

(7) $d_g(i) = d_g(j)$, $d_g(k) = d_g(l)$;

(8) Edges $\{i,j\}, \{k, l\} \notin E(g)$;

(9) The permutation $(ij)(kl)(s_1)(s_2)...(s_{p-4}) \in A(g)$., where i, j, k, l, $s_1, s_2, ..., s_{p-4}$ are vertices of g.

Therefore, in this case the coefficient of λ^{p-2} is N_{2d}.

We have proved the following

Theorem 2. $(a_{p-2}/a_p) = C(q, 2) - A - E + 2N_3 + N_{2d}$, where the meanings of q, A, E, N_3, and N_{2d} are the same as before.

Example 2. To calculate the chromatic coefficient of λ^2 in example 1, $C(q, 2) = C(3, 2) = 3$, $A = 0$, $E = 2 \cdot 3 = 6$, $N_3 = 1$, $N_{2d} = 0$, thus $a_2/a_4 = 3 - 0 - 6 + 2 \cdot 1 + 0 = -1$, $a_2 = -a_4$, and $|G| = 4$, then $a_4 = |G|/4! = 1/6$. Therefore, the chromatic cofficient of λ^2 in $\chi(G, \lambda)$ is -(1/6).

We next discuss a_{p-3}/a_p in the following seven cases:

Case 1. The coefficient of λ^{p-3} in $\chi(g, I, \lambda)$ is $-C(q, 3) + (q-2)A + B - 2C$, where A, B and C are the number of subgraphs of g which are triangles, pure quadrilaterals and complete graphs with four vertices, respectively.

Case 2. If the cyclic form of the permutation π in A(g) is $(ij)(k_1)(k_2)...(k_{p-2})$ and edge $\{i, j\} \notin E(g)$, then the coefficient of λ^{p-3} in $\chi(g, \pi, \lambda)$ is $C(|E(g{:}\pi)|, 2) - A_{(g:\pi)}$, where $A_{(g:\pi)}$ is the number of subgraphs of $g{:}\pi$ which are triangles. There are N_2 such permutations in A(g), therefore we have $T = \sum_{i=1}^{N_2} [C(|E(g{:}\pi_i)|, 2) - A_{(g:\pi_i)}]$

Case 3. Let the cyclic form of the permutation p in A(g) be $(ijk)(l_1)(l_2)...(l_{p-3})$ and edges $\{i, j\}, \{j, k\}, \{k, l\} \notin E(g)$, then the coefficient of λ^{p-3} in $\chi(g, \pi, \lambda)$ is $-|E(g{:}\pi)|$. There are $2N_3$ such permutations in A(g) in which the quotient graphs of every pair of permutations are the same. Let $E_1 = \sum_{j=1}^{N_3} |E(g, \pi_j)|$.

Therefore, in this case, the coefficient of λ^{p-3} is $-2E_1$.

Case 4. If the cyclic form of the permutation π in A(g) is $(ij)(kl)(s_1)(s_2)...(s_{p-4})$ and edges $\{i, j\}, \{k, l\} \notin E(g)$, the coefficient of λ^{p-3} in $\chi(g, \pi, \lambda)$ is $-|E(g{:}\pi)|$. There are N_{2d} such permutations in A(g). Let $E_2 = \sum_{j=1}^{N_{2d}} |E(g{:}\pi_j)|$.

Therefore, in this case, the coefficient of λ^{p-3} is $-E_2$.

Case 5. If the cyclic form of the permutation π in A(g) is $(ijkl)(s_1)(s_2)...(s_{p-4})$ and there is no edge of g between any two of the vertices i, j, k, l, then the coefficient of λ^{p-3} in $\chi(g, \pi, \lambda)$ is +1.

Let N_4 be the number of vertex subsets {i, j, k, l} of g satisfying the following conditions:

(10) $d_g(i) = d_g(j) = d_g(k) = d_g(l)$;

(11) Edges {i,j}, {i, k}, {i, l}, {j, k}, {j, l}, {k, l} $\notin$ E(g);

(12) The permutations $(ijkl)(s_1)(s_2)...(s_{p-4})$, $(ilkj)(s_1)(s_2)...(s_{p-4})$ $\in$ A(g),where i, j, k, l, s_1, ...,s_{p-4} are vertices of g.

In this case, the coefficient of λ^{p-3} is $2N_4$.

Case 6. If the cyclic form of the permutation π in A(g) is $(ij)(kl)(st)(u_1)...(u_{p-6})$ and edges {i, j}, {k, l}, {s, t} $\notin$ E(g), then the coefficient of λ^{p-3} in $\chi(g, \pi, \lambda)$ is +1.

Assume that N_{3d} is the number of the sets of triples of pairs of vertices $\{\{i, j\}, \{k, l\}, \{s, t\}\}$ of g satisfying the following conditions:

(13) $d_g(i) = d_g(j)$, $d_g(k) = d_g(l)$, $d_g(s) = d_g(t)$;

(14) Edges {i, j} , {k, l}, {s, t} $\notin$ E(g).

(15) The permutation $(ij)(kl)(st)(u_1)(u_2)...(u_{p-6}) \in$ A(g), where i, j, k, l, s, t, u_1, u_2, ..., u_{p-6} are vertices of g.

In this case, the coefficient of λ^{p-3} is N_{3d}.

Case 7. Let the cyclic form of the permutation π in A(g) be $(ijk)(ls)(t_1)(t_2)...(t_{p-5})$ and edges {i, j}, {i, k}, {j, k}, {l, s} $\notin$ E(g), then the coefficient of λ^{p-3} in $\chi(g, \pi, \lambda)$ is + 1.

Assume that N_m is the number of sets containing a triple and a pair of vertices, $\{\{i, j, k\}, \{l, s\}\}$ of g satisfying the following conditions:

(16) $d_g(i) = d_g(j) = d_g(k)$, $d_g(l) = d_g(s)$;

(17) Edges $\{i, j\}, \{i,k\}, \{j, k\}, \{l, s\} \notin E(g)$;

(18) The permutations $(ijk)(ls)(t_1)(t_2)\dots(t_{p-5})$, $(ikj)(ls)(t_1)(t_2)\dots(t_{p-5})$ $\in A(g)$, where $i, j, k, l, s, t_1, t_2, \dots, t_{p-5}$ are vertices of g.

In this case the coefficient of λ^{p-3} is $2N_m$.

Because there are only three cases, Case 5, 6, and 7, for which p vertices of g are partitioned into p-3 distinct cycles, we have

Theorem 3. $(a_{p-3}/a_p) = -C(q, 3) + (q-2)A + B - 2C + T - 2E_1 - E_2 + 2N_4 + N_{3d} + 2N_m$, where the meanings of q, A, B, C, T, E_1, E_2, N_4, N_{3d}, and N_m are the same as before.

REFERENCES

[1] Li, Nian Zu, *The Theory of Chromatic Polynomials*, Unpublished Lecture Notes, Lanzhou University, June, 1988.

[2] Hanlon, P., *The chromatic polynomial of an unlabeled graph*, J. Comb. Th. (B), 1986.

[3] Farrell, E. J., *On chromatic coefficients,* Discrete Math. 29(1980), 257-264.

[4] Li, Nian Zu, I. Wen, & H. S. H. Sun, *Properties of Chromatic Polynomials of Unlabeled Graphs*, J. Comb. Math. & Comb. Comp., V. 17, 1995, pp. 21-32.

Modular Threshold Graphs are Perfect

Chi Wang

(Department of Mathematics, University of Louisville, Louisville, Kentucky 40292, USA)

Abstract We study a generalization of threshold graphs. Let N be the set of non-negative integers. A graph $G = (V, E)$ is modular threshold if there are function $f : V \to N$ and $m, t \in N, m > t$, such that $uv \in E$ if and only if $[f(u) + f(v)] \pmod{m} \geq t$. A modular threshold graph can be represented as the symmetric difference of three threshold graphs. We show that threshold modular graphs are perfectly orderable and weakly triangulated.

Key words perfect graph; threshold graph; modular threshold graph; perfectly orderable; weakly triangulated.

1. Introduction

Given a finite simple graph $G = (V, E)$, a subset S of V is a *clique* if all of the vertices in S are mutually adjacent. The *clique number* $\omega(G)$ is the maximum size of a clique of G. A subset S of V is a *stable set* if all of the vertices in S are mutually non-adjacent. A *coloring* of G is a partition of the vertex set V into stable sets. The *chromatic number* $\chi(G)$ is the minimum number of stable sets in a coloring of G. G is *perfect* if $\omega(H) = \chi(H)$ for all induced subgraphs H of G.

A natural scheme to color a graph is to give a linear order to the vertices of the graph and then according to the order successively assigning each vertex the smallest positive integer which has not been assigned to any of the neighbors of the vertex. Obviously the largest integer used in this approach is at least the chromatic number of the graph. A graph is *perfectly orderable* if there is a linear order $\prec$ of its vertices such that the prescribed greedy scheme always gives an optimal coloring for each of its induced subgraphs.

Berge and Duchet call a graph *strongly perfect* if each of its induced subgraphs H contains a stable set meeting all the maximal cliques in H [1]. They showed that strongly perfect graphs are perfect. Chvátal proved that perfectly orderable graphs are strongly perfect and he characterized perfectly orderable graphs as graphs which permit an acyclic order

$\prec$ such that no induced path $uvxy$ of four vertices in the graph is ordered as $u \prec v$ and $y \prec x$ [2]. Comparability graphs, triangulated graphs, complements of triangulated graphs, complements of tolerance graphs are some known classes of perfectly orderable graphs [2], [5].

A graph G is *weakly triangulated* if both G and its complement do not have induced cycle of more than 4 vertices. Hayward showed that weakly triangulated graphs are perfect [8]. Perfectly orderable graphs and weakly triangulated graphs are classes of perfect graphs that none of them properly contain the another. The graph F in Figure 1 is a weakly triangulated graph but not perfectly orderable. On the other hand any even cycles of length larger than 4 is perfectly orderable but not weakly triangulated.

Figure 1: The graphs F and H.

Let N be the set of all non-negative integers. A graph $G = (V, E)$ is *threshold* if there are function $f : V \to N$ and $t \in N$ such that $uv \in E$ if and only if $f(u) + f(v) \geq t$. Threshold graphs were studied by Hammer and Chvátal [3] and many others. Hammer, Mahadev and Peled showed that unions of two threshold graphs are weakly triangulated [6]. Chvátal, Hoang, Mahadev and De Werra showed that they are also perfectly orderable [4]. In this paper, we study a generalization of threshold graphs which include threshold graphs as a subclass and can be represented as the symmetric difference of three threshold graphs. We show that those graphs are weakly triangulated and perfectly orderable. Therefore they are perfect.

2. Modular Threshold Graphs

For non-negative integers $p, m \in N, m > 0$, denote $p \,(\text{mod } m) = r$ if $p = qm + r$, $0 \leq r < m$. For $p, m, t \in N$ with $m > t$, denote $p \,(\text{mod } m) \geq t$ if $p = qm + r, 0 \leq r < m$, and $0 \leq t \leq r$. Inequalities $p \,(\text{mod } m) \leq t$, $p \,(\text{mod } m) < t$, $p \,(\text{mod } m) > t$ are similarly defined.

McMorris and Wang introduced modular intersection graphs [9]. Given a family of subsets $\{S_i | i \in I\}$ of a set S, a graph $G = (V, E)$ is the *modular intersection graph of*

$\{S_i | i \in I\}$ if there are $f : V \to \{S_i | i \in I\}$ and $m, t \in N, m > t$, such that $uv \in E$ if and only if $|f(u) \cap f(v)| \pmod{m} \geq t$. Here, we introduce a similar generalization. We call a graph $G = (V, E)$ *modular threshold* if there are function $f : V \to N$ and $m,\ t \in N,\ m > t$, such that

$$uv \in E \text{ if and only if } [f(u) + f(v)] \pmod{m} \geq t. \tag{1}$$

The triple (f, m, t) is called a *representation* of G. Obviously, threshold graphs are modular threshold (by choosing a sufficiently large m).

Lemma 1 *If $G = (V, E)$ is a modular threshold graph, then G has a representation (f, m, t) such that $0 \leq f(u) \leq m - 1\ \forall u \in V$.*

Proof Suppose (f^*, m, t) is a representation of G. Suppose $f^*(u) = q_u m + r_u,\ 0 \leq r_u < m, u \in V$. Let $f(u) = r_u\ \forall u \in V$. Then (f, m, t) is a representation of G such that $0 \leq f(u) \leq m - 1\ \forall u \in V$. □

We call the representation in Lemma 1 a *major representation* of G. If (f, m, t) is a major representation of G, then $uv \in E$ if and only if $t \leq f(u) + f(v) < m$ or $f(u) + f(v) \geq m + t$, or equivalently, $uv \notin E$ if and only if $f(u) + f(v) < t$ or $m \leq f(u) + f(v) < m + t$. We will use this fact throughout the remaining of this paper.

Given k subsets $S_1, ..., S_k$ of a set S, the symmetric difference $S_1 \Delta ... \Delta S_k$ consists elements of S who is in exactly odd number of S_i's. Given k graphs $G_1 = (V, E_1), ..., G_k = (V, E_k)$ on the same vertex set, define the symmetric difference $G_1 \Delta ... \Delta G_k$ be the graph $G = (V, E_1 \Delta ... \Delta E_k)$.

Theorem 2 *If G is a modular threshold graph, then it is the symmetric difference of three threshold graphs (with possibility that some of them may be empty or identical).*

Proof Let (f, m, t) be a major representation of G. Then two vertices u, v are adjacent if and only if $m - 1 \geq f(u) + f(v) \geq t$ or $f(u) + f(v) \geq m + t$. Let $G_1 = (V, E_1)$ be the graph such that $uv \in E_1$ if and only if $f(u) + f(v) \geq t$, let $G_2 = (V, E_2)$ be such that $uv \in E_2$ if and only if $f(u) + f(v) \leq m + t - 1$, and let $G_3 = (V, E_3)$ be such that $uv \in E_3$ if and only if $f(u) + f(v) \leq m - 1$. Clearly G_i are threshold graphs, $i = 1, 2, 3$, and G is the symmetric difference of G_1, G_2 and G_3. □

A bipartite graph $G = (B_1, B_2, E)$ with bipartition (B_1, B_2) of its vertex set is *modular difference* if there are $f : B_1 \cup B_2 \to N$ and $m, t \in N, m > t$, such that for $u \in B_1, v \in$

$B_2, uv \in E$ if and only if $[f(u)+f(v)] \pmod m \geq t$. Modular difference graph is an analogue of difference graphs. A bipartite graph is a *difference graph* if there are $f: B_1 \cup B_2 \rightarrow N$ and $t \in N$ such that $uv \in E$ if and only if $f(u)+f(v) \geq t$. Difference graphs were studied by Hammer, Peled and Sun [7]. They are the bipartite parallel of threshold graphs. Obviously, similar to Theorem 2, modular difference graphs can be represented as the symmetric difference of three difference graphs.

Theorem 3 *If $G=(V,E)$ is a modular threshold graph, then there is a partition (V_1, V_2) of V such that the subgraphs induced by V_1 and V_2 are threshold graphs and the subgraph with edges one end in V_1 and one in V_2 is a modular difference graph.*

Proof Suppose G is a modular threshold graph and (f,m,t) is a major representation of G. Let $V_1=\{u \in V | 0 \leq f(u) \leq \lfloor \frac{m}{2} \rfloor - 1\}$ and let $V_2=\{u \in V | f(u) \geq \lceil \frac{m}{2} \rceil\}$. If $u,v \in V_1, uv \in E$ if and only if $f(u)+f(v) \geq t$. If $u,v \in V_2$, then $uv \in E$ if and only if $f(u)+f(v) \geq m+t$. □

To end this section, we mention another structure related to the partition of the edge set of a modular threshold graph. A graph is *equi-modular* if there is $f: V \rightarrow N$ and $t, m \in N$, $t < m$, such that $uv \in E$ if and only if $[f(u)+f(v)] \pmod m = t$. Clearly the edge set of a modular threshold graph is the union of a family of edge disjoint equi-modular graphs. The following characterizes equi-modular graphs.

Theorem 4 *A graph G is equi-modular if and only if G is the vertex disjoint union of some complete bipartite graphs and at most two disjoint cliques.*

Proof Suppose that G is equi-modular. Let (f,m,t) be a representation of G. For $i = 0, \ldots, m-1$, let $V_i = \{v \in V \,|f(v) \pmod m = i\}$. (Some of the V_i's may be empty.) For $u \in V_i, v \in V_j, uv \in E$ if and only if $i+j=t$ or $m+t$. Therefore subsets in the form of $V_0 \cup V_t, V_1 \cup V_{t-1}, \ldots, V_{\lfloor \frac{t}{2} \rfloor -1} \cup V_{\lfloor \frac{t}{2} \rfloor +1}, V_{t+1} \cup V_{m-1}, V_{t+2} \cup V_{m-2}, \ldots, V_{m+\lfloor \frac{t}{2} \rfloor -1} \cup V_{m+\lfloor \frac{t}{2} \rfloor +1}$ induce disjoint complete bipartite graphs while $V_{\frac{t}{2}}$ and $V_{\frac{m-t}{2}}$ induce disjoint cliques when t is even or $m+t$ is even.

Now suppose that $G_i = (B_{i_1}, B_{i_2}, E_i), i = 1, \ldots, k$ are k complete bipartite graphs and C_1, C_2 are two disjoint cliques, where C_1 and C_2 may be empty. Suppose that G is the disjoint union of G_i and C_j, $i = 1, ..., k, j = 1, 2$. Let $f : \cup_{i=1}^{k} (B_{i_1} \cup B_{i_2}) \to N$ be such that

$$f(v) = \begin{cases} i, & v \in B_{i_1} \\ 2k - i & v \in B_{i_2} \\ k & v \in C_1 \\ 4k & v \in C_2 \end{cases}$$

and let $m = 6k, t = 2k$. Then (f, m, t) is an equi-modular representation of G. □

3. Perfection of Modular Threshold Graphs

Theorem 2 shows that modular threshold graphs are symmetric difference of 3 threshold graphs. The union of two threshold graphs are called *2-threshold graphs.* Chvátal, Hoang, Mahadev and De Werra showed that 2-threshold graphs are perfectly orderable [4] and Hammer, Mahadev and Peled showed 2-threshold graphs are weakly triangulated [6]. Their results are best possible in the sense that the graph H in Figure 1 is the union of three threshold graphs, but it is neither perfectly orderable nor weakly triangulated. Symmetric difference of two threshold graphs may not necessarily be perfect orderable, nor weakly triangulated. For example, the H graph in Figure 1 is not weakly triangulated since it is the complement of C_6, chordless cycle of 6 vertices. Figure 2 illustrates that it is the symmetric difference of two threshold graphs but not perfectly orderable. Now we show that modular threshold graphs are actually weakly triangulated and perfectly orderable.

Figure 2: The symmetric difference decompositon of the H graph.

Lemma 5 *Suppose G is a modular threshold graph and (f, m, t) is a major representation of G.*

1. *If $uv \in E, f(u) \geq f(v)$ and $f(u) + f(v) \geq m + t$, then $ux \in E$ for any x such that $f(x) \geq f(v)$.*

2. *If $uv, vx \in E$, $ux \notin E$ and $f(u) > f(v) > f(x)$, then $f(u) + f(v) \geq m + t$, $m > f(v) + f(x) \geq t$.*

3. *If* $uv, xy \in E, vy \notin E$, $f(u) > f(y)$ *and* $f(v) > f(x)$, *then* $f(u)+f(v) \geq m+t, m > f(x)+f(y) \geq t$.

4. *If* $uv \notin E$, $f(u)+f(v) < t$, *then* $m > f(u)+f(x) \geq t$ *for any* $ux \in E$.

Proof (1) follows from that $f(u)+f(x) \geq f(u)+f(v) \geq m+t$.

For (2), if $f(v)+f(x) \geq m+t$, then $f(u)+f(x) > f(v)+f(x) \geq m+t$, contrary to $ux \notin E$. If $f(u)+f(v) < m$, then $m > f(u)+f(v) \geq f(u)+f(x) \geq f(v)+f(x) \geq t$, contrary to $ux \notin E$.

To show (3), if $f(x)+f(y) \geq m+t$, then $f(v)+f(y) > f(x)+f(y) \geq m+t$, contrary to $vy \notin E$. If $f(u)+f(v) < m$, then $m > f(u)+f(v) > f(y)+f(v) > f(y)+f(x) \geq t$, contrary to $vy \notin E$.

Finally, suppose that $uv \notin E, ux \in E, f(u)+f(v) < t$. Since f is major, $f(u) \leq t-1$ and $f(x) < m$. Therefore $t \leq f(u)+f(x) \leq t-1+m-1 < m+t$ which implies $t \leq f(u)+f(x) < m$. □

Recall that a graph G is weakly triangulated if both G and its complement do not have any induced cycle of length larger than 4.

Theorem 6 *Let G be a modular threshold graph and let (f, m, t) be a major representation of G. Suppose $u_0u_1...u_n, n \geq 4$, is an induced path or cycle such that $u_iu_{i+1} \in E, i = 0, ..., n-1$ and $f(u_1) < f(u_2) < f(u_0)$. Then*

$$f(u_1) < f(u_3) < f(u_5) < ... < f(u_n) < ... < f(u_4) < f(u_2) < f(u_0).$$

Proof Since $N(u_i) \neq N(u_j)$ and $N[u_i] \neq N[u_j]$ for $i \neq j$, it follows that $f(u_i) \neq f(u_j)$for all $i \neq j, 0 \leq i, j \leq n$.

If $f(u_3) > f(u_2)$, by (2) of Lemma 5, $f(u_2)+f(u_3) \geq m+t$. By (1) of Lemma 5, $u_3u_0 \in E$, contrary to our assumption. If $f(u_1) > f(u_3)$, by (3) of Lemma 5 $f(u_1)+f(u_0) \geq m+t$. Then $u_2u_0 \in E$ by (1) of Lemma 5, a contradiction again. Therefore $f(u_1) < f(u_3) < f(u_2) < f(u_0)$.

(3) and (1) of Lemma 5 and $u_2u_4 \notin E$ prohibit that $f(u_4) > f(u_2)$. Similarly, (2) and (1) of Lemma 5 and $u_2u_0 \notin E$ prohibit that $f(u_4) < f(u_3)$. Therefore $f(u_1) < f(u_3) < f(u_4) < f(u_2) < f(u_0)$.

If $n=4$, the lemma is proved. If $n=5$, similar argument shows that $f(u_1)<f(u_3)<f(x_5)<f(u_4)<f(u_2)<f(u_0)$. If $n\geq 6, u_2u_3...u_n$ is an induced path of length at least 5. By $f(u_3)<f(u_4)<f(u_2)$ and induction, we have $f(u_3)<f(u_5)<...<f(u_n)<...<f(u_4)<f(u_2)$. Then the lemma follows. □

Theorem 7 *If G is a modular threshold graph, then G has no induced cycle of length larger than 4.*

Proof Let (f,m,t) be a major representation of G. Suppose on the contrary that $C=x_0x_1...x_n$, $n\geq 4$, is an induced cycle such that $x_ix_{i+1}\in E, i=0,...,n-1$, and $x_nx_0\in E$. Without loss of generality, suppose $f(x_0)=\min\{f(x_i)\,|i=0,...,n\}$ and $f(x_1)<f(x_n)$. By Theorem 6,

$$f(x_0)<f(x_2)<...<f(x_{n-1})<...<f(x_3)<f(x_1)<f(x_n).$$

Since $x_{n-1}x_n\in E$, $f(x_0)<f(x_{n-1})<f(x_1)<f(x_n)$, by (3) of Lemma 5, $f(x_n)+f(x_{n-1})\geq m+t$. Then (1) of Lemma 5 implies $x_1x_n\in E$, a contradiction. □

Theorem 8 *If G is a modular threshold graph, then so is its complement.*

Proof Denote $G^c=(V,E^c)$ the complement of G. Suppose (f,m,t) is a major representation of G. We may assume that $m>t$ and $f(u)+f(v)\neq m-1$ for all $u,v\in V$. (If not, choose a large enough positive integer M and let $m^*=Mm, t^*=Mt, f^*(u)=Mf(u)\ \forall u\in V$. Then (f^*,m^*,t^*) is a such representation.) Let $g:V\longrightarrow N$ be such that $g(u)=m-f(u)-1\ \forall u\in V$. Let $m'=m, t'=m-t-1$. By definition, $uv\in E^c$ if and only if $0\leq f(u)+f(v)\leq t-1$ or $m\leq f(u)+f(v)\leq m+t-1$. If $0\leq f(u)+f(v)\leq t-1$, then $2m-t-1\leq g(u)+g(v)\leq 2m-2$. If $m\leq f(u)+f(v)\leq m+t-1$, then $m-t-1\leq g(u)+g(v)\leq m-2$. When $f(u)+f(v)\neq m-1$ for any $u,v\in V$, $g(u)+g(v)\neq m-1$. Therefore, (g,m',t') is a major representation of G^c. □

Theorem 9 *Modular threshold graphs are weakly triangulated.*

Proof It follows from Theorems 7 and 8. □

Though modular threshold graphs are weakly triangulated, it is not a proper subset of Meyniel graphs. A graph is *Meyniel* if every odd cycle of length at least 5 has at least two chords. Meyniel graphs are perfect [10]. The *house* graph in Figure 3 is modular threshold (take label on each vertex v as $f(v)$ and let $m=18$, $t=3$) but not Meyniel.

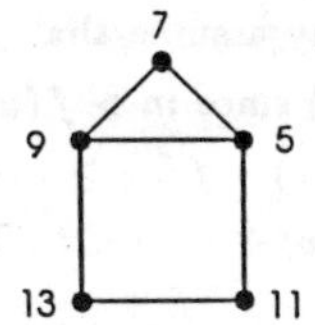

Figure 3: The house graph.

Recall that a graph is perfectly orderable if and only if there is an acyclic order $\prec$ on its vertices such that no induced path $uvxy$ of four vertices in the graph is ordered as $u \prec v$ and $y \prec x$. Such order is called *perfect.*

Theorem 10 *Modular threshold graphs are perfectly orderable.*

Proof Suppose $G = (V, E)$ is a modular threshold graph with a major representation (f, m, t). First let us assume that f is one-to-one, i.e., $f(u) \neq f(v)$ if $u \neq v$. Let $\prec$ be an order defined as follows: if $f(u) + f(v) \geq m + t$ and $f(u) \geq f(v)$ then let $u \prec v$; if $m > f(u) + f(v) \geq t$ and $f(u) > f(v)$ then let $v \prec u$.

To show that $\prec$ is acyclic, suppose that $C = \{u_1, u_2, ..., u_s\}$ is a subset of s vertices such that $u_s \prec u_1$ and $u_i \prec u_{i+1}, i = 1, ..., s-1$. Without loss of generality, suppose

$$f(u_1) = \max\{f(u_i) \,|u_i \in C\}.$$

Since $u_1 \prec u_2, u_s \prec u_1$, it must be $f(u_1) + f(u_2) \geq m + t$ and $m > f(u_1) + f(u_s) \geq t$. Let

$$C_1 = \{u_i \in C|\; m > f(u_i) + f(u_j)\; \forall u_j \in C\}.$$

Let $C_2 = C - C_1$. $C_2 \neq \emptyset$ since $u_1, u_2 \in C_2$. $u_s \in C_1$ since $m > f(u_1) + f(u_s) > f(u_i) + f(u_s)$ $\forall u_i \in C$. Obviously $f(u_i) > f(u_j)$ for any $u_i \in C_2$ and $u_j \in C_1$. Then there is no vertex $u_i \in C_2$ and $u_j \in C_1$ such that $u_i \prec u_j$ since $f(u_i) + f(u_j) < m$ for all pairs $u_i \in C_2$, $u_j \in C_1$, contrary to the assumption on C.

To show that this order is perfect, suppose on the contrary that $P = uvxy$ be an induced path such that $u \prec v$ and $y \prec x$. Suppose that $f(u) + f(v) \geq m + t$. Then $f(u) > f(v)$ by the rule of order. Since $ux, uy \notin E$ and $xy \in E$, it follows from (1) and (3) of Lemma 5 that $f(v) > \max\{f(x), f(y)\}$, $m > f(x) + f(y) \geq t$. $y \prec x$ implies $f(x) \geq f(y)$. From (2) and (1) of Lemma 5, $f(v) + f(x) \geq m + t$ and $ux \in E$, contrary to that P is induced. By symmetry, we have $m > f(x) + f(y) \geq t$ and $m > f(u) + f(v) \geq t$. Therefore $f(x) > f(y)$ and $f(v) > f(u)$.

Without loss of generality we may assume that $f(y) < f(u)$. By (3) of Lemma 5, it must be $f(x) > f(v) > f(u) > f(y)$ since $m > f(u) + f(v)$ and $m > f(x) + f(y)$. This is impossible: by (2) of Lemma 5 $f(v) + f(x) \geq m + t$; but on the other hand, by (4) of Lemma 5, $f(y) + f(v) < m$ since $f(v) + f(y) < t$. Therefore, the given order is perfect.

Next, suppose f is not one-to-one. Let $V_r = \{u \in V | f(u) = r\}$ for $m - 1 \geq r \geq 0$. Order each V_r arbitrarily as $u_1^r, ..., u_{r_s}^r$. Similar to the previous case order vertices of G according to f as follows: For $u_i^r \in V_r, u_j^s \in V_s, r \neq s$, let $u_i^r \prec u_j^s$ if $f(u_i^r) + f(u_j^s) \geq m + t$ and $f(u_i^r) \geq f(u_j^s)$; let $u_j^s \prec u_i^r$ if $m > f(u_i^r) + f(u_j^s) \geq t$ and $f(u_i^r) \geq f(u_j^s)$. For $u_i^r, u_j^r \in V_r$, let $u_i^r \prec u_j^r$ if $i > j$. Then it is easy to see that $\prec$ is also acyclic.

Notice that $N(u) = N(v)$ if $f(u) = f(v) \neq \frac{t}{2}$ or $\frac{m+t}{2}$ and $N[x] = N[y]$ if $f(u) = f(v) = \frac{t}{2}$ or $\frac{m+t}{2}$. If $P = u_1u_2u_3u_4$ is an induced path of length 4 of G, then $f(u_i) \neq f(u_j)$, $i \neq j$. The proof of perfectness for the one-to-one case is applicable here. So the order is a perfect order. □

References

[1] C. Berge and P. Duchet, Strongly perfect graphs, *Annals of Discrete Mathematics* 21 (1984) 57–61.

[2] V. Chvátal, Perfectly ordered graphs, *Annals of Discrete mathematics* 21 (1984) 63–65.

[3] V. Chvátal and P.L. Hammer, Aggregation of inequalities in integer programming, *Annals of Discrete Mathematics* 1 (1977) 142-162.

[4] V. Chvátal, C.T. Hoang, N.V.R. Mahadev and D. de Werra, Four classes of perfectly orderable graphs, Journal of Graph Theory, 11 (1987) 481–495.

[5] M.C. Golumbic, C.L. Monma, and W.T. Trotter, Jr., Tolerance graphs, *Discrete Applied mathematics* 9 (1984) 497–506.

[6] P.L. Hammer, N.V.R. Mahadev and U.N. Peled, Some properties of 2-threshold graphs, *Networks* 19 (1989) 17-23.

[7] P.L. Hammer, U.N. Peled and Xiaorong Sun, Difference graphs. *Discrete Applied Mathematics*, 28 (1989) 35–44.

[8] R.B. Hayward, Weakly triangulated graphs, *J. Combinatorial Theory* (B) 39 (1985) 200-209.

[9] F.R. McMorris and Chi Wang, Modular intersection graphs, Manuscript,1993.

[10] H. Meyniel, The graphs whose odd cycles have at least two chords, *Annals of Discreet Mathematics* 21 (1984) 115-119.

A Method for Constructing Hadamard Matrices

Wang Min, Hou Renmin, and Wang Liangtao

(*Yantai University, Yantai, Shandong Province*, 264005)

Abstract A new method for constructing Hadamard matrices is presented. By using this method we can get many Hadamard matrices. All these matrices are block symmetric circulant matrices and every block of them is a 4-order Hadamard matrix.

Key Words Hadamard Matrix

1. Introduction

Suppose $H=(h_{ij})_{n\times n}$ is a square matrix, where $h_{ij}=\pm 1\,(i,j=1,2,\cdots,n)$, then H is called a $(1,-1)$-matrix. From Hadamard's determinant theorem[1] we have

$$|\det H| \leq \prod_{i=1}^{n}\left(\sum_{j=1}^{n} h_{ij}^2\right)^{\frac{1}{2}} = n^{\frac{n}{2}} \tag{1}$$

and the equality can be established if and only if

$$HH^T = H^T H = n \cdot I_n. \tag{2}$$

Where I_n is the identity matrix of order n. We call H in (2) is a Hadamard Matrix. It isn't difficult to prove that the necessary condition of existence of n-order Hadamard matrices is $n=2$ or n is a multiple of 4. It was conjuctured that the condition also is a sufficient condition. That is the famous Hadamard's conjucture. This conjucture hasn't been proved for a hundred years, but there were many works to study constructive method for Hadamard matrices (Refer to [2], [3]). The method presented by Williamson[3] is a typical example among them. Williamson's method can be used by computer and the circulant matrices were used in it. The circulant matrix is very important in coding theory, it is closely related to pseudorandom sequence.

In this paper we present a more convenient method by using quaternion theory. By using this method we can get many Hadamard matrices. All these matrices are block symmetric circulant matrices and every block of them is a 4-order Hadamard matrix. So the storage space of these matrices can be compressed less than one percent of their original values.

2. Constructive method

Let

$$e=\begin{bmatrix}1&0&0&0\\0&1&0&0\\0&0&1&0\\0&0&0&1\end{bmatrix},i=\begin{bmatrix}0&-1&0&0\\1&0&0&0\\0&0&0&-1\\0&0&1&0\end{bmatrix},j=\begin{bmatrix}0&0&-1&0\\0&0&0&1\\1&0&0&0\\0&-1&0&0\end{bmatrix},k=\begin{bmatrix}0&0&0&-1\\0&0&-1&0\\0&1&0&0\\1&0&0&0\end{bmatrix},$$

$$\mathscr{Q}=\{c_1e+c_2i+c_3j+c_4k \mid c_i\in\mathscr{R}, i=1,2,3,4\},$$

It is evident that $\mathscr{Q}$ is isomorphic to quaternion field.

Write down

$$\left.\begin{array}{l}\omega_0=e+i+j+k,\quad \omega_1=-e+i+j+k,\quad \omega_2=e-i+j+k\\ \omega_3=e+i-j+k,\quad \omega_4=e+i+j-k.\end{array}\right\}\tag{3}$$

Let $\bar{\omega}_i$ denotes the conjugate of ω_i, evidently we have $\pm\omega_i, \pm\bar{\omega}_i\in\mathscr{Q}, (i=0,1,2,3,4)$.

Lemma 1 From (3) we can check that

1. $\omega_i\bar{\omega}_i=\bar{\omega}_i\omega_i=4e,\ i=0,1,2,3,4.$

2. $\omega_i\bar{\omega}_j+\omega_j\bar{\omega}_i=\omega_i\bar{\omega}_j+\overline{\omega_i\bar{\omega}_j}=0, \bar{\omega}_i\omega_j+\bar{\omega}_j\omega_i=0,\ i\neq j, i,j=1,2,3,4.$

3. $\omega_0\bar{\omega}_i+\omega_i\bar{\omega}_0=\omega_0\bar{\omega}_i+\overline{\omega_0\bar{\omega}_i}=4e, \bar{\omega}_0\omega_i+\bar{\omega}_i\omega_0=4e,\ i=1,2,3,4.$

4. $\bar{\omega}_i=\omega_i^T,\ i=0,1,2,3,4.$

From this Lemma we know $\pm\omega_i$ and $\pm\bar{\omega}_i (i=0,1,2,3,4)$ all are 4-order Hadamard matrices.

Defination 1 Assume A is a $n\times n$ matrix over $\mathscr{Q}$ and

$$A=\begin{bmatrix}x_0&x_1&x_2&\cdots&x_{n-2}&x_{n-1}\\x_{n-1}&x_0&x_1&\cdots&x_{n-3}&x_{n-2}\\ \cdots&\cdots&\cdots&\cdots&\cdots&\cdots\\x_1&x_2&x_3&\cdots&x_{n-1}&x_0\end{bmatrix},\ x_i\in\mathscr{Q},\ i=0,1,2,\cdots,n-1.$$

Then we call A an n-order circulant matrix over $\mathscr{Q}$. Because A can be determined by its first row, so we simply denote $A=C[x_0,x_1,x_2,\cdots,x_{n-1}]$.

Defination 2 Assume A is an n-order circulant matrix over $\mathscr{Q}$ with the form as $A=C[x_0,x_1,x_2,\cdots,x_{k-1},x_k,x_k,x_{k-1},\cdots,x_2,x_1]$, where $n=2k+1$. Then A is called an n-order symmetric circulant matrix over $\mathscr{Q}$. Simply we denote $A=CW[x_0,x_1,x_2,\cdots,x_k]$.

Let $G=C[0,e,0,\cdots,0]$ is an n-order circulant matrix over $\mathscr{Q}$, then we have $G^i=C[0,\cdots,e,0,\cdots,0]$, where e is at the i-th position and $i=1,2,\cdots,n-1$. Similarly we get $G^n=C[e,0,\cdots,0]$. Evidently $\{G,G^2,G^3,\cdots,G^n\}$ is a group of cycles. So we have $G^{n-i}=G^{-i}(i=1,2,\cdots,n-1)$.

Let $H_i=G^i+G^{n-i}=G^i+G^{-i}(i=1,2,\cdots,n-1)$, here n only can be an odd number. Then we have another lemma.

Lemma 2 $H_i(i=1,2,\cdots,n-1)$ is an n-order symmetric circulant matric over $\mathscr{Q}$, here $n=2k+1$

is an odd number. At the same time we have

1. $H_i^T = H_i$, $H_i^2 = H_{2i} + 2eE_n$, $H_{k+i} = H_{k+1-i}$, $i = 1, 2, \cdots, k$.

2. $\sum_{i=1}^{k} H_{2i} = \sum_{i=1}^{k} H_i$, $H_iH_j = H_jH_i = H_{i+j} + H_{|i-j|}$, $i \neq j$; $i, j = 1, 2, \cdots, k$.

Where $E_n = C[e, 0, \cdots, 0]$ is the n-order identity matrix over $\mathscr{Q}$.

Proof. From $H_i = G^i + G^{-i} (i = 1, 2, \cdots, n-1; n = 2k+1)$ we have

1. $H_i^2 = (G^i + G^{-i})^2 = G^{2i} + G^{-2i} + 2eE_n = H_{2i} + 2eE_n$,

$H_{k+i} = G^{k+i} + G^{2k+1-(k+i)} = G^{k+i} + G^{k+1-i} = G^{k+1-i} + G^{(2k+1)-(k+1-i)} = H_{k+1-i}$.

2. From 1. we have $H_{2k} = H_1, H_{2(k-1)} = H_3, \cdots, H_k = H_{k+1}$, then we have $\sum_{i=1}^{k} H_{2i} = \sum_{i=1}^{k} H_i$,

$H_iH_j = (G^i + G^{-i})(G^j + G^{-j}) = G^{i+j} + G^{-(i+j)} + G^{i-j} + G^{j-i} = H_{i+j} + H_{|i-j|}$.

Lemma 3 Assume $A = CW[x_0, x_1, x_2, \cdots, x_k]$, then we have

1. $A = x_0E_n + \sum_{i=1}^{k} x_iH_i$, $n = 2k+1$,

2. denote $A = (a_1, a_2, \cdots, a_n)^T$ here a_i is the i-th row vector of A. Then the necessary and sufficient condition of $a_ia_j^+ = 0 (i \neq j; i, j = 1, 2, \cdots, n)$ is $a_1a_j^+ = 0 (i = 2, 3, \cdots, k+1)$, where a_j^+ is the conjugate transpose of $a_j (j = 1, 2, \cdots, n)$.

Defination 3 If $W = CW[-\omega_0, a_1, a_2, \cdots, a_k]$ and $a_i \in \{\pm\omega_j \mid j = 1, 2, 3, 4\} (i = 1, 2, \cdots, k)$ then we call W an n-order W-type matrix over $\mathscr{Q}$, here $n = 2k+1$.

Evidently if W is an n-order W-type matrix over $\mathscr{Q}$ it is also a $4n$-order $(1, -1)$-matrix over $\mathscr{R}$ and it is also a block symmetric circulant matrix over $\mathscr{R}$, every block of W is a 4-order Hadamard matrix.

Lemma 4 Suppose an n-order W-type matrix over $\mathscr{Q}$ $A = CW[-\omega_0, a_1, a_2, \cdots, a_k]$ is also a $4n$-order Hadamard matrix over $\mathscr{R}$ (here $n = 2k+1$), if the sum of a row of A is $\delta = -\omega_0 + 2\sum_{i=1}^{4} \lambda_i\omega_i$ then we have $\sum_{i=1}^{4} \lambda_i(2\lambda_i - 1) = k$, $\lambda_i \in \mathscr{Z}, \lambda_i \in [-n, n]$, $i = 1, 2, 3, 4$.

Proof. Let $J = C[e, e, \cdots, e]$ is an n-order circulant matrix over $\mathscr{Q}$ and $A^+ = \overline{A}^T = CW[-\bar{\omega}_0, \bar{a}_1, \bar{a}_2, \cdots, \bar{a}_k]$, then we have $A^+A = AA^+ = 4neE_n$ and $A^+AJ = A^+J\delta = J\bar{\delta}\delta = \bar{\delta}\delta J$, that is $4neE_nJ = \bar{\delta}\delta J$, so we have $4ne = \bar{\delta}\delta$, that is

$$(-\bar{\omega}_0 + 2\sum_{i=1}^{4} \lambda_i\bar{\omega}_i)(-\omega_0 + \sum_{i=1}^{4} \lambda_i\omega_i) = 4ne.$$

Expand the formula above, from Lemma 1, we have

$$4e + 4\sum_{i=1}^{4} \lambda_i^2 4e - 2\sum_{i=1}^{4} \lambda_i 4e = 4(2k+1)e.$$

Simplifing the formula we know

$$\sum_{i=1}^{4} \lambda_i(2\lambda_i - 1) = k. \tag{4}$$

Lemma 5[4] Every positive integer is a sum of three triangular numbers. Where the triangular number is formed as $\frac{m(m-1)}{2}$ and m is an integer.

Lemma 6 For any natural number k, there must exist integers $\lambda_1, \lambda_2, \lambda_3$ and λ_4 such that

$$\sum_{i=1}^{4} \lambda_i (2\lambda_i - 1) = k.$$

Proof. From Lemma 5 there exists integers m_1, m_2, m_3 and m_4 such that $\sum_{i=1}^{4} \frac{m_i(m_i-1)}{2} = k$

set

$$\lambda_i = \begin{cases} m_i/2 & \text{,if } m_i \text{ is a even number} \\ (1-m_i)/2 & \text{,if } m_i \text{ is a odd number} \end{cases} ; \; i=1,2,3,4.$$

Then we have

$$\sum_{i=1}^{4} \lambda_i (2\lambda_i - 1) = k.$$

From Lemma 6, for any natural number k there must exist a $2k+1$-order W-type matrix over $\mathcal{Q}$, the sum of a row of it is $\delta = -\omega_0 + \sum_{i=1}^{4} \lambda_i \, \omega_i$ and the formula $\sum_{i=1}^{4} \lambda_i (2\lambda_i - 1) = k$ holds true.

Theorem 1 Assume $A = CW[-\omega_0, a_1, a_2, \cdots, a_k]$ is an n-order W-type matrix over $\mathcal{Q}$. Let $A = (a_1, a_2, \cdots, a_n)^T$, where $a_1, a_2, \cdots, a_n$ are row vectors of A, then A is a $4n$-order Hadamard matrix over $\mathcal{R}$ if and only if $a_1 a_i^+ = 0, i = 2, 3, \cdots, k+1$.

Proof. Evidently from Lemma 1 we have

$$a_i a_i^+ = a_1 a_1^+ = \omega_0 \bar{\omega}_0 + 2 \sum_{i=1}^{k} a_i \bar{a}_i = (2k+1) 4e = 4ne.$$

Then we have

$$A^+ A = AA^+ = 4nI_{4n} \Longleftrightarrow a_i a_j^+ = 0, \; i \neq j, i, j = 1, 2, \cdots, n.$$

From lemma 3 we know

$$A^+ A = AA^+ = 4nI_{4n} \Longleftrightarrow a_1 a_i^+ = 0, \; i = 2, 3, \cdots, k+1.$$

Theorem 2 Assume $A = CW[-\omega_0, a_1, a_2, \cdots, a_k]$ is an n-order W-type matrix over $\mathcal{Q}$, define $\theta_i^+ = \{j \mid a_j = \omega_i\}$ and $\theta_i^- = \{j \mid a_j = -\omega_i\}, i = 1, 2, 3, 4$. Then A is a $4n$-order Hadamard matrix if and only if

$$\sum_{m=1}^{4} \Big[\sum_{\substack{i,j \in \theta_m^+ \\ 1 \leqslant i < j \leqslant k}} H_i H_j + \sum_{\substack{i,j \in \theta_m^- \\ 1 \leqslant i < j \leqslant k}} H_i H_j - \sum_{\substack{i \in \theta_m^+ \\ i \in \theta_m^-}} H_i H_j \Big] + \sum_{m=1}^{4} \sum_{i \in \theta_m^-} H_i = 0$$

Proof. Evidently we have

$$\begin{aligned} A^+ A &= (-\bar{\omega}_0 E_n + \sum_{i=1}^{k} \bar{a}_i H_i)(-\omega_0 E_n + \sum_{i=1}^{k} a_i H_i) \\ &= \bar{\omega}_0 \omega_0 E_n - \sum_{i=1}^{k} (\bar{\omega}_0 a_i + \bar{a}_i \omega_0) H_i + \sum_{i=1}^{k} \sum_{j=1}^{k} (\bar{a}_i a_j H_i H_j) \end{aligned}$$

$$= 4eE_n + 4e\sum_{i=1}^{k} H_i^2 + \sum_{1\leqslant i<j\leqslant k} (\bar{a}_j a_i + \bar{a}_i a_j) H_i H_j - \sum_{i=1}^{k} (\bar{\omega}_0 a_i + \bar{a}_i \omega_0) H_i.$$

From Lemma 2 and defination of θ_m^+, θ_m^- $(m=1,2,3,4)$ we have

$$\sum_{i=1}^{k} H_i^2 = \sum_{i=1}^{k} H_{2i} + 2keE_n = \sum_{i=1}^{k} H_i + 2keE_n,$$

$$\sum_{1\leqslant i<j\leqslant k} (\bar{a}_j a_i + \bar{a}_i a_j) H_i H_j = \sum_{m=1}^{4} \Big(\sum_{\substack{i,j\in\theta_m^+ \\ 1\leqslant i<j\leqslant k}} 8H_iH_j + \sum_{\substack{i,j\in\theta_m^- \\ 1\leqslant i<j\leqslant k}} 8H_iH_j - \sum_{\substack{i\in\theta_m^+ \\ j\in\theta_m^-}} 8H_iH_j \Big),$$

$$\sum_{i=1}^{k} (\bar{a}_i\omega_0 + \bar{\omega}_0 a_i) H_i = 4\sum_{m=1}^{4} \Big(\sum_{i\in\theta_m^+} H_i - \sum_{i\in\theta_m^-} H_i \Big).$$

Then we have

$$A \text{ is a Hadamard matrix} \Longleftrightarrow A^+A = 4neE_n = 4(2k+1)eE_n$$

$$\Longleftrightarrow 4\sum_{i=1}^{k} H_i + \sum_{1\leqslant i<j\leqslant k} (\bar{a}_j a_i + \bar{a}_i a_j) H_i H_j - \sum_{i=1}^{k} (\bar{\omega}_0 a_i + \bar{a}_i \omega_0) H_i = 0$$

$$\Longleftrightarrow \sum_{m=1}^{4} \Big[\sum_{\substack{i,j\in\theta_m^+ \\ 1\leqslant i<j\leqslant k}} H_iH_j + \sum_{\substack{i,j\in\theta_m^- \\ 1\leqslant i<j\leqslant k}} H_iH_j - \sum_{\substack{i\in\theta_m^+ \\ j\in\theta_m^-}} H_iH_j \Big] + \sum_{m=1}^{4} \sum_{i\in\theta_m^-} H_i = 0$$

Corollary let $\Omega = \{1,2,\cdots,k\} = \bigcup_{m=1}^{4} (\theta_m^+ \cup \theta_m^-)$, where θ_m^+ and θ_m^- $(m=1,2,3,4)$ are subsets of Ω, they are not intersect each other, we define sets φ_m^+, φ_m^- and φ_m $(m=1,2,3,4)$ as follows (the elements in these sets can be repeated)

$$\left.\begin{aligned} \varphi_m^+ &= \{ j\pm i \mid i,j\in\theta_m^+ \text{ and } i<j \} \\ \varphi_m^- &= \{ j\pm i \mid i,j\in\theta_m^- \text{ and } i<j \} \\ \varphi_m &= \{ |i\pm j| \mid i\in\varphi_m^+, j\in\varphi_m^- \} \end{aligned}\right\} \quad m=1,2,3,4.$$

If $i+j>k$ replace $i+j$ by $2k+1-(i+j)$. There exist $4(2k+1)$-order W-type Hadamard matrices over $\mathscr{R}$ if and only if

$$\Big(\bigcup_{m=1}^{4} (\varphi_m^+ \cup \varphi_m^-)\Big) \cup \Big(\bigcup_{m=1}^{4} \theta_m^-\Big) = \bigcup_{m=1}^{4} \varphi_m.$$

3. Computation for finding W-type Hadamard matrices

From theorem 1 we can search for all W-type Hadamard matrices by computer. Limited to our condition we have only computed the cases with $k\leqslant 11$ by a computer (386), we got all W-type Hadamard matrices with $k\leqslant 11$. From the computing result the number of W-type Hadamard matrices show an increasing tendency when their orders increase. The computing method is as follows.

There only $a_1, a_2, \cdots, a_k$ are variables in a W-type matrix $CW[-\omega_0, a_1, a_2, \cdots, a_k]$. In the steps below we replace $\pm\omega_i$ by $\pm i$ $(i=1,2,3,4)$.

1. construct a permutation $l_1, l_2, \cdots, l_k$ here $l_i \in \{\pm 1, \pm 2, \pm 3, \pm 4\}, i=1,2,\cdots,k$. From this permutation we can construct a W-type matrix.

2. If the matrix doesn't satisfy Lemma 4, then return to step 1 to construct a new permutation.

3. Define a operation $\otimes$ as follows

$i \otimes j = 0$, here $|i| \neq |j|$, $i, j = \pm 1, \pm 2, \pm 3, \pm 4$;

$-0 \otimes i = -1$, $-0 \otimes (-i) = 1, (\pm i) \otimes (\pm i) = 2$, $(\pm i) \otimes (\mp i) = -2$, $i = 1,2,3,4$.

From the permutation, the first row of the W-type matrix is $(-0, l_1, l_2, \cdots, l_k, l_k, \cdots, l_2, l_1)$, and the i-th row of the matrix also can be given. If the first row and the i-th row are not orthogonal, then return to step 1 to construct a new permutation, here $i = 2, 3, \cdots, k+1$. Otherwise all rows of the matrix are orthogonal each other, we get a Hadamard matrix.

According to the results above we conjucture there must exist $4n$-order W-type Hadamard matrices over $\mathscr{R}$ for any odd number $n > 1$.

As follows we give some 92-order (that is $k=11$) W-type Hadamard matrices for reference.

$$CW[-\omega_0, \omega_1, \omega_1, \omega_2, -\omega_3, \omega_4, -\omega_4, -\omega_1, -\omega_4, -\omega_3, \omega_4, -\omega_2]$$

$$CW[-\omega_0, \omega_1, \omega_1, -\omega_1, \omega_2, \omega_3, -\omega_3, -\omega_2, -\omega_1, \omega_3, -\omega_4, -\omega_4]$$

$$CW[-\omega_0, \omega_1, \omega_3, -\omega_2, -\omega_1, -\omega_4, -\omega_4, \omega_2, -\omega_3, \omega_2, -\omega_1, \omega_1]$$

$$CW[-\omega_0, \omega_1, -\omega_3, -\omega_2, -\omega_1, -\omega_3, \omega_3, \omega_4, \omega_4, -\omega_2, -\omega_4, \omega_3]$$

The authors wish to express their deep gratitude to Professor Pu Deqian for his guidance and help.

Reference

[1] Hall, Jr., M., Combinatorial Theory, 2nd Edition, Wiley (New York), 1986.

[2] Paley, R. E. A. C., On orthogonal matrices, J. Math. Phys., 12(1933), 311—320.

[3] Williamson, J., Hadamard's determinant theorem and the sum of four squares, Duke math. J., 11(1944), 65—81.

[4] J. P. Serre, A course in arithmetric, Springer-Verlag New York Inc. 1973.

Two Summation Formulas for Basic Bilateral Series

Wang Tianming*
(*Dalian University of Technology, Dalian, 116023*)

Zhang Xiangde*
(*Northeastern University, Shenyang, 110006*)

Abstract Two summation formulas for basic bilateral series are derived by using the difference method.

Key words basic bilateral series; summation formulas; difference method.

1 Introduction

Defining the q-shifted factorial by

$$(a;\ q)_{-n} = \frac{(-q/a)^n}{(q/a;\ q)_n}\ q^{\binom{n}{2}}$$

where $(a;\ q)_0 = 1,\quad |q| < 1, (a;\ q)_n = \Pi_{i=0}^{n-1}(1 - aq^i)$ and $\binom{n}{2} = \frac{n(n-1)}{2},\quad n \geq 1$, and letting $(a_1,\ a_2,\ ...,\ a_m;\quad q)_n = (a_1;\ q)_n(a_2;\ q)_n \cdots (a_m;\ q)_n$, an ${}_r\Psi_s$ basic bilateral series in base q may be defined by [1].

$$\begin{aligned} &{}_r\Psi_s\left(\begin{matrix} a_1, & a_2, & ..., & a_r \\ b_1, & b_2, & ..., & b_s \end{matrix}\ ;\quad q,\quad x\right) \\ &= \sum_{-\infty}^{+\infty} \frac{(a_1,\ a_2,\ ...,\ a_r;\ q)_n}{(b_1,\ b_2,\ ...,\ b_s;\ q)_n}\ x^n \end{aligned} \tag{1}$$

Using the difference method, we obtain the following summation formulas for basic bilateral series.

$$\begin{aligned} &{}_2\Psi_2\left(\begin{matrix} a, & a/bc \\ aq/b, & aq/c \end{matrix}\ ;\quad q,\quad q\right) \\ &= \frac{(1-a)(bc-a)}{a(1-b)(c-1)}\left(\frac{(aq,\ aq/bc;\ q)_\infty}{(aq/b,\ aq/c;\ q)_\infty} - \frac{(b/a,\ c/a;\ q)_\infty}{(1/a,\ bc/a;\ q)_\infty}\right) \end{aligned} \tag{2}$$

*Supported by the Natural Science Foundation of China and Liaoning Province.

and

$$
{}_6\Psi_6\left(\begin{matrix} aq/\sqrt{bcd}, & -aq/\sqrt{bcd}, & a, & a/bc, & a/bd, & a/cd, \\ a/\sqrt{bcd}, & -a/\sqrt{bcd}, & aq/b, & aq/c, & aq/d, & aq/bcd, \end{matrix} ;\ q,\ q\right)
$$

$$
= \frac{(1-a)(bc-a)(1-a/bd)(1-a/cd)}{a(1-b)(c-1)(1-1/d)(1-a^2/bcd)}
$$

$$
\times\left(\frac{(aq,\ aq/bc,\ aq/bd,\ aq/cd;\ q)_\infty}{(aq/b,\ aq/c,\ aq/d,\ aq/bcd;\ q)_\infty} - \frac{(b/a,\ c/a,\ d/a,\ bcd/a;\ q)_\infty}{(1/a,\ bc/a,\ bd/a,\ cd/a;\ q)_\infty}\right) \tag{3}
$$

2 Derivation of the summation formulas

Set

$$
r_n = \frac{(aq,\ aq/bc;\ q)_n}{(aq/b,\ aq/c;\ q)_n} \tag{4}
$$

and

$$
s_n = \frac{(aq,\ aq/bc,\ aq/bd,\ aq/cd;\ q)_n}{(aq/b,\ aq/c,\ aq/d,\ aq/bcd;\ q)_n} \tag{5}
$$

for $n = 0, \pm 1, \pm 2, \ldots$, and define the difference operator Δ by

$$
\Delta r_n = r_n - r_{n-1}
$$

and

$$
\Delta s_n = s_n - s_{n-1}
$$

Then

$$
\sum_{k=-m}^{n} \Delta r_k = r_n - r_{-m-1}
$$

and

$$
\sum_{k=-m}^{n} \Delta s_k = s_n - s_{-m-1}
$$

Note that

$$
\begin{aligned}
\Delta r_k &= r_k - r_{k-1} \\
&= \frac{a(1-b)(c-1)q^k}{(1-a)(bc-a)} \frac{(a,\ a/bc;\ q)_k}{(aq/b,\ aq/c;\ q)_k} \\
&= \frac{(aq,\ aq/bc;\ q)_k}{(aq/b,\ aq/c;\ q)_k} - \frac{(aq,\ aq/bc;\ q)_{k-1}}{(aq/b,\ aq/c;\ q)_{k-1}}
\end{aligned} \tag{6}
$$

and

$$\begin{aligned}\Delta s_k &= s_k - s_{k-1}\\ &= a(1-\frac{1}{d})(1-b)(c-1)q^k(1-\frac{a^2q^{2k}}{bcd})\\ &\times\frac{1}{(1-a)(bc-a)(1-a/bd)(1-a/cd)}\times\frac{(a,\ a/bc,\ a/bd,\ a/cd;\ q)_k}{(aq/b,\ aq/c,\ aq/d,\ aq/bcd;\ q)_k}\\ &= \frac{(aq,\ aq/bc,\ aq/bd,\ aq/cd;\ q)_k}{(aq/b,\ aq/c,\ aq/d,\ aq/bcd;\ q)_k}-\frac{(aq,\ aq/bc,\ aq/bd,\ aq/cd;\ q)_{k-1}}{(aq/b,\ aq/c,\ aq/d,\ aq/bcd;\ q)_{k-1}},\end{aligned}\tag{7}$$

and sum (6) and (7) over k from $-\infty$ to $+\infty$, respectively, we have

$$\begin{aligned}&\sum_{k=-\infty}^{+\infty}\frac{a(1-b)(c-1)}{(1-a)(bc-a)}\frac{(a,\ a/bc;\ q)_k}{(aq/b,\ aq/c;\ q)_k}\ q^k\\ &=\frac{(aq,\ aq/bc;\ q)_\infty}{(aq/b,\ aq/c;\ q)_\infty}-\frac{(b/a,\ c/a;\ q)_\infty}{(1/a,\ bc/a;\ q)_\infty}\end{aligned}\tag{8}$$

and

$$\sum_{k=-\infty}^{+\infty}\frac{a(1-b)(c-1)(1-1/d)(1-a^2q^{2k}/bcd)}{(1-a)(bc-a)1-a/bd)(1-a/cd)}\times\frac{(a,\ a/bc,\ a/bd,\ a/cd;\ q)_k}{(aq/b,\ aq/c,\ aq/d,\ aq/bcd;\ q)_k}\ q^k$$

$$=\frac{(aq,\ aq/bc,\ aq/bd,\ aq/cd;\ q)_\infty}{(aq/b,\ aq/c,\ aq/d,\ aq/bcd;\ q)_\infty}-\frac{(b/a,\ c/a,\ d/a,\ bcd/a;\ q)_\infty}{(1/a,\ bc/a,\ bd/a,\ cd/a;\ q)_\infty}\tag{9}$$

so, (2) and (3) follow immediately from (8) and (9), respectively.

References

[1] H.Exton. q-Hypergeometric Functions and Applications, John Wiley & Sons, 1983.

[2] G.Gasper. Summation,Transformation and Expansion Formulas for Bibasic Series, Trans. Amer. Math. Soc. 312(1989), 257-277.

[3] X.D.Zhang. On Some q-Analysis Problems in Combinatorics. Ph.D. Thesis,Dalian University of Technology, 1994.1.

The Number of Isomeric Types of G(4)

Wu Qiqi
(Foshan University, Guangdong, 528000, China)

Abstract In this paper, three formulas for the number of isomeric types of the graphs with degree 4 are obtained. Then we shall cite some examples to illustrate the applications on graph theory and molecule structure.

Key Words graphs with degree 4; distribution domain; integral point; isomeric type; distribution law; hydrocarbon.

1. Introduction

The study of the enumeration and the structure of hydrocarbons by using graph theory is one of the most interesting and the most meaningful problems. During 1874～1875, Cayley (1821～1895) approached the problem of determining the number of different isomers of the saturated hydrocarbons C_nH_{2n+2} for all values of n. ([1]). During 1932～1934, Blair and Henze also approached the problem([2]). But the carbon chain graph obtained by a molecule from the formula C_nH_{2n+2} is a tree, i.e, its cycle rank is zero. Therefore, this is a simplest case of hydrocarbons (including the saturated and the nonsaturated).

Polya also approached the formula C_nH_{2n}, and obtained rather an approximate formula for the number of isomers. ([2]) We have proved that the cycle rank of any carbon chain graph from C_nH_{2n} is one ([3]), therefore, it is rather difficult to study the above problems in the cases of cycle rank $v \geqslant 1$.

All the above problems may be regarded as being based on the parameter n, the number of carbon atoms.

It is noted that the maximal degree in the carbon chain graph obtained by any hydrocarbon is not exceed 4. Thus, the author made a study of such graphs with maximal degree Δ, the set of the above graphs is denoted by $G(\Delta)$, in which the vertices n and the cycle rank υ are regarded as parameters in integers.

In [4—5], we obtained the existence theorems for G(3) and G(4), respectively. In [5], we also obtained the three theorems of distribution domain of G(4). In this paper we shall give the three formulas of the numbers of isomeric types of G(4), and cite an example to illustrate the applications on graph theory and molecule structure.

In this paper, let $[x]$ be the integer part of the real number x, and we denote the number $\{x\}=x-[x]$.

1. Fundamental Definitions

We first provide the following definitions from [5]:

Definition 1. Let $G=(X,U)$ be any loopless undirected connected planar graph with fixed maximal degree $\Delta(\geqslant 3)$ and the fixed number of vertices, $|X|=n$. Such a type of graph is called *the graph with degree* Δ, the set of all the above graphs is denoted by $G(\Delta)$.

Let $G\in G(\Delta)$, $|U|=l$. Let $l=v-n+1$ be the cycle rank of G and $\delta(i)$ the number of vertices of degree i in G, $i=1,2,\cdots,\Delta$. Clearly,

$$\sum_{i=1}^{\Delta} i\delta(i) = 2n + 2v - 2, \tag{1}$$

$$\sum_{i=1}^{\Delta} \delta(i) = n, \tag{2}$$

where, $\delta(\Delta)\geqslant 1, \delta(\Delta-1)\geqslant 0,\cdots,\delta(2)\geqslant 0, \delta(1)\geqslant 0$.

In particular, when $\Delta=4$, the $G(4)$ is namely the set of all the above *graphs with maximal degree* 4, from (1) and (2) We obtain:

$$\delta(1) + 2\delta(2) + 3\delta(3) + 4\delta(4) = 2n + 2v - 2, \tag{3}$$

$$\delta(1) + \delta(2) + \delta(3) + \delta(4) = n, \tag{4}$$

$$\delta(4) \geqslant 1, \quad \delta(3) \geqslant 0, \quad \delta(2) \geqslant 0, \quad \delta(1) \geqslant 0. \tag{5}$$

$$\delta(2) = n + 2v - 2 - 3\delta(4) - 2\delta(3) \geqslant 0, \tag{6}$$

$$\delta(1) = 2\delta(4) + \delta(3) + 2 - 2v \geqslant 0. \tag{7}$$

Definition 2. For any $v\geqslant 0$, the minimal integer n from which there exists at least one graph in $G(4)$ is called *the initial value* of $G(4)$, and denoted by $N_4(v)$.

In [5], we have proved the following theorems:

Theorem 1. A graph with vertices n and cycle rank v belongs to $G(4)$ if and only if:

$$n \geqslant \begin{cases} 5, & \text{when } v = 0; \\ 4, & \text{when } v = 1; \\ 3, & \text{when } v = 2; \\ v-1, & \text{when } v \geqslant 3. \end{cases}$$

That is, $N_4(0)=5, N_4(1)=4, N_4(2)=3$, and $N_4(v)=v-1$ for $v\geqslant 3$.

2. Distribution Domains of G(4)

Definition 3. Given n, v and $n\geqslant N_4(v)$. From (5), (6) and (7) the following inequalities are obtained:

$$1 \leqslant \delta(4) \leqslant [(n + 2v - 2)/3], \tag{8}$$

$$0 \leqslant \delta(3) \leqslant [(n + 2v - 5)/2], \tag{9}$$

$$3\delta(4) + 2\delta(3) \leqslant n + 2v - 2, \tag{10}$$

$$2v - 2 \leqslant 2\delta(4) + \delta(3). \tag{11}$$

Then, any solution in integers $(\delta(4),\delta(3),\delta(2),\delta(1))$ obtained by (6)～(11) is called an *Isomeric type* of G(4), which is represented by $(\delta(4),\delta(3),\delta(2),\delta(1))$ yet.

Definition 4. Let $\delta(4)=t_1$, $\delta(3)=t_2$. And, the t_1 is regarded as *the axis of abscissas* and t_2 *the axis of ordinates* respectively. Then, the domain bounded by (8)～(11) on the plane $t_1 0 t_2$ is called *the Distribution domain* of the isomeric types of G(4) corresponding to the number pair (v,n), and denoted by $R(v,n)$, or by R briefly. That is,

$$R(v,n):\begin{cases}1\leqslant t_1\leqslant[(n+2v-2)/3], & (12)\\ 0\leqslant t_2\leqslant[(n+2v-5)/2], & (13)\\ 3t_1+2t_2\leqslant n+2v-2, & (14)\\ 2v-2\leqslant 2t_1+t_2. & (15)\end{cases}$$

3. Classification of Distribution Domain

For the sake of convenience, we first give the following notations:

The two straight lines are defined as follows: (See (14) and (15))

$$l_1:\quad 2t_1+t_2=2v-2, \tag{16}$$

$$l_2:\quad 3t_1+2t_2=n+2v-2. \tag{17}$$

Some key points are defined as follows: B=(1,0).

M: the intersection of the line l_1 and the line $t_1=1$,

N: the intersection of the line l_1 and the t_1-axis;

The segment OC equals the integer part of abscissas of the intersection of the line l_2 and the t_1-axis;

The segment AB equals the integer part of ordinates of the intersection of the line l_2 and the line $t_1=1$;

That is,

$$M=(1,2v-4),\qquad N=(v-1,0); \tag{18}$$

$$C=([(n+2v-2)/3],0),\qquad OC=[(n+2v-2)/3]; \tag{19}$$

$$A=(1,[(n+2v-5)/2]),\qquad AB=[(n+2v-5)/2]. \tag{20}$$

Also, let D be the intersection of l_2 and the parallel to t_2-axis passing through the point C; and E the intersection of l_2 and the parallel to t_1-axis passing through the point A; finally, let T be the intersection of l_1 and l_2. Clearly, D or E may be not integral point. It is easily seen that

$$T=(2v-n-2,2n-2v+2). \tag{21}$$

All the above points, A,B,C,D,E,M,N and T, are all *key points* by which the domain $R(v,n)$ is defined.

In [5], we proved that $R(v,n)$ can be classified in the following three forms:

Theorem 2. (Distribution Law I) When $v=0,1,2,n\geqslant N_4(v)$. Then, the domain R nonempty and is bounded by the above inequalities (12),(13) and (14). By using the above key points, it is easily seen that the R is shown as the polygon ABCDE in Fig. 1, which is the A-

Form.

Theorem 3. (Distribution Law Ⅱ) When $v \geqslant 3, n \geqslant 2v-3$. Then, the domain R nonempty and is bounded by the above inequalities (12), (13), (14) and (15).

Similarly, the domain R is shown as the polygon AMNCDE in Fig. 2, for the sake of convenience, which is called the B-Form.

Theorem 4. (Distribution Law Ⅲ) When $v \geqslant 3, v-1 \leqslant n \leqslant 2v-4$. Then, the domain R nonempty and is bounded by the above inequalities (14) and (15) and the following inequalities:

$$2v - n - 2 \leqslant t_1 \leqslant [(n + 2v - 2)/3], \tag{22}$$

$$0 \leqslant t_2 \leqslant 2n - 2v + 2. \tag{23}$$

Similarly, it is shown as the polygon TNCD in Fig. 3, for the sake of convenience, which is called the C-Form.

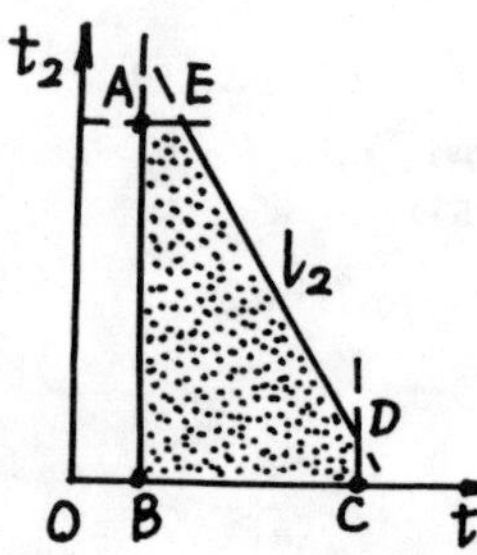

Fig. 1

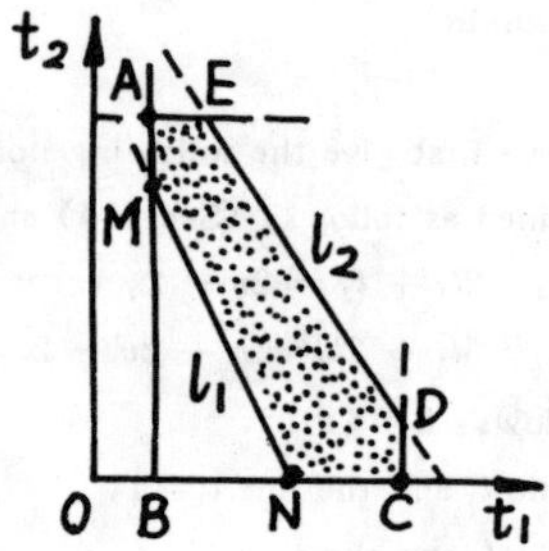

Fig. 2

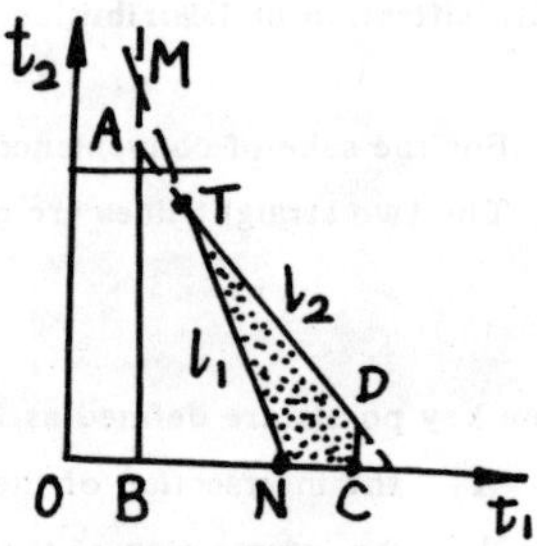

Fig. 3

4. The Numbers of Isomeric Types of G(4)

Let $I = I(v, n)$ be the number of isomeric types of G(4) with v and n. As we have seen that an isomeric type of G(4) is one-to-one corresponding to an integer point on $R(v, n)$. In other words, the number I equals that of the integral points on $R(v, n)$.

We denote

$$K = K(v, n) = (n + 2v - 2)/3, \tag{24}$$

which plays an important role in the number $I(v, n)$.

Now, we give the following three theorems for $I(v, n)$.

Theorem 5. For A-Form, i. e, for $v = 0, 1, 2$ and $n \geqslant N_4(v)$,

$$I(v, n) = \frac{3}{4}[K](K + \{K\}) + (-1)^{n+1} \cdot \frac{1}{2}\left\{\frac{[K]}{2}\right\} \tag{25}$$

Theorem 6. For B-Form, i. e, for $n \geqslant 2v - 3 \geqslant 3$.

$$I(v, n) = \frac{3}{4}[K](K + \{K\}) + (-1)^{n+1} \cdot \frac{1}{2}\left\{\frac{[K]}{2}\right\} - (v-1)(v-2). \tag{26}$$

Theorem 7. For C-Form, i. e, for $v \geqslant 3,\ v - 1 \leqslant n \leqslant 2v - 4$,

$$I(v,n) = \frac{3}{4}((2n - 2K + 1)^2 - \{K\}^2) - (n - v + 2)(n - v + 1) + O(n,K) \quad (27)$$

where, the number $O(n,K)$ is given by

$$O(n,K) = 1/4, \text{ if and only if } n \equiv [K] \pmod 2, \quad (28)$$

$$O(n,K) = O(zero), \text{ if and only if } n \not\equiv [K] \pmod 2. \quad (29)$$

As we have seen that the C-Form is the most complex case. For the sake of convenience, by the condition of C-Form we give the following table.

Table 1. Correspondent table of number pairs n and v for C-Form

v	$v-1$	$2v-4$	n
3	2	2	2
4	3	4	3 4
5	4	6	4 5 6
6	5	8	5 6 7 8
7	6	10	6 7 8 9 10
8	7	12	7 8 9 10 11 12
9	8	14	8 9 10 11 12 13 14
10	9	16	9 10 11 12 13 14 15 16

5. Proofs of Theorems

Proof for Theorem 6. From Fig. 2 and the Theorem 3, the number of integer points on the polygon ABCDE is denoted by the notation I_1, and the number of integer points on the domain R^* bounded by the following inequalities,

$$R^*: \begin{cases} 2t_1 + t_2 < 2v - 2, \\ 1 \leqslant t_1 \leqslant v - 2, \\ 0 \leqslant t_2 \leqslant 2v - 5, \end{cases} \quad (30)$$

is denoted by the notation I_2.

It is easily seen that

$$I(v,n) = I_1 - I_2 \quad (31)$$

The condition $2t_1 + t_2 < 2v - 2$, is namely $2t_1 + t_2 \leqslant 2v - 3$. Hence, for any straight line $t_1 = i$ $(1 \leqslant i \leqslant v - 2)$ on which there are $[2v - 3 - 2i] + 1$ (namely, $2v - 2 - 2i$) integer points belonging to R^*. We can hence write

$$I_2 = \sum_{i=1}^{v-2} (2v - 2 - 2i) = (v - 1)(v - 2). \quad (32)$$

Second, we consider the number I_1. For any straight line $t_1 = i(1 \leqslant i \leqslant [K])$ on which there are $[(n + 2v - 2 - 3i)/2] + 1$ integer points belonging to the polygon ABCDE. We can hence write that

$$I_1 = \sum_{i=1}^{[K]} \left(\left[\frac{n + 2v - 2 - 3i}{2}\right] + 1 \right) = \sum_{i=1}^{[K]} \left[\frac{n + 2v - 3i}{2}\right] \quad (33)$$

Below, we have need to consider the odevity of the number n in (33) for completing this proof (the detailed process has been omitted). We can hence prove that

$$I_1 = \frac{3}{4}[K](K + \{K\}) + (-1)^{n+1} \cdot \frac{1}{2}\left\{\frac{[K]}{2}\right\}. \tag{34}$$

Finally, from the equalities (31), (32) and (34), the proof is hence completed.

The Theorem 5 and Theorem 7 can be proved in a similar way to the Theorem 6, hence, the proofs have been omitted.

6. Examples

Now, we cite an example for B-form to illustrate the conclusions of this paper and their meaning on graph theory and molecule structure.

Let $n=6$, $v=3$. By the Theorem 6,

$K=(n+2v-2)/3=10/3$, $[K]=3$, $\{K\}=1/3$, $\{[K]/2\}=1/2$.

By using the formula (26), we obtain:

$$I(3,6) = \frac{9}{4}\left(\frac{10}{3} + \frac{1}{3}\right) - \frac{1}{4} - 2 = 6\text{(types)}$$

On the other hand, the domain $R(3,6)$ is bounded by the following conditions: (See. (12), (13), (14) and (15))

$$R\begin{cases} 1 \leqslant t_1 \leqslant 3, \\ 0 \leqslant t_2 \leqslant 3, \\ 3t_1 + 2t_2 \leqslant 10, \\ 4 \leqslant 2t_1 + t_2. \end{cases}$$

The $R(3,6)$ is shown as the polygon AMNCDE in Fig. 4. We observe that there are 6 integral points on this domain, which are: (2,0), (3,0), (2,1), (2,2), (1,2) and (1,3). Using the above equalities (6) and (7), all isomeric types as follows: (See the following Table 2).

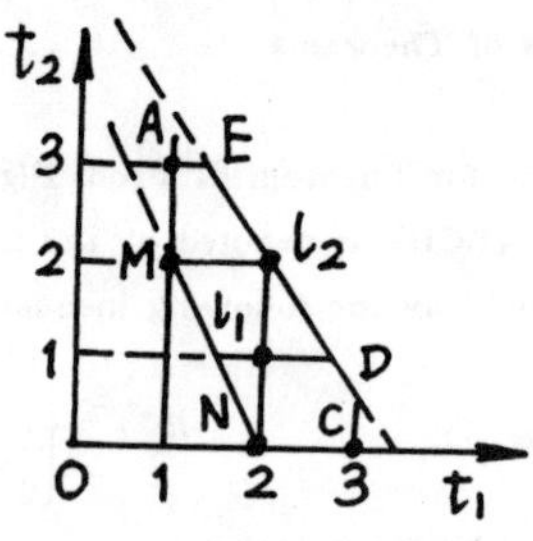

Fig. 4 The domain R(3,6)

Table 2. The isomeric types of G(4) with $n=6, v=3$

Integral point	Type	Integral point	Type
(2,0)	$T_1=(2,0,4,0)$	(2,2)	$T_4=(2,2,0,2)$
(3,0)	$T_2=(3,0,1,2)$	(1,2)	$T_5=(1,2,3,0)$
(2,1)	$T_3=(2,1,2,1)$	(1,3)	$T_6=(1,3,1,1)$

Now, we give a graph belonging to G(4) for the each type in Table 2 by using the way given by the existence theorems in [5].

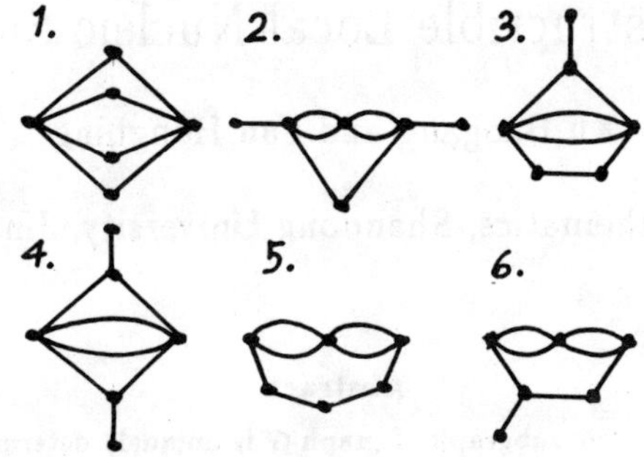

Fig. 5 Some graphs of isomeric types from $R(3,6)$

7. Meanings

As we have seen from Fig. 5 that if every point is used to represent a carbon atom, then the graphs are exactly corresponding to 6 formulas of hydrocarbon. Take the sixth graph, (fig. 5). It is namely the following Fig. 6, which also corresponds to the Fig. 7. Oppositely, the Fig. 7 can be simplified as the Fig. 6, which is corresponding to the sixth graph from the type T_6. The T_6 corresponds to the point (1,3) from $R(3,6)$. And, the domain $R(v,n)$ is completely determined by v and n.

Finally, from Fig. 5 we have seen that the graphs among G(4) may be not simple graphs, also can be the multiple graphs. Clearly, the graphs among G(4) is different from the graphs discussed in Chapter 6 of [6].

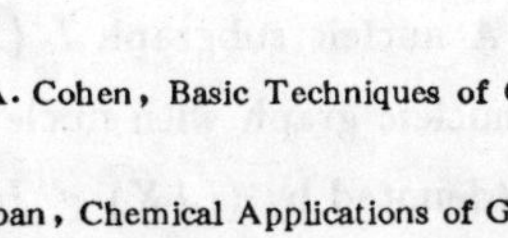

Fig. 6

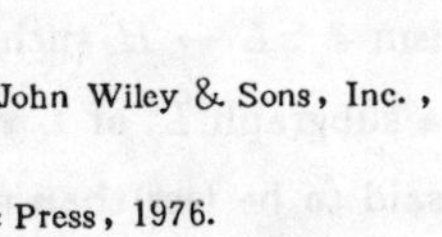

Fig. 7

References

[1] Daniel I. A. Cohen, Basic Techniques of Combinatorial Theory, John Wiley & Sons, Inc., Copyright © 1978.

[2] A. T. Balaban, Chemical Applications of Graph Theory, Academic Press, 1976.

[3] Wu Qiqi, Derive Some Formulas of Unification of Molecules by means of Euler's Formula of Planar Graphs, Journal of Central China Normal Univ., Monograph(Ⅱ), Graph Theory, 1989.

[4] Wu Qiqi, On the Connected Planar Graphs with $\Delta(G)=3$, Journal of Zhengzhou Univ., (1)1985.

[5] Wu Qiqi, On the Existence of Connected Planar Graphs with $\Delta(G)=4$ and its Distribution Domains, Journal of Huazhong (Central China) Univ. of Sci. & Tech., (1)1990.

[6] Harary, F. Graph Theory, Addison-Wesley, 1969.

On Reconstructible Local-Nucleic Subgraphs *

Xu Baogang and Fan Hongbing

(Institute of Mathematics, Shandong University, Jinan,250100, P.R.C)

Abstract

A d-local-nucleic subgraph of graph G is uniquely determined by certain preassigned properties. In this paper, it is proved that the number of d-local-nucleic subgraph of graph G isomorphic to a given d-local-nucleic subgraph is reconstructible(resp., edge reconstructible) if the number of vertices (resp., edges) of every d-local-nucleic subgraph of the same kind is less than the number of vertices (resp., edges) of G minus one.

Introduction

There are various reconstructible and edge reconstructible parameters related to subgraphs of a simple graph $G^{[1-5]}$. In [8],Xie Litong and Fan Hongbing have show several results which supply a conditional answer to the question raised in [4]: How to reconstruct the third-order degree sequence of G. In the present paper, a new kind of subgraph called local-nucleic subgraph is introduced, accompanied by the concept of "nucleus" of a graph G. The purpose of the present paper is to extend the results in [8] to local-nucleic subgraph.

Let L be a graph without isolate vertex. X is a proper subgraph of L and $V(X) \neq \emptyset$, if $V(X) \bigcap V(C) \neq \emptyset$ for each connected component C of L,then L is called a nucleic graph with a nucleus X and denoted by $L\{X\}$. Two nucleic graphs $L\{X\}$ and $H\{R\}$ are isomorphic ($L\{X\} \cong H\{R\}$) if there is a isomorphism $\phi : L \to H$ such that $\phi(X) = R$. A nucleic subgraph $L'\{X\}$ of $L\{X\}$ is a subgraph L' of L where L' is also a nucleic graph with nucleus X. $L\{X\}$ is said to be less than or equal to $H\{R\}$ (denoted by ($L\{X\} \preceq H\{R\}$)

***This research was supported by the National Natural Science Foundation of China**

if $L\{X\}$ is isomorphic to a nucleic subgraph of $H\{R\}$. If $(L\{X\} \preceq H\{R\}$ and $(L\{X\} \not\cong H\{R\}$, then we say $(L\{X\}$ is less than $H\{R\}$ (denoted by $(L\{X\} \prec H\{R\}$.

Suppose $L\{X\}$ is a nucleic graph, $e = ab \in E(L)$ and $e' = a'b' \in E(L)$. We define the distance from e to e' is $min\{d_L(a,a'), d_L(a,b'), d_L(b,a'), d_L(b,b')\}+1$, the distance from e to a' is the minimum of the distances from e to edges incident to a'. If $e \in E(X)$, then we define the distance from e to X is zero. Otherwise, the distance from e to X is $min\{d_1, d_2+1\}$, where d_1 is the minimum of distances from e to edges of X and d_2 is the distances from e to isolate vertices of X. The *peripheral distance of* $L\{R\}$ *is the maximum of distances from edges in* $E(L)\backslash E(X)$ *to* X. *While the* peripheral distance of $L\{R\}$ is d, $L\{R\}$ is denoted by $L^d\{R\}$.

Suppose $L\{X\}$ is a nucleic graph, $A(L)$ is the automorphism group of L. Then for each $\pi \in A(L)$, $\pi(X)$ is a subgraph of L being isomorphic to X. Denote by $O(L\{X\})$ the number of subgraph in the orbit of $A(L)$, containing X. Let G be a graph without isolate vertex, Y is a proper subgraph of G and $V(Y) \neq \emptyset$. For $d \geq 1$ an integer, the subgraph of G induced by edges having distance to Y no more than d is called a d-local-nucleic subgraph of G with nucleus Y, denoted by $L_G^d\{Y\}$. A d-local-nucleic subgraph of G is called a maximal d-local-nucleic subgraph of G if it is not less than any other d-local-nucleic subgraph of G .

Let $L\{X\}$ be a nucleic graph, we use $s(L,G)$ to denote the number of subgraphs of G which are isomorphic to L (Without causing confusion, we also use $s(L,G)$ to denote the set of subgraph of G which are isomorphic to L), $S(L\{X\},G)$ the number of nucleic subgraphs of G which are isomorphic to $L\{X\}$, $\dot{S}(L^d\{X\}, L_G^{d'}\{Y\})$ the number of local-nucleic subgraphs of $L_G^{d'}\{Y\}$ which are ismorphic to $L^d\{X\}$, $\dot{S}(L\{X\},G)$ the number of local-nucleic subgraphs of G which are isomorphic to $L\{X\}$.

All graphs considered in this paper are finite and simple. We use $\nu(G)$ to denote the number of vertices of G, $\varepsilon(G)$ the number of edges of G. Undefined terminologies are from [6].

1 Results

Lemma 1 *Let* $L^d\{X\}$ *be a nucleic graph with peripheral distance* d. *Then*

$$S(L^d\{X\},G) = O(L^d\{X\})s(L,G) = \sum_{Y \in s(X,G)} \dot{S}(L^d\{X\}, L^d_G\{Y\}).$$

Proof: Because each subgraph of G isomorphic to L when L is considered as unrooted graph, corresponds $O(L^d\{X\})$ nucleic subgraphs of G isomorphic to $L^d\{X\}$ and each $Y \in s(X,G)$ corresponds $\dot{S}(L^d\{X\}, L^d_G\{Y\}$ d-local-nucleic subgraphs of G isomorphic to $L^d\{X\}$. □

According to *Kelly's lemma* and lemma 1, the following corollarys immediately follows.

Corollary 2 *Let $L\{X\}$ be a nucleic graph and $\varepsilon(L) < \varepsilon(G)$. Then $S(L\{X\},G)$ is edge-reconstructible.*

Corollary 3 *Let $L^d_G\{X\}$ be a maximal d-local-nucleic subgraph of G. Then $S(L^d_G\{X\},G) = O(L^d\{X\})s(L,G) = \bar{S}(L^d_G\{X\},G)$.*

Lemma 4 *Let $L^d_G\{X\}$ be a maximal d-local-nucleic subgraph of G. If $\varepsilon(L^d_G\{X\}) \leq \varepsilon(G) - 2$. Then $L^d_G\{X\}$ is also a maximal d-local-nucleic subgraph of each edge reconstruction of G, and $\bar{S}(L^d_G\{X\},G)$ is edge reconstructible.*

Proof: By contradiction. Suppose $L^d_G\{X\}$ is not a maximal d-local-nucleic subgraph of an edge reconstruction G' of G. From corollary 2, there must be a d-local-nucleic subgraph $L^d_{G'}\{Y\}$ such that $L^d_G\{X\} \prec L^d_{G'}\{Y\}$. If $\varepsilon(L^d_{G'})\{Y\} \leq \varepsilon(G) - 1$, choose $e \in E(G') \backslash E(L^d_{G'}\{Y\})$, otherwise choose $e \in E(L^d_{G'}\{Y\})$ such that $L^d_G\{X\} \prec L^d_{G'-e}\{Y\}$. Let $G" = G' - e + f \cong G$, then $L^d_G\{X\}$ would be less than $L^d_{G"}\{Y\}$, a contradiction to that $L^d_G\{X\}$ is a maximal d-local-nucleic subgraph of G, and $\bar{S}(L^d_G\{X\},G)$ is edge reconstructible by corollary 2 and 3. □

Theorem 5 *Let X be a proper subgraph of graph G and $d \geq 1$ an integer. If $\varepsilon(L^d_G\{Y\}) \leq \varepsilon(G) - 2$ for each $Y \in s(X,G)$, then each $\bar{S}(L^d_G\{Y\},G)$ is edge reconstructible.*

Proof: Suppose $L^d_G\{Z\}$ is a d-local-nucleic subgraph of G such that $L^d_G\{Y\} \prec L^d_G\{Z\}, Z \in s(X,G)$. Define a $(L^d_G\{Y\}, L^d_G\{Z\})$-chain of length n to be the sequence

$$(L^d_G\{Y\}, L^d_G\{Y_1\}, ..., L^d_G\{Y_{n-1}\}, L^d_G\{Z\})$$

where $\{Y_1, Y_2, ..., Y_{n-1}\} \subset s(X, G)$ such that

$$L_G^d\{Y\} \prec L_G^d\{Y_1\}, \prec ... \prec L_G^d\{Y_{n-1}\} \prec L_G^d\{Z\})$$

The grade of a d-local-nucleic subgraph $L_G^d\{Y\}$ is defined to be the maximum length of all $(L_G^d\{Y\}, L_G^d\{Z\})$-chain for all d-local-nucleic subgraph $L_G^d\{Z\}$ of G with $L_G^d\{Y\} \preceq L_G^d\{Z\}$. While $L_G^d\{Y\} \cong L_G^d\{Z\}$, $(L_G^d\{Y\}, L_G^d\{Z\})$-chain is of length 0. The grade of maximal d-local-nucleic subgraph is defined to be 0. Next we prove the theorem by induction on the grade of $L_G^d\{Y\}$.

By Lemma 4, the theorem is certainly true while the grade is 0. Suppose that it is true for each of grade less than r. Now suppose $L_G^d\{Y\}$ is of grade $r \geq 1$. By Lemma 1, we have

$$S(L^d\{Y\}, G) = \sum_{Z \in s(X,G)} \dot{S}(L^d\{Y\}, L_G^d\{Z\} = \sum \dot{S}(L^d\{Y\}, L_G^d\{Z\})\bar{S}(L_G^d\{Z\}, G)$$

where the last sum is taken over all isomorphism types $L_G^d\{Z\}$ of d-local-nucleic subgraph of G with $L_G^d\{Y\} \preceq L_G^d\{Z\}$. Inverting this identity, we obtain

$$\bar{S}(L_G^d\{Y\}, G) = S(L_G^d\{Y\}, G) +$$

$$\sum_{n=1}^{r} \sum (-1)^n \dot{S}(L_G^d\{Y\}, L_G^d\{Y_1\}) ... \dot{S}(L_G^d\{Y_{n-1}\}, L_G^d\{Z\}) S(L_G^d\{Z\}, G)$$

where the inner sum is taken over all isomorphism types of $(L_G^d\{Y\}, L_G^d\{Z\})$-chain

$$(L_G^d\{Y\}, L_G^d\{Y_1\}, ..., L_G^d\{Y_{n-1}\}, L_G^d\{Z\})$$

of length n of G with $L_G^d\{Y\} \prec L_G^d\{Z\}$. Then each of $L_G^d\{Y_i\}$ $i = 1, ..., n-1$ has grade less than r, so by induction hypothesis, $\bar{S}(L_G^d\{Y_i\}, G) > 0, i = 1, 2, ..., n-1$, are edge reconstructible. Then $L_G^d\{Y_i\}, i = 1, 2, ..., n-1$ are also the d-local-nucleic subgraph of every edge reconstructible of G. Therefore

$$\dot{S}(L_G^d\{Y\}, L_G^d\{Y_1\}), \dot{S}(L_G^d\{Y_1\}, L_G^d\{Y_2\}), ..., \dot{S}(L_G^d\{Y_{n-1}\}, L_G^d\{Z\})$$

are edge reconstructible. By Crrollary 2, $S(L_G^d\{Y\}, G)$ and $S(L_G^d\{Z\}, G)$ are also edge reconstructible. Therefore all terms in right hand side of above identity are edge reconstructible, so that the left hand side is edge reconstructible. □

Using similar proof to the above theorem, we have

Theorem 6 *Let X be a proper subgraph of graph G and $d \geq 1$ an integer. If $\nu(L_G^d\{Y\}) \leq \nu(G) - 2$ for each $Y \in s(X,G)$, then each $\bar{S}(L_G^d\{Y\},G)$ is reconstructible.*

The following corollary also supply a conditional answer to the question raised by W.L. Kocay [4]: How to reconstruct the third-order degree sequence of G.

Corollary 7 *Let $d > 1$ and G be a graph. Then, $\bar{S}(L_G^d\{x\},G)$ is reconstructible if $\nu(L_G^d\{x\}) \leq \nu(G) - 2$ for each $x \in V(G)$; $\bar{S}(L_G^d\{x\},G)$ is edge reconstructible if $\epsilon(L_G^d\{x\}) \leq \epsilon(G) - 2$ for each $x \in V(G)$.*

2 Some Extensions

Let $d \geq 1$ be an integer and G be a graph without isolate vertex. X is a proper subgraph of G with $\nu(X) > 0$. Given $Y \in s(X,G)$, if $\nu(L_G^d\{Y\}) \geq \nu(G) - 1$, then $L_G^d\{Y\}$ is called a large d-local-nucleic subgraph of G. Given $Y_1 \in s(X,G)$, if $\dot{S}(L_G^d\{Y_1\}, L_G^d\{Y\}) = 0$ for all large d-local-nucleic subgraph of G, then $L_G^d\{Y_1\}$ is called a free d-local-nucleic subgraph of G. Edge-large d-local-nucleic subgraph and edge-free d-local-nucleic subgraph of of G are defined idem.

Corollary 8 *Let $d \geq 1$. Then $\bar{S}(L_G^d\{Y\},G)$ is reconstructible if $L_G^d\{Y\}$ is a free d-local-nucleic subgraph of G; and it is edge reconstructible if it is a edge free d-local-nucleic subgraph of G.*

Let $L^d\{X\}$ be a nucleic graph. If $S(L^d\{X\},G) \neq 0$ and $\dot{S}(L^d\{X\}, L_G^d\{Y\}) = 0$ for all large d-local-nucleic subgraph of G, then $L^d\{X\}$ is called a free nucleic graph from G.

Theorem 9 *Let $d \geq 1$ be an integer and H be a reconstruction of G. Then $L^d\{X\}$ is a free nucleic graph from H iff $L^d\{X\}$ is a free nucleic graph from G.*

Proof: It's enough to prove the sufficiency. Since $L^d\{X\}$ is a free nucleic graph from G, then $\nu(L^d\{X\}) \leq \nu(G) - 2$, so $S(L^d\{X\},G)$ is reconstructible. By lemma 1, $S(L^d\{X\},G) = \sum_{Y \in s(X,G)} \dot{S}(L^d\{X\}, L^d_G\{Y\})$

$$= \sum_{\substack{L^d_G\{Y\}\ free \\ Y \in s(X,G)}} \dot{S}(L^d\{X\}, L^d_G\{Y\}) = \sum_{\substack{L^d_H\{Y\}\ free \\ Y \in s(X,H)}} \dot{S}(L^d\{X\}, L^d_H\{Y\}) = S(L^d\{X\}, H)$$

Therefore $L^d\{X\}$ must be a free nucleic graph from H. □.

Theorem 10 *Let $d \geq 1$ be an integer and H be an edge reconstruction of G. Then $L^d\{X\}$ is a free nucleic graph from H iff $L^d\{X\}$ is a free nucleic graph from G.*

bf Acknowledgement: The authors would like to give their thanks to Prof. Xie Litong for his guidance.

References

[1] P.J.Kelly, A congruence theorem for trees, Pacific J.Math., 7(1957),961-968.

[2] D.J.Greenwell and R.L.Hemminger, Reconstructing the n-connected components of a graph, Aequations Math.,9(1973),19-22.

[3] W.T.Tutte, All the King's Horses, In the Graph Theory and Related Topics, Eds. J.A.Bondy and U.S.R. Murty,Academic Press.(1979)15-33.

[4] W.L.Kocay, On reconstructing degree sequences, Utilitas Mathematica, Vol.17(1980)151-162.

[5] W.L.Kocay, Some new methods in reconstruction theory, Lecture Notes in math.,952,Springer,(1982)89-114.

[6] J.A.Bondy and U.S.R. Murty, Graph Theory with Applications, THE MACMILLAN PRESS LTD.,1976.

[7] J.A.Bondy, A Reconstructor's Manual, University of Waterloo.1991.

[8] Xie Litong and Fan Hongbing, On Reconstructible Local Subgraphs. (to appear).

Research About the Structure of PBIB Designs

Ju—Yong Xu

(Dept. of Basic Science, Wuhan Urban Construction Institute, Wuhan 430074, Hubei, P. R. China)

Abstract A necessary and sufficient conditions for dividing a given PBIB design into the Descartes product of some PBIB designs with less associate classes and with smaller parameters is given in this paper. Furthermore, the condition for dividing a PBIB design into the Descartes product of some BIB designs is also given in this paper.

Key words Partially balanced incomplete block design (PBIB design), Descartes product, Association Scheme.

In the study of PBIB designs[1][2], one always wishes that the studied PBIB design is able to be reduced to some simple PBIB designs with less associate classes and with smaller parametes. In [3] , we have given a method, which is called Descartes product of PBIB designs, for constructing new PBIB designs by using some known PBIB designs. We call a PBIB design dividable if it can be expressed as the Descartes product of some PBIB designs. Naturally, the following two problems arise. (1) What is the necessary and sufficient conditions for dividing a given PBIB design into the Descartes product of some PBIB designs? (2) How to decompose a given dividable PBIB designs? In the present paper we deal with the two problems and give complete solutions to them. It is possible to obtain some new PBIB designs by using the method in this paper.

Let $\mathscr{B}$ be a PBIB design with w associate classes and with the parameters

$$b,\ v,\ r,\ k,\ \lambda_i,\ n_i,\ p^i_{jl},\ i,\ j,\ l \in [1,w], \tag{1}$$

where $[a,b]$ is the set of the integers neither smaller than a nor larger than b . Let S be its treatment set and R_1, $\cdots$, R_w its w associate classes.

Suppose S can be divided into Descartes product of t subsets

$$S = S_1 \times \cdots \times S_t \tag{2}$$

It is easy to see that

$$v = \prod_{q=1}^{t} v_q,\ where\ v_q = |S_q|,\ q = 1,\cdots,\ t, \tag{3}$$

and for any $s \in S$, we have $s = (s_1, \cdots, s_t)$, $s_q \in S_q$ is said to be q−th component of s. Suppose $\mathscr{B}$ as the collection of k−subsets B of S can be divided into the Descartes product of the collection $\mathscr{B}_q$ of k_q−subsets B_q of S_q

$$\mathscr{B} = \mathscr{B}_1 \times \cdots \times \mathscr{B}_q \tag{4}$$

and for any $B\in\mathscr{B}$, we have

$$B=(B_1,\cdots,B_t),\ B_q\in\mathscr{B}_q,\ |B_q|=k_q,\ q=1,\cdots,t \tag{5}$$

It is clear that

$$b=\prod_{q=1}^{t} b_q,\ where\ b_q=|\mathscr{B}_q|,\ q=1,\cdots,t, \tag{6}$$

Let

$$R_{qi_q}=\{(s_q,s_q')\in S_q\times S_q|((s_1,\cdots,s_q,\cdots,s_t),(s_1',\cdots,s_q',\cdots,s_t'))\in R_i,$$

$$s_p,s_p'\in S_p,\ p=1,\cdots,t,\ p\neq q\},\ q=1,2,\cdots,t.$$

It is easy to see that for any $i\in[1,w]$ there exists ordered t−tuples $(i_1,\cdots,i_q,\cdots,i_t)$ such that

$$((s_1,\cdots,s_t),(s_1',\cdots,s_t'))\in R_i\quad iff\ (s_1,s_1')\in R_{1i_1},\cdots,(s_t,s_t')\in R_{ti_t}. \tag{7}$$

Note that for two elements $(s_1,\cdots,s_t)$, $(s_1',\cdots,s_t')\in S$, $(s_1,\cdots,s_t)\neq(s_1',\cdots,s_t')$, it is possible that $s_q=s_q'$ for some $q\in[1,t]$. Therefore ,we let $i_q=0$ when $s_q=s_q'$. Obviously, $(i_1,\cdots,i_t)\neq(o,\cdots,o)$. Let the number of different i_q be w_q+1 (including o) when i runs through $[1,w]$. Denote by $[o,w_q]$ the area of i_q, $q=1,\cdots,t$. Therefore, there is one−one mapping between $[1,w]$ and $[o,w_1]\times\cdots\times[o,w_t]\backslash\{(o,\cdots,o)\}$ and suppose α is an orbitrarily given one−one mapping from $[1,w]$ to $[o,w_1]\times\cdots\times[o,w_t]\backslash\{(o,\cdots,o)\}$. It is essy to see that

$$w=\prod_{q=1}^{t}(w_q+1)-1 \tag{8}$$

and

$$R_i=R_{1i_1}\times\cdots\times R_{ti_t},\ where\ \alpha(i)=(i_1,\cdots,i_t), \tag{9}$$

is a Descartes product of sets.

Definition. Two elements $s_q^1,s_q^2\in S_q$, $s_q^1\neq s_q^2$, are said to be the i_q−th associates of each other, if $(s_q^1,s_q^2)\in R_{qi_q}$, $q=1,\cdots,t$. The set R_{qi_q} is called the i_q−th associate class on S_q.

Obviously, $i_q\neq 0$ here. We will now prove R_{qi_q}, $i_q\in[1,w_q]$, are an association scheme on S_q.

Lemma 1.

(a) R_{qi_q} is symmetric, i. e. , $(s_q^1,s_q^2)\in R_{qi_q}$ if and only if $(s_q^2,s_q^1)\in R_{qi_q}$, for any $i_q\in[1,w_q]$, $q=1,\cdots,t$.

(b) $R_{qi_q}\neq\emptyset$, $i_q\in[1,w_q]$

(c) $R_{qi_q}\cap R_{qj_q}=\emptyset$, $i_q,j_q\in[1,w_q]$, $i_q\neq j_q$

(d) $S_q\times S_q\backslash\{(s_q,s_q)|s_q\in S_q\}=\bigcup_{i_q\in[1,w_q]}R_{qi_q}$

Proof By the definition of R_{qi_q}, (a) and (b) are clear.

Suppose there exist $i_q,j_q\in[1,w_q]$, $i_q\neq j_q$, such that $R_{qi_q}\cap R_{qj_q}\neq\emptyset$. Let $\alpha^{-1}(i_1,\cdots,i_q,\cdots,i_t)=i$, $\alpha^{-1}(j_1,\cdots,j_q,\cdots,j_t)=j$, where $i_p\in[1,w_p]$, $p=1,\cdots,t$, $p\neq q$, then $i\neq j$. For any $(s_q^1,s_q^2)\in R_{qi_q}\cap R_{qi_q}$ and any $(s_p^1,s_p^2)\in R_{pi_p}$ $p=1,\cdots,t$, $p\neq q$, we have

$$((s_1^1,\cdots,s_q^1,\cdots,s_t^1),(s_1^2,\cdots,s_q^2,\cdots,s_t^2))\in R_i\cap R_j$$

This is in contradiction with that R_i and R_j are two different associate classes of the PBIB design $\mathscr{B}$ on S. Hence (c).

For any $(s_q^1, s_q^2) \in S_q \times S_q \backslash \{(s_q, s_q) | s_q \in S_q\}$, we have $s_q^1 \neq s_q^2$. It is easy to see that the two elements $(s_1^1, \cdots, s_q^1, \cdots, s_t^1)$, $(s_1^2, \cdots, s_q^2, \cdots, s_t^2) \in S$ are different. Then there exists $i \in [1, w]$ such that $((s_1^1, \cdots, s_q^1, \cdots, s_t^1), (s_1^2, \cdots, s_q^2, \cdots, s_t^2)) \in R_i$. Let $\alpha(i) = (i_1, \cdots, i_t)$. Then $(s_q^1, s_q^2) \in R_{qi_q}$ and $i_q \neq 0$. We have

$$S_q \times S_q \backslash \{(s_q, s_q) | s_q \in S_q\} \subset \bigcup_{i_q \in [1, w_q]} R_{qi_q}.$$

Contrarily, for any $(s_q^1, s_q^2) \in \bigcup_{i_q \in [1, w_q]} R_{qi_q}$, then there exists $i_q \in [1, w_q]$ such that $(s_q^1, s_q^2) \in R_{qi_q}$. Since $i_q \neq 0$, so $s_q^1 \neq s_q^2$, i. e.,

$$(s_q^1, s_q^2) \in S_q \times S_q \backslash \{(s_q, s_q) | s_q \in S_q\}$$

Therefore, we have

$$\bigcup_{i_q \in [1, w_q]} R_{qi_q} \subset S_q \times S_q \backslash \{(s_q, s_q) | s_q \in S_q\}$$

Synthesizing the above, hence (d).

This completes the proof.

Lemma 2 For any $s_q \in S_q$, the number

$$n_{qi_q} = |\{s_q^* \in S_q | (s_q^*, s_q) \in R_{qi_q}\}|, \ i_q \in [1, w_q], \tag{10}$$

is independent of the choice of $s_q \in S_q$.

Proof. For and $s_q^1, s_q^2 \in S_q$, $s_q^1 \neq s_q^2$, and $i_q \in [1, w_q]$, let

$$n_{qi_q}^1 = |\{s_q^* \in S_q | (s_q^*, s_q^1) \in R_{qi_q}\}|, \quad n_{qi_q}^2 = |\{s_q^* \in S_q | (s_q^*, s_q^2) \in R_{qi_q}\}|$$

and $s_1 = (s_1^1, \cdots, s_q^1, \cdots, s_t^1) \in S$, where its q−th component is s_q^1. Let $\alpha(i) = (i_1, \cdots, i_q, \cdots, i_t)$. Since $\mathscr{B}$ is a PBIB design on S, then the number

$$\begin{aligned} n_i &= |\{s^* \in S | (s^*, s_1) \in R_i\}| \\ &= |\{(s_1^*, \cdots, s_t^*) \in S | (s_1^*, s_1^1) \in R_{1i_1}, \cdots, (s_q^*, s_q^1) \in R_{qi_q}, \cdots (s_t^*, s_t^1) \in R_{ti_t}\}| \\ &= n_{qi_q}^1 \prod_{\substack{p=1,\cdots,t \\ p \neq q}} |\{s_p^* \in S_p | (s_p^*, s_p^1) \in R_{pi_p}\}| \end{aligned} \tag{11}$$

is only dependent on the choice of i and independent of the choice of $s_1 \in S$.

Let $s_2 = (s_1^1, \cdots, s_q^2, \cdots, s_t^1)$, where its q−th component is s_q^2 and the others are the same as s_1. Similarly, we have

$$n_i = n_{qi_q}^2 \prod_{\substack{p=1,\cdots,t \\ p \neq q}} |\{s_p^* \in S_p | (s_p^*, s_p^1) \in R_{pi_p}\}| \tag{12}$$

According to (11) and (12), we have $n_{qi_q}^1 = n_{qi_q}^2$.

Hence the lemma.

Lemma 3 For any $(s_q^1, s_q^2) \in R_{qi_q}$, the number

$$p_{qj_ql_q}^{i_q} = |\{s_q^* \in S_q | (s_q^*, s_q^1) \in R_{qj_q}, (s_q^*, s_q^2) \in R_{ql_q}\}|, \tag{13}$$

$i_q, j_q, l_q \in [1, w_q]$, is independent of the choice of $(s_q^1, s_q^2) \in R_{qi_q}$

Proof. For any $(s_q^1, s_q^2), (s_q^3, s_q^4) \in R_{qi_q}$, $(s_q^1, s_q^2) \neq (s_q^3, s_q^4)$, and any $i_q \in [1, w_q]$, let

$$p_{qj_ql_q}^{i_q(1,2)} = |\{s_q^* \in S_q | (s_q^*, s_q^1) \in R_{qj_q}, (s_q^*, s_q^2) \in R_{ql_q}\}|,$$

$$p_{qj_ql_q}^{i_q(3,4)} = |\{s_q^* \in S_q | (s_q^*, s_q^3) \in R_{qj_q}, (s_q^*, s_q^4) \in R_{ql_q}\}|.$$

Let $s_1=(s_1^1, \cdots, s_q^1, \cdots, s_t^1)$, $s_2=(s_1^2, \cdots, s_q^2, \cdots, s_t^2)$, where their q−th components are s_q^1, s_q^2 respectively. Obviously, $s_1 \neq s_2$. Let $(s_1, s_2) \in R_l$, $\alpha(i)=(i_1, \cdots, i_t)$, $\alpha(j)=(j_1, \cdots, j_t)$ and $\alpha(l)=(l_1, \cdots, l_t)$. Since $\mathscr{B}$ is a PBIB design, then the number

$$\begin{aligned} p_{jl}^{i} &= |\{s^* \in S \mid (s^*, s_1) \in R_j, (s^*, s_2) \in R_l\}| \\ &= |\{(s_1^*, \cdots, s_t^*) \in S \mid ((s_1^*, \cdots, s_t^*), (s_1^1, \cdots, s_t^1)) \in R_j, \\ &\quad ((s_1^*, \cdots, s_t^*), (s_1^2, \cdots, s_t^2)) \in R_l\}| \\ &= p_{q j_q l_q}^{i_q(1,2)} \prod_{\substack{\gamma=1,\cdots,t \\ \gamma\neq q}} |\{s_\gamma^* \in S_\gamma \mid (s_\gamma^*, s_\gamma^1) \in R_{\gamma j_\gamma}, (s_\gamma^*, s_\gamma^2) \in R_{\gamma l_\gamma}\}| \quad (14) \end{aligned}$$

is independent of the choice of $(s_1, s_2) \in R_l$

Let $s_3=(s_1^1, \cdots, s_q^3, \cdots, s_t^1)$, where q−th component is s_q^3 and the others are the same as s_1, and $s_4=(s_1^2, \cdots, s_q^4, \cdots, s_t^2)$, where its q−th component is s_q^4 and the others are the same as s_2. Similarly, we have

$$p_{jl}^{i} = p_{q j_q l_q}^{i_q(3,4)} \prod_{\substack{\gamma=1,\cdots,t \\ \gamma\neq q}} |\{s_\gamma^* \in S_\gamma \mid (s_\gamma^*, s_\gamma^1) \in R_{\gamma j_\gamma}, (s_\gamma^*, s_\gamma^2) \in R_{\gamma l_\gamma}\}| \quad (15)$$

According to (14) and (15), we obtain $p_{q j_q l_q}^{i_q(1,2)} = p_{q j_q l_q}^{i_q(3,4)}$

Hence the lemma.

Combining the above lemmas, we have

Theorem 1 The S_q with all the above associate classes R_{qi_q}, $i_q \in [l, w_q]$, is an association scheme with w_q associate classes and with parameters v_q, n_{qi_q}, $p_{q j_q l_q}^{i_q}$.

Theorem 2 Let $\mathscr{B}$ be a PBIB design with w associate classes and with the parameters (1) and S the set of treatments of $\mathscr{B}$. If (2), (4) and (5) hold, then $\mathscr{B}_q$ is a PBIB design on S_q with w_q associate classes and with parameters

$$b_q, v_q, r_q, k_q, \lambda_{qi_q}, n_{qi_q}, p_{q j_q l_q}^{i_q} \; i_q, j_q, l_q \in [1, w_q], \; q=1, \cdots, t, \quad (16)$$

Proof. Based on the association scheme of Theorem 1, we will now prove $\mathscr{B}_q$ is a PBIB design on S_q.

Obviously, the parameters n_{qi_q}, $p_{q j_q l_q}^{i_q}$, are given by lemma 2 and lemma 3 respectively.

By (3), $v_q=|S_q|$, $q=1, \cdots, t$, and $v=\prod_{q=1}^{t} v_q$

By (4), $b_q=|\mathscr{B}_q|$, $q=1, \cdots, t$, and $b=\prod_{q=1}^{t} b_q$

By (5), $|B_q|=k_q$, for any $B_q \in \mathscr{B}_q$, $q=1, \cdots, t$.

For any $s_q^1 \neq s_q^2 \in S_q$ and $s_1=(s_1^1, \cdots, s_q^1, \cdots, s_t^1)$, $s_2=(s_1^1, \cdots, s_q^2, \cdots, s_t^1) \in S$, where their q−th components are s_q^1, s_q^2 respectively and the others components are equal correspondingly. Let

$$r_q^1 = |\{B_q \in \mathscr{B}_q \mid s_q^1 \in B_q\}| \text{ and } r_q^2 = |\{B_q \in \mathscr{B}_q \mid s_q^2 \in B_q\}|$$

Since $\mathscr{B}$ is a PBIB design on S, then the number

$$\begin{aligned} r &= |\{B \in \mathscr{B} \mid s_1 \in B\}| \\ &= |\{(B_1, \cdots, B_t) \in \mathscr{B} \mid (s_1^1, \cdots, s_q^1, \cdots, s_t^1) \in (B_1 \cdots, B_t)\}| \\ &= r_q^1 \prod_{\substack{p=1,\cdots,t \\ p\neq q}} |\{B_p \in \mathscr{B}_p \mid s_p^1 \in B_p\}| \quad (17) \end{aligned}$$

is independent of the choice of $s_1 \in S$.

Similarly,

$$r = r_q^2 \prod_{\substack{p=1,\cdots,t \\ p \neq q}} |\{B_p \in \mathscr{B}_p | s_p^1 \in B_p\}|. \tag{18}$$

From (17) and (18), we have $r_q^1 = r_q^2$. Therefore, for any $s_q \in S_q$, the number

$$r_q = |\{B_q \in \mathscr{B}_q | s_q \in B_q\}| \tag{19}$$

is a constant and independent of the choice of $s_q \in S_q$.

For any $(s_q^1, s_q^2), (s_q^3, s_q^4) \in R_{qi_q}$, $(s_q^1, s_q^2) \neq (s_q^3, s_q^4)$, it is clear that $\{s_q^1, s_q^2\} \neq \{s_q^3, s_q^4\}$, $s_q^1 \neq s_q^2$, $s_q^3 \neq s_q^4$. Let

$$\lambda_{qi_q}^{(1,2)} = |\{B_q \in \mathscr{B}_q | \{s_q^1, s_q^2\} \subset B_q\}| \text{ and } \lambda_{qi_q}^{(3,4)} = |\{B_q \in \mathscr{B}_q | \{s_q^3, s_q^4\} \subset B_q\}|.$$

For four elements $s_1 = (s_1^1, \cdots, s_q^1, \cdots, s_t^1)$, $s_2 = (s_1^1, \cdots, s_q^2, \cdots, s_t^1)$, $s_3 = (s_1^1, \cdots, s_q^3, \cdots, s_t^1)$ and $s_4 = (s_1^1, \cdots, s_q^4, \cdots, s_t^1)$, let their q-th components be s_q^1, s_q^2, s_q^3 and s_q^4 recpectively and the others components be equal correspondingly. Let $\alpha(i) = (i_1, \cdots, i_t)$. Noting $\mathscr{B}$ is a PBIB design on S and $\{s_p^1, s_p^1\} = \{s_p^1\}$, we have

$$\begin{aligned}
\lambda_i &= |\{B \in \mathscr{B} | \{s_1, s_2\} \subset B\}| \\
&= |\{(B_1, \cdots, B_t) \in \mathscr{B} | \{(s_1^1, \cdots, s_q^1, \cdots, s_t^1), (s_1^1, \cdots, s_q^2, \cdots, s_t^1)\} \\
&\qquad \subset (B_1 \cdots, B_t)| \\
&= |\{B_q \in \mathscr{B}_q | \{s_q^1, s_q^2\} \subset B_q\}| \prod_{\substack{p=1,\cdots,t \\ p \neq q}} (|\{B_p \in \mathscr{B}_p | \{s_p^1, s_p^1\} \subset B_p\}|) \\
&= \lambda_{qi_q}^{(1,2)} \prod_{\substack{p=1,\cdots,t \\ p \neq q}} r_p.
\end{aligned} \tag{20}$$

Similarly, we have

$$\lambda_i = \lambda_{qi_q}^{(3,4)} \prod_{\substack{p=1,\cdots,t \\ p \neq q}} r_p. \tag{21}$$

According to (20) and (21), we have $\lambda_{qi_q}^{(1,2)} = \lambda_{qi_q}^{(3,4)}$. Therefore, for any $(s_q^1, s_q^2) \in R_{qi_q}$, the number

$$\lambda_{qi_q} = |\{B_q \in \mathscr{B}_q | \{s_q^1, s_q^2\} \subset B_q\}| \tag{22}$$

is a constant and independent of the choce of $(s_q^1, s_q^2) \in R_{qi_q}$.

Synthesizing the above, we have proved that $\mathscr{B}_q$ is a PBIB design on S_q with w_q associnte classes and with the parameters (16). This completes the Proof.

Theorem 3 Let $\mathscr{B}$ be a PBIB deslgn on S with w associate classes and with the parameters (1). $\mathscr{B}$ is dividable if and only if (2), (4) and (5) hold.

Proof. let $\mathscr{B}_q$ be a PBIB design on S_q with w_q associate classes and with the parameters (16), $q=1, \cdots, t$, and $\mathscr{B}$ the Descartes product of t PBIB designs $\mathscr{B}_1, \cdots, \mathscr{B}_t$. According to the concept of Descartes product of PBIB designs[3], (2), (4) and (5) hold, i. e., the condition of this theorem is necessary.

On the other hand, if (2), (4) and (5) hold, by Theorem 2, $\mathscr{B}_q$ is a PBIB design on S_q with w_q associate classes and with the pasameters (16). Now we have to show that the given

PBIB design $\mathscr{B}$ is exactly the Descartes product of the above-mentioned PBIB designs $\mathscr{B}_1$, $\cdots$, $\mathscr{B}_t$.

Now, we research the expressions of the parameters of the given PBIB design $\mathscr{B}$ under the given conditions.

By (2) and (3), $v=\prod_{q=1}^{t} v_q$ and $v_q=|S_q|$.

By (4) and (6), $b=\prod_{q=1}^{t} b_q$ and $b_q=|\mathscr{B}_q|$.

For any $B=(B_1, \cdots, B_t)\in\mathscr{B}$, by (5), we have

$$k = |B| = \prod_{q=1}^{t} |B_q| = \prod_{q=1}^{t} k_q.$$

For any $s=(s_1, \cdots, s_t)\in S$, Since $\mathscr{B}$ is a PBIB design, we have

$$\begin{aligned} r &= |\{B \in \mathscr{B} | s \in B\}| \\ &= |\{(B_1, \cdots, B_t) \in \mathscr{B} | (s_1, \cdots, s_t) \in (B_1, \cdots, B_t)\}| \\ &= \prod_{q=1}^{t} |\{B_q \in \mathscr{B}_q | s_q \in B_q\}|. \end{aligned}$$

By (19), $|\{B_q\in\mathscr{B}_q|s_q\in B_q\}|=r_q$. Hence

$$r = \prod_{q=1}^{t} r_q.$$

For any $(s_1, s_2)=((s_1^1, \cdots, s_t^1), (s_1^2, \cdots, s_t^2))\in R_i$ and $\alpha(i)=(i_1, \cdots, i_t)$, then $(s_q^1, s_q^2)\in R_{qi_q}$, where $i_q\in[o, w_q]$. Since $\mathscr{B}$ is a PBIB design on S, we have

$$\begin{aligned} \lambda_i &= |\{B \in \mathscr{B} | \{s_1, s_2\} \subset B\}| \\ &= |\{(B_1, \cdots, B_t) \in \mathscr{B} | \{(s_1^1, \cdots, s_t^1), (s_1^2, \cdots, s_t^2)\} \subset (B_1, \cdots, B_t)\}| \\ &= \prod_{q=1}^{t} |\{B_q \in \mathscr{B}_q | \{s_q^1, s_q^2\} \subset B_q\}|. \end{aligned}$$

By (19) and (22),

$$|\{B_q \in \mathscr{B}_q | \{s_q^1, s_q^2\} \subset B_q\}| = \begin{cases} \lambda_{qi_q} & \text{for } i_q \neq 0, \\ r_q & \text{for } i_q = 0 \end{cases}$$

Hence

$$\lambda_i = \prod_{i_q\neq 0} \lambda_{qi_q} \cdot \prod_{i_q=0} r_q, \ \textit{where } \alpha(i) = (i_1, \cdots, i_t).$$

For any $s=(s_1, \cdots, s_t)\in S$, $i\in[1, w]$ and $\alpha(i)=(i_1, \cdots, i_t)$, we have

$$\begin{aligned} n_i &= |\{s^* \in S | (s^*, s) \in R_i\}| \\ &= |\{(s_1^*, \cdots, s_t^*) \in S | ((s_1^*, \cdots, s_t^*), (s_1, \cdots, s_t)) \in R_i\}| \\ &= \prod_{q=1}^{t} |\{s_q^* \in S_q | (s_q^*, s_q) \in R_{qi_q}\}| \end{aligned}$$

By (10), $|\{s_q^* \in S_q | (s_q^*, s_q) \in R_{qiq}\}| = n_{qi_q}$. Hence

$$n_i = \prod_{q=1}^{t} n_{qi_q}, \ \textit{where } \alpha(i) = (i_1, \cdots, i_t).$$

Note $i_q\in[o, w_q]$ and it is easy to verify $n_{q0}=1$. And we know that it conforms with usual convention in the PBIB design theory.

For any $(s_1, s_2)=((s_1^1, \cdots, s_t^1), (s_1^2, \cdots, s_t^2))\in R_l$ and $i, j, l \in [1,w]$. Let $\alpha(i)=(i_1, \cdots, i_t)$, $\alpha(j)=(j_1, \cdots, j_t)$ and $\alpha(l) = (l_1, \cdots, l_t)$. For PBIB design $\mathscr{B}$, we have

$$\begin{aligned} p^i_{jl} &= |\{s^* \in S \mid (s_1, s^*) \in R_j, (s_2, s^*) \in R_l\}| \\ &= |\{(s_1^*, \cdots, s_t^*) \in S \mid ((s_1^*, \cdots, s_t^*), \\ &\quad (s_1^1, \cdots, s_t^1)) \in R_j, ((s_1^2, \cdots, s_t^2), (s_1^*, \cdots, s_t^*)) \in R_l\}| \\ &= \prod_{q=1}^{t} |\{s_q^* \in S_q \mid (s_q^1, s_q^*) \in R_{qj_q}, (s_q^2, s_q^*) \in R_{ql_q}\}|. \end{aligned}$$

Noting $(s_q^1, s_q^2)\in R_{qi_q}$ and by (13), we have

$$p^i_{jl} = \prod_{q=1}^{t} p^{i_q}_{q j_q l_q}, \text{ where } i_q, j_q, l_q \in [o, w_q].$$

It is easy to verify that

$$p^0_{q j_q l_q} = n_{qi_q} \delta_{j_q l_q} \quad \text{and} \quad p^{i_q}_{q o j_q} = \delta_{i_q j_q} = p^{i_q}_{q j_q o},$$

where

$$\delta_{xy} = \begin{cases} 1, & \text{if } x = y \\ 0, & \text{if } x \neq y, \end{cases}$$

and they conform with the usual convention in the PBIB design theory.

By (8), $\quad w = \prod_{q=1}^{t} (w_q + 1) - 1.$

Synthesizing the above, we obtain that the expressions of the parameters of the given PBIB $\mathscr{B}$ by using the parameters of the PBIB designs $\mathscr{B}_q$, $q=1, \cdots, t$, *are*

$$\left.\begin{aligned} & b = \prod_{q=1}^{t} b_q \quad v = \prod_{q=1}^{t} v_q \quad k = \prod_{q=1}^{t} k_q \quad r = \prod_{q=1}^{t} r_q, \\ & \lambda_i = \prod_{i_q \neq 0} \lambda_{qi_q} \prod_{i_q=0} r_q, \quad n_i = \prod_{q=1}^{t} n_{qi_q}, \quad p^i_{jl} = \prod_{q=1}^{t} p^{i_q}_{q j_q l_q}, \end{aligned}\right\} \tag{23}$$

where $i, j, l\in[1, w]$, $\alpha(i)=(i_1, \cdots, i_t)$, $\alpha(j)=(j_1, \cdots, j_t)$, $\alpha(l)=(l_1, \cdots, l_t)$,

$$i_q, j_q, l_q \in [o, w_q], \; q = 1, \cdots, t,$$

and between the number w *of the associate classes of* $\mathscr{B}$ and the numbers w_q of the associate classes of $\mathscr{B}_q$, $q=1, \cdots, t$, there is the following relation,

$$w = \prod_{q=1}^{t} (w_q + 1) - 1$$

By the concept of the Descartes product of PBIB designs[3], we prove that the given PBIB design $\mathscr{B}$ as a design is exactly the Descartes product of the PBIB designs $\mathscr{B}_q$ on S_q, $q=1, \cdots$ t. this completes the proof.

We know that a balanced incomplete block (BIB) design can be viewed as a special PBIB design with only one associate class. Then, what is the condition for a given PBIB design to be able to be divided into the Descartes product of some BIB designs ? For the problem, we need the concept of EGD/(2^t-1)−PBIB designs, which can be found in [4], as follows.

Let $\mathscr{B}$ be a PBIB design with 2^t-1 associate classes on the treatment set S. Then $\mathscr{B}$ is an EGD/(2^t-1)—PBIB design if v elements of S and the association scheme on S are defined as follows. $v=v_1\cdots v_t$ and the elements are expressed as $(x_1, \cdots, x_t)$, $x_q \in S_q=\{1, 2, \cdots, v_q\}$, $q=1, \cdots, t$ (i. e. $S=S_1\times\cdots\times S_t$), x_q is said to be the q—th component of $(x_1, \cdots, x_t)$. Eements

$$(x_1, \cdots, x_t), (x'_1, \cdots, x'_t) \in S, (x'_1, \cdots, x'_t) \neq (x_1, \cdots, x_t),$$

belong to the $(i_1, \cdots, i_t)$—th associate class if $i_q=1$ when $x_q\neq x'_q$, and if $i_q=0$ when $x_q=x'_q$, $q=1,\cdots,t$.

Theorem 4 If a PBIB design $\mathscr{B}$ can be divided into the Descartes product of t BIB designs, then $\mathscr{B}$ is a EGD/(2^t-1)—PBIB design[4].

Proof. Let $\mathscr{B}$ be a PBIB design on S with W associate classes. Suppose $\mathscr{B}$ can be divided into the Descartes product of t BIB designs D_q on S_q, $q=1,\cdots$, t. From Theorem 2, (2), (4), and (5) hold.

By (2), $S=S_1\times\cdots\times S_t$ and $v_q=|S_q|$, $q=1, \cdots$, t. Without loss of generality, the set S_q can be taken as the set $[1,v_q]$. Therefore, any element s of S can be expressed as $(s_1, \cdots, s_t)$, $s_q\in[1,v_q]$, $q=1,\cdots$, t.

By (4) and the conditions of the theorem, $\mathscr{B}=D_1\times\cdots\times D_t$ and D_q is a BIB design on S_q. Since D_q is a BIB design, then $w_q=1$, $q=1,\cdots$, t. So, by (8), we have

$$w = \prod_{q=1}^{t}(w_q+1)-1=2^t-1,$$

i. e., $\mathscr{B}$ is a PBIB design with 2^t-1 associate classes. For any $i\in[2^t-1]$, let $\alpha(i)=(i_1, \cdots, i_t)$, obviously, $i_q\in[0,1]$, $q=1, \cdots$, t. For any $((s_1^1, \cdots, s_t^1), (s_1^2, \cdots, s_t^2))\in R_i$, without loss of generality, we can denote it as $((s_1^1, \cdots, s_t^1), (s_1^2, \cdots, s_t^2))\in R_{(i_1,\cdots,i_t)}$. And it is easy to see that $i_q=1$ when $s_q^1\neq s_q^2$, $i_q=0$ when $s_q^1=s_q^2$.

Synthesizing the above, by the definition of EGD/(2^t-1)—PBIB design[4], we have proved that $\mathscr{B}$ is a EGD/(2^t-1)—PBIB design. this completes the proof.

The general problem naturally arises: what is the necessary and sufficient conditions for a EGD/(2^t-1)—PBIB design to be able to be divided into the Descartes product of some BIB designs? This problem have been solved in [5]. The proof of the following theorem can be found in [5].

Theorem 5 An EGD/(2^t-1)—PBIB design $\mathscr{B}$ on S can be divided into the Descartes product of t BIB designs $D_1, \cdots, D_t$ if and only if $\mathscr{B}$ as a collection of k—subsets B of S can be divided into the Descartes Product of the collection D_q of k_q—subsets B_q of S_q, $q=1,\cdots,t$.

For dividing a PBIB design $\mathscr{B}$, We have the main step as follows.

1. Find all decomposition of the number $v=|S|$.
2. For any decomposition of v, $v=v_1\times\cdots\times v_t$, We find the decomposition of the set S

$$S=S_1\times\cdots\times S_t$$

such that $|S_q|=v_q$, $q=1,\cdots,t$.

3. Verify that $\mathscr{B}$ as the collection of k—subsets B of S can be divided into the Descartes

product of the collection $\mathscr{B}_q$ of k_q—Subsets B_q of S_q or can't.

Example. We know that there exists a PBIB design on Z_{77} with 3 associate classes and with the parameters

$b=v=77,\ r=k=15, \lambda_1=2,\ \lambda_2=6,\ \lambda_3=5,$

$n_1=60,\ n_2=10,\ n_3=6,$

$$P_1=\begin{bmatrix}45 & 9 & 5\\ 9 & 0 & 1\\ 5 & 1 & 0\end{bmatrix},\quad P_2=\begin{bmatrix}54 & 0 & 6\\ 0 & 9 & 0\\ 6 & 0 & 0\end{bmatrix},\quad P_3=\begin{bmatrix}50 & 10 & 0\\ 10 & 0 & 0\\ 0 & 0 & 5\end{bmatrix}$$

and $\mathscr{B}=\{B_1,\cdots,B_{77}\}$,

where $B_1=\{1, 36, 15, 71, 64, 23, 58, 37, 16, 9, 67, 25, 4, 60, 5\}$,

$B_2=\{57, 15, 71, 50, 43, 2, 37, 16, 72, 65, 46, 4, 60, 39, 32\}$,

… …

$B_{77}=\{77, 35, 14, 70, 63, 22, 57, 36, 15, 8, 66, 24, 3, 59, 52\}$.

Because $77=7\times11$ and it is easy to see that

$Z_{77}\cong Z_7\times Z_{11}$, where $Z_7\times Z_{11}$ is the direct product of the Abelian groups Z_7 and Z_{11}. It is easy to verify that mapping β,

$\beta: Z_{77}\rightarrow Z_7\times Z_{11}$

$a\rightarrow(a(\bmod 7), a(\bmod 11))$,

is a ismorphic mapping from Z_{77} to $Z_7\times Z_{11}$, So, it is easy to verify that

$$\beta(B_1)=\{(1,1),(1,3),(1,4),(1,5),(1,9),(2,1),(2,3),(2,4),(2,5),(2,9),(4,1),(4,3),(4,4),(4,5),(4,9)\}=(B_1^1,B_2^1),$$

$$\beta(B_2)=\{(1,2),(1,4),(1,5),(1,6),(1,10),(2,2),(2,4),(2,5),(2,6),(2,10),(4,2),(4,4),(4,5),(4,6),(4,10)\}=(B_1^1,B_2^2),$$

… …

$$\beta(B_{77})=\{(0,0),(0,2),(0,3),(0,4),(0,8),(1,0),(1,2),(1,3),(1,4),(1,8),(3,0),(3,2),(3,3),(3,4),(3,8)\}=(B_1^7,B_2^{11}),$$

where $B_1^1=\{1, 2, 4\}, B_1^2=\{2, 3, 5\}, B_1^3=\{3, 4, 6\}, B_1^4=\{4, 5, 0\}$,

$B_1^5=\{5, 6, 1\}, B_1^6=\{6, 0, 2\}, B_1^7=\{0, 1, 3\}$,

$B_2^1=\{1, 3, 4, 5, 9\}, B_2^2=\{2, 4, 5, 6, 10\}, B_2^3=\{3, 5, 6, 7, 0\}$,

$B_2^4=\{1, 4, 6, 7, 8\}, B_2^5=\{2, 5, 7, 8, 9\}, B_2^6=\{3, 6, 8, 9, 10\}$,

$B_2^7=\{0, 4, 7, 9, 10\}, B_2^8=\{0, 1, 5, 8, 10\}, B_2^9=\{0, 1, 6, 9, 2\}$,

$B_2^{10}=\{1, 2, 3, 7, 10\}, B_2^{11}=\{0, 2, 3, 4, 8\}$.

Let $\mathscr{B}_1=\{B_1^1,\cdots, B_1^7\}, \mathscr{B}_2=\{B_2^1,\cdots, B_2^{11}\}$. It is easy to vesify that $\mathscr{B}$ as the collection of 15—subsets B of Z_{77} can be divided into the Descartes Product of the collection $\mathscr{B}_1$ of 3—subsets B_1 of Z_7 and the collection $\mathscr{B}_2$ of 5—subsets B_2 of Z_{11}. By Theorem 3, We obtain that $\mathscr{B}_1$

is a BIB design on Z_7 with the parameters

$$b_1=v_1=7,\ r_1=k_1=3,\ \lambda_1^1=1,$$

and $\mathscr{B}_2$ is a BIB design on Z_{11} with the parameters

$$b_2=v_2=11,\ r_2=k_2=5,\ \lambda_2^1=2.$$

Of course, $\mathscr{B}_1$ and $\mathscr{B}_2$ in the example have been known already, but we think that it is possible to obtain some new PBIB desighs by using the decomposition method in this paper.

References

[1] M. Hall. jr, Combinatorial theory, Blaisdell publ. Company, walthem. Massachusetts, 1967.

[2]Wan—Di Wei, Combinatorial theory (Vol, 2), Science Press, BeiJing, 1987

[3] Wan—Di Wei and Ju—Yong Xu, Construction of PBIB designs by Using Descartes product of known PBIB designs, Chin. Ann. of Math., 14B;2(1993), 245—252.

[4] K. Hinkelmann, Extended group divisible partially bananced incomplete block designs, Ann. Math. Satist, 36 (1964) 681—695.

[5] Ju—Yong Xu, Research about the structure of EGD/(2^t-1)—PBIB designs, Combinatorics and graph theosy (Proceedings), World Scientific Publishing Co. Pte. Ltd, 1993, 81—86.

Finite Geometries and PBIB Designs (I)

Yang Benfu

(*Dept. of Math., Chengdu Teachers College, Chengdu*, 611930)

Wei Wandi

(*Dept. of Math., Sichuan Univ., Chengdu*, 610064)

Abstract We study the properties of the conjugation in a finite linear space and in a finite symplectic geometry. Then we apply them to find relations among subspaces, to construct a kind of new PBIB designs, and to give the isomorphisms between some known PBIB designs.

1. Introduction

Let q be a prime power, Fq the finite field with q elements, Vn(Fq) the n−dimensional vector space over Fq and GLn(Fq) the general linear group of degree n over Fq. GLn(Fq) can be viewed as a transformation group of Vn(Fq). Vn(Fq) with GLn(Fq) as its transformation group is called an n−dimensional linear space over Fq and denoted by LVn(Fq). Let p be an mxn matrix with rank m over Fq. The subspace of Vn(Fq) spanned by the rows of p is also denoted by the same symbol p.

Two vectors V_1 and V_2 in LVn(Fq) are said to be orthogonal if $V_1 \cdot V'_2 = 0$, where V'_2 is the transpose of V_2. For a subspace P of LVn(Fq), P^* denoteds the set of the vectors that are orthogonal to all the vectors in P. Clearly, P is an (n−m)−dimensional subspace, and is called the conjugate subspace of p or the conjugation of p.

Let $n=2\gamma$ $(\gamma \geqslant 1)$, and

$$K = \begin{pmatrix} 0 & 1^{(\gamma)} \\ -1^{(\gamma)} & 0 \end{pmatrix}.$$

The set of nonsingular n×n matrices T over Fq satisfying $TKT' = K$ with the matrix multiplication as its composition forms a group, called the symplectic group of degree 2γ over Fq (defined by K), and denoted by $Sp_{2\gamma}(Fq)$. $Sp_{2\gamma}(Fq)$ can be viewed as a transformation group of $V_{2\gamma}(Fq)$. $V_{2\gamma}(Fq)$ with $Sp_{2\gamma}(Fq)$ as its tranformation group is called a 2γ−dimensional symplectic space or symplectic geometry over Fq and denoted by $SV_{2\gamma}(Fq)$.

Two vectors V_1 and V_2 in the symplectic geometry $SV_{2\gamma}(Fq)$ are said to be orthogonal if $V_1KV'_2 = 0$. Let p be a subspace of $SV_{2\gamma}(Fq)$. Let P^* be the set of the vectors that are orthogonal to all the vectors in P. Clearly, P^* is a $(2\gamma - m)$−dimensional subspace of $SV_{2\gamma}(Fq)$, and called the conjugate subspace of P or the conjugation of P.

In the present paper, we study the properties of the conjugation in LVn(Fq) and in $SV_{2\gamma}(Fq)$. Then we apply them to find relations between subspaces, to construct a kind of new PBIB designs, and also to give the isomorphisms between some known PBIB designs.

In the finite unitary geometries and in the orthogonal geometries, the similar discussion can be conducted, which appear in some other separate papers. [2,3]

2. Properties of conjugation

For two subspaces P and Q of $Vn(Fq)$, $P\cup Q$ denotes the subspace which is spanned by P and Q, and $P\cap Q$ denotes the subspace which is the intersection of P and Q.

We first prove

Theorem 1 In $LVn(Fq)$ or in $SVn(Fq)$, for two subspaces P and Q, we have

(I) $(P^*)^*=P$;

(II) If $P\subseteq Q$, then $Q^*\subseteq P^*$;

(III) $(P\cup Q)^*=P^*\cap Q^*$;

(IV) $(P\cap Q)^*=P^*\cup Q^*$.

Proof From the definition of the conjugation, it follows that $P\subseteq(P^*)^*$. According to $\dim P=n-\dim P^*=\dim(P^*)^*$, we have (I).

(II) and (III) are the immediate consequences of the definition of the conjugation.

Replacing P and Q by P^* and Q^* in (III) respectively, one obtains (IV) by (I). This completes the proof.

Theorem 2 If P is a subspace of $LVn(Fq)$, and $T\in GLn(Fq)$, then $(PT)^*=P^*(T^{-1})'$. If P is a subspaces of $SV_{2\gamma}(Fq)$ and $T\in Sp_{2\gamma}(Fq)$, then $(PT)^*=P^*T$.

Proof In $LVn(Fq)$,

$$(PT)(P^*(T^{-1})')'=PTT^{-1}(P^*)'=P(P^*)'=0,$$

so $P^*(T^{-1})'\subseteq(PT)^*$. According to

$$\dim\ (P^*(T^{-1})')'=\dim\quad P^*=n-\dim\quad P=n-\dim\quad (PT)=\dim(PT)^*,$$

we have $(PT)^*=P^*(T^{-1})'$.

In $SV_{2\gamma}(Fq)$,

$$(PT)K(P^*T)'=PTKT'(P^*)'=PK(P^*)'=0,$$

so $P^*T\subseteq(PT)^*$. According to

$$\dim P^*T=\dim\quad P^*=n-\dim\quad P=n-\dim(PT)=\dim(PT)^*,$$

we have $(PT)^*=P^*T$. This completes the proof.

Let P be an m-dimensional subspace of $SV_{2\gamma}(Fq)$. It is well know that the rank of PKP' is an even number. If the rank of PKP' is 2s, then P is called an (m,s)—type subspace. And the set of (m,s)—type subspaces is denoted by $m\ (m,s;2\gamma)$.

In section 4, we will use the following result.

Theorem 3 Let $2s\leqslant m\leqslant\gamma+s$. In $SV_{2\gamma}(Fq)$, an (m,s)—type subspace intersects its conjugate subspace in an $(m-2s,0)$—type subspace.

Proof Let P be an (m,s)—type subspace. Without loss of generality, we assume that

$$PKP' = \begin{bmatrix} 0 & 1^{(s)} & \\ -1^{(s)} & 0 & \\ & & 0^{(m-2s)} \end{bmatrix}$$

Write

$$P = \begin{pmatrix} P_1 \\ P_2 \end{pmatrix} \begin{matrix} 2s \\ m-2s \end{matrix} .$$

Then $P_2KP'=0$, which implies that $P_2 \subseteq P^* \cap P$ and that P_2 is an $(m-2s, 0)$—type subspace. Therefore $\dim P^* \cap P \geqslant m-2s$. And then $\dim P^* \cap P = m-2s$. Otherwise, $\dim P^* \cap P > m-2s$. Writing

$$P = \begin{bmatrix} \widetilde{P}_1 \\ P^* \cap P \end{bmatrix}.$$

we have

$$PKP' = \begin{bmatrix} \widetilde{P}_1 K \widetilde{P}_1{}' & 0 \\ 0 & 0 \end{bmatrix},$$

so

rank of PKP' = rank of $\widetilde{P}_1 K \widetilde{P}_1{}' < 2s$,

a contradiction. Hence $P^* \cap P = P_2$ and is an $(m-2s, 0)$—type subspace.

This proves the theorem.

3. Conjugation mapping

Let $\mathfrak{m}(x,n)$ be the set of χ-dimensional subspaces of $LV_n(F_q)$.

The mapping which maps an m-dimensional subspace of $V_n(F_q)$ into its conjugate subspace is called the conjugation mapping from $\mathfrak{m}(m,n)$ to $\mathfrak{m}(n-m,n)$.

By (I) of Theorem 1, we know that $(P^*)^* = p$, and that if $P_1^* = P_2^*$ then $P_1 = P_2$. Therefore, we have

Theorem 4 For $LV_n(F_q)$, the conjugation mapping is a one—one mapping from $\mathfrak{m}(m,n)$ onto $\mathfrak{m}(n-m,n)$.

Similarly to Theorem 4, we have

Theorem 5 Let $2s \leqslant m \leqslant \gamma+s$. Then for $SV_{2\gamma}(F_q)$,

(1) The conjugate subspace of an (m,s)—type subspace is of $(2\gamma-m, \gamma-m+s)$—type.

(2) The conjugation mapping is a one—one mapping from $\mathfrak{m}(m,s;2\gamma)$ onto $\mathfrak{m}(2\gamma-m, \gamma-m+s; 2\gamma)$.

Proof Let

$$P = \begin{pmatrix} 1 & 0 & 0 & 0 \\ 0 & 0 & 1 & 0 \end{pmatrix} \begin{matrix} m-s \\ s \end{matrix}$$
$$\begin{matrix} \gamma-s & \gamma-m+s & s & \gamma-s \end{matrix}$$

$$P_1 = \begin{pmatrix} 0 & 1 & 0 & 0 \\ 0 & 0 & 0 & 1 \end{pmatrix} \begin{matrix} \gamma - s \\ \gamma - m + s \end{matrix}$$

$$\begin{matrix} s & \gamma-s & m-s & \gamma-m+s \end{matrix}$$

Then P is an (m,s)−type subspace, P_1 is a $(2\gamma-m, \gamma-m+s)$−type subspace, and $PKP_1' = 0$, i.e. $P_1 \subseteq P^*$. On the other hand, dim $P_1 = 2\gamma - m = \dim P^*$. Therefore $P_1 = P^*$.

For any (m,s)−type subspace Q, there must be $T \in SP_{2\gamma}(Fq)$ such that $Q = PT$, because the symplectic group $SP_{2\gamma}(Fq)$ acts trasitively on the set of subspaces of the same type. By Theorem 2, $Q^* = P^* T$. Therefor, Q^* is a $(2\gamma-m, \gamma-m+s)$−type subspace too. This proves the first assertion of the theorem. As to the second, it is obvious. This completes the proof.

4. Applications of conjugation mapping

4.1 Application to countint subspaces

Using the conjugation mapping one can easily find some relations between the numbers of some types of subspaces.

Denote by N(m,n) the number of m−dimensional subspaces in LVn(Fq), by N(K,m,n) the number of K−dimensional subspaces included in a given m−dimensional subspace of LVn (Fq) and by $N^T(K,m,n)$ the number of m−dimensional subspaces including a given K−dimensional subspace of LVn(Fq). Then we have

Theorem 6 Let $K \leqslant m \leqslant n$. For LVn(Fq), we have

(1) $N(m,n) = N(n-m,n)$;

(2) $N^T(K,m,n) = N(n-m,n-K)$;

(3) $N(m,n)N(K,m) = N(K,n)N(n-m,n-K)$.

Proof The first assertion is an immediate consequence of Theorem 4. By (Ⅰ) and (Ⅱ) of Theorem 1, a K−dimensional subspace P is included in an m−dimensional subspace Q if and only if the (n−K)−dimensional subspace P^* includes the (n−m)−dimensional subspace Q^*. Thus, the number of m−dimensional subspaces that include a given K−dimensional subspace equals the number of (n−m)−dimensional subspaces that are included in a given (n−K)−dimensional subspace. That is, $N^T(k,m,n) = N(n-m,n-k,n)$. On the other hand, $N(n-m,n-k,n) = N(n-m,n-k)$. Hence the conclusion (2). In order to prove (3), we count the pairs (P,Q):

$$P \in \mathfrak{m}(k,n), \quad Q \in \mathfrak{m}(m,n), \quad P \subseteq Q$$

in two ways: considering first P then Q, and first Q then P, respectively. This gives

$$N(k,n)N^T(k,m,n) = N(m,n)N(k,m,n).$$

By substituting (2) and $N(k,m,n) = N(k,m)$ into the above, the conclusion (3) follows. This completes the proof.

In the symplectic geometry $SV_{2\gamma}(Fq)$, Denote by $N(m,s;2\gamma)$ the number of (m,s)−type subspaces, by $N(m_1,s_1;m,s;2\gamma)$ the number of (m_1,s_1)−type subspaces included in a given

(m,s)—type subspace and by $N^T(m_1,s_1;m,s;2\gamma)$ the number of (m,s)—type subspaces including a given (m_1,s_1)—type subspace. Similarly to Theorem 5, we can prove

Theorem 7 Let $2s\leqslant m\leqslant \gamma+s$, and

$$\max\{0,-s-s_1+m_1\}\leqslant \min\{m-2s,m_1-2s_1\}.$$

Then

(Ⅰ) $N(m,s;2\gamma)=N(2\gamma-m,\gamma-m+s;2\gamma)$;

(Ⅱ) $N^T(m_1,s_1;m,s;2\gamma)=N(2\gamma-m,\gamma-m+s;2\gamma-m_1,\gamma-m_1+s_1;2\gamma)$;

(Ⅲ) $N(m,s;2\gamma)N(m_1,s_1;m,s;2\gamma)=N(m_1,s_1;2\gamma)N(2\gamma-m,\gamma-m+s;2\gamma-m_1;\gamma-m_1+s_1;2\gamma)$.

4.2 Applications to relations between subspaces and to trasitivity of symplectic group

Using the congugation mapping for $SV_{2\gamma}(Fq)$, we can derive some relations between the subspaces of higher dimensions from ones between the subspaces of lower demensions, and some transitivity of the symplectic group on the set of some subspace pairs of higher dimensions from the transitivity of symplectic group on the set of some subspace pairs of lower dimensions. And vice versa.

For example, in $SV_{2\gamma}(Fq)$, two 1—dimensional subspaces are either orthogonal or nonorthogonal. Thus, the subspace two 1—dimensional subspaces span is either a $(2,0)$—type subspace or a $(2,1)$—type subspace. Therefore, using the conjugation mapping, we have immediately.

Theorem 8 In $SV_{2\gamma}(Fq)$, two $(2\gamma-1,\gamma-1)$—type subspaces intersect either in a $(2\gamma-2,\gamma-2)$—type subspace or in a $(2\gamma-2,\gamma-1)$—type subspace.

We know that the symplectic group $Sp_{2\gamma}(Fq)$ acts transitively on the set of 1—dimensional subspace pairs the two subspaces of each of which span a $(2,0)$—type subspace, and also acts transitively on the set of 1—dimensional subspace pairs the two subspaces of each of which span a $(2,1)$—type subspace. Therefore, using the conjugation mapping, we have immediately.

Theorem 9 The symplectic group $Sp_{2\gamma}(Fq)$ acts transitively on the set of $(2\gamma-1,\gamma-1)$—type subspace pairs the two subspaces of each of which intersect in a $(2\gamma-2,\gamma-2)$—type subspace, and also transitively on the set of $(2\gamma-1,\gamma-1)$—type subspace pairs the two subspaces of each of which intersect in a $(2\gamma-2,\gamma-1)$—type subspace.

4.3 Application to construction of PBIB designs

Using the conjugation mapping, one can construct some new PBIB designs. Here we only give an illustrative example. We will need the following association scheme (see Theorem 1 in Chapter 7 of [1]).

Let $\gamma\geqslant 2$. Take as treatments the 1—dimensional subspaces of $SV_{2\gamma}(Fq)$, and define two treatments to be the first (resp. the second) associates of each other if they as subspaces are orthogonal (resp. nonorthogonal). Then one obtains an association scheme with two associate classes.

Based on this association scheme, we have

Theorem 10 Adopt the above association scheme. Take as blocks the ordered subspace pairs (W_1,W_2) of $SV_{2\gamma}(F_q)$ where W_1 are (m,s)-type subspaces, W_2 1-dimensional subspaces and $W_2\subseteq W_1^*\setminus W_1$. And define a treatment V to be arranged in a block (W_1,W_2) if the subspace V is orthogonal to W_1 and nonorthogonal to W_2. Then we obtain a PBIB design with two associate classes and with the parameters

$$v=\frac{q^{2\gamma}-1}{q-1},\qquad n_1=\frac{q^{2\gamma-1}-q}{q-1},\qquad n_2=q^{2\gamma-1},$$

$$P_{11}^1=\frac{q^{2\gamma-2}-1}{q-1}-2,\qquad P_{11}^2=\frac{q^{2\gamma-2}-1}{q-1},\qquad P_{22}^1=(q-1)q^{2\gamma-2},$$

$$b=N(m,s;2\gamma)\cdot[N(1,2\gamma-m)-N(1,m-2s)],$$

$$r=n_2\cdot N(m,s;2\gamma-2,\gamma-1;2\gamma),\qquad k=\frac{\gamma v}{b},$$

$$\lambda_1=P_{22}^1\cdot N(m,s;2\gamma-3,\gamma-2;2\gamma),$$

$$\lambda_2=\frac{r(k-1)-\lambda_1 n_1}{n_2}.$$

Proof Let (W_1,W_2) be a block. Without loss of generality, we may assume

$$W_1KW_1'=\begin{bmatrix}0 & 1^{(s)} & \\ -1^{(s)} & 0 & \\ & & 0^{(m-2s)}\end{bmatrix}.$$

Thus

$$\begin{pmatrix}W_1\\W_2\end{pmatrix}K\begin{pmatrix}W_1\\W_2\end{pmatrix}'=\begin{bmatrix}0 & 1^{(s)} & & \\ -1^{(s)} & 0 & & \\ & & 0^{(m-2s)} & \\ & & & 0\end{bmatrix}.$$

From this and Theorem 4 in Chapter 2 of [1], we know that the symplectic group $Sp_{2\gamma}(F_q)$ acts transitively on the set of blocks, Hence we certainly obtain a PBIB design. The parameters of the association scheme of this PBIB design have been given in [1], and the other parameters will be computed as follows.

There are $N(m,s;2\gamma)$ (m,s)-type subspaces in $SV_{2\gamma}(F_q)$, each of them can be taken as W_1 of the blocks (W_1,W_2). For a chosen W_1, W_1^* is a $(2\gamma-m,\gamma-m+s)$-type subspace. By Theorem 3. There are exactly

$$N(1,2\gamma-m)-N(1,m-2s)$$

ways of choosing W_2. Therefore, the number of block is

$$b=N(m,s;2\gamma)\cdot[N(1,2\gamma-m)-N(1,m-2s)].$$

Let V be a given 1-dimensional subspace. We now count the blocks (W_1,W_2) which the treatment V is arranged in. First consider W_2 then W_1. There are exactly n_2 ways of choosing W_2. For a chosen W_2 the number of choosing W_1 is the number of (m,s)-type subspaces that are orthogonal to V and W_2, because W_2 is nonorthogonal to V. Since $\begin{pmatrix}V\\W_2\end{pmatrix}$ is a $(2,1)$-type subspace, its conjugation is a $(2\gamma-2,\gamma-1)$-type subspace. Therefore, the parameter r is $r=n_2\cdot N(m,s;2\gamma-2,\gamma-1;2\gamma)$.

Let V_1 and V_2 be two 1-dimensional orthogonal subspaces. We count the blocks (W_1,W_2) which the treatments V_1 and V_2 are arranged in. First consider W_2, then W_1. There are exactly P^1_{22} ways of choosing W_2. For a chosen W_2, the number of choosing W_1 is the number of (m, s)-type subspaces that are orthogonal to V_1, V_2 and W_2. Since

$$\begin{pmatrix} V_1 \\ V_2 \\ W_2 \end{pmatrix}$$

is a (3,1)-type subspace and its conjugation is a $(2\gamma-3,\gamma-2)$-type subspace, it follows that

$$\lambda_1=P^1_{22}\cdot N(m,s;2\gamma-3,\gamma-2;2\gamma).$$

This completes the proof.

4.4 Application to isomorphisms between some known PBIB designs

We now point out that some known PBIB designs constructed in Chapters 6 and 7 of [1] are isomorphic to each other. The PBIB designs constructed in Theorems 4 and 7 of Chapter 7 of [1], that will be called Designs A and B, respectively, are as follows.

Design A: In the 4-dimensional symplectic geometry $SV_4(Fq)$, take the (2,0)-type subspaces as treatments, and define two treatments to be the first (resp. the second) associates of each other if they intersect in a 1-dimensional (or the 0-dimensional) subspace. Take the 1-dimensional subspaces as blocks, and define a treatment to be arranged in a block if the former includes the latter both as subspaces.

Design B: The association scheme is the same as that in Design A. Take as blocks the (3, 1)-type subspaces of $SV_4(Fq)$, and define a treatment to be arranged in a block if the latter includes the former both as subspaces.

For Designs A and B, we have

Theorem 11 Design B is isomorphic to Design A.

Proof In $SV_4(Fq)$, the conjugation of a (2,0)-type subspace is also a (2,0)-type subspace, and the conjugation of a 1-dimensional subspace (i.e. a (1,0)-type subspace) is a (3, 1)-type subspace, Hence the conclusion of the theorem.

The PBIB designs constructed in Theorems 9 and 10 of Chapter 7 of [1], that will be called Designs C and D respectively, are as follows.

Design C: Let $\gamma\geqslant 2$. In the symplectic geometry $SV_{2\gamma}(Fq)$, take the $(\gamma,0)$-type subspaces as treatments, and define two treatments to be the i-th associates of each other if they as subspaces span a $(\gamma+i)$-dimensional subspace. Let $2s\leqslant m\leqslant\gamma+s$ and $m>\gamma$. Take the (m,s)-type subspaces as blocks, and define a treatment to be arranged in a block if the latter includes the former both as subspaces.

Design D: The association scheme is the same as that in Design C. Let $m<\gamma$. Take as blocks the (m,0)-type subspaces of $SV_{2\gamma}(Fq)$, and define a treatment to be arranged in a block if the former includes the latter both as subspaces.

For Designs C and D, we have

Theorem 12 Design D is isomorpnic to Design C with m and s replaced by $2\gamma-m$ and $\gamma-m$ respectively.

Proof In $SV_{2\gamma}(Fq)$, the conjugation of a $(\gamma,0)$-type subspace is also a $(\gamma,0)$-type subspace, and for $m<\gamma$ the conjugation of a $(m,0)$-type subspace is a $(2\gamma-m,\gamma-m)$-type subspace. Therefore, by Theorem 1, the conclusion of theorem follows.

The PBIB designs constructed in Theorem 8 and 9 of Chapter 6 of [1], that will be called Designs E and F, respectively, are as follows.

Design E: Let $1\leqslant m\leqslant n$. In the n-dimensional linear space $LVn(Fq)$, take the m-dimensional subspaces as treatments, and define two treatments are the i-th associates of each other if they as subspaces span an $(m+i)$-dimensional subspace. Let $m<m_1<n$. Take the m_1 dimensional subspaces as blocks, and define a treatment to be arranged in a block if the former includes the latter both as subspaces.

Design F: The association scheme is the same as that in Design E. Let $1\leqslant m_1<m$. Take the m_1-dimensional subspaces of $LVn(Fq)$ as blocks, and define a treatment to be arranged in a block if the former includes the latter both as subspaces.

Theorem 13 Design E is isomorphic to Design F with m and m_1 replaced by $n-m$ and $n-m_1$, respectively.

Proof Note that in $LVn(Fq)$, two m-dimensional subspaces P and Q span an $(m+i)$-dimensional subspace $P\bigcup Q$ if and only if $(P\bigcup Q)^*=P^*\bigcap Q^*$ is an $(n-(m+i))$-dimensional subspace, i.e. if and only if $P^*\bigcup Q^*$ is an $((n-m)+i)$-dimensional subspace. Thus, under the conjugation mapping the design isomorphic to Desing E is as follows: In $LVn(Fq)$, take the $(n-m)$-dimensional subspaces as treatments, and define two treatments to be the i-th associates of each other if they as subspaces span an $((n-m)+i)$-dimensional subspace. Take the $(n-m_1)$-dimensional subspaces as blocks, and define a treatment to be arranged in a block if the former includes the latter both as subspaces. However, this design is nothing but Design F with m and m_1 replaced by $n-m$ and $n-m_1$, respectively. This proves the theorem.

References

[1] Wan Zhexian, Dai Zongduo, Feng Xuning and Yang Beufu, Some Studies on Finite Geometries and Incomplete Block Designs, Science Press, Beijing, 1966. (In Chinese).

[2] Yang Benfu and Wei Wandi, Finite unitary geometrics and PBIB designs (1), The Journal of Combinatorial Mathematics and Combinatorial Computing, 6(1989), 51-61.

[3] Yang Benfu and Wei Wandi, Orthogonal geometrics over finite with characteristic≠2 and PBIB designs (1), Linear Algebra and Its Applications, 134: 1-23(1990).

The Isomorphic Factorization of Complete Tripartite Graphs K (A, B, C) (III)

– The sufficiency of the divisibility condition for 18 divides complete tripartite graphs K (A, B, C)

Yang shihui

(Fuling Teacher's College Fuling Sichuan 648005 China)

Abstract F.Harary, R.W.Robinson and N.C.Wormald [1] proved that for a complete tripartite graph G = K (m, n, s), when t = 2, 4 and t | (mn + ms + ns), G has an isomorphic factorization into t isomorphic subgraphs, write t | G. They also proved that when t is odd, the condition t | (mn + ms + ns) is not sufficient for t | k (m, n, s). and conjecture tuat when t > 1 is even, the condition t | (mn + ms + ns) is sufficient for t | k (m, n, s) . The case that t = 6 was proved by S. J. Quinn [2],[3]. The case that t = 2 ^{k} (k = 1, 2, 3, ⋯) was proved in [4]. A few months ago the author proved the case that t = 10 [5], t = 12 [6] and t = 14 [7]. In this paper we shall show that the conjecture is true for the case that t = 18.

Key words Isomorphic factorization complete tripartite graph

Introduction

An isomorphic factorization of a graph G = (V, E) is a partition $\{E_1, E_2, \cdots E_t\}$ of the edge set E such that the spanning subgraphs (V, E_1), (V, E_2), ⋯ , (V, E_t) are all isomorphic to each other. If G has an isomorphic factorization into exactly t isomorphic subgraphs, we say that G is divisible by t or t divides G, write t | G. For a given positive integer t and a given graph G with exactly q edges an obvious necessary condition for t to divide G is that t divides q, write t | q. t | q is called the divisibility condition for t | G.

F.Harary, R.W.Robinson and N.C.Wormald [1] proved that for a complete tripartite graph G = K (m, n, s), when t = 2, 4 the divisibility condition t | (mn+ ms + ns) is sufficient condition for t | G. They also proved that when t is odd the divisibility condition is not sufficient for t to divide K (m, n, s), and conjecture that when t > 1 is even, G = K(m, n, s) is a complete tripartite graph and t | (mn + ms + ns), then t | G. The case that t = 6 was proved by S.J.Quinn [2],[3]. The case that t = 2 ^{k} (k = 1, 2, 3, ⋯) was proved in [4] . A few months ago the author proved the case that t = 10 [5], t = 12 [6] and t = 14 [7] .

About ten years ago, the author found a method to prove that the conjecture is true[8], but the proof is very hard to understand and need further arguements, However I feel that it is very difficult to clearly explaine the method in words. So I want to represent the decomposition method by strictly writing out all t factors in an isomorphic factorization for a few smaller positive even integers with different properties, respectively. Of course some special methods will be used and which will make the decomposition to be simpler than the general method. In this paper we will prove that the conjecture is true for $t = 18$.

Remark

Assume t is even and $t \mid (mn + ms + ns)$. Then there are at least two even numbers among m, n and s.

Theorem Assume $G = K(A, B, C)$ is a complete tripartite graph with independent vertex sets A, B and C. $t = 18$.

Then the divisibility condition is sufficient for t to divide G.

Proof Without loss of generality, by the remark, we can suppose that

$$|A| = 2m, \ |B| = 2n, \ |C| = s$$

Thus we have, by the divisibility condition,

$$18 \mid (4mn + 2ms + 2ns)$$

$$9 \mid (2mn + ms + ns)$$

we shall prove the theorem by strictly writing out all 18 factors in an isomorphic factorization of G according to each solution of the congruence

$$2mn + ms + ns \equiv 0 \pmod 9,$$

respectively. By the symmetry of m and n in the congruence, we shall identify a solution $s \equiv r$, $n \equiv h$, $m \equiv k \pmod 9$ as the same as the solution $s \equiv r$, $n \equiv k$, $m \equiv h \pmod 9$. Thus all the solutions of the congruence are as follows

(0.1) $s \equiv 0, n \equiv 0, m \equiv m_0 \pmod 9$ where $0 \leq m_0 < 9$

(0.2) $s \equiv 0 \pmod 9$; $n \equiv 0, m \equiv 0 \pmod 3$

(0.3) $s \equiv s_0, \ n \equiv 0, \ m \equiv 0 \pmod 9$ where $0 \leq s_0 < 9$

(0.4) $s \equiv s_0, \ n \equiv 3, \ m \equiv 6 \pmod 9$ where $s_0 = 1, 2, 4, 5, 7, 8$

(1.1) $s \equiv 1, \ n \equiv 2, \ m \equiv 5 \pmod 9$

(1.2) $s \equiv 1, \ n \equiv 8, \ m \equiv 8 \pmod 9$

(2.1) $s \equiv 2, \ n \equiv 1, \ m \equiv 4 \pmod 9$

(2.2) $s \equiv 2$, $n \equiv 7$, $m \equiv 7 \pmod 9$

(3.1) $s \equiv 3$, $n \equiv 3$, $m \equiv m_0 \pmod 9$ where $0 \leq m_0 < 9$

(3.2) $s \equiv 3$, $n \equiv 6$, $m \equiv 6 \pmod 9$

(4.1) $s \equiv 4$, $n \equiv 2$, $m \equiv 8 \pmod 9$

(4.2) $s \equiv 4$, $n \equiv 5$, $m \equiv 5 \pmod 9$

(5.1) $s \equiv 5$, $n \equiv 1$, $m \equiv 7 \pmod 9$

(5.2) $s \equiv 5$, $n \equiv 4$, $m \equiv 4 \pmod 9$

(6.1) $s \equiv 6$, $n \equiv 3$, $m \equiv 3 \pmod 9$

(6.2) $s \equiv 6$, $n \equiv 6$, $m \equiv m_0 \pmod 9$ where $0 \leq m_0 < 9$

(7.1) $s \equiv 7$, $n \equiv 2$, $m \equiv 2 \pmod 9$

(7.2) $s \equiv 7$, $n \equiv 5$, $m \equiv 8 \pmod 9$

(8.1) $s \equiv 8$, $n \equiv 1$, $m \equiv 1 \pmod 9$

(8.2) $s \equiv 8$, $n \equiv 4$, $m \equiv 7 \pmod 9$

For convenience we shall identify the edge set of a graph G as the same as the graph itself. i.e if E = E (G) is the edge set of a graph G, then put E = G . We shall express a graph G by indicating the adjacent relation of its vertices, Assume U and V are two disjoint nonempty subsets of the vertex set of a graph G , the notation U - V mean that every vertex in U is adjacent to each vertex in V or the edge set $\{(u, v) \mid u \in U, v \in V\}$ of the complete bipartite graph K (U,V) or the graph K (U,V) itself. If U = {u} is a singleton element set , we replace U - V by u - V, esspecially, we use uv or u - v instead of the notation {u} - {v}.

For a solution $s \equiv r$, $n \equiv k$, $m \equiv h \pmod 9$ of the congruence $2mn + ms + ns \equiv 0 \pmod 9$ we shall partition the vertex sets A, B, C as follows

$$A = A^1 U A^2 \quad A^\tau = A_1^\tau U A_2^\tau U \cdots U A_9^\tau U A_0^\tau \quad A_0^\tau = \{a_1^\tau, a_2^\tau, \cdots, a_h^\tau\}$$

$$B = B^1 U B^2 \quad B^\tau = B_1^\tau U B_2^\tau U \cdots U B_9^\tau U B_0^\tau \quad B_0^\tau = \{b_1^\tau, b_2^\tau, \cdots, b_k^\tau\}$$

$$C = C_1 U C_2 U \cdots U C_9 U C_0 \quad C_0 = \{c_1, c_2, \cdots, c_r\}$$

where $|A_i^\tau| = \left[\frac{m}{9}\right] = \frac{m-n}{9}$, $|B_i^\tau| = \left[\frac{n}{9}\right] = \frac{n-k}{9}$, $|C_i| = \left[\frac{s}{9}\right] = \frac{s-r}{9}$

$\tau = 1,2$; $i = 1, 2, \cdots, 9$

and then construct the corresponding 18 factors in an isomorphic factorization of G, respectively.

(0.1) $s \equiv 0$, $n \equiv 0$, $m \equiv m_0 \pmod 9$

The corresponding 18 isomorphic factors in an isomorphic factorization are as follows

$$F_i^1:\quad B^1 - C_i - A^1 - B_i^1 - A^2,$$
$$F_i^2:\quad B^2 - C_i - A^2 - B_i^2 - A^1,$$
$$i = 1, 2, \cdots, 9$$

(0.2) $s \equiv 0$, (mod 9); $n \equiv 0$, $m \equiv 0$ (mod 3).

In this case we partition the vertex sets A, B, C as follows

$$A = A^1 U A^2, \quad A^\tau = A_1^\tau U A_2^\tau U A_3^\tau, \quad |A_i^\tau| = \frac{m}{3}, \quad \tau = 1, 2;\ i = 1, 2, 3,$$

$$B = B^1 U B^2, \quad B^\tau = B_1^\tau U B_2^\tau U B_3^\tau, \quad |B_i^\tau| = \frac{n}{3}, \quad \tau = 1, 2;\ i = 1, 2, 3,$$

$$C = C_1 \cup C_2 \cup \cdots \cup C_9 \quad |C_1| = |C_2| = \cdots = |C_9| = \frac{s}{9}$$

The corresponding isomorphic factors are as follows

$$F_{ij}^1:\quad A^1 - C_{i+3(j-1)} - B^1 \quad A_i^1 - B_j^1 - A_i^2$$
$$F_{ij}^2:\quad A^2 - C_{i+3(j-1)} - B^2 \quad A_i^2 - B_j^2 - A_i^1$$
$$i = 1, 2, 3; \quad j = 1, 2, 3$$

(0.3) $s \equiv s_0$, $n \equiv 0$, $m \equiv 0$ (mod 9)

The corresponding isomorphic factors are as follows

$$F_j^1:\quad B^1 - A_j^1 - C - B_j^1 - A^2$$
$$F_j^1:\quad B^2 - A_j^2 - C - B_j^2 - A^1$$
$$j = 1, 2, \cdots, 9$$

(0.4) $s \equiv s_0$, $n \equiv 3$, $m \equiv 6$ (mod 9)

The corresponding isomorphic factors are as follows

$$F_1^\tau:\ \{A_1^\tau + B_1^\tau + a_1^\tau\} - C_o \quad \{a_5^1, a_6^1, a_5^2, a_6^2\} - b_2^\tau \quad B^{\tau+1} - C_1 - A^{\tau+1} - B_1^{12} \quad B_o^{\tau+1} - A_1^{12}$$
$$F_2^\tau:\ \{A_2^\tau + B_2^\tau + a_2^\tau\} - C_o \quad \{a_5^1, a_6^1, a_5^2, a_6^2\} - b_3^\tau \quad B^{\tau+1} - C_2 - A^{\tau+1} - B_2^{12} \quad B_o^{\tau+1} - A_2^{12}$$
$$F_3^\tau:\ \{A_3^\tau + B_3^\tau + a_3^\tau\} - C_o \quad \{a_5^1, a_6^1, a_5^2, a_6^2\} - b_1^\tau \quad B^{\tau+1} - C_3 - A^{\tau+1} - B_3^{12} \quad B_o^{\tau+1} - A_3^{12}$$
$$F_4^\tau:\ \{A_4^\tau + B_4^\tau + a_4^\tau\} - C_o \quad \{a_1^1, a_2^1, a_1^2, a_2^2\} - b_2^\tau \quad B^{\tau+1} - C_4 - A^{\tau+1} - B_4^{12} \quad B_o^{\tau+1} - A_4^{12}$$
$$F_5^\tau:\ \{A_5^\tau + B_5^\tau + a_5^\tau\} - C_o \quad \{a_1^1, a_2^1, a_1^2, a_2^2\} - b_3^\tau \quad B^{\tau+1} - C_5 - A^{\tau+1} - B_5^{12} \quad B_o^{\tau+1} - A_5^{12}$$
$$F_6^\tau:\ \{A_6^\tau + B_6^\tau + a_6^\tau\} - C_o \quad \{a_1^1, a_2^1, a_1^2, a_2^2\} - b_1^\tau \quad B^{\tau+1} - C_6 - A^{\tau+1} - B_6^{12} \quad B_o^{\tau+1} - A_6^{12}$$
$$F_7^\tau:\ \{A_7^\tau + B_7^\tau + b_1^\tau\} - C_o \quad \{a_3^1, a_4^1, a_3^2, a_4^2\} - b_2^\tau \quad B^{\tau+1} - C_7 - A^{\tau+1} - B_7^{12} \quad B_o^{\tau+1} - A_7^{12}$$

$$F_8^\tau:\ \{A_8^\tau + B_8^\tau + b_2^\tau\} - C_o \quad \{a_3^1, a_4^1, a_3^2, a_4^2\} - b_3^\tau \quad B^{\tau+1} - C_8 - A^{\tau+1} - B_8^{12} \quad B_o^{\tau+1} - A_8^{12}$$
$$F_9^\tau:\ \{A_9^\tau + B_9^\tau + b_3^\tau\} - C_o \quad \{a_3^1, a_4^1, a_3^2, a_4^2\} - b_1^\tau \quad B^{\tau+1} - C_9 - A^{\tau+1} - B_9^{12} \quad B_o^{\tau+1} - A_9^{12}$$

where $A_i^{12} = A_i^1 U A_i^2, \quad B_i^{12} = B_i^1 U B_i^2; \quad \tau = 1,2, \quad \tau+1 = \begin{cases} 2 & \tau = 1 \\ 1 & \tau = 2 \end{cases}$

For the solutions s ≡ r, n≡ m ≡ 9 - r (mod 9), r = 1, 2 3, ⋯ , 8, the correspondin decomposition. methods are similar to each other, so we only construct the 18 isomorphic factors in an isomorphic factorization corresponding to the solution s ≡ 2, n≡ m ≡ 7 (mod 9). Let

$$F_1^1: c_1 - A^1 - C_1 - \{B^1 - b_1^1 + b_1^2\} \quad a_1^2 - \{B^1 - b_1^1\} \quad B_o^2 - A_1^1 - B^{1*} - A_1^2 \quad A_0^1 - B_1^1 \quad a_7^2 - b_7^1$$

$$F_2^1: c_2 - A^1 - C_2 - \{B^1 - b_2^1 + b_2^2\} \quad a_2^2 - \{B^1 - b_2^1\} \quad B_o^2 - A_2^1 - B^{1*} - A_2^2 \quad A_0^1 - B_2^1 \quad c_1 - b_1^1$$

$$F_3^1: b_1^1 - A^1 - C_3 - \{B^1 - b_1^1 + b_1^2\} \quad c_1 - \{B^1 - b_1^1\} \quad B_o^2 - A_3^1 - B^{1*} - A_3^2 \quad A_0^1 - B_3^1 \quad c_2 - b_2^1$$

$$F_4^1: b_2^1 - A^1 - C_4 - \{B^1 - b_2^1 + b_2^2\} \quad c_2 - \{B^1 - b_2^1\} \quad B_o^2 - A_4^1 - B^{1*} - A_4^2 \quad A_0^1 - B_4^1 \quad a_1^2 - b_1^1$$

$$F_{2+i}: b_i^1 - A^1 - C_{2+i} - \{B^1 - b_i^1 + b_i^2\} \quad a_i^2 - \{B^1 - b_i^1\} \quad B_o^2 - A_{2+i}^1 - B^{1*} - A_{2+i}^2 \quad A_0^1 - B_{2+i}^1 \quad a_{i-1}^2 - b_{i-1}^1$$

$$F_1^2: c_1 - A^2 - C_1 - \{B^2 - b_1^2 + b_1^1\} \quad a_1^1 - \{B^2 - b_1^2\} \quad B_o^1 - A_1^2 - B^{2*} - A_1^1 \quad A_0^2 - B_1^2 \quad a_7^1 - b_7^2$$

$$F_2^2: c_2 - A^2 - C_2 - \{B^2 - b_2^2 + b_2^1\} \quad a_2^1 - \{B^2 - b_2^2\} \quad B_o^1 - A_2^1 - B^{2*} - A_2^1 \quad A_0^2 - B_2^2 \quad c_1 - b_1^2$$

$$F_3^2: b_1^2 - A^2 - C_3 - \{B^2 - b_1^2 + b_1^1\} \quad c_1 - \{B^2 - b_1^2\} \quad B_o^1 - A_3^2 - B^{2*} - A_3^1 \quad A_0^2 - B_3^2 \quad c_2 - b_2^2$$

$$F_4^2: b_2^2 - A^2 - C_4 - \{B^2 - b_2^2 + b_2^1\} \quad c_2 - \{B^2 - b_2^2\} \quad B_o^1 - A_4^2 - B^{2*} - A_4^1 \quad A_0^2 - B_4^2 \quad a_1^1 - b_1^2$$

$$F_{2+i}^2: b_i^2 - A^2 - C_{2+i} - \{B^2 - b_i^2 + b_i^1\} \quad a_i^1 - \{B^2 - b_i^2\} \quad B_0^1 - A_{2+i}^2 - B^{2*} - A_{2+i}^1 \quad A_0^2 - B_{2+i}^2 \quad a_{i-1}^1 - b_{i-1}^2$$

where $B^{1*} = B^1 - B_0^1, \quad B^{2*} = B^2 - B_0^2; \quad i = 3,4,5,6,7$

(1.1) s ≡ 1, n≡ 2, m ≡ 5 (mod 9)

The corresponding isomorphic factors are as follows

$$F_1^1:\ b_1^2 - c - A_{147}^1 \quad a_5^2 - B_{147}^1 \quad A^1 - C_1 - B^1 \quad \{A^1 - a_5^1\} - B_1^{12} \quad b_2^1 - A_1^{12} \quad a_1^1 - b_2^2 - a_1^2$$
$$F_2^1:\ b_2^2 - c - A_{258}^1 \quad a_1^2 - B_{258}^1 \quad A^1 - C_2 - B^1 \quad \{A^1 - a_1^1\} - B_2^{12} \quad b_2^1 - A_2^{12} \quad a_5^1 - b_1^2 - a_5^2$$
$$F_3^1:\ a_5^2 - c - A_{369}^1 \quad a_1^2 - B_{369}^1 \quad A^1 - C_3 - B^1 \quad \{A^1 - a_1^1\} - B_3^{12} \quad b_2^1 - A_3^{12} \quad a_4^1 - b_1^2 - a_4^2$$

F_4^1: $a_1^2 - b_1^1 - A_{147}^1$ $a_5^2 - B_{147}^2$ $A^1 - C_4 - B^2$ $\{A^1 - a_5^1\} - B_4^{12}$ $b_2^2 - A_4^{12}$ $a_1^1 - c - a_2^2$

F_5^1: $a_2^2 - b_1^1 - A_{258}^1$ $a_1^2 - B_{258}^2$ $A^1 - C_5 - B^2$ $\{A^1 - a_1^1\} - B_5^{12}$ $b_2^2 - A_5^{12}$ $a_3^1 - c - a_4^2$

F_6^1: $a_3^2 - b_1^1 - A_{369}^1$ $a_1^2 - B_{369}^2$ $A^1 - C_6 - B^2$ $\{A^1 - a_1^1\} - B_6^{12}$ $b_2^2 - A_6^{12}$ $a_2^1 - b_2^1 - a_2^2$

F_7^1: $a_1^1 - b_1^1 - A_{147}^2$ $c - B_{147}^2$ $A^2 - C_7 - B^2$ $\{A^2 - a_5^2\} - B_7^{12}$ $b_2^2 - A_7^{12}$ $a_3^2 - b_2^1 - a_3^1$

F_8^1: $a_2^1 - b_1^1 - A_{258}^2$ $c - B_{258}^2$ $A^2 - C_8 - B^2$ $\{A^2 - a_1^2\} - B_8^{12}$ $b_2^2 - A_8^{12}$ $a_4^2 - b_2^1 - a_4^1$

F_9^1: $a_3^1 - b_1^1 - A_{369}^2$ $c - B_{369}^2$ $A^2 - C_9 - B^2$ $\{A^2 - a_1^2\} - B_9^{12}$ $b_2^2 - A_9^{12}$ $a_5^2 - b_2^1 - a_5^1$

where $A_{ijk}^\tau = A_i^\tau U A_j^\tau U A_k^\tau$ $B_{ijk}^\tau = B_i^\tau U B_j^\tau U B_k^\tau$ $A_i^{12} = A_i^1 U A_i^2$ $B_i^{12} = B_i^1 U B_i^2$

The next 9 factors F_i^2 can be obtained from the first 9 factors F_i^1 by interchanging the upscripts 1 and 2 of each letter in F_i^1.

(2.1) s ≡ 2, n ≡ 1, m ≡ 4 (mod 9)

The corresponding isomorphic factors are as follows

$F_i^\tau : A^\tau - C_i - B^\tau - A_i^{12} - c_\tau - B_i^{12}$ - A_o^τ $a_i^2 - b_1^{\tau+1} - a_i^1$ i = 1,2,3,4

$F_{4+j}^\tau : A^\tau - C_{4+j} - B^\tau - A_{4+j}^{12} - c_\tau - B_{4+j}^{12}$ - A_o^τ $a_j^2 - c_{\tau+1} - a_j^1$ j = 1,2,3

$F_8^\tau : A^\tau - C_8 - B^\tau - A_8^{12} - c_\tau - B_8^{12}$ - A_o^τ $b_1^{\tau+1} - c_{\tau+1} - a_4^\tau$

$F_9^\tau : A^\tau - C_9 - B^\tau - A_9^{12} - c_{\tau+1} - B_9^{12}$ - A_o^τ $b_1^{\tau+1} - c_\tau - a_4^\tau$

where $\tau = 1,2$ $\tau+1 = \begin{cases} 2 & \tau=1 \\ 1 & \tau=2 \end{cases}$ $A_i^{12} = A_i^1 U A_i^2$ $B_i^{12} = B_i^1 U B_i^2$

(3.1) s ≡ 3, n ≡ 3, m ≡ m_0 (mod 9)

we construct the corresponding isomorphic factors as follows

$F_i^\tau : c_i - A^\tau - C_i - B^\tau$ $B_i^1 - A^\tau - B_i^2$ b_1^τ - c_{i+1} - $\{B_1^\tau + B_2^\tau + B_3^\tau\}$

$F_{3+i}^\tau : b_i^\tau - A^\tau - C_{3+i} - B^\tau$ $B_{3+i}^1 - A^\tau - B_{3+i}^2$ $b_2^{\tau+1}$ - c_i - $\{B_4^{\tau+1} + B_5^{\tau+1} + B_6^{\tau+1}\}$

$F_{6+i}^\tau : b_i^{\tau+1} - A^\tau - C_{6+i} - B^\tau$ $B_{6+i}^1 - A^\tau - B_{6+i}^2$ b_3^τ - c_i - $\{B_7^\tau + B_8^\tau + B_9^\tau\}$

where i = 1, 2, 3 $c_{3+1} = c_1$ τ = 1, 2 $\tau + 1 = \begin{cases} 2 & \tau=1 \\ 1 & \tau=2 \end{cases}$

(4.1) s ≡ 4, n ≡ 2, m ≡ 8 (mod 9)

The corresponding isomorphic factors in an isomorphic factorization can be obtained from the 18 factors in (2.1) by replacing the terms c_τ, b_1^τ, a_j^τ with sets $\{c_\tau, c_{\tau+2}\}$, $B_o^\tau, \{a_j^\tau, a_{4+j}^\tau\}$ τ=1, 2, j = 1, 2, 3, 4, respectively.

(5.1) $s \equiv 5,\ n \equiv 1,\ m \equiv 7 \pmod 7$

we construct the corresponding isomorphic factors as follows

$$F_1^1{:}B_{11} - c_1 - A_{11}\ A^1 - C_1 - B^1 - A_1^{12} - c_4 - B_1^{12} - A_0^1\ b_1^1 - a_1^2\ a_6^2 - c_5 - a_7^2$$
$$F_2^1{:}B_{11} - c_2 - A_{11}\ A^1 - C_2 - B^1 - A_2^{12} - c_4 - B_2^{12} - A_0^1\ b_1^1 - a_2^2\ a_4^2 - c_5 - a_5^2$$
$$F_3^1{:}B_{11} - c_3 - A_{11}^1\ A^1 - C_3 - B^1 - A_3^{12} - c_4 - B_3^{12} - A_0^1\ b_1^1 - a_4^2\ c_1 - a_1^2 - c_2$$
$$F_4^1{:}B_{12} - c_4 - A_{12}\ A^1 - C_4 - B^1 - A_4^{12} - c_1 - B_4^{12} - A_0^1\ b_1^1 - a_3^2\ a_4^2 - b_1^2 - c_5$$
$$F_5^1{:}B_{12} - c_2 - A_{12}\ A^1 - C_5 - B^1 - A_5^{12} - c_1 - B_5^{12} - A_0^1\ b_1^1 - a_5^2\ c_3 - a_7^2 - c_4$$
$$F_6^1{:}B_{12} - c_3 - A_{12}\ A^1 - C_6 - B^1 - A_6^{12} - c_1 - B_6^{12} - A_0^1\ b_1^1 - a_6^2\ a_2^2 - b_1^2 - a_3^2$$
$$F_7^1{:}B_{13} - c_1 - A_{13}\ A^1 - C_7 - B^1 - A_7^{12} - c_2 - B_7^{12} - A_0^1\ b_1^1 - a_7^2\ a_5^2 - b_1^2 - a_6^2$$
$$F_8^1{:}B_{13} - c_4 - \tilde{A}_{13}\ A^1 - C_8 - B^1 - A_8^{12} - c_2 - B_8^{12} - A_0^1\ b_1^1 - c_5\ a_1^2 - b_1^2 - a_7^2$$
$$F_9^1{:}B_{13} - c_5 - \tilde{A}_{13}\ A^1 - C_9 - B^1 - A_9^{12} - c_2 - B_9^{12} - A_0^1\ b_1^1 - c_4\ c_1 - b_1^2 - c_3$$
$$F_1^2{:}B_{21} - c_1 - A_{21}\ A^2 - C_1 - B^2 - A_1^{12} - c_5 - B_1^{12} - A_0^2\ b_1^2 - a_1^1\ a_5^1 - b_1^1 - a_6^1$$
$$F_2^2{:}B_{21} - c_2 - A_{21}\ A^2 - C_2 - B^2 - A_2^{12} - c_5 - B_2^{12} - A_0^2\ b_1^2 - a_2^1\ a_1^1 - b_1^1 - a_7^1$$
$$F_3^2{:}B_{21} - c_3 - A_{21}\ A^2 - C_3 - B^2 - A_3^{12} - c_5 - B_3^{12} - A_0^2\ b_1^2 - a_4^1\ c_1 - a_7^1 - c_2$$
$$F_4^2{:}B_{22} - c_4 - A_{22}\ A^2 - C_4 - B^2 - A_4^{12} - c_5 - B_4^{12} - A_0^2\ b_1^2 - a_3^1\ a_4^1 - b_1^1 - c_3$$
$$F_5^2{:}B_{22} - c_2 - A_{22}\ A^2 - C_5 - B^2 - A_5^{12} - c_5 - B_5^{12} - A_0^2\ b_1^2 - a_5^1\ c_3 - a_7^1 - c_4$$
$$F_6^2{:}B_{22} - c_3 - A_{22}\ A^2 - C_6 - B^2 - A_6^{12} - c_5 - B_6^{12} - A_0^2\ b_1^2 - a_6^1\ a_2^1 - b_1^1 - a_3^1$$
$$F_7^2{:}B_{23} - c_1 - A_{23}\ A^2 - C_7 - B^2 - A_7^{12} - c_3 - B_7^{12} - A_0^2\ b_1^2 - a_7^1\ a_4^1 - c_5 - a_5^1$$
$$F_8^2{:}B_{23} - c_4 - \tilde{A}_{23}\ A^2 - C_8 - B^2 - A_8^{12} - c_3 - B_8^{12} - A_0^2\ b_1^2 - c_2\ a_6^1 - c_5 - a_7^1$$
$$F_9^2{:}B_{23} - c_5 - \tilde{A}_{23}\ A^2 - C_9 - B^2 - A_9^{12} - c_3 - B_9^{12} - A_0^2\ b_1^2 - c_4\ c_1 - b_1^1 - c_2$$

where $B_{\tau 1} = B_1^\tau U B_2^\tau U B_3^\tau \quad B_{\tau 2} = B_4^\tau U B_5^\tau U B_6^\tau \quad B_{\tau 3} = B_7^\tau U B_8^\tau U B_9^\tau$

$$A_{\tau 1} = A_1^\tau U A_2^\tau U A_3^\tau U \{a_1^\tau, a_2^\tau, a_3^\tau\} \quad A_{\tau 2} = A_4^\tau U A_5^\tau U A_6^\tau U \{a_4^\tau, a_5^\tau, a_6^\tau\} \quad \tau = 1,2$$
$$A_{\tau 3} = A_7^\tau U A_8^\tau U A_9^\tau U \{a_4^\tau, a_5^\tau, a_6^\tau\} \quad \tilde{A}_{\tau 3} = A_7^\tau U A_8^\tau U A_9^\tau U \{a_1^\tau, a_2^\tau, a_3^\tau\} \quad \tau = 1,2$$
$$A_i^{12} = A_i^1 U A_i^2 \quad B_i^{12} = B_i^1 U B_i^2 \cdots$$

(6.2) $s \equiv 6,\ n \equiv 6, m \equiv m_0 \pmod 9$

The corresponding isomorphic factors in an isomorphic factorization can be obtained from the factors in (3.1) by replacing the terms c_i, b_i^1, b_i^2 with $\{c_i, c_{i+3}\}$, $\{b_i^1, b_{i+3}^1\}, \{b_i^2, b_{i+3}^2\}$ i =1, 2, 3, respectively.

(7.2) $s \equiv 7,\ n \equiv 5, m \equiv 8 \pmod 9$

ws construct the corresponding isomorphic factors as follows

$$F_1^1:\ c_1 - \tilde{A}^1 - C_1 - B^1\ a_1^1 - \{B^1 - b_1^1\} - A_1^{12}\ \{\tilde{A}_0^1 - a_1^1\} - B_1^{12}\ b_3^2 - a_8^1\ b_2^1 - a_2^{12} - b_2^2\ a_4^{12} - b_3^1$$

$$F_2^1:\ c_2 - A^1 - C_2 - B^1\ a_1^2 - \{B^1 - b_1^1\} - A_2^{12}\ \{A_0^1 - a_1^1\} - B_2^{12}\ b_1^2 - c_1\ b_2^1 - a_3^{12} - b_3^2\ a_5^{12} - b_3^1$$

$$F_3^1:\ c_i - A^1 - C_i - B^1\ c_{i-2} - \{B^1 - b_1^1\} - A_i^{12}\ \{A_0^1 - a_1^1\} - B_i^{12}\ b_1^2 - c_{i-1}\ b_2^1 - a_{i+1}^{12} - b_2^2\ a_{i+3}^{12} - b_3^1$$

i = 3, 4, 5

$$F_6^1:\ c_6 - A^1 - C_6 - B^1\ c_4 - \{B^1 - b_1^1\} - A_6^{12}\ \{A_0^1 - a_1^1\} - B_6^{12}\ b_1^2 - c_5\ b_2^1 - a_7^{12} - b_2^2\ a_4^{12} - b_5^1$$

$$F_7^1:\ c_7 - A^1 - C_7 - B^1\ c_5 - \{B^1 - b_1^1\} - A_7^{12}\ \{A_0^1 - a_1^1\} - B_7^{12}\ b_1^2 - c_6\ b_2^1 - a_8^{12} - b_2^2\ a_5^{12} - b_5^1$$

$$F_8^1:\ b_1^1 - A^1 - C_8 - \tilde{B}^1\ c_6 - \{B^1 - b_1^1\} - A_8^{12}\ \{A_0^1 - a_1^1\} - B_8^{12}\ b_1^2 - c_7\ b_3^1 - a_2^{12} - b_3^2\ a_6^{12} - b_5^1$$

$$F_9^1:\ b_1^2 - A^1 - C_9 - B^1\ c_7 - \{B^1 - b_1^1\} - A_9^{12}\ \{A_0^1 - a_1^1\} - B_9^{12}\ b_3^2 - a_8^2\ b_3^1 - a_3^{12} - b_2^2\ a_7^{12} - b_5^1$$

$$F_1^2:\ c_1 - \tilde{A}^2 - C_1 - B^2\ a_1^2 - \{B^2 - b_1^2\} - A_1^{12}\ \{\tilde{A}_0^2 - a_1^2\} - B_1^{12}\ b_5^1 - a_8^2\ b_4^1 - a_2^{12} - b_4^2\ a_4^{12} - b_5^2$$

$$F_2^2:\ c_2 - A^2 - C_2 - B^2\ a_1^1 - \{B^2 - b_1^2\} - A_2^{12}\ \{A_0^2 - a_1^2\} - B_2^{12}\ b_1^1 - c_1\ b_5^1 - a_3^{12} - b_4^2\ a_5^{12} - b_5^2$$

$$F_i^2:\ c_i - A^2 - C_i - B^2\ c_{i-2} - \{B^2 - b_1^2\} - A_i^{12}\ \{A_0^2 - a_1^2\} - B_i^{12}\ b_1^1 - c_{i-1}\ b_4^1 - a_{i+1}^{12} - b_4^2\ a_{i+3}^{12} - b_5^2$$

i = 3, 4, 5

$$F_6^2:\ c_6 - A^2 - C_6 - B^2\ c_4 - \{B^2 - b_1^2\} - A_6^{12}\ \{A_0^2 - a_1^2\} - B_6^{12}\ b_1^1 - c_5\ b_4^1 - a_7^{12} - b_4^2\ a_4^{12} - b_3^2$$

$$F_7^2:\ c_7 - A^2 - C_7 - B^2\ c_5 - \{B^2 - b_1^2\} - A_7^{12}\ \{A_0^2 - a_1^2\} - B_7^{12}\ b_1^1 - c_6\ b_4^1 - a_8^{12} - b_4^2\ a_5^{12} - b_3^2$$

$$F_8^2:\ b_1^2 - A^2 - C_8 - \tilde{B}^2\ c_6 - \{B^2 - b_1^2\} - A_8^{12}\ \{A_0^2 - a_1^2\} - B_8^{12}\ b_1^1 - c_7\ b_5^1 - a_2^{12} - b_5^2\ a_6^{12} - b_3^2$$

$$F_9^2:\ b_1^1 - A^2 - C_9 - B^2\ c_7 - \{B^2 - b_1^2\} - A_9^{12}\ \{A_0^2 - a_1^2\} - B_9^{12}\ b_5^1 - a_8^1\ b_4^1 - a_3^{12} - b_5^2\ a_7^{12} - b_3^2$$

where $\tilde{A}^1 = \{A^1 - a_1^1 - a_8^1 + a_1^2 + a_8^2\}$ $\quad \tilde{A}_0^1 = \{A_0^1 - a_8^1 + a_8^2\}$

$\tilde{A}^2 = \{A^2 - a_1^2 - a_8^2 + a_1^1 + a_8^1\}$ $\quad \tilde{A}_0^2 = \{A_0^2 - a_8^2 + a_8^1\}$

$\tilde{B}^1 = \{B^1 - b_1^1 + b_1^2\}$ $\quad \tilde{B}^2 = \{B^2 - b_1^2 + b_1^1\}$

$A_i^{12} = A_i^1 U A_i^2 \quad B_i^{12} = B_i^1 U B_i^2 \quad a_i^{12} = \{a_i^1, a_i^2\} \cdots$

(8.2) $s \equiv 8, n \equiv 4, m \equiv 7 \pmod 9$

Let

E_1^1: $c_{34}^1 - a_1^{12} - b_{12}^1$ $b_3^2 - c_{12}^1$ $a_{56}^2 - c_3^2$ $a_7^2 - b_{34}^1$ $b_{12}^2 - c_4^2$

E_2^1: $c_{34}^1 - a_2^{12} - b_{12}^1$ $b_4^2 - c_{12}^1$ $a_{56}^2 - c_4^2$ $a_3^2 - b_{34}^1$ $b_{12}^2 - c_3^2$

E_3^1: $c_{12}^1 - a_3^{12} - b_{12}^1$ $a_5^2 - c_{34}^1$ $c_{34}^2 - a_4^2$ $a_6^2 - b_{34}^1$ $b_{34}^2 - a_7^2$

E_4^1: $c_{12}^1 - a_4^{12} - b_{12}^1$ $a_7^2 - c_{34}^1$ $c_{34}^2 - a_3^2$ $a_5^2 - b_{34}^1$ $b_{34}^2 - c_1^2$

E_5^1: $c_{12}^1 - a_5^{12} - b_{12}^1$ $a_6^2 - c_{34}^1$ $c_{34}^2 - a_7^2$ $a_4^2 - b_{34}^1$ $b_{34}^2 - c_2^2$

E_6^1: $c_{12}^1 - a_6^{12} - b_{12}^1$ $b_1^2 - c_{34}^1$ $b_2^2 - c_{12}^2$ $c_3^2 - b_{34}^1$ $a_{34}^2 - b_3^2$

E_7^1: $c_{12}^1 - a_7^{12} - b_{12}^1$ $b_2^2 - c_{34}^1$ $b_1^2 - c_{12}^2$ $c_4^2 - b_{34}^1$ $a_{34}^2 - b_4^2$

E_8^1: $c_{12}^1 - a_1^{12} - b_{34}^1$ $a_3^2 - c_{34}^1$ $b_3^2 - c_{34}^2$ $c_1^2 - b_{12}^1$ $a_{56}^2 - b_4^2$

E_9^1: $c_{12}^1 - a_2^{12} - b_{34}^1$ $a_4^2 - c_{34}^1$ $b_4^2 - c_{34}^2$ $c_2^2 - b_{12}^1$ $a_{56}^2 - b_3^2$

E_i^2 are obtained from E_i^1 by interchanging the upscripts 1 and 2 of each letter.

Let

$$H_i^\tau:\ A^\tau - C_i - B^\tau - A_i^{12} - C_o^\tau - B_i^{12} - A_o^\tau \quad \tau = 1,2; \quad i = 1,2,\cdots,9$$

where $C_0 = \{c_1, c_2, \cdots, c_8\} = C_o^1 U C_o^2$ $\quad C_o^\tau = \{c_1^\tau, c_2^\tau, c_3^\tau, c_4^\tau\} \quad \tau = 1,2;$

$c_i^\tau = C_{4(\tau-1)+i} \quad A_j^{12} = A_j^1 U A_j^2 \quad B_j^{12} = B_j^1 U B_j^2 \cdots$

$$F_i^\tau = H_i^\tau U E_i^\tau \quad \tau = 1,2; \quad i = 1,2,\cdots,9$$

Then are 18 isomorphic factors in an isomorphic factorization of K (A, B, C).

Up to now we have constructed an isomorphic factorization of K (A, B, C) for each solution of the congruence 2mn + ms + ns ≡ 0 (mod 9). The proof is complete.

References:

[1] Frank. Harary, Robert. W. Robinson and Nicgolas.C.Wormald: Isomorphic Factorization III: Complete multipartite graphs in "Lecture Notes in Mathematics 086" 47 - 58.

[2] S.J.Quinn On the existence and enumeration of isomorphic factorizations and graphs, Master's Thesis University of Newcastle, Australia (1984).

[3] F.Harary and R.W.Robinson: Isomorphic factorization X: Unsolved problems Journal of Graph. Theory 9 (1985) 67 - 86.

[4] Yang Shihui Some sufficient conditions for complete tripartite graph K (A, B, C) to have isomorphic factorization ACTA Mathematicae Applicatae Sinica Vol 6 No.4 (1983) 393 - 405.

[5] Yang Shihui The sufficiency of the divisibility condition for 10 divides K (m, n, s) Journal of Taiyuan Institute of Machinery Vol. 15 (1994) 95 - 103.

[6] Yang Shihui The Isomorphic Factorization of Complete tripartite graph K (A,B,C) (I), The necesity and sufficiency for 12 divides K (m, n, s) to appear in Journal of Qu - fu Normal Universty china.

[7] Yang Shihui The Isomorphic factorization of Complete tripartite graph K (A, B, C) (II), The necesity and sufficiency for 14 divides K (m, n, s) to appear in Journal of Qu - fu Normal University china.

[8] Yang Shihui The Isomorphic Factorization of Complete Tripartite graph K (m,n,s) to appear in Dis Math.

On Coverings of DK_v by Even Circuits

J. Yin
Department of Mathematics, Suzhou University, Suzhou 215006, P. R. China

Abstract Let DK_v be the complete directed graph with v vertices. A covering of DK_v by m−circuits, directed cycle of length m, is a family F of m−circuits whose union is DK_v. The covering problem is to determine the minimum number of m−circuits, c(v,m), in such a covering for every integer $v\geqslant m$. In this paper, the problem is reduced to the case $m+3\leqslant v\leqslant 2m-2$ for any fixed even integer $m\geqslant 4$. In particular, the values of c(v,m) are completely determined for the case m=6,8 and 10.

1. Introduction

Let DK_v denote the complete directed graph with v vertices, where any two distinct vertices x and y are joined by exactly two arcs (x,y) and (y,x). By an m−circuit we mean a directed cycle of length m.

A covering of DK_v by m−circuits, denoted by (v,m)−DCC, is defined to be a family F of m−circuits whose union is DK_v. If no such covering has fewer circuits, the covering is said to be minimum, and the number of circuits in a minimum covering is called the covering number, denoted by c(v,m). As a general covering problem in graph theory, the main problem here is to determine the value of c(v,m) for every integer $v\geqslant m$. If we define $[v(v-1)/m]=T(v,m)$, where [x] is the least integer y satisfying $y\geqslant x$, then it is easy to see that

$$c(v,m)\geqslant T(v,m) \tag{1.1}$$

In view of (1.1), we refer to a minimum (v,m)−DCC as an optimal (v,m)−DCC if c(v, m)=T(v,m). We wish to remark that an optimal (v,m)−DCC is essentially an arc−disjoint decomposition of DK_v into m−circuits whenever $v(v-1)\equiv 0 \pmod m$. Many researchers have been involved in decomposing DK_v into arc−disjoint m−circuits. Mendelsohn [7] was among the first to study decompositions of DK_v into 3−circuits. He showed that the necessary condition, $v\equiv 0$ or 1 (mod 3), is also sufficient unless v=6 (see also Bermond [2] for an independent −proof). Schonheim [8] showed that the necessary condition for the decomposition of DK_v into 4−circuits, $v\equiv 0$ or 1 (mode 4), is also sufficient except for v=4. Bermond et al. (see, for example, [3,4,5]) later made a further investigation for the decomposition of DK_v into m−circuits. They also developed many other results. Quite recently, F. E. Bennett and the author [1] determined the values of c(v,m) for m=3 and 4. They also obtained similar results for the complete symmetric directed multigraph.

In this paper, we are interested in finding optimal (v,m)−DCCs for every integer $v\geqslant m$. We shall restrict ourselves to m even and present some constructions to obtain the optimal (v,m)−DCCs. As a consequence, the determination of the values of c(v,m) is reduced to the case where $m+3 \leqslant v \leqslant 2m-2$ for any fixed even integer m. In particular, the values of $c(v,m)$ are completely determined for the cases $m=6,8$ and 10.

2. Constructions

Through out the sequel, $m \geqslant 4$ is always a fixed even integer unless otherwise stated.

The object of this section is to present some constructions for optimal (v,m)-DCCs. We denote by $DK_{r,s}$ the complete symmetric bipartite directed graph with vertex set $x=x_1 \cup x_2$, where x_1 and x_2 are disjoint sets with $|x_1|=r$, $|x_2|=s$ and where two vertices x and y from sets x_1 and x_2 respectively are linked by exactly two arcs (x,y) and (y,x). For convenience we also refer to an arc-disjoin decomposition of DK_v into m-circuits as an m-circuit system of DK_v. A similar terminology applies to $DK_{r,s}$.

The following result is due to D. Sotteau [9], which is quite useful for our constructions.

Lemma 2.1 An m-circuit system of $DK_{r,s}$ exists if and only if $r,s \geqslant m/2$ and $2rs \equiv 0 \pmod{m}$.

We shall also make use of the following two lemmas which are contained in [3,10].

Lemma 2.2 For every even integer $m \geqslant 4$, an m-circuit system of DK_{m+1} exists.

Lemma 2.3 For every even integer $m \geqslant 8$, an m-circuit system of DK_m exists, whereas an m-circuit system of DK_m does not exist if $m=4$ or 6.

Note that we can think of an m-circuit system of $DK_{r,s}$ as being an m-circuit system of DK_{r+s} based on the same vertex set as that of $DK_{r,s}$, leaving two holes of size r and s separately. Moreover, the number of arcs (counting multiplicities) which occur more than once in an optimal (v,m)-DCC only depends on the congruity of v (mod m). From these observations we may use known technique of "filling in holes" (see, for example, [3,4,6]) to establish the following recursive construction.

Lemma 2.4 If an optimal (n,m)-DCC exists, then so does an optimal $(n+m,m)$-DCC, where $n \geqslant m$.

Proof Let F_1 be an m-circuit system of $DK_{n-1,m}$ based on vertex set $X=X_1 \cup X_2$ where $|X_1|=n-1$ and $|X_2|=m$, which exists by Lemma 2.1. Using the hypothesis, let F_2 be the optimal (v,m)-DCC based on vertex set $X_1 \cup \{w\}$ where $w \notin X$. We now use Lemma 2.2 to construct an m-circuit system F_3 of DK_{m+1} based on $X_2 \cup \{w\}$. Let $F=F_1 \cup F_2 \cup F_3$. It is a routine matter to check F forms an $(n+m,m)$-DCC based on $X \cup \{w\}$. Moreover, by a simple calculation we can know that $|F|=T(n+m,m)$. Therefore the conclusion holds.

Iteration of Lemma 2.4 gives the following.

Lemma 2.5 If $c(v,m)=T(v,m)$ for $m \leqslant v \leqslant 2m-1$, then $c(v,m)=T(v,m)$ for all integers $v \geqslant m$.

As already mentioned earlier, an m-circuit system of DK_v is indeed an optimal (v,m)-DCC. From Lemmas 2.2 and 2.3, we further have the following.

Lemma 2.6 If $c(v,m)=T(v,m)$ for $m+2 \leqslant v \leqslant 2m-1$, then $c(v,m)=T(v,m)$ for every integer $v \geqslant m$ with the exception of $(v,m) \in \{(4,4),(6,6)\}$.

It is worth mentioning that corollaries 2.5 and 2.6 say that the determination of $c(v,m)$ for all integers $v \geqslant m$ can be reduced to a finite number of values of v. In what follows, we will further reduced the problem to the cases where $m+3 \leqslant v \leqslant 2m-2$ by two constructive proofs utilizing difference set techniques. In aid of these, we denote by $(a_0, a_1, \cdots, a_{m-1})$ an m−circuit consisting of arcs (a_i, a_{i+1}) for $0 \leqslant i \leqslant m-1$, where subscripts are reduced modulo m. We need to remark that the method using difference sets was used in the construction of m−circuit systems (see, for example, [3,4]). Definitions not given here can be found in [3,4].

Lemma 2.7 There exists an optimal $(m+2,m)$−DCC.

Proof For $m=4$, the result was established in [1]. For $m \geqslant 6$, let the vertex set of DK_{m+2} be $Z_m \cup \{x,y\}$. An optimal $(8,6)$−DCC is obtained by taking the 6−circuit family $F_1 \cup F_2$ where $F_1 = \{(0,1,2,y,x,3),(x,y,2,5,4,3),(2,3,4,1,0,5),(4,5,0,3,2,1)\}$ and F_2 is obtained by developing the 6−circuit $(0,2,x,3,1,y)$ modulo 6. We need to make the observation that the extra points x and y are always fixed whenever we develop some base circuit.

To handle the case for $m \geqslant 8$, we define an m−circuit D by $D=(x,-1,1,-2,2,\cdots,-(m+2e)/4,(m+2e)/4,y,-(m+2e+4)/4,(m+2e+4)/4,\cdots,-(m-2)/2,(m-2)/2)$ where $e=0$ or 1 depending on whether $m \equiv 0 \pmod 4$ or $m \equiv 2 \pmod 4$. Since among the differences arising from the arcs of D all nonzero residues modulo m, except for $r \in \{1, m-1, (m-2e-2)/2\}$, occur exactly once, developing the m−circuit D modulo m produces m m−circuits. On the other hand, since r is relatively prime to m we have also the following 3 m−circuits: $(0,r,2r,\cdots,(m-1)r)$, where $r=1$, $m-1$ and $(m-2e-2)/2$. The above $m+3$ m−circuits contain every arc of DK_m exactly once, except for the arcs (x,y) and (y,x) which do not occur in any of these m−circuits. To cover the arcs (x,y) and (y,x), we replace the circuit $(0,1,2,\cdots,m-1)$ obtained above with the following two m−circuits:

$$(x,y,2,3,4,5,\cdots,m-1),$$
$$(0,1,2,y,x,5,\cdots,m-1).$$

It is easily checked that the above construction yields an optimal $(m+2,m)$−DCC and the proof is complete.

Lemma 2.8 There exists an optimal $(2m-1,m)$−DCC.

Proof The existence of an optimal $(2m-1,m)$−DCC implies that we must have $T(2m-1,m)=2(2m-3)+1$ m−circuits. We proceed as in Lemma 2.7. Take the vertex set of DK_{2m-1} to be $Z_{2m-3} \cup \{x,y\}$ and define two m−circuits D_1 and D_2 by

$$D_1 = (x,0,1,-1,2,-2,\cdots,(m-2)/2,-(m-2)/2),$$
$$D_2 = (y,0,-1,1,-2,2,\cdots,-(m-2)/2,(m-2)/2).$$

It is easy to see that all nonzero residues modulo $2m-3$ occur precisely once in the differences arising from the arcs of D_1 and D_2. We then develop D_1 and D_2 to obtain $2(2m-3)$ m−circuits. In these $2(2m-3)$ m−circuits, the arcs (x,y) and (y,x) do not occur and the other arcs of DK_{2m-1} occur exactly once. We now distinguish one of the $2(2m-1)$ m−circuits obtained above, say

$$(y,0,-1,1,-2,2,\cdots,-(m-2)/2,(m-2)/2),$$

and replace it with the following two m−circuits:

$$(y,x,-1,1,-2,2,\cdots,-(m-2)/2,(m-2)/2),$$
$$(x,y,0,-1,-2,2,\cdots,-(m-2)/2,(m-2)/2),$$

This creates an optimal $(2m-1,m)$−DCC.

Summarizing the results of Lemmas 2.4~2.8, we have proved the following theorem.

Theorem 2.9 Let $m\geqslant 4$ be an even integer. If $c(v,m)=T(v,m)$ for $m+3\leqslant v\leqslant 2m-2$, then $c(v,m)=T(v,m)$ for all integers $v\geqslant m$ with the exception of $(v,m)\in\{(4,4),(6,6)\}$.

3. The case m = 6, 8 and 10

In this section, we turn our attention to determination of $c(v,m)$ for the special case $m\leqslant 10$. We proceed with the following two known results which can be found in [1,4].

Lemma 3.1 The equality $c(v,4)=T(v,4)$ holds for all integers $v\geqslant 4$ except for $v=4$. Moreover, $c(4,4)=T(4,4)+1=4$.

Lemma 3.2 Let $v\geqslant m$ be an integer and $m\in\{4,6,8,10\}$. Then an m−circuit system of DK_v exists if and only if $v(v-1)\equiv 0 \pmod m$ and $(v,m)\neq(4,4)$ and $(6,6)$.

As an immediate consequence of Lemma 3.2, we have the following.

Corollary 3.3 Let $v\geqslant m$ be an integer and $m\in\{6,8,10\}$. Then $c(v,m)=T(v,m)$ if $v(v-1)\equiv 0 \pmod m$ and $(v,m)\neq(6,6)$.

We now work towards determining $c(v,m)$ for $6\leqslant m\leqslant 10$.

Lemma 3.4 Let $v\geqslant 6$ be an integer. Then $c(v,6)=T(v,6)$ if $v\neq 6$. Moreover, $c(6,6)=T(6,6)+1=6$.

Proof In view of Theorem 2.9 and Corollary 3.3, it remains to handle the outstanding case for $v=6$. From Lemma 3.2, we know that $c(6,6)\geqslant 6$. A $(6,6)$−DCC with six 6−circuits is given by developing the circuit $(0,1,5,2,4,3)$ modulo 6.

Lemma 3.5 For all integers $v\geqslant 8$, we have $c(v,8)=T(v,8)$.

Proof From Theorem 2.9, we can establish the result by constructing an optimal $(v,8)$−DCC for each value of $v\in\{11,12,13,14\}$. When $v=14$, we take $(Z_6\times Z_2)\cup\{x,y\}$ as the vertex set of DK_{14}. Start with a $DK_{6,6}$ having the vertex set $(Z_6\times\{0\})\cup(Z_6\times\{1\})$. By Lemma 2.1, there exists an 8−circuit system S of $DK_{6,6}$. Let F_i be an 8−circuit of DK_8 based on $(Z_6\times\{i\})\cup\{x,y\}$, which exists by Lemma 2.3. It is easy to see that $F=S\cup F_0\cup F_1$ is the required optimal $(14,8)$−DCC. For the other values of v, the required optimal $(v,8)$−DCCs are provided below, where the vertex set of DK_v is given along with the required $T(v,m)$ m−circuits of DK_v.

For $v=11$, the vertex set is $Z_8\cup\{x,y,z\}$ and the 8−circuits are

(x,y,2,6,z,5,0,3),(x,y,6,0,z,1,4,7),(y,x,6,2,z,3,7,1)
(y,x,0,4,z,7,2,5),(x,z,0,2,y,3,6,1),(y,z,2,4,x,1,5,3)
(z,y,4,0,x,5,1,7),(z,x,4,6,y,7,3,1),(2,x,7,0,1,6,4,5)
(6,x,2,1,3,0,5,7),(5,x,3,z,4,2,0,7),(0,y,5,z,6,3,4,1)
(4,y,1,2,3,5,6,7),(7,y,0,6,5,4,3,2).

For v=12, the vertex set is $Z_{10} \cup \{x,y\}$ and the 8−circuits are

(x,y,0,3,6,9,2,5),(y,x,5,8,1,4,7,0)
(x,i,9+i,y,6+i,3+i,8+i,4+i),
(j,2+j,6+j,4+j,5+j,3+j,7+j,9+j)

where $i \in Z_{10}$, $j \in \{0,2,4,6,8\}$ and the sums are reduced modulo 10.

For v=13, the vertex set is $Z_9 \cup \{x,y,z,w\}$ and the 8−circuits are

(x,y,0,7,4,3,8,6),(x,z,3,1,7,6,2,0),(x,w,6,4,1,0,5,3),
(y,z,0,2,5,6,1,3),(y,w,3,5,8,0,4,6),(y,x,6,8,2,3,7,0),
(z,w,0,8,7,2,6,3),(z,x,3,2,1,5,0,6),(z,y,6,5,4,8,3,0),
(w,x,0,1,2,7,3,6),(w,y,3,4,5,1,6,0),(w,z,6,7,8,4,0,3),
(x,1,4,7,2,5,w,8),(x,2,y,7,z,1,w,5),(x,7,y,2,z,8,w,1),
(x,4,y,8,5,7,w,2),(x,8,y,5,z,2,w,4),(x,5,y,1,z,4,2,7),
(2,4,w,7,1,8,z,5),(7,5,2,8,1,y,4,z).

Lemma 3.6 For all integers $v \geqslant 10$, we have $c(v,10)=T(v,10)$.

Proof By Theorem 2.9 and Corollary 3.3, we need only to construct an optimal (v,10)−DCC for each value of $v \in \{13,14,17,18\}$ in order to complete the proof.

For v=13, let the vertex set of DK_{13} be $Z_{12} \cup \{x\}$. Then there required circuits are provided by taking the orbit of (0,1,2,3,5,7,9,4,11,6) under the action of the cyclic group (of order 4) generated by the permutation (x) (0,3,6,9) (1,4,7,10) (2,5,8,11) and the orbit of (x,0,11,9,6,10,1,7,3,8) under the action of the cyclic group generated by the permutation which fixes x and sends $i \rightarrow i+1 \pmod{12}$ for $i \in Z_{12}$.

For $v \in \{14,17\}$, the proof is similar to that of Lemmas 2.7 and 2.8. Take $Z_{v-2} \cup \{x,y\}$ as the vertex set of DK_v. By the following procedure, we first construct $T(v,10)-1$ m−circuits such that they contain every arc of DK_v exactly once, but the arcs (x,y) and (y,x) are both not covered. When v=14, the circuits are obtained by taking three orbits, each having six 10−circuits, under the action of the cyclic group (of order 6) generated by the permutation (x) (y) (0 2 4 6 8 10) (1 3 5 7 9 11). Representatives of these orbits are (0,6,4,8,x,1,7,5,9,y), (0,1,3,4,6,9,5,2,7,x) and (1,0,8,5,10,9,4,11,2,y). For v=17, the circuits are constructed by taking the orbit of (0,1,2,3,4,5,7,9,11,13), (0,13,11,9,7,5,4,3,2,1), (0,7,14,6,13,5,1,12,8,4) and (0,4,8,12,1,5,13,6,14,7) under the action of the cyclic group (of order 3) generated by the permutation (x) (y) (0 5 10) (1 6 11) (2 7 12) (3 8 13) (4 9 14), and the orbit of (x,0,3,8,2,12,9,y,1,7) under the action of the cyclic group generated by the permutation which fixes x,y and sends $i \rightarrow i+1 \pmod{15}$ for $i \in Z_{15}$. We then distinguish one m−circuit from the above $T(v,10)-1$ circuits and replaced it by two suitable m−circuits, as we did in

Lemmas 2.7 and 2.8. The result is an optimal $(v,10)$—DCC.

For $v=18$, we define the vertex set of DK_{18} to be $Z_{14}\cup\{x,y,z,w\}$. Then the desired circuits are

$$(1,2,3,4,5,6,x,y,w,z),(0,13,6,9,10,11,y,z,x,w),$$
$$(5,6,7,8,9,1,w,x,z,y),(11,12,13,0,1,5,z,w,y,x),$$
$$(x,11+i,y,5+i,z,1+i,w,i,13+i,6+i),$$
$$(j,4+j,6+j,2+j,7+j,1+j,13+j,8+j,5+j,11+j),$$

where $i\in Z_{14}\setminus\{0\}$ and $j\in Z_{14}$ and the sums are reduced modulo 14.

Summarizing the above, the values of $c(v,m)$ are determined completely for all even integers $m\leqslant 10$. For any fixed even integer $m\geqslant 12$ we have reduced the determination of $c(v,m)$ to the case where $m-3\leqslant v\leqslant 2m-3$. We now conclude this paper by making the following.

Conjecture: For any fixed even integer $m\geqslant 12$ and $m-3\leqslant v\leqslant 2m-3$, the equality $c(v,m)=T(v,m)$ holds, and hence so does for all integers $v\geqslant m$.

References

[1] F. E. Bennett and J. Yin, Packings and Coverings of the complete directed graph with k—circuits, preprint.

[2] J. C. Bermond, An application of the solution of Kirkman's schoolgirl problem: the decomposition of the symmetricoriented complete graph into 3—circuits, Discrete Math. 8(1974), 301—304.

[3] J. C. Bermond and V. Faber, Decomposition of the complete directed graph into k—circuits, J. Combin. Theory (B) 21(1976), 146—155.

[4] J. C. Bermond, C. Huang and D. Sotteau, Balanced cycle and circuit designs: even cases, Ars Combinatoria 5(1978), 293—318.

[5] J. C. Bermond and D. Sotteau, Cycle and circuit designs: odd case, Beitrage Zur Graphen theorie and deren Anwendungen, Vorgetragen auf dem int. Kolloq Oberhof (DDR) vom 10(1977), 11—32.

[6] C. C. Linder and C. A. Rodger, Decomposition into cycles Ⅱ: cycle systems, in "Contemporary design theory" (J. H. Dinitz and D. R. Stinson, Eds), PP. 325—369, Wiley, 1992.

[7] N. S. Mendelsohn, A natural generalization of Steiner triple systems, Computers in Number Theory (edited by A. O. L. Atkin and B. J. Birch, Academic Press, New York, 1971), 323—338.

[8] J. Schonheim, Partition of the edges of the directed complete graph into 4—cycles, Discrete Math. 11(1975), 67—70.

[9] D. Sotteau, Decomposition of $K_{m,n}$ ($K^*_{m,n}$) into cycles (circuits) of length 2k, J. Combin. Theory (B) 30(1981), 75—81.

[10] T. Tillson, A hamiltonian decomposition of K^*_{2m}, $2m\geqslant 8$, J. Combin. Theory (B) 29(1980), 68—74.

VERTEX TRANSITIVE GRAPHS AND THEIR AUTOMORPHISM GROUPS

Hong Zhang

Department of Mathematical Sciences

Indiana University-Purdue University at Fort Wayne

Fort Wayne, IN 46805

U.S.A.

Abstract. Certain vertex transitive graphs are classified based on their automorphism groups. It is shown that a nontrivial circulant graph has a primitive automorphism group if and only if it has a prime number of vertices. In this case, the classification of edge transitive circulant graphs is obtained by a result on symmetric graphs with a prime number of vertices. In the case of imprimitive automorphism groups, the family of strongly edge transitive graphs with strongly edge transitive complements is classified.

Key Words. vertex transitive and edge transitive graphs, circulant graphs, primitive and imprimitive groups, doubly transitive groups, Burnside groups.

1. Introduction

In this paper we classify several special families of vertex transitive graphs based on the properties of their automorphism groups. Two cases are considered here. If the automorphism group is primitive, we show that a nontrivial circulant graph must have a prime number of vertices. Consequently symmetric circulant graphs with primitive automorphism groups can be classified using a result by Chao on symmetric graphs of a prime number of vertices. In the second case, the automorphism groups are imprimitive. We show that a strongly edge transitive graph with a strongly edge transitive complement is either mK_n, a disjoint union of m copies of the complete graph K_n, or $T_{m,mn}$, a complete m-partite graph with n vertices in each partite set, where $m, n > 1$.

As corollaries to the results above, we shall classify the family of edge transitive circulant graphs with edge transitive circulant complements and, in particular, all self-complementary symmetric circulant graphs. Similar results for digraphs are also obtained.

The graphs and digraphs we consider here are assumed to be finite and simple. A graph G is said to be vertex transitive if its automorphism group $A(G)$ is transitive on the vertex set. A graph with v vertices is circulant if its automorphism group contains a v-cycle. A circulant graph is clearly vertex transitive. The automorphism group also induces an action on the edge set. A graph is called edge transitive if the action of the automorphism group on the edge set is transitive. An undirected graph is called strongly edge transitive if for any two edges ab and cd there exists an automorphism θ such that $\theta a = c$, $\theta b = d$. A graph is symmetric if it is both vertex transitive and edge transitive. The complement $\bar{G}$ of a graph G has the same vertex set as G and the edges of $\bar{G}$ are the non-edges of G. A graph G is called self-complementary if it is isomorphic to its complement $\bar{G}$.

A permutation group A acting on a set X is said to be doubly transitive if for any two ordered pairs (a, b), (c, d), where $a, b, c, d \in X$, $a \neq b$, and $c \neq d$, there exists an $\theta \in A$ such that $\theta a = c$, $\theta b = d$.

An abstract group H is called a Burnside group (or B-group) if any primitive group containing the regular representation of H as a transitive subgroup is doubly transitive (see [1] and [6]).

Let A be a transitive permutation group acting on a set X. A subset B of X is called a block if for any $\theta \in A$, $\theta B = B$ or $\theta B \cap B = \emptyset$. The sets $\emptyset$, X, and the singleton subsets of X are called trivial blocks. The transitive group A is said to be primitive if it has no nontrivial blocks. It is said to be imprimitive if it has nontrivial blocks. Let B be a block of A. The elements of A that leave B fixed setwise form a subgroup of A which is denoted by $A_{(B)} = \{\theta \in A \mid \theta B = B\}$.

2. Primitive Groups

The automorphism group of the complete graph K_n or the null graph with

n vertices is the symmetric group S_n, which is clearly primitive. We call a graph nontrivial if it is neither a complete graph nor a null graph. For a nontrivial circulant graph with a primitive automorphism group, the following theorem shows that it has a prime number of vertices.

Theorem 1. *Let G be a circulant graph which is neither complete nor null. The automorphism group $A(G)$ is primitive if and only if G has a prime number of vertices.*

Proof. If the number of vertices of G is a prime, clearly $A(G)$ is primitive since any transitive permutation group of a prime degree is primitive.

By a theorem of Schur [5] (see [6], Theorem 25.3), a cyclic group of composite order is a B-group. If the number of vertices v of G is not a prime, then the cyclic group of order v is a B-group. Since G is circulant, the primitive group $A(G)$ contains the regular representation of the cyclic group as a transitive subgroup. Hence $A(G)$ is doubly transitive and G is either complete or null, which is a contradiction. ∎

A set of symmetric circulant graphs can be constructed as follows. Let p be an odd prime and $\mathbf{Z}_p$ the finite field with p elements. Let d be an even integer with $d|(p-1)$ and $d < (p-1)$. Then the multiplicative group $\mathbf{Z}_p^*$ contains a unique subgroup H of order d. The graph $X(p,d)$ is defined as the graph with the vertex set $\mathbf{Z}_p$ and the edge set $\{ab \mid b-a \in H\}$.

In the special case that $p \equiv 1 \pmod 4$ and $d = (p-1)/2$, the graph $X(p,d)$ is also called a Paley graph and is denoted by $P(p)$.

It is well-known that the graph $X(p,d)$ is symmetric and circulant (see [2]). The Paley graph $P(p)$ is also self-complementary (see [7]).

In [2], Chao classified the family of symmetric graphs with a prime number of vertices. It was shown that a nontrivial symmetric graph with p vertices is isomorphic to one of the graphs $X(p,d)$ defined above and the symmetric graph with p vertices and degree d is unique up to isomorphism.

By Theorem 1 and Chao's result, we obtain the following classification of symmetric circulant graphs with primitive automorphism groups.

Theorem 2. *A nontrivial symmetric circulant graph has a primitive automorphism group if and only if it is isomorphic to a graph $X(p,d)$, where p is an odd prime and d is an even integer with $d|(p-1)$ and $d<(p-1)$.*

3. Imprimitive Groups

For vertex transitive graphs with imprimitive automorphism groups, we shall classify the family of strongly edge transitive graphs with strongly edge transitive complements.

Consider the disjoint union of m copies of complete graph on n vertices mK_n. Clearly the complement of mK_n is the complete m-partite graph $T_{m,mn}$, which is also a special type of Turán graphs. We first show that mK_n and $T_{m,mn}$ are circulant and edge transitive graphs and their automorphism groups are imprimitive if $m,n>1$.

Lemma 1. *If G is circulant and edge transitive, then the disjoint union of m isomorphic copies of G, denoted by mG, is also circulant and edge transitive.*

Proof. Label the vertices of the ith copy of G by $v_1^i, v_2^i, \ldots, v_n^i$ such that the permutation $(v_1^i, v_2^i, \ldots, v_n^i)$ is an automorphism of G. Then it is easy to verify that

$$(v_1^1, v_1^2, \ldots, v_1^m, v_2^1, v_2^2, \ldots, v_2^m, \ldots, v_n^1, v_n^2, \ldots, v_n^m)$$

is an automorphism of mG. Hence mG is circulant. mG is clearly edge transitive.

Lemma 2. *The graph mK_n and its complement $T_{m,mn}$ are circulant and edge transitive. If $m>1$ and $n>1$, the automorphism groups are imprimitive.*

Proof. K_n is clearly circulant and edge transitive. By lemma 1, mK_n is also circulant and edge transitive. Since a graph and its complement have the same automorphism group, $T_{m,mn}$, being the complement of mK_n, is also circulant.

We now show that $T_{m,mn}$ is edge transitive. It is easy to see that the automorphism group of mK_n and $T_{m,mn}$ is the wreath product (see [4]) $S_n \wr S_m$. Let ab and cd be two edges of $T_{m,mn}$. By vertex transitivity, there is an automorphism θ_1 that maps a to c. Define $b_1 = \theta_1 b$. Clearly by selecting an appropriate permutation in S_m there is an automorphism θ_2 in $S_n \wr S_m$ that fixes c and maps b_1 to b_2 which is in the same partite set as d in $T_{m,mn}$. Finally, using a permutation in S_n, there is an automorphism θ_3 that fixes c and maps b_2 to d. Hence, $\theta_3\theta_2\theta_1$ is an automorphism that maps the edge ab to cd and $T_{m,mn}$ is edge transitive.

If $m > 1$, clearly the vertices in a component of mK_n form a block. If $n > 1$ the block is nontrivial. ∎

The following theorem gives a classification for the strongly edge transitive graphs with strongly edge transitive complements and imprimitive automorphism groups.

Theorem 3. *Let G be a vertex transitive graph. If both G and $\bar{G}$ are strongly edge transitive and the automorphism group $A(G)$ is imprimitive, then either G or $\bar{G}$ is isomorphic to mK_n for some positive integers $m, n > 1$.*

Proof. Let B be a nontrivial block of $A(G)$. Let $S = \{s \mid as \in E(G)\}$ and $S' = \{s \mid as \in E(\bar{G})\}$. Since B is a nontrivial block, B intersects either S or S'. Suppose that S intersects B and $b \in B \cap S$. Since G is strongly edge transitive, for each $s \in S$, there exists an automorphism θ such that $\theta a = a$ and $\theta b = s$. This implies $s \in B$ since $a, b \in B$ and B is a block. Hence $S \subset B$. Similarly, if S' intersects B, then $S' \subset B$. Therefore, we have either $B = \{a\} \cup S$ or $B = \{a\} \cup S'$. If $B = \{a\} \cup S$, consider the action of $A_{(B)}$ on B. Since A is transitive and B is a block, $A_{(B)}$ is transitive on B. By the strong edge transitivity of G, the stabilizer of $A_{(B)}$ at a is transitive on S. Hence the action of $A_{(B)}$ on B is doubly transitive. Therefore B induces a complete graph K_n. Since G is regular, B must be a component of G. By the vertex transitivity, each component of G is a complete graph K_n and $G = mK_n$. If $B = \{a\} \cup S'$, by a similar argument, we have $\bar{G} = mK_n$. ∎

4. Corollaries and Generalizations

Using the results in section 2 and 3, we can obtain a classification of edge transitive circulant graphs with edge transitive complements.

Lemma 3. *If a graph G is circulant and edge transitive, then G is strongly edge transitive.*

Proof. Identify the vertices of G as the elements of $\mathbf{Z}_v$ in such a way that $\rho(x) = x + 1$ is an automorphism. Let a be a vertex adjacent to the vertex 0. It suffices to show that there is an automorphism θ such that $\theta(0) = a$ and $\theta(a) = 0$. The permutation $\sigma(x) = -x$ is an automorphism of G since any two vertices s and t of G are adjacent if and only if the vertices $\rho^{-s-t}(s) = -t$ and $\rho^{-s-t}(t) = -s$ are adjacent. Let $\theta = \sigma\rho^{-a}$. Then $\theta(0) = -(0 - a) = a$ and $\theta(a) = -(a - a) = 0$. ‖

Corollary 1. *Let G be a circulant graph and both G and $\bar{G}$ be edge transitive. Then G is isomorphic to one of the following.*

1. *mK_n, a disjoint union of m copies of complete graph K_n.*
2. *$T_{m,mn}$, a complete m-partite graph with n vertices in each partite set.*
3. *$P(p)$, a Paley graph on p vertices, where p is a prime and $p \equiv 1 \pmod 4$.*

Proof. By lemma 3, G and $\bar{G}$ are strongly edge transitive. If the automorphism group $A(G)$ is imprimitive, then by theorem 3, G is isomorphic to mK_n or $T_{m,mn}$ for some m and n. If G is complete or null, then G is also isomorphic to mK_n for $m = 1$ or $n = 1$. Otherwise, $A(G)$ is primitive and G is neither complete nor null. By theorem 2, G is isomorphic to $X(p, d)$ for some prime p and even integer d, where d is the degree of G. Since both G and $\bar{G}$ are symmetric, we have $d|(p-1)$ and $(p-1-d)|(p-1)$. Therefore, $d = (p-1)/2$ is even and $p \equiv 1 \pmod 4$. By the uniqueness of the symmetric graph, G must be the Paley graph $P(p)$. ‖

In [7], it is shown that the number of vertices of a self-complementary symmetric graph must be a prime power. By the above theorem, we obtain a classification of self-complementary symmetric circulant graphs.

Corollary 2. *A graph with at least two vertices is self-complementary, symmetric,* *and circulant if and only if it is isomorphic to the Paley graph $P(p)$, where p is a prime and $p \equiv 1 \pmod 4$.*

We may generalize the results to digraphs. The definitions of circulant graphs, vertex transitive graphs, edge transitive graphs, and symmetric graphs can be naturally extended to digraphs. We omit the proofs here since they are similar to those in the undirected case.

Theorem 4. *Let G be a circulant digraph which is neither complete nor null. The automorphism group $A(G)$ is primitive if and only if G has a prime number of vertices.*

The following result uses the classification of symmetric digraphs of a prime number of vertices by Chao and Wells [3].

Theorem 5. *A nontrivial symmetric circulant digraph has a primitive automorphism group if and only if it is isomorphic to a digraph $X(p, d)$ defined in Chao and Wells [3], where p is an odd prime and d is an integer with $d|(p-1)$.*

Theorem 6. *Let G be a vertex transitive digraph. If both G and $\bar{G}$ are edge transitive and the automorphism group $A(G)$ is imprimitive, then either G or $\bar{G}$ is isomorphic to mK_n for some positive integers $m, n > 1$.*

Similarly, we have the following corollaries that classify the edge transitive circulant digraphs with edge transitive complements and the self-complementary circulant symmetric digraphs.

Corollary 3. *Let G be a circulant digraph and both G and $\bar{G}$ be edge transitive. Then G is isomorphic to one of the following.*

1. *mK_n, a disjoint union of m copies of complete digraph K_n.*
2. *$T_{m,mn}$, a complete m-partite digraph with n vertices in each partite set.*
3. *A Paley digraph $P(p)$, where p is an odd prime.*

Corollary 4. *A digraph with at least two vertices is self-complementary, symmetric, and circulant if and only if it is isomorphic to the Paley digraph $P(p)$, where p is an odd prime.*

5. References

[1] W. Burnside, *Theory of groups of finite order*, Cambridge University Press, Cambridge (1911).

[2] C. Y. Chao, On the classification of symmetric graphs with a prime number of vertices, *Trans. Amer. Math. Soc.*, 158 (1971) 247–256.

[3] C. Y. Chao and J. G. Wells, A class of vertex-transitive digraphs, *J. Combin. Theory*, Ser. B, 14 (1973), 246-255.

[4] D. S. Passman, Permutation groups, W. A. Benjamin, New York (1968).

[5] I. Schur, Zur Theorie der einfach transitiven Permutaionsgruppen, *S. B. Preuss. Akad. Wiss., Phys.-Math. Kl.* 1933, 598–623.

[6] H. Wielandt, Finite Permutation Groups, Academic, New York (1964).

[7] H. Zhang, Self-complementary symmetric graphs, *J. Graph Theory*, 16-1 (1992), 1–5.

On q-Derangement Polynomials

Zhang Xiangde*
(*Northeastern University, Shenyang, 110006*)

Abstract The q-deranqement polynomials are studied and a conjecture of Brenti is proved.

Key words derangement; q-derangement polynomials; Sturm sequence.

Let $P = \{1, 2, \cdots, n \cdots\}$ and S_n denote the symmetric group on n elements. For $n \in P$ and $\sigma \in S_n$, an element $i \in [n]$ (where $[n] = \{1, 2, \cdots, n\}$) is called an excedance of σ if $\sigma(i) > i$; denote the number of excedances of σ by $e(\sigma)$. The permutation $\sigma \in S_n$ is called a derangement if $\sigma(i) \neq i$ for $i = 1, 2, \cdots, n$. Denote the set of all derangement of S_n by D_n, F. Brenti [2] defined polynomials $d_n(q)$ by

$$d_n(q) = \sum_{\sigma \in D_n} q^{e(\sigma)}$$

for $n \in P$ (so that $d_1(q) = 0$), and let $d_0(q) = 1$. Since $d_n(1) = |D_n|$, Brenti considered $d_n(q)$ as q-derangement polynomials.

In [2], F. Brenti posed the following conjecture: for $n \in P$, the polynomials $d_n(q)$ have only real zeros. We have verified the conjecture. The main result of this paper is as follows:

Theorem The polynomials $d_n(q)$ satisfy the following properties:

1)

$$d_n(q) = \sum_{k=0}^{n} (-1)^{n-k} \binom{n}{k} A_k(q)$$

where $A_0(q) = q$, for $n \in P$, $A_n(q)$ is the nth Eulerian polynomials.

2) $d_n(q) = (n-1)qd_{n-1}(q) + q(1-q)d'_{n-1}(q) + (n-1)qd_{n-2}(q), n \geq 2.$

3) The polynomials $d_n(q)$, i.e. $\sum_{k=0}^{n}(-1)^{n-k}A_k(q)$ form a Sturm sequence, that is $d_n(q)$ has $n-1$ distinct real roots (≤ 0), separated by the roots of $d_{n-1}(q)$.

---

*Supported by the Natural Science Foundation of China and Liaoning Province.

Proof Recall the following relation between $d_n(q)$ and $A_n(q)$ [2]:

$$A_n(q)/q = \sum_{k=0}^{n} \binom{n}{k} d_k(q), \qquad n \in P \tag{1}$$

Let $A_0(q) = q$, then when $n = 0$ (1) also holds. Using the following inversion formulas

$$f_n = \sum_{k=0}^{n} \binom{n}{k} g_k \iff g_n = \sum_{k=0}^{n} (-1)^{n-k} \binom{n}{k} f_k,$$

we obtain $d_n(q) = \sum_{k=0}^{n} (-1)^{n-k} \binom{n}{k} A_k(q)/q$.

Since the Eulerian polynomials $A_n(q)$ satisfy the following recurrence relation:

$$A_n(q) = q(1-q)A'_{n-1}(q) + nqA_{n-1}(q), n \geq 2.$$

Noting that when $n = 1$ and $A_0(q) = q$ the recurrence relation also holds, we have

$$\begin{aligned}
d_n(q) &= \sum_{k=0}^{n} (-1)^{n-k} \binom{n}{k} A_k(q)/q \\
&= \sum_{k=0}^{n} (-1)^{n-k} \left(\binom{n-1}{k} + \binom{n-1}{k-1} \right) A_k(q)/q \\
&= -d_{n-1}(q) + \sum_{k=1}^{n} (-1)^{n-k} \binom{n-1}{k-1} A_k(q)/q \qquad \left(\binom{n-1}{-1} = 0 \right) \\
&= -d_{n-1}(q) + \sum_{k=1}^{n} (-1)^{n-k} \binom{n-1}{k-1} ((1-q)A'_{k-1}(q) + k\, A_{k-1}(q)) \\
&= -d_{n-1}(q) + (1-q)\,(q\, d_{n-1}(q))' + \sum_{k=1}^{n} (-1)^{n-k} \binom{n-1}{k-1} k\, A_{k-1}(q) \\
&= -q\, d_{n-1}(q) + q\,(1-q)\, d'_{n-1}(q) + \sum_{k=1}^{n} (-1)^{n-k} \left(\binom{n}{k} - \binom{n-1}{k} \right) k\, A_{k-1}(q) \\
&= (n-1)\, q\, d_{n-1}(q) + q\,(1-q)\, d'_{n-1}(q) + (n-1)\, q\, d_{n-2}(q),
\end{aligned}$$

where $n \geq 2$. Hence, 2) is proved.

In order to prove 3), let us use induction on n. For $n = 1, 2, 3$ we have $d_1(q) = 0$, $d_2(q) = q$ and $d_3(q) = q + q^2$, so, 3) is easily checked for $n \leq 3$. Assume that it holds for $n - 1 (n \geq 3)$, we proceed to show that it holds for n. Using the following property [2]:

$$\sum_{n \geq 0}^{\infty} d_n(q) \frac{t^n}{n!} = \frac{1}{1 - \sum_{k \geq 2} (q + q^2 + \cdots + q^{k-1}) \frac{t^k}{k!}},$$

we obtain that zero is a simple root of $d_n(q)$ and that the polynomials $d_n(q)$ have no positive real roots. By assumption $d_{n-1}(q)$ and $d_{n-2}(q)$ have only distinct real roots, denote the roots of $d_{n-1}(q)$ and $d_{n-2}(q)$ by $\alpha_1 = 0, \alpha_2, \cdots, \alpha_{n-2}$ and $\beta_1 = 0, \beta_2, \cdots, \beta_{n-3}$, respectively, where $0 = \alpha_1 > \alpha_2 > \cdots > \alpha_{n-2}$ and $0 = \beta_1 > \beta_2 > \cdots > \beta_{n-3}$, by assumption, we have

$$\alpha_{i+1} \leq \beta_i \leq \alpha_i, \quad i = 1, 2, \cdots, n-3. \tag{2}$$

By Rolle's theorem, we get that $d'_{n-1}(q)$ has $n-3$ distinct real roots and that the roots of $d'_{n-1}(q)$ separate those of $d_{n-1}(q)$. Denote the roots of $d'_{n-1}(q)$ by $\gamma_1, \gamma_2, \cdots, \gamma_{n-3}$, where $\gamma_1 > \gamma_2 > \cdots > \gamma_{n-3}$, then we have

$$\alpha_{i+1} < \gamma_i < \alpha_i, \quad i = 1, 2, \cdots, n-3. \tag{3}$$

Hence, we obtain

$$\begin{aligned} d_n(q) &= (n-1)\, q^2\, (q-\alpha_2) \cdots (q-\alpha_{n-2}) \\ &\quad + (n-2)\, q\, (1-q)\, (q-\gamma_1)\, (q-\gamma_2) \cdots (q-\gamma_{n-3}) \\ &\quad + (n-1)\, q^2\, (q-\beta_2) \cdots (q-\beta_{n-3}) \end{aligned} \tag{4}$$

Letting

$$a = -\frac{n-2}{n-1} |\gamma_1/2| \prod_{i=2}^{n-3} (\frac{\gamma_1}{2} - \gamma_i)/N,$$

where

$$N = (1 + |\alpha_{n-2}|) \prod_{i=2}^{n-3} |\alpha_i + \gamma_i| + \prod_{i=2}^{n-3} |\beta_i + \gamma_i|,$$

then $\alpha_2 < \gamma_1/2 < a < 0$; using (4), we have

$$d_n(a) < 0 \tag{5}$$

By (2), (3) and (4), The following inequalities can be easily proved

$$(-1)^k d_n(\alpha_k) > 0, \quad k = 2, ..., n-2.$$

Since $\lim_{q \to -\infty} (-1)^{n-1} d_n(q) = +\infty$, there exists $M > 0$ ($M > 1 + |\alpha_{n-2}|$) such that if $q \leq -M$ then $(-1)^{n-1} d_n(q) > 0$ so in particular $(-1)^{n-1} d_n(-M-1) > 0$. Hence, by the intermediate value theorem of continuous functions, we get that $d_n(q)$ has only one root in each open interval (α_2, a), (α_{i+1}, α_i) and $(-M-1, \alpha_{n-2})$, where $i = 2, 3, \cdots, n-3$. Recall that zero is a simple root of $d_n(q)$, therefore $d_n(q)$ has $n-1$ distinct real roots and its roots are obviously separated by those of $d_{n-1}(q)$. Hence, we have proved 3) by induction.

Corollary 1 The conjecture is true.

Corollary 2 The polynomials $d_n(q)$ are log-concave.

Notes (1) The polynomials $d_n(q)$ are different from the q-derangement numbers for the major index that have been previously considered in the literature (see, e.g., [3], [5], [6] , [7] and [8])

(2) The claim 3) has been generalized, the details will be published elsewhere.

Acknowledgements.

I thank Prof. W. Y. C. Chen, L. C. Hsu and X. H. Xu for valuable suggestions and generous help.

References

[1] F. Brenti, Unimodal, Log-concave, and Polya Frequency Sequences in Combinatorics, Memoirs Amer. Math. Soc. no. 413. 1989.

[2] F. Brenti, Unimodal Polynomials Arising from Symmetric Functions, Proc. Amer. Math. Soc. 108(1990), 1133-1141.

[3] W. Y. C. Chen and G. -C. Rota, q-Analogs of the Inclusion - Exclusion Principle and Permutation with Restricted Position, Discrete Math. 104(1992), 7-22.

[4] L. Comet, Advanced Combinatorics, Reidel Dordrecht/Boston, 1974.

[5] A. M. Garsia and J. B. Remmel, A Combinatorial Interpretation of q-Derangement and q-Laguerre numbers, European J. Combin, 1(1980), 47-59.

[6] I. Gessel and C. Reutenauer, Counting Pernutations with Given Cycle Structure and Dessent Set, J. Combina. Theory. Ser. A, 64(1993), 189-215.

[7] M. L. Wachs, on q-Derangement Numbers, Proc. Amer. Math. Soc. 106(1989) 273-278.

[8] X. D. Zhang, On Some q-Analysis Problems in Combinatorics, Ph. D. Thesis, Dalian University of Technology, January 1994.

Combinatorial Algebraic Properties of Continuous Mapping Set $C(M, E^n)$

Zhang Zhiyu
Wang Xiaoxia and Zhang Meiyu
(*North China Institute of Technology, Taiyuan, Shanxi, P. R. China*, 030008)

Abstract This paper by introducing four kinds of operations on continuous mapping from a topological space X to a n-dimensional Banach algebra E^n obtains the result that, for any subset $M \subset X$, the continuous mapping set $C(M, E^n)$ has the combinatorial algebraic properties of group, ring, module and Banach algebra. As far as the references the author can reach are concerned, these results look completely new.

Key words continuous mapping; group; ring; field; module; combinatorial algebra; Banach algebra.

1. Introduction

It is well-known that the study of the existence and uniqueness of the solution to a differential equaion can usually be transformed into that of the existence of a fixed point of a mapping on certain space. This space is generally considered to be the so-called continuous functional space $C(\Omega)$. Because of the varying problems, it is required that $C(\Omega)$ is not only a Banach space but also a Banach algebra sometimes (see paper[1−4]). The purposes of this paper are to build the additive, subtractive and multiplicative operations of continuous mapping from a general topological space X to a n-dimensional Banach algebra E^n, and for any subset $M \subset X$, to obtain the combinatorial algebraic results that, under a certain condition, the continuous mapping set $C(M, E^n)$ becomes a group, a ring, a module or a Banach algebra. These results are very important to the study of the solutions to differential equations (especially those in abstract space), integral equations or operator equations. As space is limited, however, the application in this respect will be discussed in another paper. This paper provides a definition called the multiplication on $C(X, E^n)$. Moreover, it defines division in passing, and establishes the theorem about the divisibility of the set $C(X, E^n)$.

2. Main Result

Unless a special statement is given, in this paper, X, Y are always supposed as the general topological spaces, E as a Banach algebra and

$$E^n = \prod_{i=1}^{n} E_i = \{(x_1, x_2, \cdots, x_n) \mid x_i \in E_i = E,\ i = 1, 2, \cdots, n\}$$

as a n -dimensional Banach algebra.

Definition 1 Suppose that f, g are mappings from X to E^n , write as

$$f(x)=(f_1(x),f_2(x),\cdots,f_n(x)),\text{for all } x\in X \tag{1}$$

and

$$g(x)=(g_1(x),g_2(x),\cdots,g_n(x)),\text{for all } x\in X \tag{2}$$

where $f_i, g_i : X\to E$ are mappings $(i=1,2,\cdots,n)$. Define

$$f\pm g: X\to E^n$$

by

$$(f\pm g)(x)=(f_1(x)\pm g_1(x),\cdots,f_n(x)\pm g_n(x)),\forall\ x\in X \tag{3}$$

and

$$f\cdot g: X\to E^n$$

by

$$(f\cdot g)(x)=(f_1(x)\cdot g_1(x),\cdots,f_n(x)\cdot g_n(x)),\forall\ x\in X. \tag{4}$$

Then, $f\pm g$ and $f\cdot g$ are called respectively the sum, the difference and the product of the mappings f and g . In the same way, we can define the quotient of f and g , which is denoted f/g . It goes if only we change " $\cdot$ " in (4) to "/", in which $f_i(x)\cdot g_i(x)$ is a product on E , $f_i(x)/g_i(x)=f_i(x)\cdot[g_i(x)]^{-1}$, $[g_i(x)]^{-1}$ is the multiplicative inverse of $g_i(x)$ in $E(i=1,2,\cdots,n)$.

Notation: In Definition 1, we often abbreviate respectively (1)—(4) to

$$f=(f_1,f_2,\cdots,f_n);\ g=(g_1,g_2,\cdots,g_n);$$
$$f\pm g=(f_1\pm g_1,f_2\pm g_2,\cdots,f_n\pm g_n);$$
$$f\cdot g=(f_1\cdot g_1,f_2\cdot g_2,\cdots,f_n\cdot g_n);$$

and similarly

$$f/g=(f_1/g_1,f_2/g_2,\cdots,f_n/g_n).$$

Theorem 1 If mappings $f,g:X\to E$ are continuous, then $f\pm g:X\to E$ is continuous. If E also satisfies multiplicative inequality, then $f\cdot g:X\to E$ is continuous, too.

Proof Prove first $f\pm g: X\to E$ is continuous. $\forall\ x_0\in X$ and $\varepsilon>0$, since $f,g:X\to E$ are continuous, there exist two neighbourhoods $U_1(x_0)$ and $U_2(x_0)$ such that

$$\|f(x)-f(x_0)\|<\frac{\varepsilon}{2},\ \text{for all } x\in U_1(x_0);$$

$$\|g(x)-g(x_0)\|<\frac{\varepsilon}{2},\ \text{for all } x\in U_2(x_0).$$

Thus, there exists a neighbourhood $U(x_0)\subset U_1(x_0)\cap U_2(x_0)$ of x_0 such that for all $x\in U(x_0)$,we always have

$$\begin{aligned}&\|(f(x)\pm g(x))-(f(x_0)\pm g(x_0))\|\\&=\|(f(x)-f(x_0))\pm(g(x)-g(x_0))\|\\&\leqslant\|f(x)-f(x_0)\|+\|g(x)-g(x_0)\|<\frac{\varepsilon}{2}+\frac{\varepsilon}{2}=\varepsilon.\end{aligned}$$

So, $f\pm g$ is continuous at $x_0\in X$. Because x_0 is an arbitrary element of X , $f\pm g:X\to E$ is continuous.

Prove next that $f\cdot g:X\to E$ is continuous. $\forall\ x_0\in X$ and $\varepsilon>0$, as $f,g:X\to E$ are continuous, we know there exists $U(x_0)$ such that for all $x\in U(x_0)$, the following three forms hold simultaneously

$$\|f(x)\|\leqslant 1+\|f(x_0)\|,$$

$$\|f(x)-f(x_0)\| < \frac{\varepsilon}{2(1+\|g(x_0)\|)},$$
$$\|g(x)-g(x_0)\| < \frac{\varepsilon}{2(1+\|f(x_0)\|)}.$$

Therefore,

$$\begin{aligned}&\|f(x)\cdot g(x)-f(x_0)\cdot g(x_0)\|\\ =\ &\|f(x)\cdot(g(x)-g(x_0))+(f(x)-f(x_0))\cdot g(x_0)\|\\ \leqslant\ &\|f(x)\|\|g(x)-g(x_0)\|+\|f(x)-f(x_0)\|\|g(x_0)\|\\ <\ &\frac{(1+\|f(x_0)\|)\varepsilon}{2(1+\|f(x_0)\|)}+\frac{\|g(x_0)\|\varepsilon}{2(1+\|g(x_0)\|)}<\varepsilon.\end{aligned}$$

So, $f\cdot g: X\to E$ is continuous at $x_0\in X$. As $x_0\in X$ is arbitrary, we know $f\cdot g: X\to E$ is continuous. The proof is complete.

Lemma 1[5] Let X, Y be general topological spaces and let mapping $f: X\to Y$ be continuous. If x_n, $x\in X$ and $x_n\to x(n\to\infty)$, then $f(x_n)\to f(x)(n\to\infty)$.

If X is also a first countable topological space, then mapping $f: X\to Y$ is continuous iff for every sequence $\{x_n\}$ converged to x in X, there always is the sequence $\{f(x_n)\}$ converged to $f(x)$ in Y.

Lemma 2[6] All the multiplicative invertible elements in a Banach algebra E set up an open set $G(E)$. If we define $\varphi(x)=x^{-1}$, $\forall\ x\in G(E)$, then the mapping $\varphi: G(E)\to G(E)$ is continuous.

Theorem 2 Given that X is a first countable topological space, E is a divisible Banach algebra (i.e. each non-null element of E has a multiplicative inverse) and the mapping $f: X\to E$ is continuous. If $f(x)\neq\theta$, $\forall\ x\in X$, and let $\bar{f}(x)=[f(x)]^{-1}$, $\forall\ x\in X$, then the mapping $\bar{f}: X\to E$ is continuous.

Proof Since X is first countable, E is divisible, and the mapping $f: X\to E$ is continuous with $f(x)\neq\theta$, $\forall\ x\in X$, we have that if x_n, $x\in X$ and $x_n\to x(n\to\infty)$, then $f(x_n)\to f(x)(n\to\infty)$ by Lemma 1. Since E is divisible, so $G(E)=E\backslash\{\theta\}$, also $f(x)\neq\theta, f(x_n)\neq\theta$, and hence, $f(x_n), f(x)\in G(E)$. By Lemma 2,

$$[f(x_n)]^{-1}\to[f(x)]^{-1}(n\to\infty).$$

That is

$$\bar{f}(x_n)\to\bar{f}(x)\ (n\to\infty).$$

Also, by Lemma 1, we know $\bar{f}: X\to E$ is continuous. The proof is complete.

Theorem 3. Suppose that X is first countable, the Banach algebra E satisfying multiplicative inequality is divisible, and mappings $f, g: X\to E$ are continuous. If $f(x)\neq\theta$, $\forall\ x\in X$, then the mapping $g/f: X\to E$ is continuous.

The proof can directly be given by Theorem 1 and Theorem 2. Omitted here.

Lemma 3 If the mapping $f: X\to E^n$ is

$$f=(f_1, f_2, \cdots, f_n),$$

then a necessary and sufficient condition of f being continuous is that each mapping $f_i: X\to E$ is continuous $(i=1,2,\cdots,n)$.

The proof is similar to paper[5]. Omitted.

Theorem 4 If continuous mappings $f, g: X\to E^n$ are $f=(f_1, f_2, \cdots, f_n)$, and $g=(g_1, g_2, \cdots, g_n)$. Then,

(i) the mapping $f\pm g: X\to E^n$ is continuous;

(ii) if E also satisfies multiplicative inequality, then the mapping $f \cdot g: X \to E^n$ is continuous;

(iii) if X is first countable, E is divisible and satisfies multiplicative inequality, also $f_i(x) \neq \theta$, $\forall\, x \in X$, then mapping $f/g: X \to E^n$ is continuous.

Proof Because the mappings $f, g: X \to E^n$ are continuous, and $f = (f_1, f_2, \cdots, f_n)$, $g = (g_1, g_2, \cdots, g_n)$, we know by Lemma 3 that f_i, $g_i: X \to E$ are all continuous $(i = 1, 2, \cdots, n)$. By the hypotheses of Theorem 4 and the conclusions of Theorem 1 and Theorem 3, we have $f_i \pm g_i$, $f_i \cdot g_i$ and $f_i/g_i: X \to E (i = 1, 2, \cdots, n)$ are all continuous. Hence, according to Definition 1 and Lemma 1, the mappings $f \pm g$, $f \cdot g$ and $f/g: X \to E^n$ are all continuous. The proof is complete.

Definition 2 Let $f: X \to E^n$ be a mapping from the topological space X to the n-dimensional Banach space E^n, where

$$f(x) = (f_1(x), f_2(x), \cdots, f_n(x)), \forall\, x \in X.$$

Let A be a number field. Then, λf is called a scalar product of the number λ and the mapping f, where

$$\lambda f(x) = (\lambda f_1(x), \lambda f_2(x), \cdots, \lambda f_n(x)), \text{ for all } \lambda \in A,\ x \in X.$$

For the mapping $f: X \to Y$, it is not difficult to know that a necessary and sufficient condition of $f: X \to Y$ being continuous is that for every subset $M \subset X$, $f: M \to Y$ is continuous. Hence, the above operation on the continuous mapping set $C(X, E^n)$ can be translated onto $C(M, E^n)$. From Definition 1, Definition 2 and Theorem 4, we can obtain the following combinatorial algebraic results about the continuous mapping set $C(M, E^n)$ from any subset $M \subset X$ to E^n.

Theorem 5 Let M be any subset of the topological space X, $E^n = \prod_{i=1}^{n} E_i (E_i = E, i = 1, 2, \cdots, n)$ be a n-dimensional Banach algebra, then

(i) $C(M, E^n)$ is an Abelian group, and also is a linear space on any number field A.

(ii) $C(M, E^n)$ is an exchangeable ring with null divisor, thus it is not a field.

(iii) $C(M, E^n)$ is an algebra on any number field A. Moreover, when M is a compact set and E satisfies multiplicative inequality, $C(M, E^n)$ becomes a Banach algebra satisfying multiplicative inequality with respect to the norm $\|f\|_2 = \sum_{i=1}^{n} \|f_i\|_1$, where $f = (f_1, f_2, \cdots, f_n)$ and $\|\cdot\|_1$ is any norm which makes $C(M, E)$ a Banach algebra satisfying multiplicative inequality.

(iv) $C(M, E^n)$ is a $C(M, E^n)$-module.

Proof In view of paper [7], it is easy to prove the conclusions (i), (ii) and (iv). Next, to prove the conclusion (iii). Judging by (i) and (ii), $C(M, E^n)$ is a linear space on any number field A, and a ring too. Moreover, on the basis of Difinition 1 and Definition 2,

$$(\lambda f) \cdot g = f \cdot (\lambda g) = \lambda (f \cdot g), \forall\, \lambda \in A; f, g \in C(M, E^n).$$

Thus, $C(M, E^n)$ is an algebra on any number field A.

In order to show that $C(M, E^n)$ is a Banach algebra with respect to the norm $\|\cdot\|_2$, it suffices only to prove that when M is a compact set, with respesct to $\|\cdot\|_2$, $C(M, E^n)$ is a Banach space, the left multiplication and the right muliplication defined by Definition 1 are continuous.

Because $M \subset X$ is a compact set and E is a Banach algebra satisfying multiplicative inequality, the norm $\|\cdot\|_1$, which makes $C(M, E)$ a Banach algebra satisfying multiplicative inequality, is in existence. For example

$$\|f\|_1 = \sup_{x\in M}\{\|f(x)\| \mid \|\cdot\| \text{ is } E\text{'s norm}\}$$

is in the case.

For any $f, g \in C(M,E^n)$, that is

$$f = (f_1, f_2, \cdots, f_n),\ g = (g_1, g_2, \cdots, g_n)$$

where $f_i,\ g_i \in C(M,E)(i = 1,2,\cdots,n)$. Since $C(M,E)$ satisfies multiplicative inequality with respect to the norm $\|\cdot\|_1$, then

$$\|f\cdot g\|_2 = \sum_{i=1}^{n}\|f_i\cdot g_i\|_1 \leqslant \sum_{i=1}^{n}\|f_i\|_1\|g_i\|_1$$
$$\leqslant (\sum_{i=1}^{n}\|f_i\|_1)(\sum_{i=1}^{n}\|g_i\|_1) = \|f\|_2\|g\|_2.$$

Thus, $C(M,E^n)$ satisfies multiplicative inequality with respect to the norm $\|\cdot\|_2$.

Again given f^n, $g \in C(M,E^n)$, also $\|f^n - f\|_2 \to 0(n\to\infty)$, since $C(M,E^n)$ satisfies multiplicative inequality with respect to the norm $\|\cdot\|_2$, then

$$\|f^n\cdot g - f\cdot g\|_2 = \|(f^n - f)\cdot g\|_2$$
$$\leqslant \|f^n - f\|_2\|g\|_2 \to 0(n\to\infty).$$

Thus, left multiplication is continuous on $C(M,E^n)$. In the same way, right multiplication can be proved to be continuous, too.

Because $C(M,E)$ is a Banach space with respect to the norm $\|\cdot\|_1$, according to the structure of $C(M,E^n)$ and the definition of $\|\cdot\|_2$, similar to paper [6], we can prove that $C(M,E^n)$ is also a Banach space with respect to the norm $\|\cdot\|_2$.

To sum up, we have proved that $C(M,E^n)$ is a Banch algebra satisfying multiplicative inequality with respect to the norm $\|\cdot\|_2$. The proof is complete.

References

[1] R. Reissig, G. Sansone, R. Conti, *Nonlinear Differential Equation of Higher Order*, Noordhoff international publishing leydan, 1974.

[2] T. A. Burton, *Volterra Integral and Differential Equations*, Academic Press, New York, 1983.

[3] V. Lakshmikantham, S. Leela, V. Moauro, Existence and Uniqueness of Solutions of Delay Differential Equations on a Closed Subset of a Babach Space, *Nonlinear analysis*, 2(1978), 311—327.

[4] R. W. Leggett, On Centain Nonlinear Integral Equations, *J. Math. Anal. Appl.*, 57(1977), 462—468.

[5] J. R. Munkres, *Topology: A First Course*, Prentice-Hall, 1975.

[6] Ding Guanggui, *Introduction to Banach Space*, Science Press, 1984.

[7] L. C. Grove, *Algebra*, Academic Press, 1983.

A Note on Win's Conjecture

Zhao Bingxin

(*Economic Management Department, Shandong University, Jinan, China, 250100*)

Abstract

S.win posed the following conjecture in paper [2]: Let G be a graph of order 2n. If $d_G(u) + d_G(v) = 2n + k$ for all pairs of nonadjacent vertices u and v of G, then G contains $k + 2$ disjoint 1-factors. In this paper, we give a further conjectue on this topics: let G be a connected graph of order 2n. If $d_G(u) + d_G(v) = 2n + k$ for all pairs of vertices u and v which are at distance two. Then G contains $k + 2$ disjoint 1-factors. We prove it when $k = 1$ and improve the win's result.

Key words: hamilton cycle, 1-factor, connected graph.

1 Introduction

In this paper, we only consider finite simple graph. Let G be a graph with vertex set $V(G)$ and edge set $E(G)$. For a vertex v of G, we denote by $d_G(v)$ the degree of v in G. The number of odd components of G is denoted by $\omega(G)$, let $[S, T]$ denote the set of edges joining the vertex sets S and T of G. A spanning subgraph H of G is called a k-factor if $d_H(v) = k$ for every vertex v of H. A k-factor H of G is called a connected k-factor if H is connected. In paper [2], Win get the following result.

Theorem 1 *Let G be a graph of order $2n$. If $d_G(u) + d_G(v) \geq 2n + 1$ for all pairs of nonadjacent vertices u and v of G, then G contains a hamilton cycle C and a 1-factor F, both of which are edge disjoint.*

Furthermore, Win give the following conjecture.

Conjecture 1 *Let G be a graph of order $2n$. If $d_G(u) + d_G(v) \geq 2n + k$ for all pairs of nonadjacent vertices u and v of G. Then G contains $k + 2$ disjoint 1-factors.*

Obviously, the conjecture holds when $k = 1$. In this paper, we get the following result.

Theorem 2 *Let G be a connected graph of order* $2n$, *if* $d_G(u) + d_G(v) \geq 2n + 1$ *for all pairs of vertices u and v which are at distance two. Then G contains a hamilton cycle C and a l-factor F, both of which are edge disjoint.*

According to the above theorem, we may get the following result easily.

Corollary 1 *If G satisties the same conditions of the above theorem, then G contains a connected 3-factor.*

According to the win's conjecture, we give the following conjecture.

Conjecture 2 *Let G be a graph of order* $2n$, *if* $d_G(u) + d_G(v) \geq 2n + k$ *for all pairs of vertices u and v which are at distance two. Then G contains* $k+2$ *disjoint 1-factors.*

It is obvious that the theorem improves theorem 1, conjecture 1 would hold if conjecture 2 hold and conjecture 2 hold when $k = 1$.

2 Proof of Theorem 2

Let G be a graph which satisfies the conditions of theorem 2, then G contains hamilton cycles. Assume that theorem 2 is not true, then for any hamilton cycle C of G there is no 1-factor in $G - E(C)$. *Let C be a hamilton cycle of G,* $G' = G - E(C)$. *From the Tutte's 1-factor theorem, there exists a nonempty set S of* $V(G')$ *satisfying* $\omega(G' - S) \geq s + 1$, *where s denotes the size of S. In fact, we may get that* $\omega(G' - S) \geq s + 2$ *when the order of G is a even number. We choose a nonempty vertex set S of G', such that* $\omega(G' - S) \geq s + 2$ *and the number of components of* $G' - S$ *is as large as possible. We shall denote the odd components of* $G' - S$ *by* $H_1, H_2, \cdots, H_\omega$, *and the even components of* $G' - S$ *by* $H_{\omega+1}, H_{\omega+2}, \cdots, H_{\omega+\lambda}$. *The size of the set* $V(H_i)$ *will be denoted by* $h_i, i = 1, \cdots, \omega + \lambda$. *Let* S_i *be the subset of S, whose elements are joined to at lease one vertex of* $V(H_i), i = 1, \cdots, \omega + \lambda$. *For a subset X of* $V(G)$, $\partial_G(X)$ *denotes the edge set* $[X, V(G) - X)]$. $\partial_c(X)$ *will be defined similarly.*

Lemma 1 Lemma 1 *a)* $S = S_1 \cup S_2 \cup \cdots \cup S_{\omega+\lambda}$,

b) For any nonempty subset X of $V(G)$, $|\partial_c(X)| \geq 2$.

From the choice of the subset S of $V(G)$ *and G contains hamilton cycle we may prove it easily.*

Lemma 2 *There exist two vertices* x_i *of* $V(H_i)$ *and* x_j *of* $V(H_j)$ *such that the distance between the two vertices is two.*

Proof: *Suppose that the lemma is false. We only need to consider the following cases.*

Case 1. There exists a component H_α such that $h_\alpha \geq 3$. Let A be the union set of all vertices of $V(G'-S)$ except $V(H_\alpha)$, then it is obvious that $[A, V(h_\alpha)]$ is an empty set. (Otherwise, there exist a vertex x of A and a vertex x_α of $V(H_\alpha)$ such that $xx_\alpha \in E(G)$, according to the hypothesis, we may get that $|[\{x\}, V(H_\alpha)]| = h_\alpha \geq 3$, so $2 = |\partial_c(X)| \geq 3$, contradiction). Similarly, we may get that $S_i \cap S_\alpha$ is an empty set for $1 \leq i \neq \alpha \leq \omega + \lambda$. Thus there exists a vertex y_j of S_j, which is adjacent to a vertex y_α of S_α, let y_j is joined to vertex x_j of $V(H_j)$, then the vertices x_j and y_α are at distance two. So we may have $2n+1 \leq d_G(x_j) + d_G(y_\alpha) \leq d_{G'}(x_j) + 2 + d_G(y_\alpha) \leq h_j - 1 + 2 + h_\alpha + s - 1 + 2$, i.e., $h_j + h_\alpha \geq 2n - 2s + 2$. Hence, $\omega - 2 \leq \sum_{k \neq j,\alpha} h_k \leq 2n - s - (h_j + h_\alpha) \leq 2n - s - 2n + 2s - 2 \leq s - 2 \leq \omega - 4$, a concradition.

Case 2. For any $i, h_i \leq 2, i = 1, \cdots, \omega + \lambda$, then all odd components contain only one vetex. We only need to consider the following two subcases.

Subcase 2.1. There is a compent H_α satisfying $[H_\alpha, H_i]$ is an empty set for all of the other components H_i of $G' - S (1 \leq i \neq \alpha \leq \omega + \lambda)$. The proof would be found in the proof of case 1.

Subcase 2.2. For any component H_i of $G' - S$, there exists another component H_j satifying that $[V(H_i), V(H_j)]$ isn't empty. From this we may get that there exists at least one element in $\{H_i, H_j\}$ which contains only one vertex (otherwise, $h_i \geq 2, h_j \geq 2$, because there are not vertices x_i of $V(H_i)$ and x_j of $V(H_j)$ which are at distance two, so $|[H_i, H_j]| = h_i h_j$. let $Z = V(H_i) \cup V(H_j)$ then $\partial_c(Z)$ is empty, a contradiction). We assume $h_i = 1$. Then based on first hypothesis, we may get that both $[H_i \cup H_j, H_\alpha]$ and $S_\alpha \cap (S_i \cup S_j)$ are empty sets for $1 \leq \alpha \neq i, j \leq \omega + \lambda$. Then there exists a vertex y_k of S_k which is adjacent to a vertex y of $S_i \cup S_j$. Assume that the vertex x_k of $V(H_k)$ is joined by y_k in G, then x_k and y are at distance two. Therefore $2n + 1 \leq d_G(x_k) + d_G(y) = d_{G'}(x_k) + 2 + d_G(y) \leq h_k - 1 + s - 1 + h_i + h_j + s - 1 + 2 \leq h_k + h_j + 2s$, $h_k + h_j \geq 2n - 2s + 1$. So $\omega - 2 \leq \sum_{t \neq k,j} h_t \leq 2n - s - (h_k + h_j) \leq 2n - s - 2n + 2s - 1 + 2 = s - 1 \leq \omega - 3$, a contradition.

Corollary 2 *There exist two integers i and $j (1 \leq i \neq j \leq \omega + \lambda)$ satisfying $\sum_{k \neq i,j} h_k \leq \omega - 1$.*

In fact, based on the Lemma 2, we know that there exist two vertices x_i of $V(H_i)$ and x_j of $V(H_j)$ which are at distance two. So, $2n + 1 \leq d_G(x_i) + d_G(x_j) = d_{G'}(x_i) + d_{G'}(x_j) + 4 \leq h_i - 1 + h_j - 1 + 4 + 2s$, i.e., $h_i + h_j \geq 2n - 2s - 1$. Therefore, $\sum_{h_k} h_k = 2n - s - (h_i + h_j) \leq 2n - s - 2n + 2s + 1 = s + 1 \leq \omega - 1$.

Corollary 3 *There exists at least one odd component H_i satisfying $h_i = 1$.*

Otherwise, $h_i \geq 3$ *for* $1 \leq i \leq \omega$*,then there exist two integers* i *and* j *satisfying* $3(\omega - 2) \leq \sum_{k \neq i,j} h_k \leq \omega - 1$*, i.e.,* $2\omega \leq 5$*,So on the one hand,*$\omega \leq 2$*, on the other hand,* $\omega \geq s + 2 \geq 3$*.There is a contradiction.*

Without lossing the generalily, we suppose that $h_1 = h_2 = \cdots h_{\omega_1} = 1$*, then we may have*

Corollary 4 *a)*$0 \leq \omega - \omega_1 \leq 2 - \lambda$
b)$0 \leq \lambda \leq 1$ *if* $\omega = \omega_1$

Proof: *a) Based on the corrollary 1, there exist two integers* i *and* $j(1 \leq i \neq j \leq \omega + \lambda)$ *satisfying that*

$$\omega_1 + 3(\omega - \omega_1 - 2) + 2\lambda \leq \sum_{k \neq i,j} h_k \leq \omega - 1, i.e., 2(\omega - \omega_1 + \lambda) \leq 5$$

So $\omega - \omega_1 + \lambda \leq 2$*. Therefore,* $0 \leq \omega - \omega_1 \leq 2 - \lambda$*.*

Similarly we may prove b).

Based on the corrollary 3, we only require considering the following cases when there are no 1-factors contained in G*.*

Case A $\omega = \omega_1, \lambda = 0$.
Case A_2 $\omega = \omega_1, \lambda = 1$.
Case B_1 $\omega = \omega_1 + 1, \lambda = 0$.
Case B_2 $\omega = \omega_1 + 1, \lambda = 1$.
Case C $\omega = \omega_1 + 2, \lambda = 0$.

Lemma 3 *The case* A_1 *can not occur when there are no 1-factors contained in* G'*.*

Proof: *Assume that the case* A_1 *can occur. Then* $s = n - 1, \omega = \omega_1 + 1$*. In fact, based on the Lemma 2, we may have that there exist two vertices* x_i *of* $V(H_i)$ *and* x_j *of* $V(H_j)$ *whose distance is two. So,* $2n + 1 \leq d_G(x_i) + d_G(x_j) \leq 2s + 4$*,* $s \geq n - 1$*,* $\omega \geq s + 2 \geq n + 1$*. On the other hand,* $2s + 2 \leq \omega + s \leq 2n$*, so,* $s \leq n - 1$*, therefore,* $s = n - 1$*, and* $n + 1 \leq \omega \leq 2n - s \leq n + 1$*, i.e.,* $\omega = n + 1$*. Let* $G^* = G[V(G) - S]$*,* $V(H_i) = \{x_i\}$*,* $1 \leq i \leq n + 1$*, then* $1 \leq \delta(G^*) \leq \Delta(G^*) \leq 2$*. If not, there exists a vertex* x_i *of* $V(G)$*,* $d_{G^*}(x_i) = 0$*, so for any* $k \neq i$*,* $d_G(x_i) + d_G(x_k) \leq s + s + 2 \leq n - 1 + n - 1 + 2 = 2n$*, i.e, the distance of any two vertices* x_i *and* x_k *does not equal two for* $1 \leq k \neq i \leq n+1$*. Furthmore,* $S_i \cap S_k$ *is empty for all* $k \neq i$*. Because* G *is connected, there exists a vertex* y_j *of* S_j*, satisfying* $y_i y_j \in E(G)$*. So, the two vertices* x_j *and* y_i *are at distance two. Therefore,* $2n + 1 \leq d_G(y_i) + d_G(x_j) \leq s + s + 2 = 2n$*, a contradiction.*

Let $V_r = \{x \in V(G^*) | d_{G^*}(x) = r\}$*,* $r = 1, 2$*, then we may get the following conclusions:*

(1) The distance of any two vertices contained in V_1 doesn't equal two. (because for any two vertices x_i and x_j of V_1, $d_G(x_i) + d_G(x_j) \le s + s + 1 = 2n$)

(2) There are no cycles contained in $G[V_2]$. If not, let Z be the set of $V(C)$, then $\partial_c(Z) = \emptyset$.

(3) There are no edges contained in $G[V_1]$. Otherwise, if there is one edge $x_l x_m$ contained in $G[V_1]$, then $[\{x_l, x_m\}, V(G_1^*) - \{x_i, x_j\}] = \emptyset$ and $|S_1 \cup S_m| \ge 2$. From $\omega = \omega_1 \ge 3$, we may get $|V_1| \ge 4$. Furthermore, Based on the conclusion (1), we may get that $S_r \cap (S_1 \cup S_m) = \emptyset$ for any vertex of $x_r \in V_1, r \ne 1, m$. So for any vertex x_j of V_2, the distance of x_j and x_1 (or x_m) does not equal two. (because $d_G(x_e) + d_G(x_j) \le 1 + s - 1 + s + 2 = 2n$ for any $x_j \in V_2$). Therefore $S_j \cap (S_l \cup S_m) = \emptyset$, for any $x_j \in V_2$. Based on the choice of S, and G being connected, we may have that there exists an integer $k \ne l, m, [S_k, S_l \cup S_m] = \emptyset$, let $y_k y_0 \in [S_k, S_l \cup S_m]$, then the vertices y_0 and x_k are at distance two. So $2n + 1 \le d_G(y_0) + d_G(x_k) \le s + 2 - 1 + s - 2 + 2 = 2n - 1$, a contradiction.

Let V_1 be the set $\{x_1, x_2, \cdots, x_{2m}\}$, V_2 be the set $\{x_{2m+1}, x_{2m+2}, \cdots, x_{n+1}\}$, then the sets $S_1, S_2, \cdots, S_{2m}$ are not intersection each other, and $|V_1| \ge 2$, $|V_2| \ge 2$. So $d_G(x_i) \le 1 + s - 1 = s = n - 1, 1 \le i \le 2m, d_G(x_j) \le s + 2 = n + 1$. Therefore, $d_G(x_i) + d_G(x_j) \le n - 1 + n + 1 = 2n$. Thus, the vertices x_i and x_j and are not at distance two, i.e.,there are not vertices x_i of V_1 and x_j of V_2 whose distance equal two, the connected compoments of G^* have to be the path P_3, contradiction to the conclusion(1).

Lemma 4 *The cases A_2, B_1, and C cannot emerge when there are no 1-factors contained in G'.*

We only prove the case A_2, the proofs of other cases may be found in the proof of case A_2.

Assume that the case A_2 can emerge .According to the discussing same as the above, we may get that any two vertices x_i and x_j are not at disdance two for $1 \le i \ne j \le \omega$.

So there exists one odd component H_i satisfying that the two vertices x_i and $x_\omega + 1$ of $H_{\omega+1}$ are at distance two. Let $\alpha_{\omega+1}$ denote the number of vertices contained in S which are not adjacent to $x_{\omega+1}, \alpha_i$ denote the number of vertices contained in S which are not adjacent to $x_i, \beta_{\omega+1}$ denote the number of vertices contained in $H_{\omega+1}$ which do not adjacent to $x_{\omega+1}$. Then $2n + 1 \le d_G(x_{\omega+1}) + d_G(x_i) \le d_{G'}(x_{\omega+1}) + d_{G'}(x_i) + 4 \le s - \alpha_i + h_\omega + 1 - \alpha_{\omega+1} + s - \beta_{\omega+1} - 1 + 4$, i.e., $h_{\omega+1} \ge 2n + 1 - 2s + (\alpha_i + \alpha_{\omega+1} + \beta_{\omega+1}) - 3$. So, $\omega = \omega_1 = \sum_{k \ne \omega+1} = 2n - 2s - h_{\omega+1} \le 2n - s - 2n + 2s - (\alpha_i + \alpha_{\omega+1} + \beta_{\omega+1} + 2 = s + 2 - (\alpha_i + \alpha_{\omega+1} + \beta_{\omega+1})$, therefore, $\alpha_i + \alpha_{\omega+1} + \beta_{\omega+1} \le 0$. Thus,$\alpha_i = \alpha_{\omega+1} = \beta_{\omega+1} = 0$, i.e., $|[x_i, S]| = s$, $|[x_{\omega+1}, S]| = s$.

We only need to discuss the following cases.

Case 1. There exists a vertex $x_j, x_i x_j \in E(G)(1 \le j \ne i \le \omega)$*, then there exists a vertex* $x \in S \cup H_\omega$ *and for any vertex* $x_r (1 \le r \ne i, j \le \omega)$, $[\{x_r\}, H_i \cup H_j \cup S] = \emptyset$*. So there is an odd component* $H_t = \{x_t\}$ *satisfying that* $[H_t, H_{\omega+1}] \ne \emptyset$*, furthermore,* $|[H_t, H_{\omega+1}]| = h_{\omega+1}$*. Therefore, we get* $h_{\omega+1} = 2$ *from* $|\partial_c(x_t)| = 2$*. Let B be the vertex set of* $\bigcup_{i=1, i\ne t}^{\omega} V(H_i)$*, then we may prove that* $[H_t, B] = \emptyset$*. Suppose that* $x_t x_r \in [H_t, B]$*, i.e.,* $x_t x_r \in E(G)$*. Because of* $d_G(x_r, x_{\omega+1}) \ne 2$, $[\{x_r\}, \{x_i, x_j\}] = \emptyset$*, the set* $[\{x_r\}, H_{\omega+1}] \ne \emptyset$*, so* $|[\{x_r\}, H_{\omega+1}]| = h_{\omega+1} = 2$*. Therefore* $|\partial_c(x_r)| = 3$*, a contradiction. Thus,* $d_G(x_t) = 2$*. It is obvious that there is a vertex y of S satisfying that* $d_G(x_t, y) = 2$ *, so* $2n + 1 \le d_G(x_t) + d_G(y) \le 2 + 2n - 2 = 2n$*, a contradiction.*

Case2. For any odd component $H_r = \{x_r\}, [H_r, H_i] = \emptyset$*. Then* $[H_r, S] = \emptyset$*. Because G is connected,* $[H_r, H_{\omega+1}] \ne \emptyset$*, for any vertex* $x_{\omega+1}$ *of* $H_{\omega+1}, d_G(x_r, x_{\omega+1}$ $\ne 2$*. So* $|[x_r, H_{\omega+1}]| = h_{\omega+1} \ge 2$*. Let Z be the set of* $\bigcup_{j=1, j\ne i}^{\omega} V(H_j)$*, then* $Z \ne \emptyset$ *and* $\partial_c(Z) = \emptyset$*, a contradiction.*

References

[1] J.A.Bondy and U.S.R.Murty, Graph Theory with Applications, American Elsevier, NewYork,1976.

[2] Sein.Win, A Sufficient Condition for a Graph to Contain Three Disjoint 1-factors, J. of Graph Theory, No.6(1982), 489-492.

Construction of Circulant Graphs with the Best Connectivity

Zhou Yongsheng

(Guangdoug Institute for Nationalities Guangzhou 510633)

ABSTRACT It is known in document [1] that determing the sufficient and necessary condition for a connected circulant graph which has the best connectivity is very difficult. In this paper, after getting some natures about connected circulant graphs relating to connectivity, we deduce the sufficient and necessary condition of circulant graphs with the best connectivity. Thereby the mathod of constructing the graphs with the best connectivity has been obtained and the method of constructing connected graph with m - connectivity has been also (m is a number which is smaller than the smallest degree of vertex of the graphs).

Key words circulant graphs atomic part atomic numbers

1. DEFINITIONS AND LAMMAS

Definition 1. Let G be a connected graph, the number

$$K(G) = \min \{ |T| ; T \text{ is vertex cut of } G \}$$

is called vertex connectivity.

Let S (G) = {T: | T | = K (G), T is vertex cut of G}, a subgraph P of G which is a component of G [V (G) - C] for some C ∈ S (G) is called a part of G with respect to C, Each vertex in C is adjacent to some vertex of p and some vertex in V (G) - (C∪ v (P)).

Definitioa 2. If a part P has

$$|V(P)| = \min \{ \min \{ |V(H)| ; H \text{ is a part with respect to } C \} ; C \in S(G) \},$$

then the part P is called an atomic part of G.

Definition 3. If K (G) = δ (G) (the smallest degree of vertex of G) for graph G, then graph G is called the graph with the best connectivity.

Lemma 1 [2]. Distinct atomic parts in a connected graph are disjoint.

Lemma 2 [2]. Let G be a vertex - transitive graph and suppose 0 < K (G) < ρ (G) (ρ (G) denotes the degree of each vertex), let P is a atomic part of G, then

(1) P is a vertex transitive graph.

(2) G is isomorphic to a disjoint union of two or more copies of P together with some edges joining them.

Lemma 3 [2]. Let G be vertex-transitive graph and suppose $0 < K(G) < \rho(G)$, then

$$K(G) = m\,|V(P)|$$

for some integer $m \geq 2$

Lemma 4 [2] If G is connected vertex-transitive graph then $K(G) \geq \frac{2}{3}\rho(G)$.

Because circulant graph must be vertex-transitive, lemma 2, 3, 4 must be true after vertex - transitive graphs in lemma 2, 3, 4 are changed into circulant graph.

2. NATURES RELATING TO CONNECTIVITY OF CONNECTED CIRCULANT GRAPH

Theorem 1. Let G be a cnnected circulant graph and suppose $0 < K(G) < \rho(G)$. Let P be an atomic part of G and $|V(P)| = s$. Then G has $\frac{n}{s}$ atomic parts and these $\frac{n}{s}$ atomic parts are isomorphic circulant graphs for each other, their vertex labels are

$$\{h, h+\frac{n}{s}, h+2\frac{n}{s}, \cdots, h+(s-1)\frac{n}{s}\}(h=0,1,\cdots,\frac{n}{s}-1)\}$$

Proof. Let $V(P) = \{P_1, P_2, \ldots, P_s\}$, label the vertices of circulant graph G so that its abjacent matrix is circulant. Suppose this labeling method labels vertex $P_1, P_2, \cdots P_s$ in G with $t_1, t_2, \cdots, t_s$ respectively, not losing generality, suppose $t_i < t_{i+1}$, therefore

$$t_i - t_{i-1} = t_{i+1} - t_i \quad (\text{i is moded by s})\ 2 \leq i \leq s$$

otherwise

$$t_k - t_{k-1} \neq t_{k+1} - t_k \qquad 2 \leq k \leq s$$

Thus s vertices labeled with

$$t_1 + (t_{k+1} - t_k),\ t_2 + (t_{k+1} - t_k), \cdots, t_{k-1} + (t_{k+1} - t_k),\ t_k + (t_{k+1} - t_k), \cdots t_s + (t_{k+1} - t_k)$$

forms a vertex set of another atomic parts of G. But

$$t_k + (t_{k+1} - t_k) = t_{k+1},\ t_{k-1} + (t_{k+1} - t_k) \neq t_k$$

This means this atomic part has some same vertices as $V(P) = \{P_1, P_2, \cdots, P_s\}$. This is contradictory with lemma 1. So S vertex labels of P can be represented by

$$h,\ h+\frac{n}{s}, h+2\frac{n}{s}, \cdots, h+(s-1)\frac{n}{s}$$

According to natures of circulant graph, G has $\frac{n}{s}$ atomic parts and their labels are

$$V_h(P)=\{h,h+\frac{n}{s},h+2\frac{n}{s},\cdots,h+(s-1)\frac{n}{s}\} \quad h=0,1,\cdots,\frac{n}{s}-1$$

By definition of circulant graph, the sufficient and necessary condition that vertex $h + i_1\frac{n}{s}$ is adjacent to vertex $h + i_2\frac{n}{s}$ is vertex $h + (i_1 + t)\frac{n}{s}$ is adjacent to vertex $h + (i_2 + t)\frac{n}{s}$ $(0 \leqslant i_1 \leqslant i_2 \leqslant s - 1)$. So vertex-induced subgraph P of G is circulant graph. #

Let $G = C_n < j_1, j_2, \cdots, j_r >$ be a connected circulant graph and P be an atomic part of G with respect to C for C∈S (G). Graph $H_1[V(P) \cup C]$ represents a vertex - induced subgraph whose vertices $V(P) \cup C$ is in G. $H_2[C]$ represents a vertex-induced subgraph whose vertices C is in G. Graph $H = H_1 - E(H_2)$ is constructed by P and connecting vertices C with some edges of P in G. Suppose $|V(P)| = s$, $P = \{P_1, P_2, \cdots, P_s\}$, label of vertex P_i is denoted by b_i, and let $b_{i+1} > b_i$ according to theorem 1, substraction between label of vertex $P_{i+1}$and P_i is equal to $\frac{n}{s}$ for all i. If there exists t vertices with lables between b_i and $b_{i+1}$for vertex P_i and P_{i+1}, then we call that there are t vertices between vertex P_i and P_{i+1}.

Theorem 2. There must be $\frac{|c|}{s}$ vertices between vertex P_i and $P_{i+1}$of graph H. If label of vertex P_i is denoted by b_i and $\frac{|c|}{s} = t$ vertex labels are labels are denoted by $a_1, a_2 \cdots, a_t$ $(a_1 < a_2 < \cdots < a_t)$ then

$$b_{i+1} - a_{i+1-h} = a_h - b_i \quad | \quad (h = 1,2,\cdots,t)$$

Proof: Notlosing generality, suppose labels of vertex $b_1, b_2, \cdots, b_s$ are $0, \frac{n}{s}, \frac{2n}{s}, \cdots, \frac{(s-1)n}{s}$ respectivily. Suppose that there are 1 vertices between vertex 0 and $\frac{n}{s}$ and their labels are denoted by $a_1', a_2', \cdots, a_l'$ $(a_1' < a_2' < \cdots < a_l')$. If a_h' $(1 \leq h \leq l)$ is adjacent to 0, then

$\frac{n}{s} \pm a'_h$ is adjacent to $\frac{n}{s}$, $\frac{2n}{s} - a'_h$ to $\frac{2n}{s}$ If a_h is disadjacent to 0, then there must exist a k so that a'_k in adjacent to $k\frac{n}{s}$. Thus vertex $a'_h + \frac{n}{s}$ is adjacent to $(K+1)\frac{n}{s}$. Therefore there must be vertex $a'_1 + \frac{n}{s}$, $a'_2 + \frac{n}{s}$, $\cdots$, $a'_L + \frac{n}{s}$ between vertex $\frac{n}{s}$ and $\frac{2n}{s}$, As same as above, if there are L vertices between vertex $\frac{n}{s}$ and $\frac{2n}{s}$, then there are 1 vertices between vertex 0 and $\frac{n}{s}$ also and between vertex P_i and P_{i+1}, so there are $l = \frac{|c|}{s} = t$ vertices between vertex P_i and P_{i+1}.

If vertex a_h ($1 \leqslant h \leqslant t$) is adjacent to vertex $b_i = (i-1)\frac{n}{s}$, because H is a subgraph of circulant graph G and ρ (H) = Δ (G), then according to natures of circulant graph, it is known that vertex a_k ($1 \leqslant k \leqslant t$) in vertex $a_1, a_2, \cdots, a_t$ is adjacent to vertex $b_{i+1} = i\frac{n}{s}$ and $a_h - b_i = b_{i+1} - a_k$. If vertex a_h and vertex $b_i = (i-1)\frac{n}{s}$ are not adjacent, because a_h ($1 \leqslant h \leqslant t$) is cut vertex there must exist k so that vertex a_h and vertex $k\frac{n}{s}$ are adjacent. There must exist one vertex a_k which is adjacent to vertex $\frac{ln}{s}$ in among vertex $a_1, a_2, \cdots, a_t$ and $a_k - \frac{ln}{s} = \frac{kn}{s} - a_h \pmod{n}$. According to results discussed as adove, it is known that for any vertex a_h in among vertex $a_1, a_2, \cdots, a_t$, there exists one vertex a_k in $a_1, a_2, \cdots, a_t$ so that $a_h - b_i = b_{i+1} - a_R$ or $a_R - \frac{ln}{s} = \frac{kn}{s} - a_h \pmod{n}$. Thus $R = t + 1 - h$. So $b_{i+1} - a_{t+1-h} = b_i - a_h$ ($h = 1, 2, \cdots, t$). #

Theorem 3. If | C | < ρ (G), then each vertex of C in a graph H has at least 2 degrees.

Proof: P is an atomic part of G. So it is a circulant graph. Let $p = C_s < i_1, i_2, \cdots, i_b > \alpha \leqslant b \leqslant [\frac{s-1}{2}]$) and V (P) = $\{P_1, P_2, \cdots, P_s\}$. If label of P_1 is o, then labels of $P_2, P_3, \cdots, P_s$

are $\frac{n}{s}, \frac{2n}{s}, \cdots, \frac{(s-1)n}{s}$ respectively. Suppose the labels of $t = \frac{|C|}{s}$ vertices between vertex o and $\frac{n}{s}$ are denoted by $a_1, a_2, \cdots, a_t$. If vertex with label 0 is not adjacent to vertex with label a_i ($1 \leqslant i \leqslant t$), then there must exist at least 2 vertices in vertices with label $\frac{n}{s} + a_i, \frac{2n}{s} + a_i, \cdots, [\frac{s-1}{2}] \frac{n}{s} + a_i$ and these two vertices are adjacent to vertex with label 0. Otherwise $\rho(G) \leqslant |C| - s + 2 + s - 1 - 1 =. |C|$. This is contradictory with assumption of this theorem. If vertex $\frac{xn}{s} + a_i$ and $\frac{yn}{s} + a_i$ ($1 \leqslant x < y \leqslant [\frac{s-1}{2}]$) are both adjacent to vertex 0, then vertex a_i is adjacent to vertex $n - \frac{xn}{s}$ and $n - \frac{yn}{s}$. So vertex a_i has at least two degrees.

If vertex with label 0 is adjacent to vertex with label a_i, then there must exist as least one vertex in vertices with label $\frac{n}{s} + a_i, \frac{2n}{s} + a_i, \cdots, [\frac{(s-1)}{2}] \frac{n}{s} + a_i$ and it is adjacent to vertex with label 0. Otherwise $\rho(G) \leqslant |C|$. This is contradictory with supposion of this theorem. By th same proof. method as above, we can prove that vertex a_i has at least two degrees.

So each vertex of C in graph H has at least two degrees.

3. SUFFICIENT AND NECESSARY CONDITION FOR CIRCULANT GRAPHS WITH THE BEST CONNECTIVITY

Let G is a connected circulant graph and $K_1(G), K_2(G), \cdots, K_d(G)$ be all imprime integers. between $\frac{2}{3}\rho(G)$ and $\rho(G)$, If λ_g is a number which satisfies

$$\left\{\begin{array}{l} \lambda_g | n \\ K_g(G) = m_g \lambda_g \, (m \geq 2,\ \mathrm{g} = 1,2,\cdots,\mathrm{d}) \\ \lambda_g + K_g(G) > \rho(G) \\ when\ \rho(\mathrm{G})\ \text{is odd}\ .\ \lambda_g + K_g(G)\ is\ even \end{array}\right\}$$

then λ_g is called an meat atomic number of G.

Let $\lambda_{g.1}, \lambda_{g.2}, \cdots, \lambda_{g.2(g)}$ be all meat atomic numbers of G with respect to C_g (C_g is vertex cut of G. $|C_g| = K_g(G)$).

Theorem 4. If $G = C_n < j_1, j_2, \cdots, j_r >$ is a Connected graph. $\rho(G) < n-1$, $\lambda_{g.i}$ ($g = 1, 2, \cdots q$, $i = 1, 2, \cdots, \partial(g)$) arc all meat atomic numbers of G, $b_1, b_2, \cdots, b_r$ is a Permutation of $\{1, 2 \cdots r\}$.. then Connectivity $K(G) = \rho(G)$ if and only if at least one expression in each following group of expressions does not hold.

$$\left\{\begin{array}{l} J_{b_f} = \dfrac{n}{\lambda_{g,i}} \dfrac{2n}{\lambda_{g,i}} \cdots, or \left[\dfrac{\lambda_{g,i}-1}{2}\right] \\ (f = 1,2\cdots t, 1 \leq t \leq \left[\dfrac{\lambda_{g,i}-1}{2}\right]) \\ J_{b_{t+2\beta+1}} \pm J_{b_{t+2\beta+z}} = 0\,(\mathrm{mod}\,\dfrac{n}{\lambda_{g,i}})(\beta = 0,1,\cdots\left[\dfrac{r-t}{2}\right]-1, amd\ \text{if}\ \mathrm{b}_{t+2\beta+s} > r, then\ \mathrm{b}_{t+2\beta+2} \in \\ g = 1,2,\cdots\beta, i = 1,2,\cdots,\partial(g). \qquad \{\mathrm{b}_{t+1}, b_{t+2}, \cdots, b_r\}) \end{array}\right. \tag{A}$$

Proof : Now we prove sufficient condition

Let K (G) be connectivity of G and P be an atomic part of G. By lemma 4, we know that K (G) $> \frac{2}{3}\rho(G)$. Suppose that $K_1(G), K_2(G), \cdots, K_d(G)$ are all imprime integers between $\frac{2}{3}\rho(G)$. and $\rho(G)$, C_g is a vertex cut with respect to $K_g(G)$.

If $K(G) = K_g(G) < \rho(G)$ ($g = 1, 2, \cdots, d$) and suppose that vertex-cut of $K_g(G)$ is C_g. By lemma 3, we know

$$K_g(G) = m_g |V(P)| \qquad (m_g > 2,\ g = 1, 2, \cdots d) \tag{1}$$

H_g represents the graph which is cosnstructed by p and by connecting vertex C_g with all edges of P in G. Since P is an atomic part of G and the maximum degree Δ (H_g) of H equals to $p(G)$, $|V(H_g)| > \rho(G)$, that is

$$|V(P)| + K_g(G) > \rho(G) \tag{2}$$

Suppose $P_{g,i}$ ($g = 1, 2, \cdots d$, $i = 1, 2 \cdots \alpha(g)$) is all atomic part which satisfies equation (1) and (2). Suppose $|P_{g.i}| = \lambda_{gi}$ obviously $\lambda_{g,i}$ are all meatatomic parts of G.

Because we suppose that $K(G) = K_g(G) < \rho(G)$ ($g = 1, 2, \cdots d$) and P is an atomic part of G them $P_{g,i}$ ($g = 1, 2, \cdots d$, $i = 1, 2, \cdots \alpha(g)$)
is an atomic part of G. According to theorem 1. We know that $P_{g,i}$ ($g = 1, 2, \cdots d$, $i = 1, 2, \cdots \alpha(g)$)

are circulant graphs, Let $V(P_{gi}) = \left\{p_{gi}^{(1)}, P_{gi}^{(2)}, \cdots, P_{gi}^{(\lambda_{gi})}\right\} = \left\{0, \frac{n}{\lambda_{g,i}}, 2\frac{n}{\lambda_{g,i}} \cdots, (\lambda_{g,i} - 1)\frac{n}{\lambda_{g,i}}\right\}$. If

$P_{g,i}$ is ∂t--degree circulant graph, them there are t numbers among $j_{b_1}, j_{b_2}, \cdots, j_{b_r}$ which belong to

set $\left\{\frac{n}{\lambda_{g,i}}, \frac{2n}{\lambda_{g,i}}, \cdots, \left[\frac{\lambda_{g,i} - 1}{2}\right]\frac{n}{\lambda_{g.i}}\right\}$ without losing generality let

$$j_{b_f} = \frac{n}{\lambda_{g,i}}, \frac{n}{\lambda_{g,i}} \cdots, or \left[\frac{\lambda_{g.i} - 1}{2}\right]\frac{n}{\lambda_{g.i}} \quad \left(f = 1, 2, \cdots, t, 1 \le t \le \left[\frac{\lambda_{g,i} - 1}{2}\right]\right)$$

If $P_{g.i}$ is (2t + 1) - degree cirulant graph, then there are t numbers among $j_{b_1}, j_{b_2}, \cdots j_{b_{r-1}}$

which belong to set $\left\{\frac{n}{\lambda_{g.i}}, \frac{n}{\lambda_{g.i}}, \cdots, \left[\frac{\lambda_{g.i} - 1}{2}\right]\frac{n}{\lambda_{g.i}}\right\}$。 without losing geuerality. Le

$$J_{b_f} = \frac{n}{\lambda_{g.i}}, \frac{n}{\lambda_{g.i}}, \cdots, or \left[\frac{\lambda_{gi} - 1}{2}\right]\frac{n}{\lambda_{g.i}} \left(f = 1, 2, \cdots, t, 1 \le t \le \left[\frac{\lambda_{g.i} - 1}{2}\right]\right)$$

then there are r - t vertices among c_g whose lables are $j_{b_{t+1}}, j_{b_{t+2}}, \cdots j_{b_r}$. According to theorem 3 following expressions are correct

$$j_{b_{t+2\beta+1}} \pm j_{b_{t+2\beta+2}} = 0 \pmod{\frac{n}{\lambda_{g.i}}}$$

$$(\square\beta = 0, 1, \cdots, \{\frac{r-t}{2}\}, \text{when } b_{t+2\beta+1} > r, b_{t+2\beta+2} \in \{b_{t+1}, b_{t+2}, \cdots, b_r\})$$

$$g = 1, 2, \cdots, d; i = 1, 2, \cdots, \partial(g)$$

These arc contradictory with supposion of this theorem. So necessory condition of this theorem has been proven.

If one group of expressions in this theorem are cosrrect, when $g = \theta$ ($1 \leqslant \theta \leqslant q$), $i = w$ ($1 \leqslant w. \leqslant \partial(g)$) this (A) are correct. Obviously C_θ is a vertex cut of graph G. Thus K (G) < ρ (G), This is contradictorg with K (G) = ρ (G). So necessory conditon of this theorem has been proven.

Corollary Let G connected graph $C_n < j_1, j_2, \cdots, j_r >$, $\lambda_{g,i}$ ($g = 1, 2, \cdots d$; $i = 1, 2, \cdots, \partial(g)$) be all meat atomic numbers of G and $b_1, b_2, \cdots, b_r$ be one permutation of { 1, 2, ⋯ r, }. If fo $C_{.g}$ ($P_{g.i}$ is an atomic part with respect to C_g and $|P_{g.i}| = \lambda_{g.i}$) there are

$$j_{b_f} = \frac{n}{\lambda_{g,i}}, \frac{2n}{\lambda_{g,i}}, \ldots, \text{ or } \left[\frac{\lambda_{g.i-1}}{2}\right]\frac{n}{\lambda_{g.i}}$$

$$(f = 0, 1, \cdots, \left\{\frac{r-t}{2}\right\} - 1, \text{when } b_{t+2\beta+1} > r, b_{t+2\beta+2} \in \{ b_{t+1}, b_{t+2}, \cdots, b_r\})$$

$$j_{b_{t+2\beta+1}} \pm j_{b_{t+2\beta+2}} = 0(\mathrm{mod} \frac{n}{\lambda_{g.i}})$$

$$(\beta = 0, 1, \cdots, \left\{\frac{r-t}{2} - 1, \textit{ when } b_{t+2\beta+1} > r, b_{t+2\beta+2} \in \{b_{t+1}, b_{t+2}, \cdots, b_r\right\})$$

$(g = 1, 2, \cdots d, \quad i = 1, 2, \cdots \partial(g))$

then $K(G) = \min \{|C_g|; g = 1, 2, \cdots, d\}$

The steps of constructing Circulant graph $G = C_n < j_1, j_2, \cdots j_r >$ with the best connectivity are descript as following.

(1) Find out all imprime integers K (G): K_1 (G), K_2 (G), ⋯ , K_d (G) so $\frac{2}{3}\rho(G) < K(G) < \rho(G)$ holds.

(2) Find out all meatatomic numbers of G.

(3) Find out $j_1, j_2, \cdots j_r$ So that at least one expression in each group of expressions (A) does not hold.

According to corollary graphs with K (G) connectivity ($\frac{2}{3}$ ρ (G) $\leqslant$ K (G) $\leqslant$ ρ (G)) can be constructing by using connected circulant graph $G = C_n < j_1, j_2, \cdots, j_r >$.

Reference

[1] D.Bauer, C.Suffl, F.Boesch, R.Tindell. Proceedings-Fourth International Graph Theory Conference, 45-54.

[2] Mar E.Watkins. Counectivity of Transitive Graphs, Journal of Combinatorial Theory, 8(1970), 23-29.

Recent progress on Room frames and related designs

L. Zhu
Department of Mathematics
Suzhou University
Suzhou, 215006, China

Abstract The existence of Room frames of type h^n and $1^n u^1$ has been studied for more than ten years until recently it becomes close to the end. Recent work solves the two types and the type $2^n u^1$, leaving one possible exception for each case, that is 14^4, $1^{46}21^1$ and $2^{19}18^1$. Similar progress has also been made for skew Room frames of type h^n. On the other hand, (skew) Room frames are recently found useful in constructing weakly 3-chromatic linear spaces, almost resolvable directed k-cycle systems, 4-frames and resolvable (K_4 - e)-designs. In this paper, we survey the existence of Room frames and their recent applications to other designs. We also mention some related open problems.

1. Introduction

Let S be a finite set, and let $\{S_1, S_2, ..., S_n\}$ be a partition of S. An $\{S_1, S_2, ..., S_n\}$-*Room frame* is an $|S| \times |S|$ array, $\boldsymbol{F}$, indexed by S, which satisfies the following properties:

1. Every cell of $\boldsymbol{F}$ either is empty or contains an unordered pair of symbols of S.
2. The subarrays $S_i \times S_i$ are empty, for $1 \le i \le n$ (these subarrays are referred to as *holes*).
3. Each symbol $x \notin S_i$ occurs once in row (or column) s, for any $s \in S_i$.
4. The pairs in $\boldsymbol{F}$ are those {s, t}, where $(s, t) \in (S \times S) \setminus \cup_{1 \le i \le n} (S_i \times S_i)$.

The *type* of the frame is defined to be the multiset $\{|S_i|: 1 \le i \le n\}$. We usually use an "exponential" notation to describe types: a type $t_1^{u_1} t_2^{u_2} ... t_k^{u_k}$ denotes u_i occurrences of t_i, $1 \le i \le k$. We briefly denote a frame of type $t_1^{u_1} t_2^{u_2} ... t_k^{u_k}$ by $RF(t_1^{u_1} t_2^{u_2} ... t_k^{u_k})$.

An $\{S_1, S_2, ..., S_n\}$-Room frame $\boldsymbol{F}$ is called *skew* if one cell is occupied, then its

[1] Research supported by NSFC grant 19231060-2.

symmetric cell is empty. A skew RF($t_1^{u_1} t_2^{u_2} \ldots t_k^{u_k}$) is denoted by SRF($t_1^{u_1} t_2^{u_2} \ldots t_k^{u_k}$).

Example 1.1 An SRF(1^7) is shown in Table 1.1

	2 6	4 5		1 3		
		3 0	5 6		2 4	
			4 1	6 0		3 5
4 6				5 2	0 1	
	5 0				6 3	1 2
2 3		6 1				0 4
1 5	3 4		0 2			

Table 1.1 An SRF(1^7)

It is well known (see [6] for example) that the existence of an RF(1^n) is equivalent to the existence of a Room square of side n. In Table 1.1 if we put the pair {i, ∞} in the i-th diagonal cell, we get a Room square of side 7. This is also true in general.

After Room's 1955 paper, the existence problem of a Room square attracted the interest of many researchers. It was finally solved in 1973 when W. D. Wallis reported a Room square of side 257. The following theorem can be found in a survey paper [12] by Mullin and Wallis.

Theorem 1.2 A Room square of side n exists if and only if n is odd and $n \neq 3$ or 5.

Since its introduction by Wallis, the notion of a Room frame has played an important role not only in the investigation of Room squares but also in many other combinatorial designs. In this paper, we describe the recent progress on the existence of Room frames of type h^n, $1^n u^1$, $2^n u^1$, and the existence of a skew Room frame of type h^n. Since several applications to problems of nesting of Steiner triple systems, nesting group divisible designs and partial Steiner triple systems, nesting cycle systems, construction of almost resolvable cycle systems, construction of decompositions of K_n and $2K_n$ into wheels, and construction of BIBDs with block size four admitting a blocking set have already been mentioned in [6], [11], [16], we focus on the recent applications to problems of constructing weakly 3-chromatic linear spaces, almost resolvable directed k-cycle systems, 4-frames and resolvable $(K_4 - e)$-designs. We also mention some related open problems.

2. Existence of Room frames

In [19], the following results are presented.

Theorem 2.1 Suppose h and n are positive integers, $n \geq 4$, and $(h, n) \neq (1, 5), (2, 4)$. Then there exists a Room frame of type h^n if and only if $h(n - 1)$ is even, except possibly $h = 14$ and $n = 4$.

Theorem 2.2 Suppose v and u are odd positive integers, $v \geq 3u + 2$, and $(v, u) \neq (5, 1)$. Then there exists a Room frame of type $1^{v-u}u^1$, except possibly for $(v, u) = (67, 21)$.

Theorem 2.3 Suppose n and u are positive integers. If $u \geq 4$, then there exists a Room frame of type $2^n u^1$ if and only if u is even and $n \geq u + 1$, except possibly when $n = 19$ and $u = 18$.

Similar results for skew Room frames are presented in [3].

Theorem 2.4 Suppose h and n are positive integers, $n \geq 4$, and $(h, n) \neq (1, 5), (2, 4)$. Then there exists a skew Room frame of type h^n if and only if $h(n - 1)$ is even, except possibly when (1) $n = 5$ and $h \in \{11, 13, 17, 19, 23, 29, 31, 41, 43\}$, or (2) $n = 4$ and $h \in \{8, 24\} \cup \{k > 2: k \equiv 2 \bmod 4\}$.

To prove these results, various kinds of recursive constructions are used such as Weighting construction, various Filling in Holes constructions and Inflation constructions (see [6], [19], [18]). The direct constructions used are mainly starter-adder and intransitive starter-adder techniques.

Let G be an abelian group, written additively, and let H be a subgroup of G. Denote $g = |G|$, $h = |H|$ and suppose that $g - h$ is even. A *frame starter* in $G \setminus H$ is a set of unordered pairs $S = \{\{s_i, t_i\}: 1 \leq i \leq (g - h)/2\}$ satisfying

(1) $\cup_{1 \leq i \leq (g-h)/2} (\{s_i\} \cup \{t_i\}) = G \setminus H$, and

(2) $\cup_{1 \leq i \leq (g-h)/2} \{\pm (s_i - t_i)\} = G \setminus H$.

An *adder* for S is an injection A: S $\rightarrow$ G \ H, such that

$$\cup_{1\leq i\leq (g-h)/2} (\{s_i + a_i\}\cup\{t_i + a_i\}) = G\setminus H,$$

where $a_i = A(s_i, t_i)$, $1\leq i\leq (g-h)/2$. An adder A is *skew* if, further,

$$\cup_{1\leq i\leq (g-h)/2} (\{a_i\}\cup\{-a_i\}) = G\setminus H.$$

From a starter S and a skew adder A, we can construct a skew Room frame in which the cell (f, $-a_i + f$) is occupied by $\{s_i + f, t_i + f\}$ for $1\leq i\leq (g-h)/2$ and any $f\in G$.

Example 2.5 Let $G = Z_{10}$ and H = {0, 5}. Let S = {{4, 8}, {1, 2}, {7, 9}, {3, 6}} and A = {8, 7, 4, 1}. An SRF(2^5) can be constructed from the starter S and the skew adder A shown in Table 2.1.

		48	12			79			36
47			59	23			80		
	58			60	34			91	
		69			71	45			02
13			70			82	56		
	24			81			93	67	
		35			92			04	78
89			46			03			15
26	90			57			14		
	37	01			68			25	

Table 2.1 An SRF(2^5)

Let G and H be as above. A *2k-intransitive frame starter-adder* in G \ H is defined to be a quadruple (S, C, R, A), where

$S = \{\{s_i, t_i\}: 1\leq i\leq (g-h)/2 - 2k\}\cup\{u_r: 1\leq r\leq 2k\}$

$C = \{\{p_j, q_j\}: 1\leq j\leq k\}$

$R = \{\{p_j', q_j'\}: 1\leq j\leq k\}$

A: S $\rightarrow$ G \ H is an injection that satisfies the following properties:

(i) $\{s_i\}\cup\{t_i\}\cup\{u_r\}\cup\{p_j\}\cup\{q_j\} = G\setminus H$

(ii) $\{s_i + A(s_i, t_i)\}\cup\{t_i + A(s_i, t_i)\}\cup\{u_r + A(u_r)\}\cup\{p_j'\}\cup\{q_j'\} = G\setminus H$

(iii) $\{\pm (s_i - t_i)\}\cup\{\pm (p_j - q_j)\}\cup\{\pm (p_j' - q_j')\} = G\setminus H$.

(iv) Any element $p_j - q_j$ or $p_j' - q_j'$ has even order, $1 \le j \le k$.

An adder A is *skew* if, further

(v) $\{\pm A(s_i, t_i)\} \cup \{\pm A(u_r)\} = G \setminus H$, and for each j, $1 \le j \le k$, there exists a $d \ge 1$ such that $p_j - q_j$ has order $2^d m$ and $p_j' - q_j'$ has order $2^d m'$, where m and m' are odd.

From a 2k-intransitive (skew) frame starter-adder we can construct a (skew) Room frame in which the cells $(f, -A(s_i, t_i) + f)$ and $(g, -A(u_r) + f)$ are occupied by $\{s_i + f, t_i + f\}$ and $\{u_r + f, \infty_r\}$ respectively, for $1 \le i \le (g - h)/2$, $1 \le r \le 2k$ and any $f \in G$. At the same time the square contains some extra rows and columns constructed from R and C.

Example 2.6 Let $G = Z_{10}$ and $H = \{0, 5\}$. Let $S = \{\{1, 3\}, \{2, 6\}\} \cup \{8, 9\}$, $C = \{4, 7\}$, $R = \{8, 9\}$, and $A(1, 3) = 3$, $A(2, 6) = 1$, $A(8) = 4$ and $A(9) = 2$. An SRF(2^6) can be constructed from (S, C, R, A), shown in Table 2.2

						8x	13	9y	26	47	
37							9x	24	0y		58
1y	48							0x	35	69	
46	2y	59							1x		70
2x	57	3y	60							81	
	3x	68	4y	71							92
		4x	79	5y	82					03	
			5x	80	6y	93					14
				6x	91	7y	04			25	
					7x	02	8y	15			36
	90		12		34		56		78		
89		01		23		45		67			

Table 2.2 An SRF(2^6).

We mention some applications of Room frames to other designs in the following sections.

3. Weakly 3-chromatic linear spaces

In this section, a linear space is restricted to be a BIB design with block size four and index unity. The *weak chromatic number* of a linear space is the least number of colours required to colour the points in such a way that there are no monochromatic lines.

The existence problem for linear spaces with weak chromatic number 2 is settled by Hoffman et al. [9]. In [4], Colbourn and Rosa suggested the existence problem for weakly 3-chromatic linear spaces. It is known that there are no weakly 3-chromatic linear space of order 13 or 16. They do exist of orders 25 and 28. In a recent survey [16], Rodger presented a construction which uses a skew Room frame with at least one hole of size 4 and other holes of size 2 or 4. The construction can be described as follows.

Suppose a skew Room frame $\boldsymbol{F}$ is based on set S with hole set $\boldsymbol{H}$, where at least one hole has size 4 and other holes have size 2 or 4. The linear space will be based on set $(S \times Z_6) \cup \{\infty\}$.

(1) Construct a GDD based on $S \times Z_6$ with groups $H \times Z_6$ for $H \in \boldsymbol{H}$ and blocks

$$\{\{(x, g), (y, g), (c, 1 + g), (r, 4 + g)\}: (x, y) \text{ occurs in cell } (r, c) \text{ of } \boldsymbol{F}, g \in Z_6\}.$$

(2) On each set $(H \times Z_6) \cup \{\infty\}$, construct a linear space of order 13 or 25.

Colour the points in $(S \times Z_6) \cup \{\infty\}$ such that there are no monochromatic lines and one of the linear spaces of order 25 is weakly 3-chromatic. This construction solves the case $v \equiv 1 \pmod{12}$. For the case $v \equiv 4 \pmod{12}$, the linear space is based on set $(S \times Z_6) \cup \{\infty_1, \infty_2, \infty_3, \infty_4\}$ and (2) should be replaced by (2'), where a weakly 3-chromatic linear space of order 28 is needed.

(2') On each set $(H \times Z_6) \cup \{\infty_1, \infty_2, \infty_3, \infty_4\}$, construct a linear space of order 16 or 28 such that $\{\infty_1, \infty_2, \infty_3, \infty_4\}$ is a line.

These constructions have been used to solve the existence problem completely in a recent paper [15], where the following result is proved.

Theorem 3.1 A weakly 3-chromatic linear space of order v with all lines having size 4 exists if and only if $v \geq 25$ and $v \equiv 1, 4 \pmod{12}$.

In the process, the following result is also established.

Theorem 3.2 For any even integer $n \geq 16$, there is an integer $b = b(n) \geq 1$ such that a skew Room frame of type $4^b 2^{(n - 4b)/2}$ exists.

The construction suggested by Rodger has other applications too, which we describe below.

4. New results on 4-frames

A *group divisible design* (or GDD), is a triple (X, *G*, *B*) which satisfies the following properties:

(1) G is a partition of a set X (of *points*) into subsets called *groups*,

(2) B is a set of subsets of X (called *blocks*) such that a group and a block contain at most one common point,

(3) every pair of points from distinct groups occurs in a unique block.

The group type of the GDD is the multiset $\{|G|: G \in \mathcal{G}\}$. A GDD (X, *G*, *B*) will be referred to as a K-GDD if $|B| \in K$ for every block B in *B*.

A *frame* is a group divisible design (X, *G*, *B*) whose block set admits a partition into *holey* parallel classes, each holey parallel class being a partition of $X \setminus G$ for some $G \in \mathcal{G}$. The groups of a frame are usually referred to as *holes*. The *type* of the frame is defined to be the group type of the GDD. A frame of block size k is denoted by k-frame. A frame of type h^u has u holes of size h and is called *uniform*.

It is known [13] that if there is a k-frame of type h^u (with $u > 1$), then $u \geq k + 1$, $h \equiv 0 \mod (k - 1)$ and $h(u - 1) \equiv 0 \mod k$. The necessary conditions for the existence of a uniform k-frame have been proved to be sufficient for $k = 3$ [17]. A partial solution for $k = 4$ has been provided in [13]. This is improved in [5] by using certain skew Room frames.

In Rodger's construction mentioned above, a skew Room frame *F* of type h^n, based on X and having hole set $\mathcal{H} = \{H_1, ..., H_n\}$, leads to a GDD of type $(6h)^n$. The blocks are generated modulo 6 from the initial blocks

$$\{\{(x, 0), (y, 0), (c, 1), (r, 4)\}: (x, y) \text{ occurs in cell } (r, c) \text{ of } \mathcal{F}\}.$$

If all the quadruples (x, y, c, r) can be partitioned into some sets such that each set forms a partition of $X \setminus H_i$ for some i and each H_i coresponds to 2h of such sets, we call such a skew Room frame *partitionable*. It is clear that from a partitionable skew Room frame, each set which partitions $X \setminus H_i$ will result in a holey parallel class in the resulting GDD. We state the fact in the following lemma.

Lemma 4.1 If there is a partitionable skew Room frame of type h^n, then there exists a 4-frame of type $(6h)^n$.

Example 4.2 A partitionable skew Room frame of type 3^5. From [7] we have a starter S and a skew adder A of type 3^5 as follows.

$G = Z_{15}$ and $H = \{0, 5, 10\}$.

$S = \{\{1, 7\}, \{2, 3\}, \{4, 8\}, \{6, 14\}, \{9, 12\}, \{11, 13\}\}$.

$A = \{12, 14, 4, 8, 9, 13\}$.

We can partition all the quadruples $(s_i + g, t_i + g, -a_i + g, g)$ as follows. First, translate the initial quadruples:

(1, 7, 3, 0) --- add 1 --- (2, 8, 4, 1)

(2, 3, 1, 0) --- add 1 --- (3, 4, 2, 1)

(4, 8, 11, 0) --- add 3 --- (7, 11, 14, 3)

(6, 14, 7, 0) --- add 2 --- (8, 1, 9, 2)

(9, 12, 6, 0) --- add 2 --- (11, 14, 8, 2)

(11, 13, 2, 0) --- add 1 --- (12, 14, 3, 1)

It is easily seen that each quadruple (u, v, w, z) on the right covers the four non-zero residues modulo 5, and hence will give one partition of $G \setminus H$: $\{(u + g, v + g, w + g, z + g): g \in H\}$. Altogether we get six partitions of $G \setminus H$. Under the action of group G, we get further partitions so that $\boldsymbol{F}$ is partitionable. Hence, from Lemma 4.1 there exists a 4-frame of type 18^5.

Using this construction and other recursive constructions the result in [13] has been improved in [5] which we state as follows.

Theorem 4.3 There exists a 4-frame of type h^u, if and only if $u \geq 5$, $h \equiv 0 \bmod 3$ and $h(u - 1) \equiv 0 \bmod 4$, except possibly where

(i) $h = 9$ and $u \in \{13, 17, 29, 93, 113, 153, 173, 193\}$;

(ii) $h \equiv 12 \bmod 24$ and $u \in \{8, 12\}$,

$h = 36$ and $u \in \{7, 18, 23, 28, 38, 43, 48\}$;

(iii) $h \equiv 0 \bmod 24$ and $u \in \{12\}$,

$h = 24$ or 120 and $u \in \{7\}$; or

(iv) $h \equiv 6 \bmod 12$ and $u \in \{n: n \equiv 3 \bmod 4 \text{ and } n \le 267\}$,

$h = 18$ and $u \in \{17\}$.

5. Resolvable (K_4 - e)-designs

A *(K_4 - e)-design* of order n is a pair (X, ***B***) where ***B*** is an edge-disjoint decomposition of the edge set of K_n (the complete undirected graph on n vertices) with vertex set X into copies of

$K_4 - e =$ 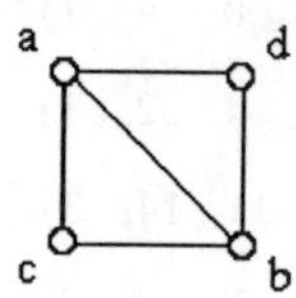

We denote K_4 - e by any one of (a, b, c, d), (a, b, d, c), (b, a, c, d), or (b, a, d, c) and call it a *block* of the design. A (K_4 - e)-design is called *resolvable* if the block set ***B*** can be partitioned into *parallel classes*, each forming a partition of X. Simple counting shows that if a resolvable (K_4 - e)-design of order n exists, then $n \equiv 16 \pmod{20}$, $n \ge 16$. A. Street posed the existence question at Auburn Conference (1994). A. Rosa (private communication) provided the first example of order 16 as follows.

Example 5.1 A resolvable (K_4 - e)-design of order 16. Let $X = \{\infty, 0, 1, ..., 14\}$. Let ***B*** be the set of 24 blocks shown below, which is partitioned into six parallel classes, where each row forms a parallel class.

(1, ∞, 0, 2), (5, 14, 4, 6), (9, 10, 3, 7), (11, 13, 8, 12);

(5, 10, 0, ∞), (4, 8, 1, 2), (13, 6, 7, 9), (3, 14, 11, 12);

(6, 11, ∞, 1), (2, 5, 3, 9), (8, 14, 7, 10), (0, 4, 12, 13);

(7, 12, ∞, 2), (4, 6, 3, 10), (0, 9, 8, 11), (1, 13, 5, 14);

(∞, 3, 8, 13), (7, 11, 4, 5), (1, 12, 9, 10), (0, 2, 6, 14);

(∞, 9, 4, 14), (8, 12, 5, 6), (2, 10, 11, 13), (3, 7, 0, 1).

Then (X, ***B***) is the desired resolvable (K_4 - e)-design of order 16.

A *(K_4 - e)-group divisible design* (GDD) of type h^n is a triple (X, ***G***, ***B***), where X is the vertex set, ***G*** is a partition of X into groups (*holes*) of size h each, ***B*** is an edge-disjoint decomposition of the edge set of $K_{n, n, ..., n}$ (the multipartite complete undirected graph with ***G*** as the partition of the vertex set X) into copies of (*blocks*) K_4 - e. A *(K_4 - e)-frame* of type h^n is a (K_4 - e)-GDD of type h^n (X, ***G***, ***B***) in which the bolcks can be partitioned into *holey parallel classes*, each forming a partiton of X \ G for some G ∈ ***G***.

In [5], a construction for (K_4 - e)-frames using partitionable skew Room frames is described. Let ***F*** be a partitionable skew Room frame of type h^n with hole set ***H*** and let {x, y} be a pair in the cell (r, c) of ***F***. We first construct a (K_4 - e)-GDD of type $(5h)^n$ (X, ***G***, ***B***), where ***G*** = {H × Z_5: H ∈ ***H***}, X = $\cup_{G \in \boldsymbol{G}}$ G and ***B*** contains all the 5 graphs ((x, g), (y, g), (c, 1 + g), (r, 2 + g)) for {x, y}∈ ***F*** and g ∈ Z_5. This construction has been used implicitly in [10]. Since ***F*** is partitionable, for each H' ∈ ***H*** the quadruples which partition the set $\cup_{H \in \boldsymbol{H} \setminus H'}$ H will result in a holey parallel class in the resulting (K_4 - e)-GDD of type $(5h)^n$. Therefore, the (K_4 - e)-GDD is actually a (K_4 - e)-frame.

Lemma 5.2 If there exists a partitionable skew Room frame of type h^n, then there exists a (K_4 - e)-frame of type $(5h)^n$.

Using the partitionable skew Room frames and some recursive constructions for (K_4 - e)-frames, the following existence result is shown in [5].

Lemma 5.3 There exist (K_4 - e)-frames of types 15^u for u ∈{n: n ≥ 5, n ≡ 1 (mod 4), n ≠ 33, 49}.

By adding a new point ∞ to the (K_4 - e)-frame and construct a resolvable (K_4 - e)-design of order 16 on set $H\cup\{\infty\}$ for each hole H of the (K_4 - e)-frame, we get a resolvable (K_4 - e)-design of order 15u + 1. We then have the following.

Theorem 5.4 There exists a resolvable (K_4 - e)-design of order n for $n \equiv 16 \pmod{60}$, $n \geq 16$ and $n \neq 496, 736$.

6. Almost resolvable directed 5-cycle systems

A *directed k-cycle system* of order n is a pair (S, ***B***), where S is an n-set and ***B*** is a collection of arc disjoint directed k-cycles that partition the complete directed graph K_n^*. An *almost parallel class with deficiency* x is a set of directed k-cycles which form a partition of $S \setminus \{x\}$. An *almost resolvable directed k-cycle system* of order n, denoted by ARDkCS(n), is a directed k-cycle system of order n in which the cycles can be partitioned into almost parallel classes. Simple counting shows that $n \equiv 1 \pmod{k}$ is a necessary condition for the existence of such a system. It has been shown that this necessary condition is also sufficient in the case when k = 3 by Bennett and Sotteau [1] and in the case when k = 4 by Bennett and Zhang [2]. In a recent paper [8], the case k = 5 has also been solved by using a kind of skew Room frames.

A skew Room frame ***F***, based on S, is called *strong* if each unordered pair {x, y} in ***F*** can be replaced either by (x, y) or by (y, x) such that if an ordered pair (a, b) appears in row r, then r must appear in ***F*** as the second element in column a and as the first element in column b. A strong skew Room frame of type T is denoted by SSRF(T).

Example 6.1 A strong skew Room frame of type 1^7 is shown in Table 1.1.

Let $\{e_i\}$ and $\{d_i\}$ be two sequences, where for $0 \leq i \leq [k/2]$

$$e_i = (-1)^{i+1}[(i+1)/2] \pmod{k},$$

$$d_i = [k/2] + 1 + (-1)^i [(i+1)/2] \pmod{k}.$$

Let the given SSRF(h^u) F be based on S with partition $\{S_1, S_2, ..., S_u\}$. We construct an ARDkCS(huk +1) on $X = \{\infty\} \cup (S \times Z_k)$. In this construction, all additions are computed modulo k. Define a collection of directed k-cycles B as follows:

(1) for each S_i, $1 \le i \le u$, define an ARDkCS(hk + 1) on the set $\{\infty\} \cup (S_i \times Z_k)$ and place these directed k-cycles in B;

(2) for each pair (x, y) of row r and column c in F and for each j in Z_k, place in B two directed k-cycles:

$(x, y, r, d, j) = ((x, d_0 + j), ..., (y, d_{2t} + j), (r, d_{2t+1} + j), (x, d_{2t} + j), ..., (y, d_0 + j))$ and

$(y, x, c, e, j) = ((y, e_0 + j), ..., (x, e_{2t} + j), (c, e_{2t+1} + j), (y, e_{2t} + j), ..., (x, e_0 + j))$

if k = 4t + 3, or

$(x, y, r, d, j) = ((x, d_0 + j), ..., (x, d_{2t-1} + j), (r, d_{2t} + j), (y, d_{2t-1} + j), ..., (y, d_0 + j))$ and

$(y, x, c, e, j) = ((y, e_0 + j), ..., (y, e_{2t-1} + j), (c, e_{2t} + j), (x, e_{2t-1} + j), ..., (x, e_0 + j))$

if k = 4t + 1. Then (X, B) is a directed k-cycle system and also it is almost resolvable. This leads to the following.

Theorem 6.2 Suppose there exist an SSRF(h^u) and an ARDkCS(hk +1) for odd k. Then there exists an ARDkCS(huk +1).

When h = 1 or 2, an ARDkCS(hk +1) can easily be constructed. So, we have the following corollaries.

Corollary 6.3 Suppose there is an SSRF(1^u). Then there exists and ARDkCS(uk + 1) for any odd $k \ge 3$.

Corollary 6.4 Suppose there is an SSRF(2^u). Then there exists and ARDkCS(2uk + 1) for any odd $k \ge 3$.

A frame starter $S = \{\{s_i, t_i\}: 1 \le i \le (g - h)/2\}$ in $G \setminus H$ is called a *skew frame starter* if $\{\pm(s_i + t_i): 1 \le i \le (g - h)/2 \} = G \setminus H$. It is known that the existence of a skew frame starter of type h^u implies the existence of an SSRF(h^u). The following skew frame starters are known (see [8]).

Theorem 6.5 There are skew frame starters of the following types:

(1) type 1^n where $n = 2^k t + 1$ is a prime power and $t > 1$ is odd.

(2) type 1^n where $n = 16k^2 + 1$ and k is any positive integer.

(3) type h^n where $h = 2^u$, $u \geq 1$, and $n \equiv 1 \pmod 4$ is a prime power.

(4) type 4^4 and type 1^{35}.

Using the known skew frame starters and strong skew Room frames along with some recursive constructions, the existence of ARD5CS(5v + 1) is completely solved in [8]. However, the existence of ARDkCS(n) for general k is still open.

7. Further questions

For Room frames and related designs, there has been much progress made recently. However, there are still some problems remaining open. We mention the following several ones.

1. (Section 2) Construct Room frames of types 14^4, $1^{46}21^1$, and $2^{19}18^1$.

2. (Section 2) Construct skew Room frames of type h^n for $n = 5$ and $h \in \{11, 13, 17, 19, 23, 29, 31, 41, 43\}$, or for $n = 4$ and $h \in \{8, 24\} \cup \{k > 2: k \equiv 2 \bmod 4\}$.

3. (Section 4) Construct more partitionable skew Room frames, especially for types 3^{17}, 4^7, 4^{12}, 6^7, to improve the existence results on 4-frames.

4. (Section 5) Construct resolvable $(K_4 - e)$-designs of order n for n = 496, 736, or $n \equiv 36, 56 \pmod{60}$, $n \geq 36$.

5. (Section 6) Find more skew frame starters (see Theorem 6.5), especially for types 1^u and 2^u.

6. (Section 6) Solve the existence of almost resolvable directed k-cycle systems for $k > 5$.

Acknowledgment Part of the work was done while the author was visiting Waterloo University during the summer of 1994 with a Foreign Researcher Award from NSERC. He appreciates the hospitality and support accorded him.

References

[1] F. E. Bennett and D. Sotteau, Almost resolvable decompositions of K_n*, J. Combin. Theory (B) 30 (1981) 228-232.

[2] F. E. Bennett and X. Zhang, Resolvable Mendelsohn designs with block size 4, Aequationes Math. 40 (1990) 248-260.

[3] K. Chen and L. Zhu, On the existence of skew Room frames of type t^u, Ars Combinatoria, to appear.

[4] C. J. Colbourn and A. Rosa, Colorings of block designs, in: Dinitz and Stinson, eds., Contemporary Design Theory, Wiley, 1992, 401-430.

[5] C. J. Colbourn, D. R. Stinson and L. Zhu, More frames with block size four, preprint.

[6] J. H. Dinitz and D. R. Stinson, Room squares and related designs, in: Contemporary Design Theory, Wiley, 1992, 137-204.

[7] J. H. Dinitz and D. R. Stinson, The construction and uses of frames, Ars Combinatoria 10(1980) 31-54.

[8] G. Ge and L. Zhu, Existence of almost resolvable directed 5-cycle systems, Australasian J. Combinatorics, to appear.

[9] D. G. Hoffman, C. C. Lindner and K. T. Phelps, Blocking sets in designs with block size 4, Europ. J. Combin. 11(1990) 451-457.

[10] D. G. Hoffman, C. C. Lindner, M. J. Sharry and A. P. Street, Maximum packings of K_n with copies of K_4 - e, Aequationes Math., to appear.

[11] C. C. Lindner and C. A. Rodger, Decomposition into cycles II: cycle systems, in: Contemporary Design Theory, Wiley, 1992, 325-369.

[12] R. C. Mullin and W. D. Wallis, The existence of Room squares, Aequationes Math. 13(1975) 1-7.

[13] R. Rees and D. R. Stinson, Frames with block size four, Can. J. Math. 44(1992) 1030-1049.

[14] C. A. Rodger, Linear spaces with many small lines, Discrete Math. 129(1994) 167-180.

[15] C. A. Rodger, E. B. Wantland, K. Chen and L. Zhu, Existence of certain skew Room frames with application to weakly 3-chromatic linear spaces, J. Combinatorial Designs, to appear.

[16] C. A. Rodger, Graph decompositions, Le Mathematiche XLV(1990) 119-140.

[17] D. R. Stinson, Frames for Kirkman triple systems, Discrete Math. 65(1987) 289-300.

[18] D. R. Stinson and L. Zhu, Towards the spectrum of Room squares with subsquares, J. Combin. Theory (A) 63(1993) 129-142.

[19] D. R. Stinson, L. Zhu and J. H. Dinitz, On the spectra of certain classes of Room frames, Electronic J. Combinatorics 1(1994), #R7, 21pp.

LIST OF PARTICIPANTS

Prof.A.K.Agarwal
Department of Mathematics
Birla Institute of Technology & Science
Pilani(Rajasthan) 333 031
India

Prof.Y.Alavi (Alavi, Yousef)
Department of Mathematics & statics
Western Michigan Vniversity
Kalamazoo, MI 49008-5152
USA
Tel:001-616-387-4558 or 4513

Prof.A.Altshuler
Dean,Faculty of Natural Sciences
The Hyman Kreitman Chair of Bioenergetics
Ben-Gurion University of The Negev
P.O.B.653,Beer-Sheva 84105
Israel
Tel:972-57-30 919 & 461-633
Fax:972-57-38 717 & 31 340

Prof.L.D.Andersen
Department of Mathematics and Computer Science,
Institute of Electronic Systems
Aalborg University
Fr. Bajers Vej 7
DK 9220 Aalborg O
Denmark
Tel:0045-98-158522
e-mail:lda@iesd.auc.dk

Philbert Banani(P.Banani)
Department of Statistics Operations Research,
Fudan University
Shanghai 200433
P.R.China
Tel:021-5488440(O)

Prof.A.Barlotti
Universita Degli studi difirenze
Dipartimento Di Matematica"U.Dini"
Viale Morgagni 67/A-50134 Firenze
Italy
Tel: 0039-55-411985-411986
Fax: 0039-55-4222695
e-mail: UDINI@IFIIDG.BITNET

Prof.L.Batten
Department of Mathematics
University of Manitoba
Winnipeg Manitobà
Canada R3T 2N2
Tel:001-204-474-8703
001-204-474-9693
Fax:001-204-261-0177
e-mail: BATTEN@CCM.UMANITOBA.CA

Prof.D.Bedford
University of Keele
Department of Mathematics
Keele, Staffordshire, ST5 5BG
Fax:0782 715194

Prof.K.A.Berman(Kenneth A.Berman)
Dept.of Electrical & Computer Engin.& Computer Science,
University of Cincinnati
Cincinnati OH
USA

Prof.N.L.Biggs (Norman Biggs)
Department of Statistical and Mathematical Sciences,
Honghton Street
London WCZA ZAE
Tel:071-405 7686
Fax:071-955-7416

Prof.F.Brenti
Dipartmento Di Matematica
Universita Di Perugia
Via Vanvitelli, 1
I-06123 Perugia
Italy
Tel: 0039-75-585 5016
Fax: 0039-75-585 5024

N.Budharsanam
Department of Mathematics
National College
Trichirapalli-T.N
Tel:620001
India

Assoc.Prof.Cai Junliang(J.L.Cai)
Basic Science Department
Taiyuan Heavy Machinery Institute
Taiyuan, Shanxi Province, 030024
P.R.China
Tel:0351-6129611(H)

Prof.Beifang Chen
Department of Mathematics
Hong Kong University of Science and Technology,
Clear Water Bay, Kowloon
Hong Kong

Prof.CHen Chuanping(C.P.Chen)
Institute of System Science
Academia Sinica
Beijing 100080
P.R.China

Chen Dewu(D.W.Chen)
Room 404
14 Xinyi Xinchun, Shiquan Road
Shanghai 200061
P.R.China
Tel:021–3779439–Math.Dept.(O)
021–2148834(H)

Chen Hao(H.Chen)
The Hefei Branch Research Centre of
Combinatorial Mathematics and
Computer Science, Academia Sinica
P.O.Box 1110 Hefei, 230031 Anhui
P.R.China
Tel:0086–551–5591289(O)
Fax:0086–551–5591270

Prof.H.Chen(Chen Hong)
Dept. of Math. & Computer Science
Central Missouri State Univ.
Warrensborg MO 64093
USA
Tel:001–816–543–8907
Fax:001–816–543–8006

Prof.Chen WenLi(W.L.Chen)
Department of Mathematics
Southwest Normal University
Chongqing, Beibei 630715
P.R.China
Tel:0811–8863901–2350(O)
0811–8863907–2097(H)

Assoc.Prof.Chen Xiebin(X.B.Chen)
Department of Mathematics
Zhangzhou Teachers' College
Zhangzhou 363000, Fujian
P.R.China

Prof. Chen Yongchuan(Y.C.Chen)
Nankai University
Tianjin 300071
P.R.China
Nankai Institute of Mathematics
Tel: 022–3501029
Fax: 022–3501532

Prof.Zhi–Hong Chen
Dept.of Math.and Computer Science
Butler University
4600 Sunset Avenue
Indianapolis, Indiana 46208–3485
U.S.A
Tel:001–317–920–6540
Fax:001–317–283–9519
e–mail: Chen@Butler.Edu

Assoc.Prof.Cheng Jiangang(J.G.Cheng)
Department of Mathematics
Yantai University
Yantai 264005 Shandong
P.R.China

Prof. A.Chernyak
ul Russiyanova, 10, kv 299
Minsk, 220141,
Republic Belarus

Prof.D.W.Crowe
Department of Mathematics
University of Wisconsin
Madison, WI 53706
U.S.A
Tel: 001–608–233–3604
Fax: 001–608–263–8891

Cui Guoqin(G.Q.Cui)
Institute of Applied Mathematics
Academia Sinica
Beijing 100080
P.R.China
Tel:01–2541693(O)
Fax:01–2541689

Assoc.Prof.Dai Hongtu(H.T.Dai)
182/15/201 Zhenwu Rol.
Wu Song, Shanghai 200940
P.R.China
Tel:021–6671153(O)
021–6161282(H)

Prof.I.Dejter
Mathe. Dept.
University of Pverto Rico
P.O.Box 23355
San Juan PR 00931–3355
Tel:(089)–789–5796
Fax:(089)–751–0625

Prof.A.Deza
c/o Prof.Masakazu Kojima
Dept.of Math.and Computing Science
Tokyo Institute of Technology
2–12–1 Ohokayama,Meguro–ku
Tokyo 152
Japan

Prof.M.Deza (Deza, Michel Marie)
C.N.R.S.,Ecoie Normale Superieure
Dept.Math.et Informatique
75010 Paris
France
Tel:331–47–70–36–73
Fax:331–47–70–40–44

Assoc.Prof.Ding Jiyu(J.Y.Ding)
Dept. of Math.
Qiqihar Teacher's College
Qiqihar City 161005
P.R.China
Tel:0452–711810–286(O)
0452–713592(H)

Dong Jinquan(J.Q.Dong)
Department of Mathematics

Inner Mongolia University
Huhhot, 010021
P.R.Chian
Tel:0471–454433–947(O)
0471–454433–729(H)
Fax:0471–451761

Prof.A.W.M.Dress
Universitat Bielefeld
Fakultat fur Mathematik
Universitatsstra e
Postfach 8640
4800 Bielefeld 1
Germany
Tel:(0521)106–5034
Fax:106–4743

Prof.D.Z.Du(Du Ding–Zhu)
Department of Computer Science
Institute of Technology
Univ of Minnesota
4–192 EELCS Buliding
200 Union Street S.E.
Minneapolis, MN55455–0159
USA

Prof.G.P.Egorychev
Computer Centre, Akademgorodok
660036, Krasnoyarsk
Russia
Fax:007–3912–448625

Prof.P.Erdos
Mathematical Institute
Hungarian Academia of Sciences
Realtanoda U 13–15,
Budapest, 1053
Hungary

Prof.R.Euler
Department d'Informatique
Universite de Bretagne Occidentale
Faculte des Sciences
6, Avenue Victor Le Gorgeu
B.P.809
F–29285 Brest Cedex
France
Tel:0033–98016209
Fax:0033–98016131

Prof.S.Fajtlowicz
Department of Mathematics
University of Houston
Houston, TX 77204–3476
U.S.A
Tel: 001–713–723–3458

Fan Guimei(G.M.Fan)
Qingdao Institute of Architectural Engineering
Qingdao Shandong 266033
P.R.China

Dr.Fan Suohai(S.H.Fan)
Department of Mathematics
Jinan University
Guangzhou, 510632
P.R.China
Tel:020–5516511–4521(O)
020–5516511–2602(H)
Fax:020–5516941

Prof. L.Filus
Dept. of Mathematics
Northeastern Illinois University
Board of Governors Universities
5500 North St. Louis Avenue
Chicago, Illinois 60625–4699
U.S.A.

Prof. Damir N. Gainanov
"Data–Centre" Research Firm,
Ekaterinburg
620059 Postbox 65 Ekaterinburg,
Russia
Fax: 007–3432–373997
e–mail: damir@data–c.e–burg.su

Gao Liyan(L.Y.Gao)
Institute of Applied Mathematics
Academia Sinica
Beijing 100080
P.R.China
Tel:01–2541695(O)

Gao Weidong(W.D.Gao)
Institute of Mathematical Sciences
Dalian University of Technology
Dalian, 116024
P.R.China

Gao Xuefeng(X.F.Gao)
208, No.1 Building
Xibianmen Xili
Beijing 100053
P.R.China
Tel:01–3022255–913(O)

Gao Xinhui(X.H.Gao)
208, No.1 Building
Xibianmen Xili
Beijing 100053
P.R.China
Tel:01–302.2255–913(O)

Prof.R.Giudici
Depto. de Matematicas
Universidad Simon Bolivar
Apartado Postal 89000
Caracas,
Venezuela
Tel:0058–2–963 8875
Fax:0058–2–962 1695

Prof.M.Goldberg
Department of Computer Science
RPI
Troy, NY 12180
USA
Tel: 001–581–2762609
Fax: 001–581–2764033

Prof.H.Gropp
Muhling str. 19
D–69121 Heidelberg
Germany

Prof.W.Z.Gu(Weizhen Gu)
Dept.of Mathematics
Southwest Texas State University
San Marcos, Tx 78666
U.S.A.
Tel: 001–512–245–8019

Prof.Gu Yiming
Department of Decision Science
Faculty of Business Administration
National University of Singapore
#10 Kent Ridge Crescent
Singapore 0511
Tel:0065–4927769(pg)
e–mail: fbap4127@leonis.nus.sg

Prof.Y.B.Guo
Lehrstuhl II fur Mathematik
RWTH Aachen
52056 Aachen
Germany
Fax:0049–241–8888136
e–mail:guo@math2.rwth–aachen.de

Prof.C.K.Gupta
Dept.of Mathematics and Astronomy
The University of Manitoba
winnipeg, Manitoba
Canada R3T 2N2
Tel:(204)–474 –9376
Fax: (204)–275–5011

Prof.F.Harary(Harary, Frank)
Department of Mathematics and
Astronomy,
New Mexico State Univ.
Las Cruces, NM 88003–0001
USA

Assoc.Prof.He Chao(C.He)
(1). 82 Chongren Road, Hankou Wuhan
Hubei 430033
P.R.China
(2). 11 Yong Fengli of Zhizi Jie
Hankou Wuhan Hubei 430022
P.R.China
Tel:027–5862749(O)
027–5811600(H)

Assoc.Prof.He Guoliang(G.L.He)
Department of Mathematics
Universitatis Normalis Hunaensis
Changsha 410081
P.R.China
Tel:0731–8883131–2515(O)
0731–8854101(H)

He Wei(W.He)
Room 101 No.24 Lane 450
Da Mu Qiao Road
Shanghai, 200032
Tel:021–4160443

Prof.H.Hector
Instituto de Matematicas
Universidad Catolica de
Valparaiso
Casilla 4059, Valparaiso
Chile
Tel:0056–32–258860
Fax:0056–32–258847
e–mail: hhevia@aix1.ucv.cl

Hou Renmin(R.M.Hou)
Mathematics Department
Yantai University
Yantai 264005 , Shandong
P.R.China
Tel:0535–6888995–2776(O)

Prof.Leetsch C.Hsu(L.C.Hsu)
Applied Mathematics Institute
Dalian University of Technology
Dalian 116024
P.R.China
Tel:0086–411–4709361(H)
0086–411–4709360(O)
Fax:011–86–411–4671009

Prof.Hu Yuda(Y.D.Hu)
Dept.of Applied Mathematics
Shanghai Jiao Tong University
Shanghai,200030
P.R.China
Tel:021–4810437(H)
Fax:86–21–4330892

Huang Liying(L.Y.Huang)
Dept.of Math.
Cuanzhou Teacher's College
Cuanzhou,Fujian 362000
P.R.China
Tel:0595–2785106(O)
0595–2581771(H)

Assoc.Prof.Huang Qiongxiang
(Q.X.Huang)
Dept. of Math.
Xingjiang Univ.
Xingjiang Urumqi 830046
P.R.China

Dr.Huang Yuanqiu(Y.Q.Huang)
Institute of Applied Mathematics
Academia Sinica
Beijing, 100080
P.R.China

Huang Zhenjie(Z.J.Huang)
Department of Mathematics
Zhangzhou Teacher's College
Zhangzhou, Fujian 363000
P.R.China
Tel:0596–2026033–240(O)

Prof.F.Hwang (Frank Hwang)
AT&T Bell Laboratories 2 C379
Murray Hill, NJ 07974–2070
USA
Tel:908–5824733
Fax:908–5822379

Prof.Yuan–Ji Huo(Y.J.Huo)
Dept.of Math.
Zhanggiakou Teacher's College
Zhangjiakou,Hubei 075028
P.R.China

Prof.W.Imrich
Applied Mathematics
Montan Universitaet
A–8700 Leosen
Austria
Tel:++43–03842–402 420
Fax:++43–03842–402 740
e–mail:
imrich@grz08u.unileoben.ac.at

Assoc.Prof.Jiang Zhi–Ming(Z.M.Jiang)
Department of Management
Shanghai College of Petrochemical
Technology,
Shanghai 200540
P.R.China
Tel:021–7942100–80

Prof.S.Kageyama (Kageyama, Sanpel)
Faculty of School Education,
Hiroshima University
1–33 Shinonome
3–Chome
Minami–ku
Hiroshima 734
Japan

Kang Liying(L.Y.Kang)
Dept. of Basic Courses
Shijiazhuang Railway Institute
Shijiazhuang , Hebei 050043
P.R.China
Tel:0311–6039031–Math.Dept.

Prof. Aloxandor A. Kolokolov
The Institute of Information
Technologies and Applied
Mathematics,
Omsk
28 Andrianov Str. Omsk 644077
Russia
e–mail: iitpm@intskh.omsk.su

Assoc.Prof.Ku Tung–Hsin(T.H.Ku)
The Hefei Branch Research Centre of
Combinatorial Mathematics and
Computer Science, Academia Sinica
P.O.Box 1110 Hefei, 230031 Anhui
P.R.China
Tel:0086–551–5591289(O)
0086–551–5591166(H)
Fax:0086–551–5591270

Dr.Lai Chunhui(C.H.Lai)
Department of Mathematics
Zhangzhou Teachers College
Zhangzhou, Fujian, 363000
P.R.China
Tel:0596–2026033–240(O)
0596–2067879(H)

Prof.C.Lam
Dept. of Computer Science
Concordia University
1455 de Maisonneuve Blvd. W.
Montreal, Quebec
H3G 1M8, Canada
Tel:001–514–848–3002
Fax:001–514–848–2830
e–mail: lam@vax2.concordia.ca

Dr.Serge Lawrencenko (Russia)
Dept.of Math. & Computer Centre
Northern Jiaotong University
Room2001, Computer Center ,
c/o Foreign Affdirs Office
Northern Jiaotong University
Beijing 100044.
Tel:(0861)3240486 (ask Room2001)
Fax:(0861)3240301(Room 4201)

Prof.V.M.Levchuk
Department of Mathematics
Krasnoyarsk State University
Svobodnyi Pr. 79,
660062 Krasnoyarsk, Russia
Fax:007–3912–448625
e–mail: Levchuk@kgu.krasnoyarsk.su

Dr.Li Anping(A.P.Li)
Institute of Applied Mathematics
Academia Sinica
Beijing 100080
P.R.China

Prof.Li Dengxin(D.X.Li)
Department of Mathematics

Yuzhou University
Chongqing, 630033 Sichuan
P.R.China
Tel:0811-9863746-35
0811-9863746-66

Prof.Ko-WeiLih(Lih, Ko-Wei)
Institute of Mathematics,
Academia Sinica
Nankang, Taipei
Taiwan 11529
e-mail:makwlih@twnas886.bitnet

Prof.Li Jiong-Sheng(J.S.Li)
Department of Mathematics
University of Science and
Technology of China,
Hefei, Anhui 230026
P.R.China
Tel:0551-3601044(H)

Prof.Li Nian Zu(N.Z.Li)
Dept.of Math.
The Second Technology Univ.of
Shanghai,
Shanghai,200122
P.R.China
Tel:021-8772377(H)

Li Rongyan(R.Y.Li)
Institute of Applied Mathematics
Academia Sinica
Bejing 100080
P.R.China

Li Xuechao(X.C.Li)
Department of Mathematics
Central China Normal University
Wuhan, Hebei 430070
P.R.China
Tel:027-7818444-2317

Prof.Li Xueliang(X.L.Li)
Institute of Mathematics & Physics
Xinjiang University
Urumchi, Xinjiang 830046
P.R.China

Prof. D.Liang
Department of Computer Science
Indiana University
Purdue University
at Fort Wayne
USA
Tel: 001-219-481-6184
e-mail:
liangy@panda.ipfw.indiana.edu

Liang Haijiang(H.J.Liang)
Basic Education Department
Zhengzhou Electric Prower College
Zhengzhou 450004 Henan
P.R.China
Tel:0371-6229188-3012(O)
0371-6327225(H)

Assoc.Prof.Lin Cuiqin(C.Q.Lin)
Dept.of Applied Mathematics
Qinghua University
Beijing 100080
P.R.China

Prof.Lin Hua-Yi(H.Y.Lin)
Department of Mathematics
Huazhong University of Technology
Wuhan 430074
P.R.China

Lin Yiqing(Y.Q.Lin)
Room 402, Building 26
2500 Songhuajiang Rol.
Shanghai 200437
P.R.China

Assoc.Prof.Liu Chunfeng(C.F.Liu)
Department of Mathematics
Jinzhou Normal College
Jinzhou,Liaoning, 121001
P.R.China
Tel:0416-4188428-3015(O)

Prof.Liu Guizhen(G.Z.Liu)
Department of Mathematics
Shandong University
Jinan, 250100
P.R.China
Tel:0531-8903861-2652(O)
0531-8903861-2641-378(H)
Fax:0531-8902167

Liu Hongmei(H.M.Liu)
Department of Mathematics
Central China Normal University
Wuhan Hubei 430070
P.R.China

Prof.Liu Mulan(M.L.Liu)
Institute of Systems Science
Academia Sinica
Beijing 100080
P.R.China

Prof.Liu Ruying(R.Y.Liu)
Dept.of Math.
Qinghai Normal Univ.
Xining,Qinghai 810008
P.R.China

Assoc.Prof.Liu Ying(Y.Liu)
Institute of Applied Mathematics
Academia Sinica
Beijing 100080
P.R.China
Tel:01-2541695(O)

Prof.Liu Yan-Pei(Y.P.Liu)
Dept.of Math.
Northern Jaotong University
Beijing, 100044
P.R.China
Tel:01-2256622-3394(H)
01-3240500-3394(H)
Fax:01-2255671
e-mail:mx%"yLiu@bepc2.ihep.ac.cn"

Liu Yuji(Y.J.Liu)
Dept.of Math.
Yueyang Teacher's College
Yueyang,Hunan 414000
P.R.China

Liu Zhixiong(Z.X.Liu)
No.2 Staff and Workers Hospital
Liao River Oil Field
Liaoning 124120
P.R.China
Tel:7847325(O)

E.K.Lloyd(Lloyd, E.Keith)
Faculty of Mathematical Studies
University of Soutyampton
Southampton SO9 5NH
Great Britain

Prof.Lu Kai-Cheng(K.C.Lu)
Department of Computer Science
Qinghua University
Beijing, 100080
P.R.China
Tel:01-2594401(H)

Lu Qinglin(Q.L.Lu)
Department of Mathematics
Xuzhou Normal College
Xuzhou 221009 ,Jiangsu
P.R.China

Lu Xiangfen(X.F.Lu)
Tonghui Bridge Primarg School
P.R.China
Tel:0510-2723676(O)

Lu Xinzhong(X.Z.Lu)
Lanzhou Railway Institute
Lanzhou,Gansu 730070
P.R.China

Prof.Luo Jianjin(J.J.Luo)
Dept.of Math.
Northwest university
Xian,Shanxi 710069
P.R.China

Prof.Luo Youxue(Y.X.Luo)
Dept.of Math.
Fuzhou University
Fuzhou,Fujian 350002
P.R.China
Tel:0591-3710845-Math.Dept.(O)
0591-7530972(H)

Prof.A.A.Makhner
Institute of Mathematics and
Mechanics,
UB RAS 620219
Yekaterinburg Kovalevskoi 16
Russia

Prof.V.Manevich
CMSR Laboratory for Computer
Graphics & CAD,
Technion-Israsl Institute of
Technology,
Faculty of Mechanical Engineering
Technion City, Haifa 32000
Israel

Prof.Mao Jingzhong(J.Z.Mao)
Department of Mathematics
Central China Normal University
Wuhan 430070 Hubei
P.R.China
Tel:027-7818444-3295(O)
027-7818444-2118(H)

Prof.Andrey O. Matveev
Dept. of Automation and
Information Technologies,
Urals State Technical University
Mira Str. 32
620002, Ekaterinburg
"Data Contre" Research Firm,
Ekaterinburg
Mamin-Sibiryak Str. 193, box 132,
620055, Ekaterinburg,
Russia
Fax: 007-3432-373997
e-mail: ddoo@quorus.ru

Assoc.Prof.Meng Jixiang(J.X.Meng)
Department of Mathematics
Xinjiang University
Urumqi 830046 Xinjiang
P.R.China

Miao Zhengke(Z.K.Miao)
Department of Mathematics
Xuzhou Techer's College
Xuzhou 221009
P.R.China
Tel:0516-5719166

Prof.V.Mironov
Dept.of Algebra and Math.Logic
Moscow Institute of
Electronic and Mathematics,
Bolshoy Vusovski Per.
3/12, Moscow, 109028

Russia
Tel: 007–95–245–9837
Fax: 007–95–121–9060
e-mail: mir@avt.energy.msk.su

Prof.R.N.Mohan
Reader & Head,
Deptt. of Mathematics
D.A.R.College
Bank Street
Nuzvid–521 201
A.P.India
Tel: 0091–8656–758

Prof.F.Moth
Northeastern illinois University
Board of Governors Universities
5500 North Street Louis Avenue
Chicago Illinois 60625–4699
USA

Prof.G.L.Mullen
Department of Mathematics
Eberly College of Science
The Pennsylvania State University
218 Mcallister Building
University Park PA 16802–6401
USA

Prof.C.Mynhardt
1580 York Place
Victoria BC
Canada V8R 5X2
Tel:001–604–598–8880
Fax:001–604–598–6763

Prof.Melvyn B.Nathanson
The Graduate School and
University Center,
The City University of New York
Ph.D Program in Mathematics
Box 330
33 West 42 Street,
New York, N.Y.10036–8099
USA

Prof.J.N.Nuzhin
Deprtment of MODUS
Krsnoyarsk State Technical Univ.
Kirenskogo Str. 26,
660074 Krasnoyarsk
Russia
Fax:007–3912–430692
e-mail:
vsid@depfc.kgtu.krasnoyarsk.su

Prof.P.J.Owens
Math.& Computing Sciences Dept.
University of Surrey
Guildford Surrey GU2 5XH
United Kingdom
Tel:0044–01483–300800 Ext 2628,
0044–01483–259640
Fax: National: 01483–259385
International: 0044–1483–259385

Dr.Ou Yangkeyi(Y.K.Y.Ou)
Institute of Applied Mathematics
Academia Sinica
Beijing 100080
P.R.China
Tel:01–2541695(O)

Dr.K.R.Pardasani
Deptt. of Maths & MCA
Maulana Azad College of Technology
(Regional Engineering College)
BHOPAL–462007
India

Prof.C.M.Pareek
Maths. Depat.
Kuwait University
P.O.Box 5969
Safat 13060
Kuwait
Tel:00965–4849153
Fax:00965–4817201

Prof.M.Pennisi
Dipartimento Di Matematica
Citta Universitarla
Viale A.Doria 6–I 95125–Catania
Tel:330–533
E-Mail: PENNISI@DIPMAT.UNICT.IT

Prof.N.Polat
I.A.E. DELYON
Universite Jean Moulin (LYON 3)
15, quai Claude Bernard.B.P.0638
69239 LYON Cedex 02
France
Tel: 0033–72–72–21–60
Fax: 0033–72–72–20–81

Prof.A.Pott
Lehrstuhl fur Angewandte
Mathematil ll,
Universitat Augsburg
UniversitatsstraBe 8
86135 Augsburg
Germany
Tel:0821–598–2216
Fax:0821–598–2200

Prof.E.Prisner
Mathematisches Seminar
Universitat Hamburg
Bumdesstr.55
20146 Hamburg
Germany
e-mail:
ms3a505@math.uni–hamburg.de

Pu Changjiang(C.J.Pu)
Mathematics and Mechanical Dept.
Harbin Shipbuilding Engin.Institute
Harbin 150001
P.R.China
Tel:0451–2519753(O)
0451–4506634(H)

Qin Yiming(Y.M.Qin)
c/o Liu Xiaofeng
Department of Foreign Trade
Sichuan University
Chengdu 610064
P.R.China

Qiu Hesheng(H.S.Qiu)
Henan Research Institute of Math.
Zhengzhou, Henan 450003
P.R.China
Tel:0371–5955803–2080(H)

Prof.Qiu Weisheng(W.S.Qiu)
Department of Mathematics
Peking University
Beijing 100871
P.R.China
Tel:01–2501807(O)
01–2585641(H)
Fax:01–2501801
e-mail:
mathpu%BEPC2@scs.SLAC.STANFORD.EDU

Prof.L.V.Quintas
Mathematics Department
Pace University
New York
NY10038–1502
USA
e-mail:Quintas @pacevm.bitnet
or
@Pacevm.dac.pace.edu

Prof.S.Ramachandran
Department of Mathematics
Vivekananda College
Agasteeswaram–629701
Cape Comorin
Tamil Nadu State
India

Prof.A.Recski
Technical University of Budapest
Faculty of Electrical Engineering
Budapest , Stoczek u.2–4
H–1111
Hungary
Tel:36–1–166–5446
Fax:36–1–1812366

Ren Han(H.Ren)
Department of Mathematics
Wuhan Iron steel University
Wuhan Hubei 430081
P.R.China
Tel:027–6858783(H)

Prof.R.W.Robinson
(Robinson, Robert W)
Department of Computer Science
University of Georgia
415 Boyd Grad Studies Bidg.
Athens, GA 30602–7404
USA
Tel:001–706–542–2911
Tax:001–706–542–2966
e-mail:
rwr@pollux.cs.uga.edu(internet)
robinson@uga(bitnet)

Prof.H.Sachs
Technische Universitat Ilmenau
Institut fur Mathematik
D–98684 ILMENAU
Germany

Prof.N.Z.Salvi
Dip. di Matehatica
Politecnico Di Milano
P.za L.DA vinci 32
20133 Milano
Italy
Tel: 0039–02–9516779
0039–02–23994579
Fax: 0039–02–23994568

Prof.B.Sasan
Deprtment of Computer Science
Monash University
Clayton,
Victoria 3168
Australia
Tel:(03)565 5207
Fax:(03)565 5146

Prof.J.Schoenheim
School of Mathematics Science
Faculty of Exact Science
TEI Aviv University
Ramat Aviv, 69978 Tel Aviv
Israel

Shan Erfang(E.F.Shan)
Department of Mathematics
Shijiazhuang Normal College
Shijiazhuang 050041
P.R.China
Tel:0311–6039759–3096(O)

Assoc.Prof.Shao Xuncai(X.C.Shao)
Department of Computer Science
Beijin Polytechnic University
Beijin 100022
P.R.China
Tel:01–7711177–470(O)

Prof.Prof.A.Shastri
Deptment of Mathematics and Computer Science.
Banasthali University
P.O. Banasthali
Vidyapith 304 022
Raj
India
Tel:0091–43 83 32 21
Fax:0091–43 83 35

Prof.Shen Hao(H.Shen)
Department of Applied Mathematics
Shanhai JiaoTong University
Shanhai 200030,
P.R.China
Tel:021–43103101–278(O)
021–4686672(H)

Shen Xinan(X.N.Shen)
Turbocharger Division of Wuxi Engine Works,
Wuxi,Jiangsu 214023
P.R.China
Tel:0510–5754334–446

Dr.Shi Lijie(L.J.Shi)
Institute of Applied Mathematics
Acadenina Sinica
Bejing 100080
P.R.China
Tel:01–2541695(O)
e–mail:slj@ amath3.amt.ac.cn.

Prof.H.Singh
Lecturer In Mathematics
Govt College
Tarn Taran (Amritsar)
Punjab,
India–143401

Prof.V.T.Sos
Math.Institute of the Hungarian Academy of Sciences
Budapest, P.O.B.127, H–1364
Hungary
Tel:36–1–117–3151
Fax:36–1–117–7166

Dr.B.Srinivasan
Dept.of Computer Science
Monash University
clayton,Victoria 3168
Australia
Tel:0061–3–565 5207
Fax:0061–3–565 5146
e–mail:
srini@bruce.cs.monash.edu.au

Dr. N.Sudharsanam
Department of Mathematics
National College
Trichirapalli–T.N
Pin:620001–India

Prof.Hugo S.H.Sun
Department of Mathematics
School of Natural Sciences
California State University. FRESNO
FRESNO, California 93740–0108
USA
Tel:001–209–278–4633(O)
001–209–292–6828(H)
Fax:001–209–278–2872

Prof.Sun Shixin(S.X.Sun)
Department of Computer Science
Univ.of Electronic Science and Technology of China,
Chengdu 610054
P.R.China
Tel:028–3333312–2871(H)
Fax:028–3334131

Prof.G.Tallini
Dipartinento di Matematica
"G.castelnuovo",
Piazzale A.Moro, 2
1–00185, Roma
Italy
Tel:(0)6–4463181
Fax:(0)6–4991206

Tan Kehu(K.H.Tan)
Department of Mathematics
Central China Normal University
Wuhan Hubei 430070
P.R.China

Assoc.Prof.Tang Tingzai(T.Z.Tang)
Department of Mathematics
Sichuan Teachers College
Nanchong Sichuan 637002
P.R.China

Tao Changqi(C.Q.Tao)
Dept.of Information
Jiangxi Institute of Finance and Economics
Nanchang,Jiangxi,330013
P.R.China

Dr.Teng Cong(C.Teng)
Department of Mathematics
Shandong University 250100
P.R.China
Tel:0531–8905962–Dept.of Math.

J.H.van Lint
Dept.of Math. & Computer Sciences
Eindhoven University of Technology
P.O.Box 513
5600 MB Eindhoven

The Netherlands

Prof.K.Vijayan
Dept. of Mathematics
The University of Western Australia
Nedlands, West.Aust. 6009
Australia
Tel:0061–9–3803378
Fax:0061–9–3801028

Prof.A.Vince
Department of Mathematics
University of Florida
Gainesville, Florida 32611–2082
U.S.A
Tel:001–904–392–6172
Fax:001–904–392–8357

Prof.Wan Zhexian(Z.X.Wan)
Institute of System Science
Academia Sinica
Beijing,100080
P.R.China

Prof.Chi Wang
Dept.of Math.
University of Louisville
Louisville,Kentucky 40292
USA
Tel:502–852–6826(O)
Fax:502–499–5844
e–mail:
COWANGO1@ULKYVXLOUISVILLEEDU

Prof.Wang Jian–fang(J.F.Wang)
Institute of Applied Mathematics
Academia Sinica
Beijing 100080
P.R.China
Tel:01–2561441(O)
01–2565560(H)

Wang Jianjun(J.J.Wang)
Hebei Cadre College of Economic
Management 050061
P.R.China
Tel:0311–6039306–2070(O)
0311–6039306–2195(H)

Dr.Wang Jun(J.Wang)
Institute of Mathematics
Dalian University of Technology
Dalian 116024
P.R.China

Prof.Louis C.Wang
Applied Detector Corporation
2325 E.Mckiniey Avenue
Fresno,CA 93703
USA
Tel:209–485–2396
Fax:209–485–2830

Assoc.Prof.Wang Min(M.Wang)
Department of Mathematics
Yantai University
Yantai 264005 Shandong
P.R.China
Tel:0535–6888995–2779(O)
0553–6888995–2420(H)

Prof.P.Wang
Dept. of Math.& Computing Sciences
St.Francis Xavier University
Antigonish, Nova Scotia
Canada, B2G 2W5
P.O. Box 5000
Tel: 001–902–867–5166

Assoc.Prof.Wang Shaowen(S.W.Wang)
Business school
Beijing Institute of Machinery
Industry,
Beijing 100085
P.R.China
Tel:01–2913661–214(O)

Prof.Wang Tianming(T.M.Wang)
Institute of Mathematical Science
Dalian University of Technology
Dalian 116024
P.R.China
Tel:0411–4708360(O)
0411–4709057(H)
Fax:0411–4071009
e–mail:Chengqd@bepc2.ibep.ac.cn

Prof.Wang Zhemin(Z.M.Wang)
Department of Statistics
Operations Research,
Fudan University
Shanghai 200433
P.R.China
Tel:021–5491058(H)
Fax:021–5453039
e–mail:
zmwang@Solaris.fudan.edu.cn

Prof.Paul M.Weichsel
Dept. of Mathematics
University of Illinois
at Urbana–Champaign
221 Altgeld Hall
1409 West Green Street
Urbana, IL 61801
U.S.A.
Tel:001–217–333–0677
Fax:001–217–333–9576
e–mail: weichsel@math.uiuc.edu

Prof.R.J.Wilson
Department of Pure Mathematics
The Open University
Walton Hall

Milton Keynes
MK7 6AA
England
Tel:(0908)274066
Fax:(0908)653744

Wu Jianliang(J.L.Wu)
Department of Applied Mathematics
and Software Eng.,
Shandong Mining Institute
Taian 271019
P.R.China
Tel:0538-223311-258(O)

Wu shuoxin(S.X.Wu)
Shandong Kenli Huanghe Farm School
Shangdong 257509
P.R.China

Assoc.Prof.Wu Qiqi(Q.Q.Wu)
Middle Area 5, No. 304 Room
Foshan University
Foshan Guangdong , 528000
P.R.China
Tel:0757-2272105-3249(H)
Fax:0757-272233

Prof.Xia Mingyuan(M.Y.Xia)
Department of Mathematics
Central China Normal University
Wuhan 430070 Hubei
P.R.China
Tel:027-7818444-3295(O)

Xiang Kaiyao(K.Y.Xiang)
Department of Mathematics
Shiyan University
Shiyan Hubei 442000
P.R.China

Xiao Wei(W.Xiao)
Basic Sciences Department
Southwest Petroleum Institute
Nanchong 637001 Sichuan
P.R.China

Xiao Xinping(X.P.Xiao)
Department of Mathematics
Wuhan Transportation University
Wuhan 430063
P.R.China
Tel:027-6811694(O)

Assoc.Prof.Xie Keqiu(K.Q.Xie)
Department of Mathematics
Algebre Teaching & Research Section
Nanjing Institute of Education
Nanjing 210017,Jiangsu
P.R.China
Tel:025-6616461转数学系(O)
025-4587462(H)
Fax:025-6616424

Prof.Xie Litong(L.T.Xie)
Dept.of Math.
Shandong University
Jinan,Shandong 250100
P.R.China

Assoc.Prof.Xie Zheng(Z.Xie)
Dept. of System Engineering and Math.
National Univ. of Defense Technology
Changsha, 410073 Hunan
P.R.China
Tel:0731-4434601-87701(O)
0731-4434601-85360(H)

Xiong liming(L.M.Xiong)
Faculty of Mathematics and Computer
Jiangxi Normal University
Nanchang 330027
P.R.China
Tel:0791-8326060-2202(O)
0791-8326060-2141(H)

Xu Baogang(B.G.Xu)
Department of Mathematics
Shandong University
Jinan 250100
P.R.China

Xu Baogen(B.G.Xu)
Basic Dept.
East China Jiaotong University
Nanchang,330013 Jiangxi
P.R.China
Tel:0791-227024-518(O)

Xu Guanghui(G.H.Xu)
Teaching and Research Section of
Mathematics,
Zhejiang Forestry College
Linan 311300 Zhejiang
P.R.China
Tel:0571-3723544(O)
0571-3728487(H)

Assoc.Prof.Xu Jin(J.Xu)
Institute of Elect.Engin.
Xian Univ. of Sci. and Technology
Xian,Shanxi 710071
P.R.China

Assoc.Prof.Xu Juyong(J.Y.Xu)
Department of Basic Science
Wuhan Urban Construction Institute
Wuhan 430074, Hubei
P.R.China
Tel:027-7808415(H)

S.J. Xu(Xu Shao Ji)
Mathematics Department
Rutgers University
New Brunswick

N.J.
USA

Prof.Xu Yonghua(Y.H.Xu)
Basic Science Department
Taiyuan Heavy Machinery Institute
Taiyuan, Shanxi Province, 030024
P.R.China
Tel:0351-6129110(H)

Assoc.Prof.Xu Yunqing(Y.Q.Xu)
Department of Mathematics
Xinyang Teachers College
Xinyang 464000, Henan
P.R.China

Assoc.Prof.Xue Xiuqian(X.Q.Xue)
Dept.of Mathematics and Mechanics
China University of Mining and Technology,
Xuzhou, Jiangsu 221008
P.R.China
Tel:0516-3888653-443(O)
0516-38880676-2290(H)
Fax:0516-3888682

Prof.Yan Ji-Yi(J.Y.Yan)
University of Science and Technology of China,
Beijing 100039
P.R.China

Prof.Yang Benfu(B.F.Yang)
Department of Mathematics
Chengdu Techer's College
Penzhou, Sichuan 611930
P.R.China
Tel:028-871167-2140(O)
028-871167-2027(H)

Prof.Yang Lu(L.Yang)
Chengdu Branch of Academia Sinica
Chengdu,Sichuan 610041
P.R.China

Assoc.Prof.Yang Shihui(S.H.Yang)
Department of Mathematics
Fuling Teacher's College
Fuling 648005 Sichuan
P.R.China
Tel:0810-222015-2093(O)

Prof.H.P.Yap
Department of Mathematics
Faculty of Science
National University of Singapore
10 Kent Ridge Crescent
Singapore 0511

Ye Chengfu(C.F.Ye)
Department of Mathematics
Qinghai Nationalities College
Xining,Qinghai 810007
P.R.China
Tel:0971-76888(O)

Assoc.Prof.Ye Xiuming(X.M.Ye)
Department of Mathematics
Shanghai University
Jiading, Shanghai 201800
P.R.China
Tel:021-9532932-514(O)
021-9526434(H)

Assoc.Prof.Yin Jianxing(J.X.Yin)
Department of Mathematics
Suzhou University
Suzhou, 215006 Jiangsu
P.R.China
Tel:0512-5223614 ext 891(H)

Assoc.Prof.Yu Songnian(S.N.Yu)
School of Computer Science and Engineering,
Shanghai University
Shanghai 200072
P.R.China
Tel:021-6631515-2673(O)
021-6775041(H)

Assoc.Prof.Zeng Zhenbing(Z.B.Zeng)
Institute of Computer Applications
Academia Sinica Chengdu Branch
Chengdu 610041
P.R.China
Tel:028-5581260-235(O)

Zhang Chunyuan(C.Y.Zhang)
Dept.of Appl. Mathematics Info.
Engineering Institute
P.O.Box 1001-46
Zhengzhou, Henan 450002
P.R.China
Tel:0371-3941430-62664(O)
0371-3941430-63166(H)

Prof.Hong Zhang
Dept.of Mathematical Sciences
Indiana-Purdue University
at Fort Wayne,
Fort Wayne, IN 46805
USA
Tel:001-219-481-6186(O),
001-219-489-1956(H)
Fax:001-219-381-6880(O),
001-219-489-1956(H)
e-mail:
Zhang@cvax.ipfw.indiana.edu

Zhang Jianzhou(J.Z. Zhang)
Department of Applied Mathematics
University of Electronic Sci. &
Tech. of China,
Chengdu , Sichuan 610054
P.R.China

Zhang Shusheng(S.S.Zhang)
Middle School of Guhou
Ningdu, Jiangxi 342814
P.R.Chin

Assoc.Prof.Zhang Xiandi(X.D.Zhang)
Department of Applied Mathematics
University of Electronic Science and Technology of China,
Chengdu, Sichuan 610054
P.R.China
Tel:028–3333312–11 Dept.

Dr.Zhang Xiangde(X.D.Zhang)
Department of Automatic Cintrol
Notheastem University
Shengyang 110006
P.R.China
Tel:024–3893000–4339(O)
024–3893000–5446(H)

Zhang Zhizheng(Z.Z.Zhang)
Department of Mathematics
Luoyang Teacher's College
Luo yang, Henan 471022
P.R.China
Tel:0379–3937701–Math.Dept.

Prof.Zhang Zhongfu(Z.F.Zhang)
Lanzhou Railway Institute
Lanzhou 730070
P.R.China
Tel:0931–7666221–625(O)
0931–7666221–578(H)

Assoc.Prof.Zhang Zhiyu(Z.Y.Zhang)
Pox 154 Taiyuan,
Shanxi 030008
P.R.China
Tel:0351–305349–295(O)
0351–3059349–349(H)
Fax:0351–3059151

Zhao Bingxin(B.X.Zhao)
Dept.of Economic Mancgement
Shandong University
Jinan,Shandong 250100
P.R.China
Tel:0531–8904551–2792(O)
0531–8908890(H)

Zhao Haixing(H.X.Zhao)
Delilngha Farm Middle School
Qinghai 817003
P.R.China

Prof.Zhong Ji (J.Zhong)
Department of Mathematics
Huanan Normal University
Guangzhou,Guangdong 510631
P.R.China
Tel:020–7583386(H)

Zhou Sanming(S.M.Zhou)
Department of Mathematics
Huazhong University of Science and Technology,
Wuhan , Hubei 430074
P.R.China
Tel:027–7801153–3231(O)

Prof.Zhou Yongsheng(Y.S.Zhou)
Department of Mathematics
Guangdong Institute of Nationatities,
Guangzhou,Guangdong 510633
P.R.China
Tel:020–5515722–226(O)
020–5515722–214(H)

Zhu Fucheng(F.C.Zhu)
Wuxi Beitang Vocational School
Wuxi,Jiangsu 214043
P.R.China
Tel:0510–2721936(O)

Assoc.Prof.Zhu Hongyi(H.Y.Zhu)
Shanghai College of Metallurgy
121 Caobao Road, Shanghai 200233
P.R.China
Tel:021–4083051–2117(O)
021–4190190(H)

Prof.Zhu Jiacheng(J.C.Zhu)
Dept.of Computing Center
Qinghai Normal University
Xining,Qinghai 810008
P.R.China
Tel:0971–54410(H)

Prof.Zhu Lie(L.Zhu)
Department of Mathematics
Suzhou University
Suzhou,Jiangsu 215006
P.R.China
Tel:0512–5223614–985(H)

Assoc.Prof.Zuo Guangji(G.J.Guo)
Department of Appl. Math.
Qinghai Nationalities college
Xining,Qinghai 810007
P.R.China
Tel:
0971–76888(O)